Control Theory of Digitally Networked Dynamic Systems

Jan Lunze
Editor

Control Theory of Digitally Networked Dynamic Systems

Springer

Editor
Jan Lunze
Lehrstuhl für Automatisierungstechnik und
Prozessinformatik
Fakultät für Elektrotechnik und
Informationstechnik
Ruhr-Universität Bochum
Bochum
Germany

ISBN 978-3-319-01130-1 ISBN 978-3-319-01131-8 (eBook)
DOI 10.1007/978-3-319-01131-8
Springer Cham Heidelberg New York Dordrecht London

Library of Congress Control Number: 2013941663

Printed on acid-free paper

Springer is part of Springer Science+Business Media (www.springer.com)

Preface

Modern means of communication make it easy to connect system components whenever information links may contribute to improving the overall system performance. Actuators, sensors and controllers can now exchange information from nearly any place within a technological system, and wireless connections allow to implement advanced control technology for mobile objects. Due to this flexibility, a new challenge for control theory is to elaborate systematic ways for the selection of information structures that are suitable for a given control task.

This new freedom in choosing the information topology of control systems is associated with some nondeterminism of the behavior of digital networks. As many networks used for control purposes act simultaneously as communication medium for other tasks, time delays in the delivery of the information or packet loss regularly occur and the controllers have to be made tolerant with respect to such imperfections.

Both, the flexibility of the communication structure and the imperfection of the communication links pose new challenges for the design and the implementation of control systems and have brought about *networked control systems* as a new area of control theory. This book gives a concise introduction to this area and describes new modelling paradigms, analysis methods for digitally networked dynamic systems, and design methods for distributed controllers. It provides an overview of the problems, methods and application areas that have been tackled in the recent past and which will initiate new development directions in industry in the near future.

Networked control systems cannot be handled with traditional methods for two reasons. First, the *event-driven mode of operation* of networked systems violates the precondition of discrete-time control theory that data is processed and transmitted with a constant sampling rate. Several chapters of this book are concerned with control problems in which event generators rather than clocks initiate the next communication, computation and control steps.

Second, the structure of the communication network is usually not predefined but the subject of control design. The data links may even adjust themselves to the needs and the technological constraints during operation. Under these circumstances, the question "*Which information is necessary to solve a control task?*" has stimulated the second leading research line of this book.

The tight combination of communication and control has direct consequences for the development of advanced control systems in industry. Networked systems open the way towards novel application areas like multi-agent systems that are controlled over wireless communication networks or smart grids of intelligent components that are supervised remotely. It broadens the importance of control in intelligent transportation systems, health care services, smart buildings, or telerobotics, just to name a few. Some of these applications are discussed in this book.

Structure and readership. The book is structured into seven chapters, each of which is devoted to one of the research problems that are currently discussed in the international literature and at all major control conferences:

1. **Introduction to networked control systems**
   describes the motivation and surveys the main problems to be solved by the control theory of networked systems.

2. **Analysis of networked systems**
   presents important properties of dynamic systems that are controlled over digital communication networks.

3. **Distributed estimation and control**
   describes methods for state estimation and optimization of systems that are composed of subsystems with local control stations, which communicate over a data network.

4. **Distributed and networked model predictive control**
   develops methods for distributed control algorithms that tolerate time delays and packet loss.

5. **Event-based control**
   surveys six different approaches to replace the traditional periodic sampling by event-based triggering schemes.

6. **Multi-agent systems**
   proposes new methods for controlling systems that have to satisfy a common control goal.

7. **Wireless networking for control**
   investigates the requirements on the digital communication network, which is used for control purposes, and describes methods for the cross-design of control and communication.

The book is written as a **survey of recent results** on networked control systems with references to numerous journal papers and conference presentations, in which the ideas have been originally presented together with the formal proofs. Emphasis is laid on the presentation of the main results and the illustration of these results by examples. Each chapter starts with an introduction to the problems considered, which includes a statement on how the chapter extends the current knowledge about networked control systems.

The results are presented in a uniform style, notation and terminology. Many cross-references show how the different lines of research belong together. The references are organized as a comprehensive list at the end of the book.

The intended readers of the book are graduate students, PhD students and researchers in control, communication, or mathematics, as well as practicing engineers with knowledge in systems theory who want to get an introduction to the main problems of networked control together with an up-to-date survey of recent results. Interesting practical examples illustrate the applicability of the methods and **extensive bibliographical notes** point to the origins of the presented ideas and the current research lines. The evaluation of the methods and the application results should help the reader to assess the available methods and the limits of the present knowledge about networked control with respect to their particular field of interest.

Priority Program 1305 of the German Research Foundation. The book presents, in a concise way, the results obtained by the participants of the Priority Program 1305 on *Control Theory of Digitally Networked Dynamic Systems* of the German Research Foundation (DFG). In this program, 17 PhD students have been working between 2007 and 2013 in 13 different research groups situated in control, mathematics, or communication departments at German universities. The book describes the research interests, the results, and the common expertise of 42 authors. All sections are co-authored by several researchers who have contributed to the corresponding topic. The authors are mentioned in alphabetic order with the responsible author first.

Acknowledgement. The authors thank the German Research Foundation for its financial support within the Priority Program 1305, which has included grants for numerous workshops, two PhD schools and a generous guest program. Dr. ANDREAS ENGELKE as Programme Director for Engineering Sciences has continuously supported the activities of this priority program.

Mr. RENÉ SCHUH has structured the writing process by creating guidelines and templates and by retaining the communication among the co-authors. Ms. ANDREA MARSCHALL has drawn and re-drawn many figures and Ms. SUSANNE MALOW has helped to bring the book into a uniform layout. SVEN BODENBURG, ANDREJ MOSEBACH, YANNICK NKE, MELANIE SCHMIDT, RENÉ SCHUH, CHRISTIAN STÖCKER and DANIEL VEY have checked the page proof.

Finally, the close cooperation with Springer as publisher and particularly with Dr. THOMAS DITZINGER as responsible editor is gratefully acknowledged.

Bochum, April 2013 JAN LUNZE

Contents

List of Contributors

Frank Allgöwer
University of Stuttgart,
Institute for Systems Theory and Automatic Control,
Pfaffenwaldring 9,
70550 Stuttgart, Germany
frank.allgower@ist.uni-stuttgart.de
(Sections 4.1, 4.2, 4.3, 6.4, 7.3)

Rainer Blind
University of Stuttgart,
Institute for Systems Theory and Automatic Control,
Pfaffenwaldring 9,
70550 Stuttgart, Germany
rainer.blind@ist.uni-stuttgart.de
(Section 7.3)

Mathias Bürger
University of Stuttgart,
Institute for Systems Theory and Automatic Control,
Pfaffenwaldring 9,
70550 Stuttgart, Germany
mathias.buerger@ist.uni-stuttgart.de
(Section 6.4)

Alain Chamaken
University of Kaiserslautern,
Institute of Automatic Control,
Erwin Schrödinger Str. 12,
67653 Kaiserslautern, Germany
chamaken@eit.uni-kl.de
(Section 7.2)

Fritz Colonius
University of Augsburg,
Department of Mathematics,
Universitätsstr. 14,
86159 Augsburg, Germany
fritz.colonius@math.uni-augsburg.de
(Sections 2.1, 2.3)

Ozan Demir
Ruhr-Universität Bochum,
Institute of Automation and Computer Control,
Universitätsstr. 150,
44780 Bochum, Germany
demir@atp.rub.de
(Section 6.5)

Frederik Deroo
TU München,
Institute of Information-oriented Control,
Arcisstr. 21,
80333 München, Germany
fred.deroo@tum.de
(Section 3.3)

Rolf Findeisen
Otto-von-Guericke University Magdeburg,
Institute for Systems Theory and Automatic Control,
Universitätsplatz 2,
39106 Magdeburg, Germany
rolf.findeisen@ovgu.de
(Sections 4.1, 4.2, 4.3)

Jörg Fischer
Karlsruhe Institute of Technology,
Institute for Anthropomatics,
Intelligent Sensor-Actuator-Systems Laboratory,
Adenauerring 2,
76131 Karlsruhe, Germany
joerg.fischer@kit.edu
(Section 4.5)

Reinhard Gotzhein
University of Kaiserslautern,
Department of Computer Science,
Postfach 3049,
67653 Kaiserslautern, Germany
gotzhein@informatik.uni-kl.de
(Sections 7.1, 7.2)

Dominic Groß
University of Kassel,
Control and System Theory,
Wilhelmshöher Allee 73,
34121 Kassel, Germany
dgross@uni-kassel.de
(Section 4.4)

Lars Grüne
University of Bayreuth,
Chair of Applied Mathematics,
Universitätsstr. 30,
95440 Bayreuth, Germany
lars.gruene@uni-bayreuth.de
(Sections 1.3, 4.1, 4.2, 4.3, 5.1, 5.4, 5.6)

Uwe D. Hanebeck
Karlsruhe Institute of Technology,
Institute for Anthropomatics,
Chair of Intelligent Sensor-Actuator-Systems,
Adenauerring 2,
76131 Karlsruhe, Germany
uwe.hanebeck@kit.edu
(Sections 4.1, 4.5)

Andreas Haupt
University of Kaiserslautern,
Institute of Automatic Control,
Erwin Schrödinger Str. 12,
67653 Kaiserslautern, Germany
haupt@eit.uni-kl.de
(Section 7.2)

Uwe Helmke
University of Würzburg,
Department of Mathematics,
Emil-Fischer-Str. 40,
97074 Würzburg, Germany
helmke@mathematik.uni-wuerzburg.de
(Sections 2.1, 2.2, 6.1, 6.3)

Anna von Heusinger
University of Würzburg,
Department of Mathematics,
Emil-Fischer-Str. 40,
97074 Würzburg, Germany
heusinger@mathematik.uni-wuerzburg.de
(Section 6.3)

Sandra Hirche
TU München,
Institute of Information-oriented Control,
Arcisstr. 21,
80333 München, Germany
hirche@tum.de
(Sections 3.3, 5.1, 5.7)

Jens Jordan
University of Würzburg,
Department of Mathematics,
Emil-Fischer-Str. 40,
97074 Würzburg, Germany
jordan@mathematik.uni-wuerzburg.de
(Section 2.2)

Oliver Junge
TU München,
Department of Mathematics,
Boltzmannstr. 3,
85748 Garching b. München, Germany
oj@tum.de
(Sections 5.1, 5.4, 5.6)

Christoph Kawan
University of Augsburg,
Department of Mathematics,
Universitätsstr. 14,
86159 Augsburg, Germany
christoph.kawan@math.uni-augsburg.de
(Section 2.3)

Benjamin Kern
Otto-von-Guericke University Magdeburg,
Institute for Systems Theory and Automatic Control,
Universitätsplatz 2,
39106 Magdeburg, Germany
benjamin.kern@ovgu.de
(Sections 4.1, 4.2, 4.3)

Péter Koltai
TU München,
Department of Mathematics,
Boltzmannstr. 3,
85748 Garching b. München, Germany
koltai@ma.tum.de
(Section 5.6)

Marc Krämer
University of Kaiserslautern,
Department of Computer Science,
Postfach 3049,
67653 Kaiserslautern, Germany
kraemer@informatik.uni-kl.de
(Section 7.2)

Daniel Lehmann
KTH Royal Institute of Technology,
Automatic Control,
Osquldas väg 10,
100 44 Stockholm, Sweden
dlehmann@kth.se
(Section 5.2)

Lothar Litz
University of Kaiserslautern,
Institute of Automatic Control,
Erwin Schrödinger Str. 12,
67653 Kaiserslautern, Germany
litz@eit.uni-kl.de
(Sections 7.1, 7.2)

Boris Lohmann
TU München,
Institute of Automatic Control,
Boltzmannstr. 15,
85748 Garching b. München, Germany
lohmann@tum.de
(Section 3.2)

Jan Lunze
Ruhr-Universität Bochum,
Institute of Automation and Computer Control,
Universitätsstr. 150,
44780 Bochum, Germany
lunze@atp.rub.de
(Sections 1.1, 1.2, 3.1, 5.1, 5.2, 5.3, 6.1, 6.2, 6.5)

Martin Meinel
TU München,
Department of Mathematics,
Boltzmannstr. 3,
85748 Garching b. München, Germany
meinel@ma.tum.de
(Section 3.3)

Adam Molin
TU München,
Institute of Information-oriented Control,
Barer Str. 21,
80333 München, Germany
adam.molin@tum.de
(Section 5.7)

Matthias A. Müller
University of Stuttgart,
Institute for Systems Theory and Automatic Control,
Pfaffenwaldring 9,
70550 Stuttgart, Germany
matthias.mueller@ist.uni-stuttgart.de
(Section 4.3)

Jürgen Pannek
Universität der Bundeswehr München,
Institut für Mathematik und Rechneranwendung,
Werner-Heisenberg-Weg 39,
85577 Neubiberg b. München, Germany
juergen.pannek@unibw.de
(Sections 4.2, 4.3)

Peter Philipp
TU München,
Institute of Automatic Control,
Boltzmannstr. 15,
85748 Garching b. München, Germany
peter.philipp@tum.de
(Section 3.2)

Marcus Reble
University of Stuttgart,
Institute for Systems Theory and Automatic Control,
Pfaffenwaldring 9,
70550 Stuttgart, Germany
reble@ist.uni-stuttgart.de
(Sections 4.1, 4.2)

Rudolf Sailer
University of Würzburg,
Department of Mathematics,
Emil-Fischer-Str. 40,
97074 Würzburg, Germany
sailer@mathematik.uni-wuerzburg.de
(Sections 2.4, 5.5)

Manuela Sigurani
University of Bayreuth,
Chair of Applied Mathematics,
Universitätsstr. 30,
95440 Bayreuth, Germany
manuela.sigurani@uni-bayreuth.de
(Section 5.6)

René Schuh
Ruhr-Universität Bochum,
Institute of Automation and Computer Control,
Universitätsstr. 150,
44780 Bochum, Germany
schuh@atp.rub.de
(Section 6.5)

Christian Stöcker
Ruhr-Universität Bochum,
Institute of Automation and Computer Control,
Universitätsstr. 150,
44780 Bochum, Germany
stoecker@atp.rub.de
(Sections 5.3, 5.8)

Olaf Stursberg
University of Kassel,
Control and System Theory,
Wilhelmshöher Allee 73,
34121 Kassel, Germany
stursberg@uni-kassel.de
(Sections 4.1, 4.4)

Michael Ulbrich
TU München
Chair of Mathematical Optimization,
Boltzmannstr. 3,
85748 Garching b. München, Germany
mulbrich@ma.tum.de
(Section 3.3)

Paolo Varutti
Otto-von-Guericke University Magdeburg,
Institute for Systems Theory and Automatic Control,
Universitätsplatz 2,
39106 Magdeburg, Germany
paolo.varutti@ovgu.de
(Sections 4.1, 4.2, 4.3)

Fabian Wirth
University of Würzburg,
Department for Mathematics
Emil-Fischer-Str. 40,
97074 Würzburg, Germany
wirth@mathematik.uni-wuerzburg.de
(Sections 2.1, 2.4, 5.1, 5.5)

Karl Worthmann
University of Bayreuth,
Chair of Applied Mathematics,
Universitätsstr. 30,
95440 Bayreuth, Germany
karl.worthmann@uni-bayreuth.de
(Sections 4.1, 4.2, 4.3)

Notation

The symbols are chosen according to the following conventions. Lower-case letters like x, a or t denote scalar values or signals, boldface letters like $\boldsymbol{x}$ or $\boldsymbol{y}$ vectors and boldface upper-case letters such as $\boldsymbol{A}$, $\boldsymbol{B}$ matrices. Accordingly, the elements x_1,...,x_n of a vector $\boldsymbol{x}$ or a_{11}, a_{12},..., a_{mn} of a matrix $\boldsymbol{A}$ are represented by italics. Sets are symbolized by calligraphic letters like $\mathcal{F}$ and $\mathcal{Z}$. $\boldsymbol{O}$ or $\mathbf{0}$ denotes a zero matrix or a zero vector of appropriate dimension, respectively. For a matrix $\boldsymbol{E}$, e_{ij} or $[\boldsymbol{E}]_{ij}$ denotes the ij-th element. $\mathbf{1}$ is the vector of appropriate size with unity elements:

$$\mathbf{1} = \begin{pmatrix} 1 \\ 1 \\ \vdots \\ 1 \end{pmatrix}.$$

$\mathrm{diag}(a_1, ..., a_n)$ is a diagonal matrix with diagonal entries a_1,...,a_n. It is sometimes abbreviated as $\mathrm{diag}\, a_i$. In structured matrices, sometimes the vanishing blocks are suppressed for the clarity of notation.

The inequality $\boldsymbol{P} > 0$ is interpreted in two different ways, which is explicitly mentioned. In connection with optimal control or linear matrix inequalities, it states that the matrix $\boldsymbol{P}$ is positive definite. For this interpretation also the symbol $\boldsymbol{P} \succ 0$ is used. Alternatively, the sign $>$ has to be interpreted as an elementwise relation saying that all elements of the matrix $\boldsymbol{P}$ are positive ($p_{ij} > 0$ for all i, j).

Book Homepage. The book homepage at `www.atp.rub.de/buch/SPP` provides further information on the subject of this book.

1

Introduction to Networked Control Systems

J. Lunze and L. Grüne

This chapter gives an informal introduction to networked systems, explains the main problems and gives a survey of the new results reported in this monograph.

Chapter Contents

1.1 What Are Networked Control Systems?

1.1.1 Motivation and Examples

The control of dynamic systems necessitates the communication of sensor information towards a controller and of information about the control input from the controller towards the actuators. The implementation of this communication by means of digital networks has been common practice for more than 20 years. The great current interest in the field of networked control systems results from the fact that a large variety of digital networks are becoming available everywhere and can be used for the implementation of feedback loops without additional installation cost. Wireless connections facilitate the extension of the application area of automatic control towards mobile objects, because measured variables and control variables can now be transmitted to the controller from nearly every place of a technological plant.

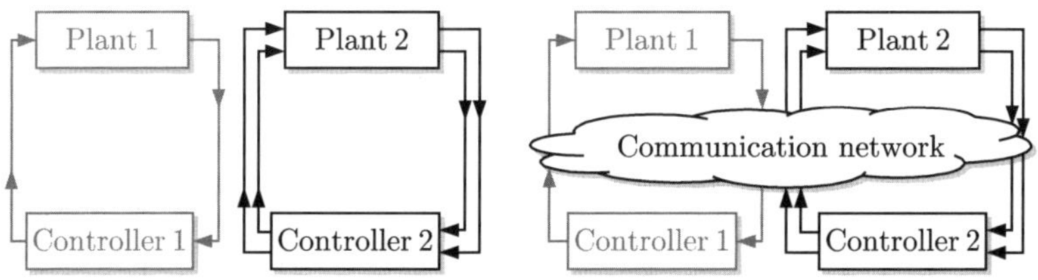

Fig. 1.1 Subject of classical control theory (left) and of the theory of digitally networked control systems (right)

A first rough comparison of the new theory of digitally networked systems and the traditional control theory is depicted in Fig. 1.1, which shows, on the left-hand side, two independent control loops. These loops may be implemented by means of a communication network, but if the network is dedicated to these control loops, the overall system behavior does not depend upon this network and both control loops can be considered as independent entities. They can be analyzed and designed by means of multivariable control theory.

If, however, the network does impose constraints on the information transmission, it influences the behavior of the overall system. Therefore, the network has to be dealt with as a separate component in the closed-loop system as shown on the right-hand side of Fig. 1.1. This new situation has severe consequences. As the figure shows, the two loops, which are separated in the classical sense, influence one another in the networked situation. The influence may deteriorate the performance of both loops, because the network introduces additional delays into the feedback loops, but the coupling may also be used to improve the behavior if the flexibility of the communication network is used to couple both control loops in favor of a quicker disturbance attenuation or set-point following.

The new control structures cannot be handled with traditional methods for two reasons. First, the event-driven mode of operation of digital networked systems violates the precondition of discrete-time control theory that data are processed and transmitted with a constant sample rate. Second, the structure of the network is usually not pre-defined but the subject of control design. The data links may even adjust themselves to the needs and the technical constraints. Research on digitally networked control systems deals with novel questions concerning the modeling of dynamic systems, the analysis of feedback systems, and the design of distributed controllers.

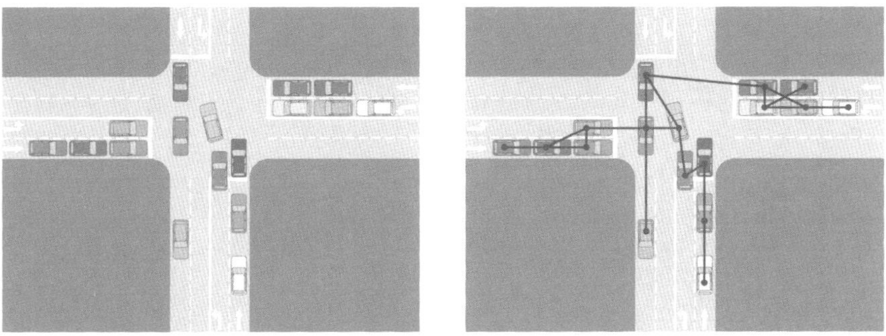

Fig. 1.2 Future crossroad management with wireless vehicle-to-vehicle communication

Examples for Networked Control Systems. In the application scenarios considered in the theory of networked control systems, the assumption of classical control theory that information can be transmitted within control loops instantaneously, reliably and with sufficient precision is no longer satisfied as the following examples show:

- **Telerobotics:** Long communication links between the (human) controller and the robots to be controlled may bring about time delays and partial loss of communication.

- **Traffic control:** If floating-car data are used for traffic supervision, state estimation and control has to use wireless components and moving sensors which do not work in accordance with a centralized sampling scheme, but they are driven by the traffic circumstances and by the tuning range of the wireless communication.

- **Future crossroad management:** Advanced communication and control systems will make it possible to bring vehicles safely over a crossroad by using vehicle-to-vehicle communication. The vehicles decide autonomously who goes next (Fig. 1.2). Only those vehicles are connected that are in the surrounding of the crossroad. The number of vehicles involved depends upon the current traffic and the communication effort changes during the flow of the traffic.

- **Smart grid:** Energy distribution networks are gaining more intelligence by the introduction of smart components at all layers from the autonomous switches towards smart measuring components [28, 112, 262]. Likewise, smart cities should be created where economy, mobility, environment, living conditions and governance are enhanced by information and communication technologies in order to reach a sustainable economic development [184].

- **Ambient intelligence:** If people are supervised by wireless sensors distributed network structures are used that have no coordinator [113, 318].

Properties of Digital Communication Networks. These examples show that the kind of networks considered in the field of "networked control systems" have the following properties:

- The networks are open and inhomogeneous with changing topology and nodes.
- The networks behave non-deterministically in dependence upon the number of nodes, the used links and the load.
- The networks provide a flexible communication structure, even for mobile objects, that can be used whenever necessary.

Hence, the behavior of the overall systems that include the physical systems to be controlled, the controllers and the communication network is severely influenced by the communication network. The theory of networked control systems should elaborate new methods for dealing with the phenomena appearing in such systems. There are, in summary, two challenges: the communication constraints imposed by the network and the flexibility to be used for solving novel control problems.

Communication Constraints. Traditional control theory uses the assumption that all communication links that are necessary for solving a control problem at hand can be implemented and used with the required quality and reliability. Moreover, the communication is assumed to be instantaneous and lossless. These assumptions are not satisfied by the networks described above. Hence, a control theory of networked systems has to take into account communication constraints with respect to timing, information loss, variable communication topology etc.

Flexibility of Communication. With the new communication technology, information can be transferred among arbitrary nodes within control loops and among different control systems. In particular, the communication topology can be adapted to the current needs. In this new situation, the value of information for solving control tasks gains a particular importance and has to be the subject of scientific research. Whereas traditional control methods start from a fixed communication topology, the flexibility brought about by modern communication networks makes it necessary to elaborate criteria for deciding which value a certain piece of information has for the solution of a control task. These criteria show under what conditions and with which quality information links within control loops and among control systems should be invoked and used.

Due to these properties, the communication network is drawn in the figures of this monograph as clouds without a well-defined boundary rather than as a block with well-known properties.

The examples above have shown that the communication constraints of the networks used will not be overcome simply by advances of technology. Networks should be realized by cheap components, which have restricted energy and computing power and, hence, provide information with restricted sampling rate and accuracy. Moving objects impose constraints on the tuning area of the components and, thus, make the communication process nondeterministic. Temporal loss of communication links cannot be avoided. To repeat sending data packets until they reach the receiver introduces time delay, which may exceed tolerable bounds. Multi-hop data transmissions are associated with large time delays because every communication step necessitates the synchronization of all senders and receivers involved.

Networked Control Systems. Having these properties of the communication network in mind, the term *networked control systems* can be defined as follows:

> Networked control systems are closed-loop systems that have to be considered as networked systems.

The term "closed-loop system" includes the fact that controllers are connected in feedforward and feedback structures with physical systems, where both components together determine the behavior of the overall system. The aspect of the "networked system" emphasizes the fact that the overall system consists of several components that are connected physically or by a digital communication network and have to be analyzed by considering the network structure. Hence, networked control systems are typically represented as an interconnection of nodes with dynamic properties.

An important aspect of using graph-theoretic representation forms lies in the fact that the overall system may change its structure during operation. Nodes may disappear or new nodes may join the network. So, the dynamics of the network do not only lie in the properties of the nodes but also in the properties of the overall system.

This definition of networked control systems leads to an answer to the question, under what conditions a feedback system has to be considered as a networked system. The communication network has to be considered as an explicit component in the control loop,

- if the task is to investigate which communication links are necessary (i.e. the communication topology is a design goal),
- if communication constraints (e.g. scheduling, time delay, packet loss) have to be taken into account,
- if control objectives have to be satisfied that necessitate coordinated actions of subsystems (e.g. multi-agent systems).

1.1.2 Cyber-Physical Systems

The close connection of the physical world with embedded (control) systems coupled through a global network is currently investigated under the buzz word of cyber-physical systems. Cyber-physical systems appear as an extension of embedded computer systems by global networks. One expects that the bandwidth of modern digital communication networks will double every two years, which leads to the question which information links are necessary. From a global viewpoint, all computer networks are expected to be coupled in the future to form the upcoming *Internet of things and services* [206, 218, 284, 298].

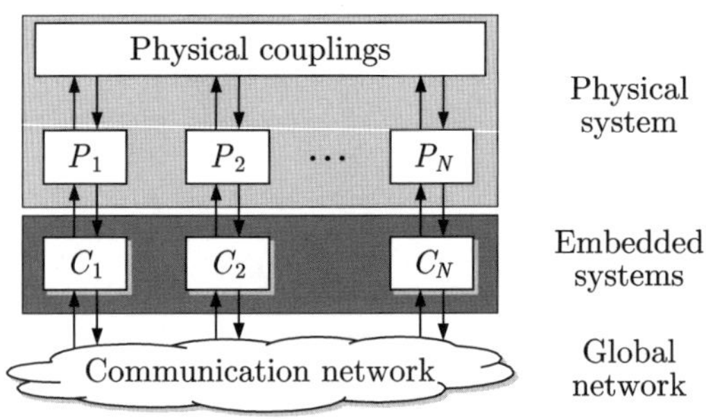

Fig. 1.3 Networked control system as cyber-physical system

The networked control systems investigated in this book are such cyber-physical systems, as Fig. 1.3 shows. The embedded systems C_i are used to implement local controllers of the subsystems P_i, $(i = 1, 2, ..., N)$. The network allows a direct information exchange among the local controllers. The novel situation provided by modern digital networks is characterized by the fact that this communication can be used extensively whenever communicated information can be utilized to improve the overall system performance.

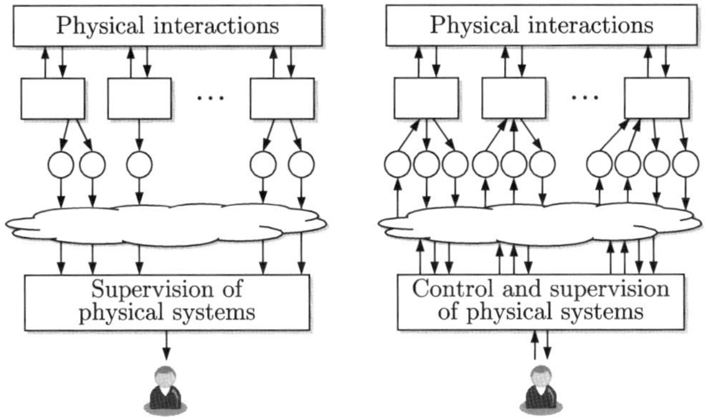

Fig. 1.4 Network of information (left) and network of action (right)

The theory of networked control systems emphasises the interaction of the cyber-system (controller, network) and the physical system (plant).

The future development of networking technology can also be seen as an evolution of the current *network of information* towards a future *network of actions*. In the network of information, the information flow goes mainly from the physical system towards the supervision system and the human operators (Fig. 1.4 (left)). The new aspect of the network of actions is the fact that the subsystems can communicate among each other and with the human operator and that this communication goes in closed loops (Fig. 1.4 (right)).

Control, Communication and Computation. A control theory of networked systems has to deal with the new challenges that originate from this new network structure and has to develop new ideas and methods for networked control systems.

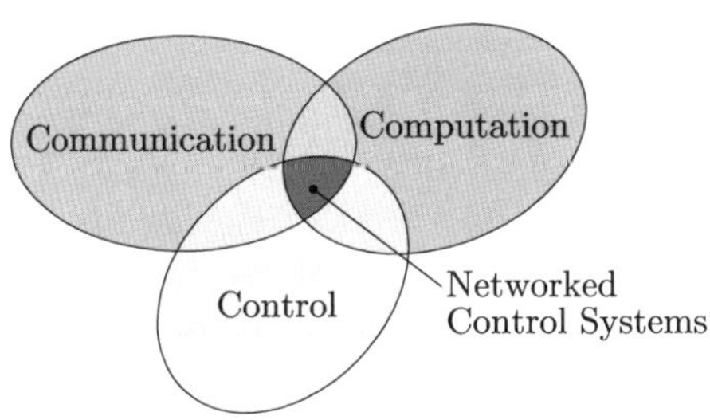

Fig. 1.5 Integration of control with communication and computation

For the scientific development, the network of actions mean that the fields of communication, computation and control have to be developed in an integrated way (Fig. 1.5). Currently, these three aspects are considered separately.

For the control loops, one can assume that communication and computation are quick enough not to bring about substantial time delays into the closed-loop systems. In contrast, in networked control systems, communication constraints and scheduling problems have to be taken into account. Hence, the models used to represent networked control systems combine aspects of the physical description of the plant with models of the communication network and of the computation schedule. Section 1.2 will summarize different classes of models together with the corresponding analysis and design methods to show how the main properties of the communication network extend the dynamic models of the plant or the closed-loop system, respectively.

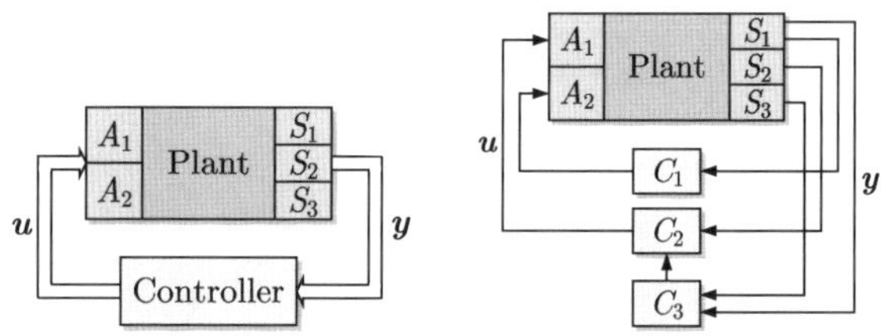

Fig. 1.6 Multivariable system vs. networked control system

Methods and tools have to be investigated to integrate the engineering view on physical systems with the computer science view on computation and communication. Consequently, new technologies have to include

- heterogeneous modeling with various model abstractions, and
- analysis and design methods for heterogeneous systems.

This combination of communication and control has direct consequences for the development of advanced control systems in industry. As long as the communication can be assumed to be ideal, the design of control systems can be separated from the design and implementation of the controllers and the communication network. Separate design decisions can be made in both fields and, if brought together, the communication network and the controllers will satisfy the overall system goals.

However, if communication and control have to be considered together, design decisions in one field have direct consequences on design decisions in the other field. Trade-offs have to be made from a global viewpoint and an inter-disciplinary cooperation between experts in communication and control is necessary.

1.1.3 Structures of Networked Control Systems

For control engineering applications, the trends outlined above lead to the question what are the main challenges for control theory if a systematic analysis and design of closed-loop systems should be facilitated. Figure 1.6 compares conventional multi-input multi-output feedback loops with networked control systems. Traditionally, a control loop is considered to consist of one plant and one controller, where all sensor information is lumped together to form the output vector $\boldsymbol{y}(t)$ and all the control inputs are considered as a single vector $\boldsymbol{u}(t)$. The individual handling of the sensor information y_i, controllers C_i and inputs u_i to the actuators results in event-driven operating modes, distributed controllers and goal-dependent communication regimes (Table 1.1).

Table 1.1 Comparison of traditional control theory and the theory of networked systems

Traditional control theory	**Theory of networked control systems**
Time-driven, sampled-data systems	Event-driven, asynchronous systems
Centralized, coordinated controllers	Decentralized, distributed controllers
Fixed communication structure	Goal-dependent communication structure

The main characteristics of networked control systems can also be seen in novel structures that are characterized by

- distributed sensing, computation, and actuation,
- decentralized and distributed control structures with intelligent nodes, and
- resource constraints and imperfect communication.

Bibliographical Notes

During the last ten years, several survey articles, special issues of journals and some monographs have appeared, which can be used as starting points to study the theory of networked control systems in more detail:

- The survey articles [8, 178] give an answer to the question what is new in networked control systems.
- Special issues of journals collect new results in this field: [7, 159, 182, 189]
- Monographs deal with special aspects of networked systems. [254] concentrate on estimation problems.
- The material developed for PhD schools and workshops give a broad and didactical introduction to the field, e.g. [26]

1.2 Theory of Networked Control Systems

1.2.1 Overview

This section describes the main problems to be tackled by the theory of digitally networked dynamic systems and gives a survey of the models and solutions described in this monograph. The methods are ordered according to the model of the networked system used.

The main question that a theory of networked control systems has to answer asks

|| Which information is necessary to solve a given control task?

This question has two important aspects:

- **Topology:** Which information links are necessary?
- **Quality:** How often and how quickly has information to be communicated and which accuracy of information is necessary?

In order to answer these questions, new modeling paradigms for networked control systems have to be elaborated, the main idea of which is to incorporate the important properties of the communication network into the model of the plant. The models combine methods that have been developed in the past in control theory, computer science, and communication theory.

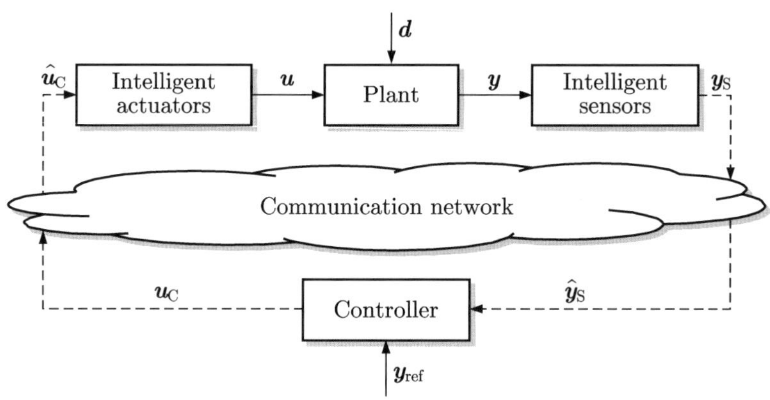

Fig. 1.7 Basic networked control system

The principal ideas used for modeling networked systems can be explained for the single control loop shown in Fig. 1.7. In order to represent effects of the communication network on the data transfer from the sensors towards the controller as well as from the controller to the actuators, the sensor signal $\boldsymbol{y}_{\mathrm{S}}(t)$ is distinguished from the controller input $\hat{\boldsymbol{y}}_{\mathrm{S}}(t)$ and the controller output $\boldsymbol{u}_{\mathrm{C}}(t)$ from the actuator input $\hat{\boldsymbol{u}}_{\mathrm{C}}(t)$. One can say that the signals $\hat{\boldsymbol{y}}_{\mathrm{S}}(t)$ and $\hat{\boldsymbol{u}}_{\mathrm{C}}(t)$ are the "network versions" of the original signals $\boldsymbol{y}_{\mathrm{S}}(t)$ or $\boldsymbol{u}_{\mathrm{C}}(t)$, respectively, sent over the corresponding communication channels.

The notions of intelligent sensors and intelligent actuators used in the figure should emphasize that these components may include computing elements, for which generally no restrictions concerning the computing power and the memory are imposed. The solid arrows show continuous-time signals, whereas the dashed arrows denote information links that are only used at certain transmission times t_k, $(k = 0, 1, ...)$.

The methods to incorporate the communication network into the model of the plant or of the closed-loop system can be classified as follows:

- **Control-theoretic approaches:** From a control-theoretic viewpoint, data transfer over the network introduces time delays, which have to be considered when analyzing the overall system and designing the controller. Furthermore, information feedback is only necessary if disturbances have to be attenuated or model uncertainties to be tolerated. Control-theoretic approaches should answer the question at which time instances communication is necessary.

- **Information-theoretic approaches:** The data transfer is restricted to a certain bit-rate, which means that the signals $\boldsymbol{y}_{\mathrm{S}}(t)$ and $\boldsymbol{u}_{\mathrm{C}}(t)$ sent over the network is restricted to a certain bit length. The question is whether this information flow is sufficient to solve the control task given for a certain application.

- **Network-theoretic approaches:** The overall system is considered as a graph, the nodes of which are dynamic elements and the edges show where information among the nodes may be exchanged. The communication topology represented by the graph may change over time if data packets get lost or if communication links are created or disappear because it cannot be used any longer. This approach is useful if the question should be answered which information links are necessary to solve a given control task.

As a result, heterogeneous models are used to represent the control system together with the communication network. The following sections give a survey of these models and the problems that can be solved by using these models. In applications, several of these approaches have to be combined to deal with all practical constraints of networked systems.

1.2.2 Control with Information Rate Constraints

The communication network may impose strong restrictions concerning the data rate, particularly if wireless networks are used. To include these restrictions into the model of the control loop, the sensor nodes are equipped with an encoder that, from a control-theoretic viewpoint, introduce a signal space partition. The encoder operation is denoted by $[\cdot]$

$$[\cdot] : \mathbb{R}^n \to \{1, 2, ..., 2^{\bar{r}}\}$$

and the symbol $[\boldsymbol{x}(t_k)]$ sent at time t_k is the number of the partition in which the state $\boldsymbol{x}(t_k)$ lies at that time t_k. $\bar{r}$ denotes the maximum bit length of the data sent over the network.

As shown in Fig. 1.8, it is usually assumed that the state $\boldsymbol{x}$ is measurable and its encoded version $[\boldsymbol{x}]$ is transmitted to the intelligent actuator, which includes a decoder together with a state observer and the controller. The fundamental question to be answered is how many bits per second have to be transmitted over the network in order to solve a given control task.

Investigations along this line have led to *data-rate theorems*, which give lower bounds on the bit rate necessary to stabilize an unstable plant. For the standard situation, where the controller should stabilize a linear discrete-time system

$$\Sigma : \begin{cases} \boldsymbol{x}(k+1) = \boldsymbol{A}\boldsymbol{x}(k) + \boldsymbol{B}\boldsymbol{u}(k), & \boldsymbol{x}(0) = \boldsymbol{x}_0 \\ \boldsymbol{y}(k) = \boldsymbol{C}\boldsymbol{x}(k) \end{cases}$$

the minimum number of bits to be sent at every sampling instant is bounded by

$$\bar{r} > \sum_{|\lambda_i|>1} \log_2 |\lambda_i|, \tag{1.1}$$

where the sum concerns all unstable eigenvalues λ_i of the matrix $\boldsymbol{A}$. It is interesting to see that the amount of data to be sent over the network directly depends upon the unstable eigenvalues of the plant.

First papers on this subject include [14, 269, 358, 389, 390] and a survey is given in [271]. The main idea to prove the inequality (1.1) is to design a state observer that reconstructs the smallest set of states in which the current plant

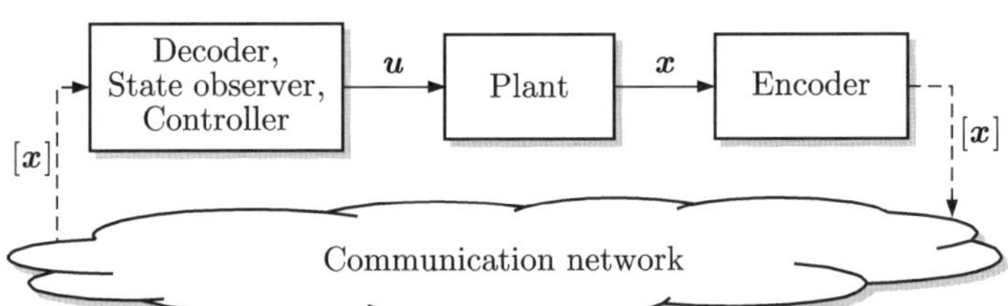

Fig. 1.8 Feedback loop with information rate constraints

state is at time k. This state set expands from one sampling time to the next time due to the unstable modes of the system but it is reduced by the new quantized measurement information. The inequality (1.1) ensures that the expansion is smaller than the reduction and, hence, after a sufficient number of sampling instants the state $\boldsymbol{x}(k)$ is known to the observer with sufficient precision. Then a state feedback can be applied to stabilize the plant.

Investigations along this line are reported in the following parts of this book:

- **Minimal bit rates and entropy for control tasks:** In Section 2.3, results on the minimal transition data rate for making a subset of the state space invariant are described.

- **Quantized event-based control:** Section 5.4 develops an event-based control strategy where the information sent from the sensor towards the controller is restricted to the quantized state. This information is communicated only whenever the state changes from one quantization region to another one. Hence, the feedback information used in the networked overall system is reduced with respect to both the time instants at which it is sent and the contents.

1.2.3 Control Subject to Networked-Induced Time Delays

If the communication network introduces temporarily or permanently severe time delays in the feedback path from the sensors via the controller towards the actuators, the network can be modeled as a time-delay system. In Fig. 1.9 the overall time delay on the feedback path is denoted by τ. Hence, the behavior of the networked system on the left-hand side of the figure can be represented as the time-delay system shown on the right-hand side, where $\tau = \tau_{\mathrm{y}} + \tau_{\mathrm{u}}$ holds.

In the structure shown, the network version $\hat{u}(t)$ of the control input $u(t)$ generated by the controller is delayed so that the overall system can bedescribed by

$$\text{Plant:} \quad \begin{cases} \dot{\boldsymbol{x}}_{\mathrm{p}}(t) = \boldsymbol{A}_{\mathrm{p}}\boldsymbol{x}_{\mathrm{p}}(t) + \boldsymbol{B}_{\mathrm{p}}\boldsymbol{u}(t-\tau) + \boldsymbol{E}\boldsymbol{d}(t) \\ \boldsymbol{y}(t) = \boldsymbol{C}_{\mathrm{p}}\boldsymbol{x}_{\mathrm{p}}(t) \end{cases}$$

$$\text{Controller:} \quad \begin{cases} \dot{\boldsymbol{x}}_{\mathrm{c}}(t) = \boldsymbol{A}_{\mathrm{c}}\boldsymbol{x}_{\mathrm{c}}(t) + \boldsymbol{B}_{\mathrm{c}}\boldsymbol{y}(t) \\ \boldsymbol{u}(t) = \boldsymbol{C}_{\mathrm{c}}\boldsymbol{x}_{\mathrm{c}}(t) \end{cases}$$

The well-known theory on time-delay systems can be applied for analyzing such feedback systems. However, in networked systems, the delay τ is usually unknown. This fact has motivated new investigations on time-delay systems with stochastic delays, on systems with delays that are described by upper bounds, and on systems for which the delay can be determined online. Overviews can be found, for example, in [236, 405].

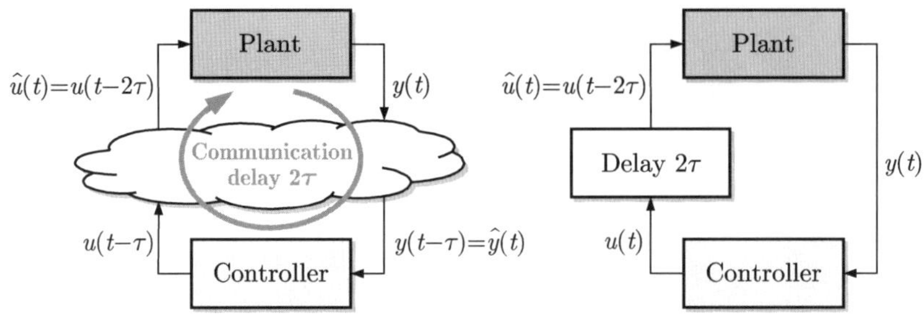

Fig. 1.9 Networked control system with delayed communication

Control with Asynchronous Communication and Computation. In a more detailed analysis, event-triggered components of the control loop work in different time schemes. Figure 1.10 shows the situation, in which the sensors work time-triggered at every clock time, whereas the controller and the actuator computes or applies the new input signal in an event-driven way as soon as it has received new information. Under the standard assumption of discrete-time control theory, all components work synchronously at the same time (Fig. 1.10 (left)). This assumption is satisfied as long as the time delays are small in comparison to the main time constants of the control loop.

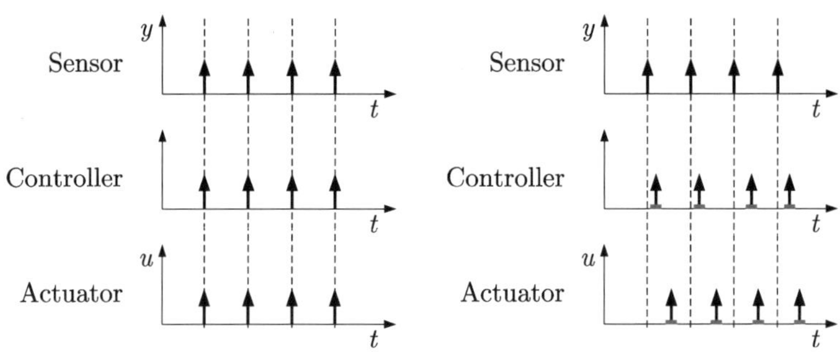

Fig. 1.10 Synchronous (left) and asynchronous behavior (right) of the components of a control loop

If the delays are larger, the missing synchrony has to be represented in the model and leads to an interesting and important extension of discrete-time models. Assume, for simplicity, that there is only a delay between the sensor and the controller in the control loop, but the actuator receives the output of the controller without delay. Then the actuator has to implement the input

$$\boldsymbol{u}(t) = \begin{cases} \boldsymbol{u}(k-1) & \text{for } kT \le t < kT+\tau \\ \boldsymbol{u}(k) & \text{for } kT+\tau \le t < (k+1)T, \end{cases}$$

where T denotes the sampling time. In this formula, $\boldsymbol{u}(t)$ denotes the continuous-time signal and $\boldsymbol{u}(k)$ the k-th value of the discrete-time version of the input $\boldsymbol{u}$. Instead of getting the usual sampled-data plant model

$$\begin{aligned} \boldsymbol{x}_{\mathrm{p}}(k+1) &= \boldsymbol{A}_{\mathrm{d}}\boldsymbol{x}_{\mathrm{p}}(k) + \boldsymbol{B}_{\mathrm{d}}\boldsymbol{u}(k) \\ \boldsymbol{y}(k) &= \boldsymbol{C}_{\mathrm{p}}\boldsymbol{x}_{\mathrm{p}}(k) \end{aligned}$$

with

$$\boldsymbol{A}_{\mathrm{d}} = \mathrm{e}^{\boldsymbol{A}_{\mathrm{p}}T}, \quad \boldsymbol{B}_{\mathrm{d}} = \int_0^T \mathrm{e}^{\boldsymbol{A}_{\mathrm{p}}\tau}\,\mathrm{d}\tau \boldsymbol{B}_{\mathrm{p}}$$

one now gets the extended plant model

$$\Sigma_{\mathrm{d}}: \begin{cases} \boldsymbol{x}_{\mathrm{p}}(k+1) = \boldsymbol{A}_{\mathrm{d}}\boldsymbol{x}_{\mathrm{p}}(k) + \boldsymbol{B}_{\mathrm{d}1}\boldsymbol{u}(k) + \boldsymbol{B}_{\mathrm{d}2}\boldsymbol{u}(k-1) \\ \boldsymbol{y}(k) = \boldsymbol{C}_{\mathrm{p}}\boldsymbol{x}_{\mathrm{p}}(k), \end{cases}$$

which has two input terms with the matrices

$$\boldsymbol{B}_{\mathrm{d}1} = \int_\tau^T \mathrm{e}^{\boldsymbol{A}_{\mathrm{p}}\tau}\,\mathrm{d}\tau \boldsymbol{B}_{\mathrm{p}} \quad \text{and} \quad \boldsymbol{B}_{\mathrm{d}2} = \int_0^\tau \mathrm{e}^{\boldsymbol{A}_{\mathrm{p}}\tau}\,\mathrm{d}\tau \boldsymbol{B}_{\mathrm{p}}.$$

In order to replace the two addends in the state equation by the usual input term, one can "lift" the model by introducing the input as a new part of the state vector leading to the model

$$\Sigma_{\mathrm{d}}: \begin{cases} \begin{pmatrix} \boldsymbol{x}_{\mathrm{p}}(k+1) \\ \boldsymbol{u}(k) \end{pmatrix} = \begin{pmatrix} \boldsymbol{A}_{\mathrm{d}} & \boldsymbol{B}_{\mathrm{d}2} \\ \boldsymbol{O} & \boldsymbol{O} \end{pmatrix} \begin{pmatrix} \boldsymbol{x}_{\mathrm{p}}(k) \\ \boldsymbol{u}(k-1) \end{pmatrix} + \begin{pmatrix} \boldsymbol{B}_{\mathrm{d}1} \\ \boldsymbol{I} \end{pmatrix} \boldsymbol{u}(k) \\ \boldsymbol{y}(k) = (\boldsymbol{C}_{\mathrm{p}} \ \ \boldsymbol{O}) \begin{pmatrix} \boldsymbol{x}_{\mathrm{p}}(k) \\ \boldsymbol{u}(k-1) \end{pmatrix}. \end{cases}$$

For both models of the discrete-time plant Σ_{d} the usual theory of sampled-data systems has to be extended to cope with the new model structure brought about by the time delay τ, which may vary in dependence upon the sampling time.

Time delays play an important role in the following sections of this book:

- **Observability of networked systems:** Section 2.2 investigates the observability of networked systems for linear and nonlinear systems with transmission time delays.

- **State estimation in networked control systems:** Section 3.2 proposes two architectures for estimating the overall system state of interconnected systems, where information about the subsystems are transmitted over a communication network. The main problem is to make the estimation result tolerate the imperfections of the network like packet delays and packet loss.

- **Stochastic model-based control based on virtual control inputs:** The idea to compensate missing or deliberately ignored information of subsystems by stochastic a-priori and a-posteriori information is explained in Section 4.5.

1.2.4 Control Subject to Packet Loss

Packet loss describes a usual phenomenon in digital networks. Some network protocols ensure that the lost packet is sent again until it reaches the receiver, but even then time has gone and, hence, packet loss may deteriorate the control performance. Therefore, several methods have been proposed to deal with this situation.

In the network-theoretic approaches to this problem, stochastic models for packet dropping are used, where p denotes the probability with which a packet does not arrive at the receiver. The critical packet drop rate is the probability $\bar{p}$ for which the average estimation error of the state estimator implemented in the intelligent actuator shown in Fig. 1.8 using the communicated information is bounded. Like in the data-rate theorems, $\bar{p}$ depends directly on the unstable eigenvalue λ of the plant to be stabilized [340]:

$$\bar{p} < \frac{1}{\lambda^2}.$$

Further methods to handle packet loss are published, for example, in [404].

Model predictive control provides a suitable means to overcome the effects of packet loss, because at every sampling time k the controller does not only determine the current control input $\boldsymbol{u}(k)$, but a sequence of future control inputs $\boldsymbol{u}(k+1)$, $\boldsymbol{u}(k+2)$,..., $\boldsymbol{u}(k+h)$ (Fig. 1.11), where h is the prediction time horizon. If this sequence is transmitted to the actuator, the actuator may use the future input values if the next packets are lost.

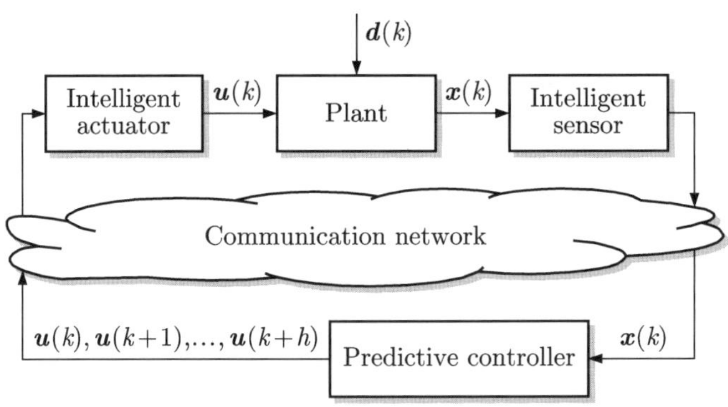

Fig. 1.11 Predictive control as a means to overcome the effect of packet loss

Although this principle seems to work easily, a more detailed analysis made in Chapter 4 shows that there are several important problems to be solved. In particular, the controller can predict the future plant behavior only if it knows which inputs the actuator has applied to the plant. In case of packet loss, the controller will use a different control input for the prediction as the actuator, which will apply "old" input values. The analysis in [120, 146, 154, 297] shows that a synchronization of the controller with the actuator is only possible if acknowledgements are sent from the actuator towards the controller whenever a new input sequence has arrived.

This control structure is investigated in this monograph also with respect to delayed information transmission:

- **Compensation of delays and packet loss by model predictive control:** Section 4.2 describes how the principle of model predictive control to determine at every time step a sequence of future control inputs can be used to compensate time delays and packet loss induced by the communication network.

- **Compensation of delays and packet loss by means of virtual control inputs:** In Section 4.5 time delays and data loss in the network connection between the controller and the actuators are compensated by buffered input sequences.

- **State estimation as part of control-communication co-design:** The state estimator developed in Section 7.2 has the aim to reconstruct the plant state even in case of packet loss between the sensor and the controller and combines this estimate with a controller that is implemented by wireless communication.

1.2.5 Control under Restrictions on the Network Access

The communication network can be used by only one sender at any time. Many network protocols include a scheduler that prescribes an order in which the different nodes have access to the network. If substantial time passes before a node is allowed to send, the communication constraints imposed by the network protocol influence the overall system performance and have to be included in the model of the feedback system.

In Fig. 1.12 only one signal of the set $\{y_1, y_2, u_1, u_2\}$ can be sent and, hence, only one signal of the set $\{\hat{y}_1, \hat{y}_2, \hat{u}_1, \hat{u}_2\}$ can be updated at a certain time t_k

$$\hat{y}_i(t_k^+) = y_i(t_k) \quad \text{or} \quad \hat{u}_i(t_k^+) = u_i(t_k) \quad \text{for one index } i \in \{1, 2\},$$

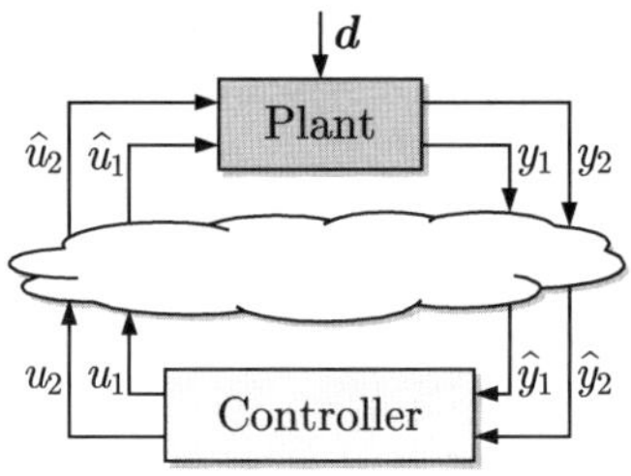

Fig. 1.12 Feedback loop in which only one signal may be transmitted at any time

where t_k^+ denotes the time just after the signal update at time t_k. The networked-induced errors

$$\begin{aligned} e_{yi}(t) &= y_i(t) - \hat{y}_i(t) \\ e_{ui}(t) &= u_i(t) - \hat{u}_i(t), \quad i = 1, 2 \end{aligned}$$

can be described by the differential equation

$$\dot{e}_{yi}(t) = \frac{\mathrm{d}}{\mathrm{d}t} y_i(t) - \underbrace{\frac{\mathrm{d}}{\mathrm{d}t} \hat{y}_i(t)}_{=0}$$

together with the update relation

$$e_{yi}(t_k^+) = \begin{cases} 0 & \text{if } \hat{y}_i(t_k) = y_i(t_k) \\ e_{yi}(t_k) & \text{else.} \end{cases}$$

Hence, the plant model and the controller

$$\text{Plant:} \quad \begin{cases} \dot{\boldsymbol{x}}_\mathrm{p}(t) = \boldsymbol{A}_\mathrm{p}\boldsymbol{x}_\mathrm{p}(t) + \boldsymbol{B}_\mathrm{p}\hat{\boldsymbol{u}}(t) + \boldsymbol{E}\boldsymbol{d}(t) \\ \boldsymbol{y}(t) = \boldsymbol{C}_\mathrm{p}\boldsymbol{x}_\mathrm{p}(t) \end{cases}$$

$$\text{Controller:} \quad \begin{cases} \dot{\boldsymbol{x}}_\mathrm{c}(t) = \boldsymbol{A}_\mathrm{c}\boldsymbol{x}_\mathrm{c}(t) + \boldsymbol{B}_\mathrm{c}\hat{\boldsymbol{y}}(t) \\ \boldsymbol{u}(t) = \boldsymbol{C}_\mathrm{c}\boldsymbol{x}_\mathrm{c}(t) \end{cases}$$

together with the model of the network-induced errors lead to the following model of the closed-loop system:

$$\begin{aligned} \begin{pmatrix} \dot{\boldsymbol{x}}_\mathrm{p}(t) \\ \dot{\boldsymbol{x}}_\mathrm{c}(t) \\ \dot{\boldsymbol{e}}_y(t) \\ \dot{\boldsymbol{e}}_u(t) \end{pmatrix} &= \begin{pmatrix} \boldsymbol{A}_\mathrm{p} & \boldsymbol{B}_\mathrm{p}\boldsymbol{C}_\mathrm{c} & \boldsymbol{O} & -\boldsymbol{B}_\mathrm{p} \\ \boldsymbol{B}_\mathrm{c}\boldsymbol{C}_\mathrm{p} & \boldsymbol{A}_\mathrm{c} & -\boldsymbol{B}_\mathrm{c} & \boldsymbol{O} \\ \boldsymbol{C}_\mathrm{p}\boldsymbol{A}_\mathrm{p} & \boldsymbol{C}_\mathrm{p}\boldsymbol{B}_\mathrm{p}\boldsymbol{C}_\mathrm{c} & \boldsymbol{O} & -\boldsymbol{C}_\mathrm{p}\boldsymbol{B}_\mathrm{p} \\ \boldsymbol{C}_\mathrm{c}\boldsymbol{B}_\mathrm{c}\boldsymbol{C}_\mathrm{p} & \boldsymbol{C}_\mathrm{c}\boldsymbol{A}_\mathrm{c} & -\boldsymbol{C}_\mathrm{c}\boldsymbol{B}_\mathrm{c} & \boldsymbol{O} \end{pmatrix} \begin{pmatrix} \boldsymbol{x}_\mathrm{p}(t) \\ \boldsymbol{x}_\mathrm{c}(t) \\ \boldsymbol{e}_y(t) \\ \boldsymbol{e}_u(t) \end{pmatrix} \\ &\quad + \begin{pmatrix} \boldsymbol{E} \\ \boldsymbol{O} \\ \boldsymbol{C}_\mathrm{p}\boldsymbol{E} \\ \boldsymbol{O} \end{pmatrix} \boldsymbol{d}(t) \end{aligned} \tag{1.2}$$

$$\begin{pmatrix} \boldsymbol{e}_y(t_k^+) \\ \boldsymbol{e}_u(t_k^+) \end{pmatrix} = \phi_k \left(\begin{pmatrix} \boldsymbol{e}_y(t_k) \\ \boldsymbol{e}_u(t_k) \end{pmatrix} \right). \tag{1.3}$$

This model shows that the closed-loop system behaves like an *impulsive system* with Eq. (1.2) describing the continuous flow and Eq. (1.3) the state jumps. The time points t_k, $(k = 0, 1, ...)$ are prescribed by the network protocol. At these time points a node has access to the network and the network-induced error of this node is set to zero. The mapping ϕ_k depends upon the network protocol. It shows which element of the vectors $\boldsymbol{e}_y(t_k^+)$ or $\boldsymbol{e}_u(t_k^+)$ are set to zero.

For a survey on hybrid dynamic systems cf. [247] and, in particular, on impulsive systems cf. [22, 179]. Further representations of control loops implemented by digital communication networks as hybrid dynamic systems can be found in [81].

From an engineering viewpoint, it is obvious that the network should not bring about large delays into the closed loop if the performance should not be deteriorated. The important aspect of systematic ways to analyze the relation between the quality of service (QoS) of the communication network and the quality of performance (QoP) of the control loop is the fact that upper bounds on the *maximum allowable transmission interval* (MATI) can be derived [230].

Reference [394] shows that the MAC protocol (medium access control protocol) is responsible for the time delay of the network. It analyzes several protocols and shows their implication for implementing control loops. Analytical results based on the hybrid representation (1.2), (1.3) of the control loop have been obtained in [63, 108, 168].

The investigations on the effect of communication constraints on the closed-loop system performance have started more than ten years ago with extensive simulation studies and the development of dedicated software tools like the MATLAB toolboxes *TrueTime* [50] and *jitterbug* [49].

Control-Communication Co-design. In the co-design of the controller and the network scheduler, the timing of the communication events and the computation steps to be performed by the controller are selected simultaneously [75]. The main question to be answered concerns the selection of the node that is allowed to send its data at a certain time point with the aim to stabilize an unstable plant or to optimize the closed-loop performance. If the sequence of sending nodes is denoted by $\sigma = (\sigma(0), \sigma(1), ..., \sigma(k_e - 1))$ and the sequence of control inputs by $U = (\boldsymbol{u}(0), \boldsymbol{u}(1), ..., \boldsymbol{u}(k_e - 1))$ the co-design problem canbe formulated as the following optimization task:

$$J(\boldsymbol{x}_0, \sigma) = \sum_{k=0}^{k_e - 1} \left(\boldsymbol{x}^{\mathrm{T}}(k) \boldsymbol{Q} \boldsymbol{x}(k) + \boldsymbol{u}^{\mathrm{T}}(k) \boldsymbol{R} \boldsymbol{u}(k) \right) + \boldsymbol{x}(k_e) \boldsymbol{Q}_e \boldsymbol{x}(k_e) \rightarrow \min_{\sigma, U}.$$

In a simplified version, the co-design problem can be reduced to a scheduling problem by considering the given state-feedback controller

$$\boldsymbol{u}(k) = -\boldsymbol{K}\boldsymbol{x}(k),$$

which should be implemented by using a digital communication network in the best possible way. With the notation for the network-induced errors introduced earlier the closed-loop system consisting of the components

$$\text{Plant:} \begin{cases} \dot{\boldsymbol{x}}_{\rm p}(t) = \boldsymbol{A}_{\rm p}\boldsymbol{x}_{\rm p}(t) + \boldsymbol{B}_{\rm p}\hat{\boldsymbol{u}}(t), & \boldsymbol{x}_{\rm p}(0) = \boldsymbol{x}_{\rm p0} \\ \boldsymbol{y}(t) = \boldsymbol{C}_{\rm p}\boldsymbol{x}_{\rm p}(t) \end{cases}$$

$$\text{Controller:} \left\{ \boldsymbol{u}_{\rm p}(t) = -\boldsymbol{K}\hat{\boldsymbol{x}}_{\rm p}(t) \right.$$

is represented by

$$\begin{pmatrix} \dot{\boldsymbol{x}}_{\rm p}(t) \\ \dot{\boldsymbol{e}}_{\rm x}(t) \end{pmatrix} = \underbrace{\begin{pmatrix} \boldsymbol{A}_{\rm p} - \boldsymbol{B}_{\rm p}\boldsymbol{K} & \boldsymbol{B}_{\rm p}\boldsymbol{K} \\ \boldsymbol{C}_{\rm p}\boldsymbol{A}_{\rm p} - \boldsymbol{C}_{\rm p}\boldsymbol{B}_{\rm p}\boldsymbol{K} & \boldsymbol{C}_{\rm p}\boldsymbol{B}_{\rm p}\boldsymbol{K} \end{pmatrix}}_{\bar{\boldsymbol{A}}} \begin{pmatrix} \boldsymbol{x}_{\rm p}(t) \\ \boldsymbol{e}_{\rm x}(t) \end{pmatrix}$$

$$\begin{pmatrix} \boldsymbol{x}_{\rm p}(t_k^+) \\ \boldsymbol{e}_{\rm x}(t_k^+) \end{pmatrix} = \boldsymbol{J}_{\sigma(k)} \begin{pmatrix} \boldsymbol{x}_{\rm p}(t_k) \\ \boldsymbol{e}_{\rm x}(t_k) \end{pmatrix},$$

where $\boldsymbol{e}_{\rm x}(t)$ denotes the network error of the state $\boldsymbol{x}(t)$. If, for simplicity, the communication events occur equidistant in time

$$t_k - t_{k-1} = T,$$

the closed-loop system has the discrete-time model

$$\text{Control loop :} \begin{cases} \begin{pmatrix} \boldsymbol{x}_{\rm p}(k+1) \\ \boldsymbol{e}_{\rm x}(k+1) \end{pmatrix} = \bar{\boldsymbol{A}}_{\sigma(k)} \begin{pmatrix} \boldsymbol{x}_{\rm p}(k) \\ \boldsymbol{e}_{\rm x}(k) \end{pmatrix} \\ \boldsymbol{y}(k) = (\boldsymbol{C}_{\rm p} \;\; \boldsymbol{O}) \begin{pmatrix} \boldsymbol{x}_{\rm p}(k) \\ \boldsymbol{e}_{\rm x}(k) \end{pmatrix} \end{cases}$$

with

$$\bar{\boldsymbol{A}}_{\sigma(k)} = {\rm e}^{\bar{\boldsymbol{A}}T} \boldsymbol{J}_{\sigma(k)}.$$

This model shows again that the overall system behaves like an impulsive system where the jump condition represented by the matrix $\boldsymbol{J}_{\sigma(k)}$ depends upon the choice $\sigma(k)$ of the node allowed to send data at time instant k. Thismodel can be used to solve the following simplified design problem

$$\hat{J}(\boldsymbol{x}_0, \sigma) = \sum_{k=0}^{k_{\rm e}-1} \boldsymbol{x}^{\rm T}(k)\hat{\boldsymbol{Q}}\boldsymbol{x}(k) + \boldsymbol{x}(k_{\rm e})\boldsymbol{Q}_{\rm e}\boldsymbol{x}(k_{\rm e}) \rightarrow \min_{\sigma}.$$

References [9, 10] describe a solution of this problem that minimizes the objective function over a restricted time horizon $k_{\rm e}$ under the assumption that after this time horizon a predefined communication sequence is periodically applied.

Results along this line of research are found in the following sections of this monograph:

- **Control and communication co-design:** Section 7.2 proposes a decomposition of dedicated wireless networked control systems into a control layer and a communication layer to investigate the connections between both layers on a common methodological basis.

- **Control requirements on network protocols:** The reduction of information sent over the communication channels is useful only if it really reduces the network load and, hence, improves the quality of service. Section 7.3 relates properties of available network protocols with requirements of the control loop to send information whenever this is necessary for ensuring a given level of performance.

Event-Based Control. Whereas the methods mentioned above have the aim to distribute the capability of the communication network in a reasonable way among the nodes of the feedback loop, event-based control aims at reducing the need for sending information. The use of the data links should be adapted to the current plant state and new information should be sent over the network only if this information is necessary for ensuring a certain level of performance of the closed-loop system.

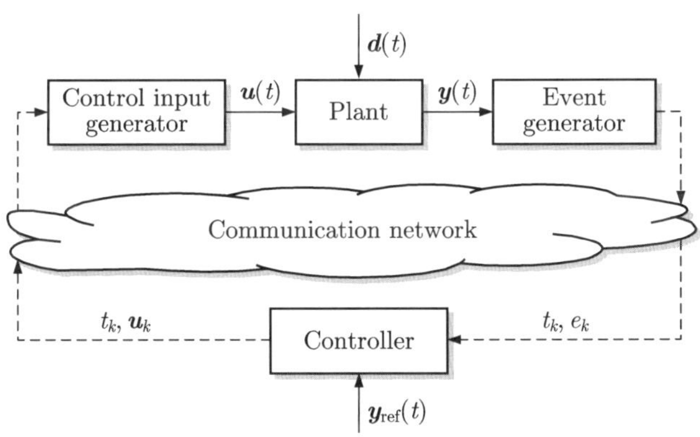

Fig. 1.13 Event-based control loop

The basic structure of event-based control loops is shown in Fig. 1.13. In comparison to the usual sampled-data control loop, the sampler is replaced by an event generator and the hold by an control input generator. The event generator decides at which instant of time, denoted by t_k, sensor information is sent to the controller and information about the control input from the controller towards the actuators. The control input generator transfers the input symbol e_{uk} obtained at time t_k into a continuous-time signal $\boldsymbol{u}(t)$ that is applied to the plant by the actuators.

Event-based control can be applied in various situations

- to reduce the data flow over the network and, hence, mitigate the constraints on the network access,
- to reduce the energy consumption of wireless sensors or actuators by reducing the data traffic from or towards these components,
- to adapt the working principle of the control loop to the event-based nature of sensors.

Furthermore, the question under what conditions information feedback is necessary for reaching a control goal is important for the principal understanding of feedback control.

Event-based control is studied extensively in Chapter 4 with the following aims:

- **Stabilization and disturbance attenuation:** Sections 5.2 and 5.4 propose new event-based control strategies that bring the system state into a bounded region around the set-point and make this region invariant under bounded disturbances.

- **Distributed event-based control:** For interconnected systems, event-based controllers consist of decentralized or distributed controllers for the subsystems. Sections 5.3, 5.5 and 5.6 explain how the performance of the overall system can be analyzed in terms of the subsystem properties.

- **Stochastic event-based control:** The event generator and the control input generator can be designed separately only if the overall system has a nested information structure. This result is described in Section 5.7 on stochastic event-based control.

1.2.6 Distributed Control of Interconnected Systems and Multi-Agent Systems

With the new communication structures, the architecture of the future control systems will change. The current hierarchically organized networks with physical subsystems at the bottom and hierarchical coordinators on higher layers (Fig. 1.14 (left)) will be replaced by distributed control structures, where the nodes represent physical subsystems together with their embedded control units (Fig. 1.14 (right)). The main question to be answered asks which groups of nodes in this network have to communicate directly.

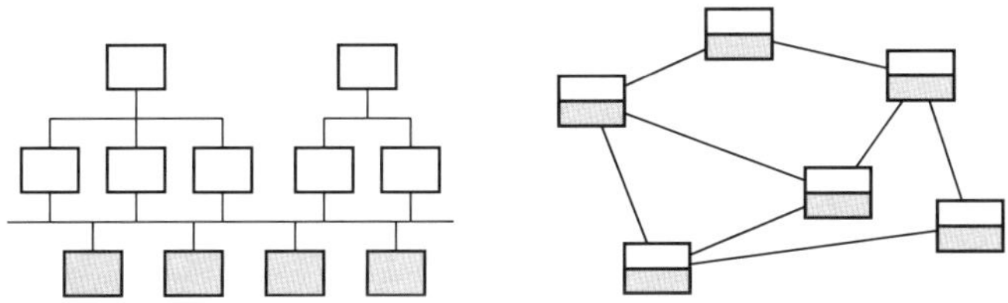

Fig. 1.14 Hierarchical vs. distributed control systems

The answer to this question depends upon the control aim and, consequently, different lines of research have been followed in the past. The following paragraphs survey the problems of distributed control and of the control of multi-agent systems, to both of which this monograph provides new methods and results.

Distributed Control. If the plant consists of several subsystems, which are physically coupled, the overall system can be described as a collection of N subsystems Σ_i and a coupling relation K as follows:

$$\Sigma_i : \begin{cases} \dot{\boldsymbol{x}}_i(t) = \boldsymbol{A}_i \boldsymbol{x}_i(t) + \boldsymbol{B}_i \boldsymbol{u}_i(t) + \boldsymbol{E}_i \boldsymbol{s}_i(t), \quad \boldsymbol{x}_i(0) = \boldsymbol{x}_{i0} \\ \boldsymbol{y}_i(t) = \boldsymbol{C}_i \boldsymbol{x}_i(t) \\ \boldsymbol{z}_i(t) = \boldsymbol{C}_{zi} \boldsymbol{x}_i(t), \quad i = 1, 2, ..., N \end{cases}$$

$$K : \left\{ \begin{pmatrix} \boldsymbol{s}_1(t) \\ \boldsymbol{s}_2(t) \\ \vdots \\ \boldsymbol{s}_N(t) \end{pmatrix} = \boldsymbol{L} \begin{pmatrix} \boldsymbol{z}_1(t) \\ \boldsymbol{z}_2(t) \\ \vdots \\ \boldsymbol{z}_N(t) \end{pmatrix} \right. .$$

The problem is to find a controller of the overall system that works predominantly locally at the subsystems. That is, the control input $\boldsymbol{u}_i(t)$ of the subsystem Σ_i should be determined mainly in dependence upon the output $\boldsymbol{y}_i(t)$ of the same subsystem and communication among the control stations of the subsystems should be reduced to a minimum.

Early research on distributed control has been concentrated on decentralized controllers, which do not at all interact directly and, hence, have the control laws of the form

$$\boldsymbol{u}_i(t) = \boldsymbol{K}_{ii} \boldsymbol{y}_i(t)$$

(cf. [245, 339]). In a more general setting, distributed control allows the control stations to communicate directly with the consequence that the control law extends towards

$$\boldsymbol{u}_i(t) = -\boldsymbol{K}_{ii} \boldsymbol{y}_i(t) - \sum_{j=1, j \neq i}^{N} \boldsymbol{K}_{ij} \boldsymbol{y}_j(t). \tag{1.4}$$

In this general setting, only the matrices $\boldsymbol{K}_{ij} \neq \boldsymbol{O}$ necessitate to communicate the state vector $\boldsymbol{x}_j(t)$ of subsystem Σ_j towards the control station of subsystem Σ_i.

Distributed control has two aspects. First, it has to be decided, which communication links among the control stations are necessary for reaching the control aim. The result is a communication topology, which can be described by a directed graph, in which each vertex represents a subsystem together with the corresponding control station and each edge a communication link. The main problem is to find the reasonable communication topology for a given control task.

The second aspect of distributed control concerns the design step. *Centralized control design* means to use a model of the overall plant to find the controllers of all subsystems together. If optimal control methods are used, the design problem is usually stated as the optimization problem

$$\min_{\boldsymbol{u}_1, \boldsymbol{u}_2, \ldots, \boldsymbol{u}_N} \sum_{i=1}^{N} \int_0^\infty \boldsymbol{x}_i^{\mathrm{T}}(t) \boldsymbol{Q} \boldsymbol{x}_i(t) + \boldsymbol{u}_i^{\mathrm{T}}(t) \boldsymbol{R} \boldsymbol{u}_i(t) \, \mathrm{d}t \tag{1.5}$$

subject to the constraint (1.4).

In *decentralized control design* the control station C_i should be found by using the model of the subsystem Σ_i only. As a corresponding formulation of the design problem

$$\min_{\boldsymbol{u}_i} \int_0^\infty \boldsymbol{x}_i^{\mathrm{T}}(t) \boldsymbol{Q} \boldsymbol{x}_i(t) + \boldsymbol{u}_i^{\mathrm{T}}(t) \boldsymbol{R} \boldsymbol{u}_i(t) \, \mathrm{d}t$$

does not take into account the couplings among the subsystems, the optimal solutions to all subsystem problems do not represent an optimal solution to the overall problem (1.5) and modifications of the subsystem problems have to be made to shift the design results of the subsystems towards an optimal solution to the overall system problem. One way of accomplishing this step is to introduce a coordinator as shown in Fig. 1.15. The coordinator communicates with the "optimizers" O_i at the subsystems in order to influence the result such that the behavior of the subsystems P_i, which include the plant subsystem Σ_i and the control station C_i, satisfies the global control goal.

Distributed control is investigated in the following sections of this book:

- **Two-layer optimal control of distributed systems:** Section 3.3 gives a survey of results on distributed optimization and the design of optimal controllers, where the controller parameter should be found by the subsystems without any complete model of the overall system.

- **Distributed model predictive control:** Chapter 4 describes several algorithms for model predictive control, where the optimization problem is solved by the subsystems with partial information about the overall system state.

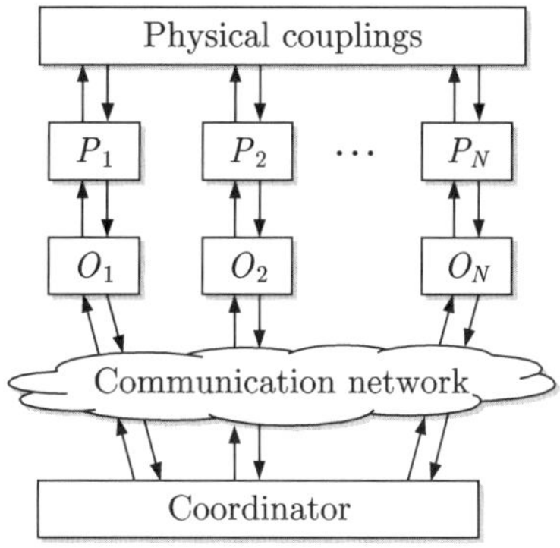

Fig. 1.15 Hierarchical structure for designing the distributed controller of an interconnected system

- **Distributed control with situation-dependent communication:** For physically interconnected systems, disturbances can be attenuated either by decentralized controllers of the subsystems or by interacting controllers. Section 6.5 shows how both methods can be combined by exchanging information among the control stations only if the current state of the system necessitates such communications.

Multi-agent Systems. The availability of wireless communication networks has increased the interest in the control of multi-agent systems, where physically uncoupled subsystem should be made to satisfy a common goal. Examples of such systems are vehicle platoons, where the vehicles should follow a leader with the same velocity and prescribed distances, or cooperating robots that should distribute over a certain region to get measurements or that should meet at a common place. Hence, multi-agent systems comprise a special class of interconnected systems, where the interconnections have to be introduced by the networked controller (Fig. 1.16).

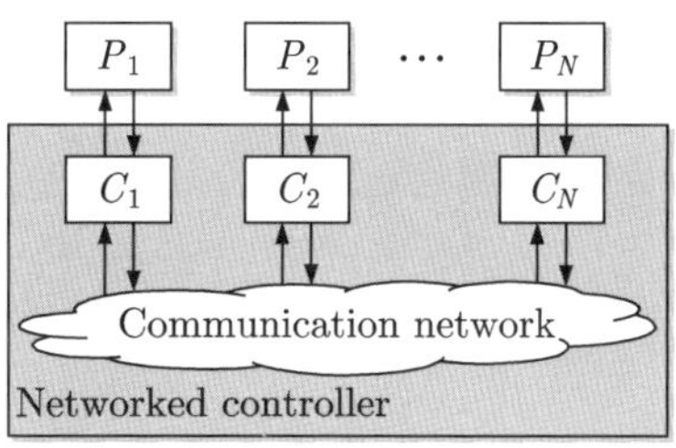

Fig. 1.16 Multi-agent system

Many control problems for multi-agent systems can be formulated as consensus or synchronization problems. The networked controller has the task to bring all agent outputs $\boldsymbol{y}_i(t)$, $(i = 1, 2, ..., N)$ onto the same trajectory $\boldsymbol{y}_\mathrm{s}(t)$, which is called the consensus trajectory or synchronous trajectory:

$$\lim_{t \to \infty} \|\boldsymbol{y}_i(t) - \boldsymbol{y}_\mathrm{s}(t)\| = 0, \quad i = 1, 2, ..., N.$$

There are two important aspects of multi-agent systems with respect to networked control. First, the coupling structure of the overall system is usually not given, but has to be found as a result of the controller design. As research on consensus and synchronization problems have shown, the choice of this structure is more important than the selection of specific controller parameters within this structure.

Second, the synchronous trajectory $\boldsymbol{y}_\mathrm{s}(t)$ appears as a result of self-organization. It depends upon the dynamics and the initial state of all agents and is not prescribed by an external entity. The exception from this general case is the leader-follower structure where one agent acts as a leader prescribing the common trajectory for all other agents, which act as followers.

The majority of papers on consensus and synchronization problems assume that the agents have identical dynamics [117, 187, 278, 319, 328, 363]. Then the synchronous trajectory is a solution to the common agent model. Under weak conditions on the agent dynamics [241], synchronization can be obtained by a static feedback, where the local components C_i of the networked controller have no dynamics (Fig. 1.16).

The problems becomes more involved if agents with individual dynamic properties are considered. Then a common trajectory only exists if the agents have some dynamics in common [243, 385]. This necessary condition on synchronizability can be satisfied by introducing dynamic components C_i into the networked controller.

Several consensus and synchronization problems are dealt with in this monograph:

- **Distributed model predictive control of multi-agent systems:** The idea of using model predictive control for cooperative control tasks is presented in Section 4.3 as an extension of distributed model predictive control.

- **Synchronizable subspaces:** The question under what conditions agents with identical linear dynamics can be synchronized is answered in Section 6.3 by deriving conditions on the network structure in terms of the Laplacian matrix of the communication graph.

- **Synchronization of agents with individual dynamics:** Whereas the literature on consensus problems and synchronization is mainly restricted to sets of agents with identical dynamics, Section 6.2 deals with agents with individual dynamics. It is shown that synchronization can only occur if the agents have some dynamics in common, which is represented by a (virtual) reference system.

- **Cluster synchronization:** Section 6.4 shows interesting cluster phenomena of nonlinear agents and develops methods to identify which agents form a group of mutually synchronized agents. Due to the nonlinearities in the interconnections, synchronization can be achieved only by groups of agents rather than by all agents in the network.

1.3 New Mathematical Concepts for Networked Control Systems

Control theory and control engineering have used sophisticated mathematical concepts at least since the times of Routh and Hurwitz and their development of easy checkable stability criteria for linear systems. It is, therefore, not surprising that also a theory of networked control systems substantially relies on mathematical concepts. In this section we give an informal overview of the main techniques used in this monograph, referring to the respective sections in the book for details. Some of these concepts, like graph theory, have primarily gained interest in control because of their applicability to networked control systems. Other concepts, like mathematical optimization or Lyapunov functions, have been successfully used in control theory for many decades. For the latter, we particularly explain which extensions are used in this monograph in order to apply them in a network setting.

1.3.1 Optimization and Optimal Control

Optimization and optimal control techniques are ubiquitous in this monograph. Already in the overview in the previous sections optimal control techniques have been mentioned in several places, either in the form of linear quadratic optimal control or in the form of model predictive control. The popularity of optimization and optimal control techniques for handling networked control problems has various reasons.

To begin with, optimization algorithms have been studied in decentralized settings for many years. Section 3.3 gives an introduction into primal and dual decomposition techniques which allow to solve optimization problems by decomposing them into several subproblems that can be solved in parallel. In the case of networked control systems with many subsystems such decompositions are naturally induced by the system structure and in this way they are used, e.g., in Sections 3.2, 3.3, 4.3 and 6.5.

Second, on-line optimal control methods form the basis of the currently very popular model predictive control (MPC) approach for controller design.

Clearly, all the possibilities to decompose optimization problems also apply to MPC and this is used in different ways in Section 4.3. An additional benefit of MPC is that instead of just a single control value a whole sequence of control inputs along with a corresponding prediction of the future system state is available in each sampling period "for free". On the one hand, this allows to use the additional control values as backup in order to compensate for packet loss or unacceptably large delays. Moreover, the prediction mechanism already available in MPC can be used in order to compensate for delays. These techniques are described in Section 4.2. On the other hand, in the case of several subsystem, MPC allows for communicating the predictions to the other subsystems. Different approaches using this technique are described in Sections 4.3–4.5.

Another benefit of optimal control based controller design is the possibility to encode different goals in both the objective and the constraints. In this monograph, this fact is used in distributed control approaches, in which the individual subsystems do not communicate but rather the other subsystems' influence is taken care of by a judicious formulation of the optimal control problem. For instance, the optimal control problem can ensure that the subsystems are dissipative (Section 4.3), robust w.r.t. errors in the predictions communicated by the other subsystems (Section 4.4) or input-to-state stable (Section 5.5). Another way to exploit the flexibility of optimal control based design is to incorporate network induced additional requirements like low communication effort directly into the optimization objectives. Such approaches are discussed in Section 5.7 in an event-based setting and in Section 7.2 in the context of a cross-design approach. A study of network effects on the performance of optimal controllers can be found in Section 7.3.

Finally, optimization techniques can not only be used for controller design but also for estimation as described in Section 3.2 and for the analysis of the systems behavior. Examples for the latter in this book are the stability analysis for MPC subject to packet loss in Section 4.2, the LMI criterion for synchronization presented in Section 6.3 and the cluster synchronization analysis in Section 6.4

1.3.2 Dynamical Properties of Control Systems: Controllability, Observability and Stability

The goal of most control-theoretic approaches is to either verify or to enforce (by design of a suitable controller) desired dynamic properties of the control system under consideration. Just like for their non-networked counterparts, for networked control systems the classical properties of controllability, observability and stability are essential for understanding the basic properties of the system.

Observability and controllability are thoroughly investigated for networked linear systems in Section 2.2. At a first glance, the mathematical tools for their verification like coprime factorization and rank conditions are similar to the analysis of non-networked systems. However, the "twist" in the mathematical analysis of networked systems comes from the special structures of the matrices under consideration. Controllability – though in the weaker form of asymptotic controllability to a desired equilibrium – also pops up in a nonlinear context in the analysis of MPC schemes without terminal constraints in Sections 4.2 and 4.3.

Stability is one of the key properties a controlled system should satisfy and it is, therefore, featured in many contributions in this monograph. In fact, even several approaches which do not consider classical stability concepts but rather invariance properties as in Section 2.3, ultimate boundedness as in Sections 5.2–5.4 or synchronization as in Section 6.3 use mathematical techniques similar to those from stability analysis.

The two main approaches for verifying stability found in this monograph are spectral theory for linear systems and and Lyapunov functions for both linear and nonlinear systems. In the time-invariant case, spectral theory boils down to computing eigenvalues of matrices. Similar to what was said for controllability above, in the networked context the matrices under consideration also reflect properties of the communication network and the coupling structure between the subsystems. Examples where such spectral information is used can be found, e.g., in those results of Section 2.3 which deal with linear systems as well as in Sections 5.3 and 6.3. In case the coupling structures are explicitly represented by a matrix $\boldsymbol{L}$ (as, e.g., in Section 5.3), respective conditions can be given in small-gain form, which essentially amounts to demanding that the couplings between the individual subsystems are sufficiently weak.

Small-gain type arguments can also be employed for nonlinear systems as in Sections 5.5 and 5.6 in an ISS Lyapunov function framework. Note that here the Lyapunov functions are not merely tools for the analysis but also used for design purposes, e.g., for determining suitable event times in Section 5.5. ISS Lyapunov functions are an appropriate tool for distributed control in which the mutual influences of the subsystems are considered as perturbations and no communication takes place.

However, Lyapunov function based stability analysis can also be employed in case the subsystems communicate, as in the MPC approaches in Sections 4.3–4.5. In order to address delays, extensions of the Lyapunov function concept in the sense of Krasovskii or Razumikhin can be used (Section 4.2).

It should finally be remarked that the analysis of dynamic properties is not necessarily restricted to the time-invariant or autonomous case, even if the underlying control system is time-invariant. This is, for instance, the case in the analysis of the topological entropy in Section 2.3, where the non autonomy is induced by the control input and methods from the theory of non-autonomous dynamic systems are applied. Another example is the networked MPC scheme in Section 4.2, in which the non autonomy is induced by possible network failures.

1.3.3 Graph Theory and Structured Matrices

In contrast to the concepts discussed in the previous two subsections, which are carried over from non-networked control systems to the networked setting, graph theory is more directly linked to the network aspect. Indeed, except for the more "exotic" use of graph theory in the set-oriented solution of quantized optimal control problems in Section 5.4, graphs in this monograph always serve for formalizing the interconnection structure between the different subsystems in a networked control system.

As an alternative to graphs - particularly in a linear setting - the interconnection structure can also be expressed via a matrix whose entries represent the coupling gains between the subsystems (Section 2.2). Formally, the graphs and these matrices are connected via the concept of the adjacency matrix (Section 3.2). More generally, interconnections are expressed by nonlinear gain matrices in Sections 5.5 and 5.6.

Regardless of the particular form, both graph-theoretic as well as matrix-valued representations have the same goal: they encode the coupling structure and allow for identifying and formulating structural conditions leading to desired system properties.

Matrix theoretic representations of coupling structures are used in Sections 2.2, 3.2, 3.3 and 5.3. The results obtained this way range from genericity results for controllability and observability via the exploitation of sparsity patterns in distributed optimal estimation and control to the formulation of small-gain conditions for stability.

Graph theoretic conditions are, on the other hand, used in Sections 4.4, 6.2 and 6.4. Here, graph theoretic concepts are used, e.g., to introduce a no cycle condition for robust MPC, to establish a relation between the existence of spanning trees and synchronization in leaderless multi-agent systems and for the formalization of clustering phenomena occuring in synchronizing systems.

This last application, which can be found in Section 6.4, combines the graph-theoretic representation with optimization approaches. Along with many other examples in this book it shows that the mathematical concepts surveyed in this section are typically not used isolated from each other but that a combination of such methods is needed in order to address analysis and design problems for networked control systems.

2

Analysis of Networked Systems

F. Colonius, U. Helmke, J. Jordan, C. Kawan, R. Sailer, and F. Wirth

This chapter deals with particular properties of dynamic systems that are important when controlling the system over digital networks. It is shown how the observability of the system depends upon the network properties. Then the minimal bit rate that is necessary to solve a control task is considered. Finally, a method for the dynamic quantization of the feedback information that is used for stabilizing an unstable plant is developed.

Chapter Contents

2.1 Overview

In this chapter fundamental properties of networked control systems are discussed that characterize obstructions to and requirements for the control of interconnections through digital channels. On the one hand, this chapter relates to general problems of the characterization of controllability and observability properties of interconnected systems, where information is distributed via communication channels. On the other hand, it is of interest to characterize the necessary bandwidth required to control a system or an interconnection of systems. The latter question is related to dynamic properties of the physical system as well as the communication channel under consideration.

Interconnected dynamic systems describe a rather broad class of networked control systems where finitely many subsystems or 'agents' are interconnected to construct a network of dynamic systems. The interconnection structure can be static or dynamic via physical couplings or through communication channels. In either case, the interconnection structure is described by a coupling graph that defines the information flow in the network. Controls are inserted into the network to be broadcasted to all nodes and may thus be non-decentralized. Structural properties of such networks refer to questions such as e.g. reachability and observability, the estimation of the amount of necessary information to be broadcasted in order to e.g. stabilize the network, or questions about quantization and stability. The control-theoretic analysis of such interconnected systems has already led to a number of fundamental insights, which are briefly described next.

The status of the theory concerning control of one or several dynamic systems using multiple sensors and actuators over a digital communication network is far from being complete. Here, in spite of the rapid growth of communication capacities, the minimal bit rate for communication may be a limit for control performance. This poses fundamental mathematical questions. Although these questions have already been realized in the 1990's, they are not yet satisfactorily solved. It is still a challenging task to determine minimal bit rates for standard control problems like stabilizing a of single dynamic system or rendering a subset of the state space invariant. It turns out that relations to the mathematical theory of (uncontrolled) dynamic systems involving topological entropy appear to be a key to the understanding of these problems. Here our work has essentially been restricted to these fundamental questions for single control systems. It is our hope that the results will contribute to the mathematical tools for understanding minimal bit rates for interconnected systems, which presently are under investigation.

A complete characterization of the reachability and observability properties of standard parallel, series and feedback interconnections of linear systems is well-known since the early 1970's. While parallel interconnections of identical linear systems can never be controllable or observable, for nonlinear systems this fact is no longer true. It is known that the parallel coupling of identical nonlinear systems can be locally accessible, at least for generic initial states. The extension of such results for general interconnection structures is an open research problem. For arbitrary homogeneous networks of linear systems, i.e., for networks of identical SISO systems, a very concise characterization of reachability and observability is known [165]. However, the extension of such results to arbitrary heterogeneous networks of control systems is an open research problem.

This chapter presents several new results on fundamental limitations for the control and estimation of interconnected systems:

- **Observability of networked systems:** Section 2.2 explores the extensions of the established observability and reachability analysis of homogeneous networks to the general case of heterogeneous networks of linear systems. A basic characterization of generic structural observability of networks of first order driftless integrators is generalized to arbitrary homogeneous networks of SISO systems. Sufficient conditions for observability of the series connection of two nonlinear systems are presented. New results on the reachability of switching networks are derived, where the interconnection parameters are considered as independent control variables.

- **Minimal bit rates for control:** Section 2.3 surveys entropy estimates for the invariance of subsets of the state space and establishes relations between invariance entropy and minimal data rates. In particular, for control sets the minimal sum of positive Lyapunov exponents along periodic solutions in the control set provides an upper bound for the strict invariance entropy. As an example, the invariance entropy for a controlled linear mathematical pendulum near the unstable position is computed. Furthermore, for nonlinear control systems, entropy estimates for the stabilization at an equilibrium with fixed exponential rate are given, which are exact in the linear case.

- **Dynamic quantization for feedback stabilization:** Section 2.4 discusses the concept of dynamic quantization and study the problem of minimal data rates in the context of this particular approach. An encoder/decoder scheme is presented for channels characterized by quantized, delayed packet-based communication with the additional possibility of packet drops. The scheme guarantees stabilization provided that the required bandwidth is available as prescribed by easily identifiable system parameters. The novel idea introduced to the classic concept of dynamic quantization is that time stamps in data packets as well as a feedback loop in the communication channel using acknowledgments are sufficient to handle unreliable packet delivery. This approach can also be analyzed for communication channels of a stochastic nature in which delays and packet delivery are described by a Markov process.

The subsequent research marks the beginning of a new theory of networked control system that is in close analogy to Shannon's theory. The development of mathematical tools for establishing fundamental limits of performance together with finding effective algorithms to achieve these limits present challenges for the future system-theoretic research.

2.2 Observability of Networked Systems

2.2.1 Motivation and Earlier Results

Large-scale networks of interconnected dynamic systems pose a number of challenging tasks for the theoretical analysis of control-theoretic issues. For example, in analyzing sensor-/actuator networks or large-scale biological networks it becomes increasingly important to estimate the state variables, to identify relevant system parameters, or to characterize the most sensitive interconnection parameters. This leads to the important questions of reachability, observability, state estimation and observer design for networks of systems.

There is a recent resurgence of interest in the analysis of controllability properties for parallel interconnection structures. The paper [39] may serve as a good starting point. In [39], controllability properties of a parallel connection of finitely many nonlinear systems are established. Somewhat surprisingly, local accessibility can hold even if the systems are identical. Early work in this area focused on the analysis of standard structures for linear time-invariant systems, such as series [46], parallel [128] and feedback interconnections [46].

In this section, our main focus is on the observability analysis of heterogeneous networks, whose interconnection structure is fixed by a coupling graph. For homogeneous networks of identical linear SISO systems a characterization of reachability has been first obtained by [166]; see also [129] for a more concise proof. The more general scenario that we consider is that of a finite number of arbitrary linear or nonlinear control systems, called *node systems*, that are connected through static or dynamic coupling laws. Since communication between the systems is mainly assumed to occur through the interconnection channels, dynamic coupling laws can model the effects of delays or time-varying topologies in the network The models for the node systems are defined by first order or higher order difference or differential equations. We show that using higher order polynomial matrix fraction decompositions, allows one to obtain very compact coprimeness conditions for reachability and observability. Our main tool is a permanence principle for strict system equivalence that implies very concise results on observability and reachability of interconnected systems. This leads to elegant new proofs for classical conditions for reachability and observability of series, parallel and feedback interconnections. We also derive new results for series or parallel connections of homogeneous networks.

Another direction of research deals with the situation of networks that are controlled by switching interconnections; see e.g. [227]. Thus the interconnections are considered as input variables that are tuned to achieve better performance of the overall network. In [40], Brockett has characterized the system Lie algebras of feedback systems, where the feedback gains act as control variables. In [103] we have generalized this characterization to homogeneous networks of SISO systems.

For graph-theoretic characterizations for generic controllability and generic observability for structured systems we refer to [100, 232]. This early work applies to networks whose node dynamics are described by simple first-order integrators. Using recent results from [129], we obtain a generalization to SISO node systems with arbitrary linear dynamics. In [401], we analyze the zero properties of blocked linear systems resulting from blocking of linear time-invariant systems. This is an important step towards analyzing zeros of networks with periodic interconnection topologies. Global observability of Morse-Smale nonlinear systems has been analyzed in [1]. In [102], we extended these results to a global observability result on real analytic systems, using weak assumptions on the dynamics of the limit sets. Our work is also connected with results on synchronizing subspaces and geometric control introduced in Section 6.3. Entropy bounds for conditioned invariant subspaces of linear systems were established in [69].

To analyze the behavior of large-scale networks of interacting dynamic systems it is of vital interest to identify those components which are highly sensitive to errors. It is well known that round-off noise and quantization errors in the implementation of linear systems in digital processing devices might lead to a substantial different performance, cf. [268]. This gave rise to a systematic sensitivity analysis for single linear systems, see [133], [173]. These investigations show that the minimum sensitivity realizations are balanced realizations. Since minimum sensitivity realizations generalize the class of balanced realizations, there is a natural relation to balanced truncation and model reduction.

In this project, we examined the sensitivity analysis from the network perspective. Precisely, we consider networks of linear discrete-time systems and develop a systematic sensitivity theory. To this end, we define a sensitivity measure that assigns to each realization of the network structure a value quantifying its sensitivity with respect to network parameter variations. In addition, new sensitivity Gramians are introduced so that the network sensitivity can be measured by the sum of their traces. In a first step we consider homogeneous networks of single-input-single-output systems and concentrate on the network sensitivity with respect to input-to-state and state-to-output parameter variations. It turns out that the network has minimal sensitivity if and only if the associated controllability and observability sensitivity Gramians are equal. This can interpreted in the way that symmetric networks are sensitivity optimal. Furthermore, we address the issue of existence and uniqueness of sensitivity optimal network realizations. Moreover, we show that sensitivity optimal realizations can be computed by a Newton algorithm which is locally quadratically convergent. Dependent on the dynamics of the node systems we derive estimates of the sensitivity Gramians in terms of the classical Gramians corresponding to the realization of the network topology and the realization of the overall network. These estimates result in some interesting perspectives on model reduction based on sensitivity optimal design, which is briefly addressed, cf. [176]. The next step is to extent these results to heterogenous networks of multi-input-multi-output linear systems. This is the content of current and future research, cf. [177].

2.2.2 Reachability and Observability of Networks

Models of Interconnected Systems. We consider heterogeneous networks described interconnecting N linear discrete-time node systems

$$\begin{aligned} \boldsymbol{x}_i(k+1) &= \boldsymbol{\alpha}_i \boldsymbol{x}_i(k) + \boldsymbol{\beta}_i \boldsymbol{v}_i(k) \\ \boldsymbol{w}_i(k) &= \boldsymbol{\gamma}_i \boldsymbol{x}_i(k), \quad i = 1, \ldots, N \end{aligned} \tag{2.1}$$

with $\boldsymbol{\alpha}_i \in \mathbb{F}^{n_i \times n_i}$, $\boldsymbol{\beta}_i \in \mathbb{F}^{n_i \times m_i}$ and $\boldsymbol{\gamma}_i \in \mathbb{F}^{p_i \times n_i}$. Here $\mathbb{F}$ is an arbitrary field. As a standard assumption, each system $(\boldsymbol{\alpha}_i, \boldsymbol{\beta}_i, \boldsymbol{\gamma}_i)$, $(i = 1, \ldots, N)$ is assumed to be observable and controllable. The systems are coupled via a static linear *coupling law*

$$\boldsymbol{v}_i(k) = \sum_{j=1}^{N} \boldsymbol{L}_{ij} \boldsymbol{w}_j(k) + \boldsymbol{B}_i \boldsymbol{u}(k) \in \mathbb{F}^{m_i} \tag{2.2}$$

with $\boldsymbol{L}_{ij} \in \mathbb{F}^{m_i \times p_j}$ and $\boldsymbol{B}_i \in \mathbb{F}^{m_i \times m}$. Here, the matrix $\boldsymbol{L} = (\boldsymbol{L}_{ij}) \in \mathbb{F}^{\overline{m} \times \overline{p}}$ with $\overline{m} := m_1 + \cdots + m_N$ and $\overline{p} := p_1 + \cdots + p_N$ models the *interconnection structure*. The *interconnection output* is given by

$$\boldsymbol{y}(k) = \sum_{i=1}^{N} \boldsymbol{C}_i \boldsymbol{w}_i(k) \ \text{ with } \ \boldsymbol{C}_i \in \mathbb{F}^{p \times p_i}, \quad i = 1, \ldots, N.$$

The *input/output interconnection* is defined by the matrices

$$\boldsymbol{C} := (\boldsymbol{C}_1, \ldots, \boldsymbol{C}_N) \in \mathbb{F}^{p \times \overline{n}} \text{ and } \boldsymbol{B} := (\boldsymbol{B}_1^{\mathrm{T}}, \ldots, \boldsymbol{B}_N^{\mathrm{T}})^{\mathrm{T}} \in \mathbb{F}^{\overline{n} \times m}$$

with $\overline{n} := n_1 + \cdots + n_N$. Let $\boldsymbol{x}(k) = (\boldsymbol{x}_1^{\mathrm{T}}(k), \ldots, \boldsymbol{x}_N^{\mathrm{T}}(k))^{\mathrm{T}} \in \mathbb{F}^{\overline{n}}$ denote the global state of the network. Of course, more complicated, dynamic couplings laws are also of interest, in particular towards modeling communication constraints, packet loss and so on. Certainly, the analysis of any such general dynamic network structures rests on a solid understanding of the influence of static coupling laws. Therefore, we focus our subsequent analysis on static interconnections.

Define

$$\begin{aligned} \boldsymbol{A}_0 &:= \operatorname{diag}(\boldsymbol{\alpha}_1, \ldots, \boldsymbol{\alpha}_N) \in \mathbb{F}^{\overline{n} \times \overline{n}}, \\ \boldsymbol{B}_0 &:= \operatorname{diag}(\boldsymbol{\beta}_1, \ldots, \boldsymbol{\beta}_N) \in \mathbb{F}^{\overline{n} \times \overline{m}}, \\ \boldsymbol{C}_0 &:= \operatorname{diag}(\boldsymbol{\gamma}_1, \ldots, \boldsymbol{\gamma}_N) \in \mathbb{F}^{\overline{p} \times \overline{n}}. \end{aligned}$$

Then the interconnected system has the following state-space representation

$$\begin{aligned} \boldsymbol{x}(k+1) &= \bar{\boldsymbol{A}}\boldsymbol{x}(k) + \bar{\boldsymbol{B}}\boldsymbol{u}(k) \\ \boldsymbol{y}(k) &= \bar{\boldsymbol{C}}\boldsymbol{x}(k) \end{aligned} \tag{2.3}$$

with

$$\bar{\boldsymbol{A}} := \boldsymbol{A}_0 + \boldsymbol{B}_0\boldsymbol{L}\boldsymbol{C}_0 \in \mathbb{F}^{\overline{n}\times\overline{n}}, \quad \bar{\boldsymbol{B}} := \boldsymbol{B}_0\boldsymbol{B} \in \mathbb{F}^{\overline{n}\times m}, \quad \bar{\boldsymbol{C}} := \boldsymbol{C}\boldsymbol{C}_0 \in \mathbb{F}^{\overline{n}\times m}.$$

As a special case we also consider *homogeneous networks*, i.e., the case where $\boldsymbol{\alpha} = \boldsymbol{\alpha}_1 = \cdots = \boldsymbol{\alpha}_N$, $\boldsymbol{\beta} = \boldsymbol{\beta}_1 = \cdots = \boldsymbol{\beta}_N$ and $\boldsymbol{\gamma} = \gamma_1 = \cdots = \gamma_N$ denote identical SISO systems. The state space representation of a homogeneous network has the tensor product form

$$\bar{\boldsymbol{A}} = I_N \otimes \boldsymbol{\alpha} + \boldsymbol{L} \otimes \boldsymbol{\beta}\boldsymbol{\gamma}, \quad \bar{\boldsymbol{B}} = \boldsymbol{B} \otimes \boldsymbol{\beta} \quad \text{and} \quad \bar{\boldsymbol{C}} = \boldsymbol{C} \otimes \boldsymbol{\gamma}. \tag{2.4}$$

In order to analyze the influence of the interconnection matrices on the reachability and observability properties it is convenient to describe the interconnected systems in terms of the transfer functions of the node systems. Define the *i-th node transfer function* as the strictly proper transfer function

$$\boldsymbol{G}_i(z) = \gamma_i(z\boldsymbol{I} - \boldsymbol{\alpha}_i)^{-1}\boldsymbol{\beta}_i. \tag{2.5}$$

of McMillan degree n_i. We define the *node transfer function* as

$$\boldsymbol{G}(z) := \operatorname{diag}(\boldsymbol{G}_1(z), \ldots, \boldsymbol{G}_N(z)) = \boldsymbol{C}_0(z\boldsymbol{I} - \boldsymbol{A}_0)^{-1}\boldsymbol{B}_0.$$

In the case that $\boldsymbol{L}$ is a square matrix we define the *interconnection transfer function* as

$$\phi(z) = \boldsymbol{C}(z\boldsymbol{I} - \boldsymbol{L})^{-1}\boldsymbol{B}.$$

The global *network transfer function* then is defined as

$$\boldsymbol{\Phi}(z) = \bar{\boldsymbol{C}}(z\boldsymbol{I} - \bar{\boldsymbol{A}})^{-1}\bar{\boldsymbol{B}}. \tag{2.6}$$

It useful to describe the network transfer function explicitly in terms of coprime factorizations of the node transfer function. Consider the case where the i-the node transfer function is strictly proper and is given in the Rosenbrock form as

$$\boldsymbol{G}_i(z) = \boldsymbol{V}_i(z)\boldsymbol{T}_i^{-1}(z)\boldsymbol{U}_i(z). \tag{2.7}$$

The pairs $(\boldsymbol{T}_i(z), \boldsymbol{U}_i(z))$ and $(\boldsymbol{T}_i(z), \boldsymbol{V}_i(z))$ are assumed to be left or right coprime polynomial matrices, respectively. Moreover, we assume that the factorizations are minimal in the sense that $\deg\det \boldsymbol{T}_i(z) = n_i$. Let

$$\begin{aligned} \boldsymbol{T}(z) &= \operatorname{diag}(\boldsymbol{T}_1(z), \ldots, \boldsymbol{T}_N(z)) \\ \boldsymbol{U}(z) &= \operatorname{diag}(\boldsymbol{U}_1(z), \ldots, \boldsymbol{U}_N(z)) \\ \boldsymbol{V}(z) &= \operatorname{diag}(\boldsymbol{V}_1(z), \ldots, \boldsymbol{V}_N(z)). \end{aligned}$$

Then we have the left- and right coprime factorization of a strictly proper transfer function

$$\boldsymbol{G}(z) = \boldsymbol{V}(z)\boldsymbol{T}(z)^{-1}\boldsymbol{U}(z).$$

Note that $\boldsymbol{T}(z) - \boldsymbol{U}(z)\boldsymbol{L}\boldsymbol{V}(z)$ is a nonsingular polynomial matrix for any constant interconnection matrix $\boldsymbol{L}$. Thus the transfer network function from $\boldsymbol{u}$ to $\boldsymbol{y}$ is well defined and given by

$$\boldsymbol{\Phi}(z) = \boldsymbol{C}\boldsymbol{V}(z)(\boldsymbol{T}(z) - \boldsymbol{U}(z)\boldsymbol{L}\boldsymbol{V}(z))^{-1}\boldsymbol{U}(z)\boldsymbol{B}.$$

This explicit representation of the network transfer function turns out to be very useful for analyzing the system-theoretic properties of the network.

Conditions for Reachability and Observability. In [129] we extended the classical notion of strict system equivalence to networks of linear systems. This was the key to proving the following extension of the classical Hautus criterion.

Theorem 2.1.

a) $(\bar{\boldsymbol{A}}, \bar{\boldsymbol{B}})$ *is reachable if and only if*

$$\operatorname{rank}(\boldsymbol{T}(z) - \boldsymbol{U}(z)\boldsymbol{L}\boldsymbol{V}(z), \boldsymbol{U}(z)\boldsymbol{B}) = \overline{n} \quad \forall z \in \mathbb{C}.$$

b) $(\bar{\boldsymbol{C}}, \bar{\boldsymbol{A}})$ *is observable if and only if*

$$\operatorname{rank}\begin{pmatrix} \boldsymbol{T}(z) - \boldsymbol{U}(z)\boldsymbol{L}\boldsymbol{V}(z) \\ \boldsymbol{C}\boldsymbol{V}(z) \end{pmatrix} = \overline{n} \quad \forall z \in \mathbb{C}.$$

In [129] this result is extended to dynamic coupling schemes. In particular, this result yields new and efficient algebraic tests for controllability and observability of networked control systems. It is a generalization of a preliminary result by Hara et. al. for homogeneous networks, i.e., from systems of type (2.4), to arbitrary heterogeneous networks. More precisely, Theorem 2.1 implies the following result.

Theorem 2.2. [129, 166] *A homogeneous network is reachable (observable) if and only if the realization* $(\boldsymbol{L}, \boldsymbol{B}, \boldsymbol{C})$ *of the interconnection transfer function* $\phi(z)$ *is reachable (observable). In particular, reachability of* $(\bar{\boldsymbol{A}}, \bar{\boldsymbol{B}}, \bar{\boldsymbol{C}})$ *is independent of the choice of the node transfer function* $g(z)$*, as long as* $g(z)$ *is scalar rational and strictly proper.*

Another application of Theorem 2.1 is to derive graph-theoretic conditions for *generic observability* of homogeneous networks, for $\mathbb{F} = \mathbb{C}$. We assume that for a given set of edges $E \subset \{1, \ldots, N\} \times \{1, \ldots, N\}$ we have $L_{ij} = 0$ if $(i, j) \notin E$. The directed graph $\mathcal{G}(E, \{1, \ldots, N\})$ is called the *coupling graph*.

The set of structured matrices $\boldsymbol{L} \in \mathbb{C}^{N\times N}$ which corresponds to a given coupling graph $\mathcal{G}(E,\{1,\ldots,N\})$ is denoted by $D_{\mathcal{G}}$. Note that $D_{\mathcal{G}}$ is a vector space of dimension $|E|$. In the following we fix the system parameters $(\boldsymbol{\alpha},\boldsymbol{\beta},\boldsymbol{\gamma},\boldsymbol{B},\boldsymbol{C})$ and the coupling graph $\mathcal{G}$ and consider the coupling parameters $L_{ij} \in \mathbb{C}$ with $(i,j) \in E$ as free parameters. Thus the set of admissible couplings is equal to $\mathbb{C}^{|E|}$. System (2.3) is *generically controllable* (*generically observable*) if it is controllable (observable) for a generic set of parameters $\boldsymbol{L}$ with. Here, we say a nonempty subset $M \subset D_{\mathcal{G}}$ is generic if M is not contained in the zero set of a nontrivial polynomial in $D_{\mathcal{G}}$.

In the special case $n = 1$ and $\alpha = 0$ and $\beta = \gamma = 1$ this concept reduces to the well-known concept of generic observability for structured matrices (see for example [100]), i.e., the system

$$\boldsymbol{x}(k+1) = \boldsymbol{L}\boldsymbol{x}(k) + \boldsymbol{B}\boldsymbol{u}(k), \quad \boldsymbol{y}(k) = \boldsymbol{C}\boldsymbol{x}(k) \tag{2.8}$$

with a structured matrix $\boldsymbol{L}$. Theorem 2.1 yields the following connection between linear structured systems and homogeneous SISO networks.

Corollary 2.1. *The homogeneous network (2.4) for $\mathbb{F} = \mathbb{C}$ is generically controllable (generically observable) if and only if the structured system* (2.8) *is generically controllable (generically observable).*

Note that there exist graph-theoretic characterizations for generic controllability and generic observability for structured systems, see [100, 232]. More precisely, a structured system is generically controllable if the corresponding structure graph fulfills the cactus-condition, i.e., in $\mathcal{G}$ there exists a vertex disjoint union of cacti that covers all state vertices.

Observability under Transmission Delays and Switching. As a special case of the general dynamic coupling laws we characterize reachability and observability for systems with delayed interconnections

$$\boldsymbol{v}_i(t) = \sum_{j=1}^{N} \boldsymbol{L}_{ij}\boldsymbol{w}_j(t-\lambda_{ij}) + \boldsymbol{B}_i\boldsymbol{u}(t),$$

where λ_{ij} are assumed to be arbitrary nonnegative integers. This model can be realized as follows using a dynamic interconnection law $\boldsymbol{E}(\sigma)\boldsymbol{v} = \boldsymbol{A}(\sigma)\boldsymbol{w} + \boldsymbol{B}(\sigma)\boldsymbol{u}$, where $\boldsymbol{E}(z)$ is a square nonsingular polynomial matrix and $\boldsymbol{A}(z), \boldsymbol{B}(z)$ appropriately sized polynomial matrices. We assume that the network transfer function is given in terms of coprime factorizations of the node transfer functions

$$\boldsymbol{G}_i(z) = \boldsymbol{Q}_i(z)^{-1}\boldsymbol{P}_i(z) = \overline{\boldsymbol{P}}_i(z)\overline{\boldsymbol{Q}}_i(z)^{-1} \tag{2.9}$$

where we assume that the factorizations are minimal in the sense that $\delta(\boldsymbol{G}_i) = \deg\det \boldsymbol{Q}_i = \deg\det \overline{\boldsymbol{Q}}_i$. We denote $\lambda_i := \max_{j=1,\ldots,N} \lambda_{ij}$, $i = 1,\ldots,N$ and define $\bar{\lambda}_{ij} = \lambda_i - \lambda_{ij} \geq 0$. Define polynomial matrices $\boldsymbol{E}(z) = \mathrm{diag}(z^{\lambda_1},\cdots,z^{\lambda_N})$, $\boldsymbol{A}(z) = (\boldsymbol{L}_{ij}z^{\bar{\lambda}_{ij}})_{i,j=1,\ldots,N} \in \mathbb{F}^{\overline{p}\times\overline{m}}[z]$, $\boldsymbol{B}(z) := \boldsymbol{E}(z)\boldsymbol{B}$ and $\boldsymbol{C}(z) = \boldsymbol{C}$.

Theorem 2.3. *The shift realization of the delayed network is reachable if and only if*

$$\operatorname{rank}\left(\boldsymbol{L}_{ij} z^{\bar{\lambda}_{ij}} \bar{\boldsymbol{P}}_j(z) - z^{\lambda_i} \bar{\boldsymbol{Q}}_j(z)\delta_{ij} \; , z^{\lambda_i}\boldsymbol{B}_i \right) = \overline{n}, \quad \forall z \in \overline{\mathbb{F}}.$$

The shift realization of the delayed network is observable if and only if

$$\operatorname{rank}\begin{pmatrix} (\boldsymbol{L}_{ij} z^{\bar{\lambda}_{ij}} \bar{\boldsymbol{P}}_j(z) - z^{\lambda_i} \bar{\boldsymbol{Q}}_j(z)\delta_{ij}) \\ \boldsymbol{C}\bar{\boldsymbol{P}}(z) \end{pmatrix} = \overline{n}, \quad \forall z \in \overline{\mathbb{F}}.$$

As a special case we consider homogeneous networks of identical SISO systems with node transfer functions $G_i(z) = \frac{p(z)}{q(z)}$ satisfying $p(0) \neq 0$. In this situation the above result implies

Corollary 2.2. *Consider a network of identical node transfer functions* $G_i(z) = \frac{p(z)}{q(z)}$ *satisfying* $p(0) \neq 0$. *Assume that* $\boldsymbol{L}$ *is invertible and* $(\boldsymbol{L}, \boldsymbol{B}, \boldsymbol{C})$ *is observable. Assume further that all delays* λ_{ij} *are equal to* $\lambda \geq 1$. *Then the delayed network is observable, independently of the value of* λ.

An interesting class of observability problems for networks of linear systems arises when the network parameters are allowed to vary with time. A special class of interest here are switching networks that are defined by time-varying state interconnection matrices $\boldsymbol{L}(t)$. Here the interconnection matrices $\boldsymbol{L}(t)$ are considered as matrix-valued control variables and the task becomes to investigate the controllability and observability properties of the resulting network. This leads to bilinear control and observation problems and therefore requires the use of techniques from geometric nonlinear control.

We begin with a brief summary of some early results on controllability of bilinear systems that are of interest here. In [40] BROCKETT characterized completely the system Lie algebras of the bilinear output-feedback control system

$$\dot{\boldsymbol{x}}(t) = (\boldsymbol{\alpha} + u(t)\boldsymbol{\beta}\boldsymbol{\gamma})\,\boldsymbol{x}(t) \tag{2.10}$$

for a SISO system $(\boldsymbol{\alpha}, \boldsymbol{\beta}, \boldsymbol{\gamma})$.

Theorem 2.4. [40] *Let* $(\boldsymbol{\alpha}, \boldsymbol{\beta}, \boldsymbol{\gamma})$ *be controllable and observable,* $g(z) = \boldsymbol{\gamma}(z\boldsymbol{I} - \boldsymbol{\alpha})^{-1}\boldsymbol{\beta}$. *The system Lie algebra of the output-feedback control system (2.10) is:*

a) $sp(n, \mathbb{R}) \iff g(z) = g(-z)$

b) $sp(n, \mathbb{R}) \oplus \mathbb{R}\, I \iff g(r + \alpha) = g(-r + \alpha)$ *for some* $r \neq 0$.

c) $sl_n(\mathbb{R}) \iff \forall r \; : g(z + r) \neq g(-z + r)$, $\boldsymbol{\gamma}\boldsymbol{\beta} = \operatorname{trace}\boldsymbol{\alpha} = 0$.

d) $gl_n(\mathbb{R}) \iff$ *else.*

These result shows that any homogeneous network of identical first order integrators, controlled by independent all-to-all interconnection functions is accessible. More generally, consider arbitrary SISO node dynamics and more general interconnection schemes. We have obtained the more general result.

Theorem 2.5. [103] *Let* $(\boldsymbol{\alpha}, \boldsymbol{\beta}, \boldsymbol{\gamma})$ *be controllable and observable. Let*

$$\mathcal{L}_{\mathcal{G}}(u(t)) := \sum_{(i,j)\in E} u_{ij}(t) L_{ij}$$

be the adjacency matrix of the graph $\mathcal{G} = (E, V)$ *with* $N > 1$ *nodes with independent input function* $u_{ij}(t)$. *Then the homogeneous network*

$$\dot{\boldsymbol{x}}(t) = (I_N \otimes \boldsymbol{\alpha} + \mathcal{L}_{\mathcal{G}}(u(t)) \otimes \boldsymbol{\beta}\boldsymbol{\gamma})\, \boldsymbol{x}(t) \tag{2.11}$$

is accessible if and only if $\mathcal{G}$ *is strongly connected. The system Lie algebra of (2.11) is either* $sl_{nN}(\mathbb{R})$ *or* $gl_{nN}(\mathbb{R})$.

Controllability and Observability of Ensembles of Systems. Spatially-invariant systems, such as the heat equation, provide interesting examples of distributed parameter systems, where control actions and measurements can take place in a spatially distributed way. Using Fourier transform techniques, spatially-invariant control systems can be identified with parameter-dependent families of linear systems; see e.g. [15] for a systematic outline of this approach. In many applications such as quantum control or the controllability of swarms, it is of interest to solve these problems using open-loop controls. Thus, given a family of desired terminal states, we attempt to construct a parameter-independent input function that steers the zero-state to these states, simultaneously for all parameter values. For simplicity we work in a Hilbert space context and focus on the notion of L^2-ensemble observability and controllability and consider the continuous-time case; the discrete-time case is treated similarly.

We begin with the controllability analysis of parameter-dependent linear systems of the form

$$\begin{aligned} \frac{\partial}{\partial t}\boldsymbol{x}(t,\theta) &= \boldsymbol{A}(\theta)\boldsymbol{x}(t,\theta) + \boldsymbol{B}(\theta)\boldsymbol{u}(t) \\ \boldsymbol{x}(0,\theta) &= \mathbf{0}. \end{aligned} \tag{2.12}$$

The system matrices $\boldsymbol{A}(\theta) \in \mathbb{R}^{n\times n}, \boldsymbol{B}(\theta) \in \mathbb{R}^{n\times m}$ are assumed to vary continuously in a compact interval $\mathbf{P} := [\theta_{\min}\ \theta_{\max}] \subset \mathbb{R}$.

Definition 2.1. *The system (2.12) is called L^2-ensemble controllable, if there exists a finite time $T > 0$ and an input function $\boldsymbol{u} : [0, T] \longrightarrow \mathbb{R}^m$ that steers the initial state $\boldsymbol{x}(0, \theta) = \boldsymbol{0}$ in time T into an ε-neighborhood of the desired state $\boldsymbol{x}_*(\theta)$, simultaneously for all parameters $\theta \in \boldsymbol{P}$; i.e. if*

$$\left(\int_{\boldsymbol{P}} \|\boldsymbol{x}(T, \theta) - \boldsymbol{x}_*(\theta)\|^2 \, \mathrm{d}\,\theta \right)^{\frac{1}{2}} < \varepsilon. \tag{2.13}$$

The following statement contains necessary and sufficient conditions that systems (2.12) is L^2-ensemble controllable.

Theorem 2.6. [174] *Let $\boldsymbol{P} := [\theta_{\min}\ \theta_{\max}] \subset \mathbb{R}$ be a compact interval. A continuous family $(\boldsymbol{A}(\theta), \boldsymbol{B}(\theta))$ of linear systems is L^2-ensemble controllable provided the following conditions are satisfied:*

(i) $(\boldsymbol{A}(\theta), \boldsymbol{B}(\theta))$ is reachable for all $\theta \in \boldsymbol{P}$.

(ii) The input Hermite indices $K_1(\theta), ..., K_m(\theta)$ of $(\boldsymbol{A}(\theta), \boldsymbol{B}(\theta))$ are independent of $\theta \in \boldsymbol{P}$.

(iii) For any pair of distinct parameters $\theta, \theta' \in \boldsymbol{P}, \theta \neq \theta'$, the spectra of $\boldsymbol{A}(\theta)$ and $\boldsymbol{A}(\theta')$ are disjoint:

$$\sigma(\boldsymbol{A}(\theta)) \cap \sigma(\boldsymbol{A}(\theta')) = \emptyset. \tag{2.14}$$

(iv) For each $\theta \in \boldsymbol{P}$, the eigenvalues of $\boldsymbol{A}(\theta)$ have algebraic multiplicity one.

This result is of significance for robust output feedback control of a given linear SISO system $(\boldsymbol{A}, \boldsymbol{b}, \boldsymbol{c})$ which is supposed to be controllable and observable. Consider a compact interval $\mathbf{P} = [\theta_{\min}, \theta_{\max}]$ of gain parameters. Then, for any $\theta \in \boldsymbol{P}$, the characteristic polynomial of the closed loop system is of the form

$$\det(z\boldsymbol{I} - (\boldsymbol{A} + \theta \boldsymbol{bc})) = q(z) + \theta p(z)$$

for some appropriate coprime polynomials p, q. This implies that for $\theta \neq \theta' \in \mathbf{P}$, there exists no complex number z so that

$$q(z) + \theta p(z) = 0 = q(z) + \theta' p(z).$$

Corollary 2.3. *[174] Let $(\boldsymbol{A}, \boldsymbol{b}, \boldsymbol{c})$ be a controllable and observable SISO system. Then output feedback system $(\boldsymbol{A}+\theta\boldsymbol{bc}, \boldsymbol{b})$ is L^2-ensemble controllable if $\boldsymbol{A}+\theta\boldsymbol{bc}$ has distinct eigenvalues for all $\theta \in [\theta_{\min}, \theta_{\max}]$.*

By duality, Theorem 2.6 can also be used to derive conditions on observability of parameter dependent linear systems. To this end, we consider the parameter-dependent system

$$\begin{aligned} \frac{\partial}{\partial t}\boldsymbol{x}(t, \theta) &= \boldsymbol{A}(\theta)\boldsymbol{x}(t, \theta), \quad \boldsymbol{x}(0, \cdot) \in L^2(\mathbf{P}, \mathbb{R}^n), \\ \boldsymbol{y}(t) &= \int_{\mathbf{P}} \boldsymbol{C}(\theta)\boldsymbol{x}(t, \theta)\, \mathrm{d}\theta. \end{aligned} \tag{2.15}$$

As before, we assume that $\boldsymbol{A}(\theta) \in \mathbb{R}^{n\times n}$ and $\boldsymbol{C}(\theta) \in \mathbb{R}^{p\times n}$ vary continuously in a compact parameter domain $\mathbf{P} \subset \mathbb{R}$.

Definition 2.2. *The system* (2.15) *is called L^2-ensemble-observable, if there exists $T > 0$ such that $\boldsymbol{y}(t) = \mathbf{0}$ on $[0, T]$ implies $\boldsymbol{x}(0, \theta) = \mathbf{0}$ for all $\theta \in \boldsymbol{P}$.*

To analyze the observability properties of system (2.15), we note that it is equivalent to the system

$$\begin{aligned} \dot{\boldsymbol{x}}(t) &= \mathcal{A}\boldsymbol{x}(t), \quad \boldsymbol{x}(0) \in L^2(\mathbf{P}, \mathbb{R}^n), \\ \boldsymbol{y}(t) &= \mathcal{C}\boldsymbol{x}(t) \end{aligned} \tag{2.16}$$

on the Hilbert space $X = L^2(\mathbf{P}, \mathbb{R}^n)$, where the bounded linear operators $\mathcal{A}$ and $\mathcal{C}$ are defined by

$$\mathcal{A} : X \longrightarrow X, \quad (\mathcal{A}\boldsymbol{x})(\theta) = \boldsymbol{A}(\theta)\boldsymbol{x}(\theta)$$

and

$$\mathcal{C} : X \longrightarrow \mathbb{R}^p, \quad (\mathcal{C}\boldsymbol{x})(\theta) = \int_{\mathbf{P}} \boldsymbol{C}(\theta)\boldsymbol{x}(\theta)\,\mathrm{d}\theta,$$

respectively. Note that the notion of L^2-ensemble observability is equivalent to the notion of approximate observability for linear systems on Hilbert spaces. In particular, (2.16) is approximate observable if and only if the dual system

$$\frac{\partial}{\partial t}\boldsymbol{x}(t, \theta) = \boldsymbol{A}(\theta)^{\mathrm{T}}\boldsymbol{x}(t, \theta) + \boldsymbol{C}(\theta)^{\mathrm{T}}\boldsymbol{u}(t), \quad \boldsymbol{x}(0, \theta) = \mathbf{0} \tag{2.17}$$

is L^2-ensemble controllable. Thus we can apply Theorem 2.6 to (2.17) and obtain.

Theorem 2.7. [174] *Let $\boldsymbol{P} = [\theta_{\min}\ \theta_{\max}] \subset \mathbb{R}$ be a compact interval. A continuous family $(\boldsymbol{A}(\theta), \boldsymbol{C}(\theta))$ of linear systems is L^2-ensemble observable, provided the following conditions are satisfied:*

(i) $(\boldsymbol{A}(\theta), \boldsymbol{C}(\theta))$ is observable for all $\theta \in \boldsymbol{P}$.
(ii) The output Hermite indices $K_1(\theta), ..., K_m(\theta)$ of $(\boldsymbol{A}(\theta), \boldsymbol{C}(\theta))$ are independent of $\theta \in \boldsymbol{P}$.
(iii) For any pair of distinct parameters $\theta, \theta' \in \boldsymbol{P}, \theta \neq \theta'$, the spectra of $\boldsymbol{A}(\theta)$ and $\boldsymbol{A}(\theta')$ are disjoint:

$$\sigma(\boldsymbol{A}(\theta)) \cap \sigma(\boldsymbol{A}(\theta')) = \emptyset. \tag{2.18}$$

(iv) For each $\theta \in \boldsymbol{P}$, the eigenvalues of $A(\theta)$ have algebraic multiplicity one.

Global Observability of Nonlinear Systems. As a preliminary step to investigate interconnections of nonlinear systems we discuss observability of nonlinear systems, with special emphasis on state constraints. Using a generalization of an approach by [1], we derive observability conditions for real analytic vector fields, i.e.,

$$\dot{\boldsymbol{x}}(t) = \boldsymbol{f}(\boldsymbol{x}(t)), \quad \boldsymbol{y}(t) = \boldsymbol{h}(\boldsymbol{x}(t)) \tag{2.19}$$

on a manifold M. This work is done as a preliminary step to a subsequent observability analysis of interconnected nonlinear systems. Necessary and sufficient condition for observability of analytic systems are due to [74], who showed that observability holds if and only if the observation space $\mathcal{O}$, spanned by iterated Lie derivatives $L_f^j \boldsymbol{h}$, separates points on M. However, this condition is hard to test, since the observation space $\mathcal{O}$ is infinite dimensional.

If the limiting dynamics on an attractor of $\boldsymbol{f}$ is known a-priori, then a sensible observability condition might consist of combining local observability criteria together with some global information on the phase portrait on the attractor. Aeyels[1] has been the first who proposed such an asymptotic dynamics based observability condition for the special class of nonlinear Morse-Smale systems. A lot of examples of particular interest, such as Riccati equations do not satisfy this condition. The following result extends that of [1] to a wide class of systems.

Theorem 2.8. [102] *Let $\boldsymbol{f} : M \to TM$ be a real analytic complete vector field on a manifold M and let $\boldsymbol{h} : M \to \mathbb{R}^p$ be real analytic. Assume:*

A1: The ω-limit set $\omega(\boldsymbol{x})$ of any $\boldsymbol{x} \in M$ is a nonempty compact subset of M.

A2: The system is observable on each ω-limit set $\omega(\boldsymbol{x})$.

A3: For any points $\boldsymbol{x}_1, \boldsymbol{x}_2 \in M$ with $\omega(\boldsymbol{x}_1) \neq \omega(\boldsymbol{x}_2)$ one has $\boldsymbol{h}(\omega(\boldsymbol{x}_1)) \neq \boldsymbol{h}(\omega(\boldsymbol{x}_2))$.

A4: For any $\boldsymbol{x} \in M$ there exists some $p \in \omega(\boldsymbol{x})$ such that Σ is locally observable at p.

Then system (2.19) is globally observable on an arbitrary time interval $[0, T]$.

As an application towards interconnected systems we consider the **series connection** scheme

$$\dot{\boldsymbol{x}}(t) = \boldsymbol{f}(\boldsymbol{x}(t), \boldsymbol{z}(t)),\ \dot{\boldsymbol{z}}(t) = \boldsymbol{g}(\boldsymbol{x}(t)),\ \boldsymbol{y}(t) = \boldsymbol{h}(\boldsymbol{x}(t)) + \boldsymbol{k}(\boldsymbol{z}(t)) \tag{2.20}$$

where M is a compact manifold and $\boldsymbol{f} : M \times \mathbb{R}^m \to TM$, $\boldsymbol{g} : \mathbb{R}^m \to \mathbb{R}^m$, $\boldsymbol{h} : M \to \mathbb{R}^p$, $\boldsymbol{k} : \mathbb{R}^m \to \mathbb{R}^p$ are real analytic. The zero dynamics and the driving system are then defined as

$$\dot{\boldsymbol{x}}(t) = \boldsymbol{f}(\boldsymbol{x}(t), \boldsymbol{0}),\ \boldsymbol{y}(t) = \boldsymbol{h}(\boldsymbol{x}(t)) \tag{2.21}$$

$$\dot{\boldsymbol{z}}(t) = \boldsymbol{g}(\boldsymbol{z}(t)),\ \boldsymbol{w}(t) = \boldsymbol{k}(\boldsymbol{z}(t)), \tag{2.22}$$

respectively.

Theorem 2.9. [102] *Assume that* $\mathbf{0} \in \mathbb{R}^m$ *is a globally asymptotic stable equilibrium point of* $\dot{\boldsymbol{z}}(t) = \boldsymbol{g}(\boldsymbol{z}(t))$ *and (2.22) is observable. Assume further that the zero dynamics (2.21) satisfy*

(i) Every solution of (2.21) converges to a compact invariant subset $A \subset M$ *and there exist finitely many* ω*–limit sets* $\omega(x_i)$*,* $x_1, \ldots, x_k \in M$ *with* $A := \bigcup_{i=1}^{k} \omega(x_i)$*.*

(ii) Each $\omega(x_i)$ *is minimal, nonempty, compact and distal.*

(iii) $A = \bigcup_{i=1}^{k} \omega(x_i)$ *satisfies the non cycle condition, i.e. for any indices* $1 \leq j_1 < \cdots < j_r \leq k$ *and* $\Lambda_j := \omega(x_j)$ *there exist no* $q_1, \cdots, q_r \notin A$*, such that*

$$\alpha(q_i) \subset \Lambda_{j_i} \quad , \quad \omega(q_i) \subset \Lambda_{j_{i+1}}, i = 1, \cdots, r,$$

where $\Lambda_{j_{r+1}} := \Lambda_{j_1}$*.*

(iv) For each $i = 1, \cdots, k$ *there exists* $a_i \in \omega(x_i)$ *such that* $h|_{\omega(x_i)}^{-1}(h(a_i)) = \{a_i\}$*.*

(v) $h(\omega(x_i)) \neq h(\omega(x_j))$ *holds whenever* $i \neq j$*.*

(vi) For any $i = 1, \cdots, k$ *there exists* $p_i \in \omega(x_i)$ *such that (2.21) is locally observable at* p_i*.*

Then system (2.20) is globally observable on an arbitrary time interval $[0, T]$*.*

As another application of Theorem 2.8 we consider linear perspective system on the projective space $\mathbb{P}(\mathbb{F}^n)$ of all real, and respectively, complex lines through the origin. Let $\langle \boldsymbol{X} \rangle$ denote teh image space of a matrix $\boldsymbol{X}$. The system is given by

$$\dot{\boldsymbol{X}} = \boldsymbol{A}\boldsymbol{X}, \quad \boldsymbol{Y} = \boldsymbol{C}\langle \boldsymbol{X} \rangle, \tag{2.23}$$

whose output is the image space $\langle \boldsymbol{X} \rangle$ under $\boldsymbol{C} \in \mathbb{F}^{p \times n}$. Our basic assumption is $p := \text{rank}\boldsymbol{C} \geq 2$. Equation (2.23) can be regarded as a real analytic system that acts on $\mathbb{P}(\mathbb{F}^n)$ via

$$\langle \boldsymbol{X} \rangle \mapsto e^{\boldsymbol{A}t} \langle \boldsymbol{X} \rangle \in \mathbb{P}(\mathbb{F}^n). \tag{2.24}$$

The corresponding output map is given by

$$\langle \boldsymbol{X} \rangle \mapsto \boldsymbol{Y} = \boldsymbol{C}\langle \boldsymbol{X} \rangle \in \mathbb{P}(\mathbb{F}^p). \tag{2.25}$$

Note that the image of (2.25) is not entirely contained in $\mathbb{P}(\mathbb{F}^p)$. However, $\boldsymbol{C}\langle X \rangle$ belongs to $\mathbb{P}(\mathbb{F}^p)$ for almost all $\langle \boldsymbol{X} \rangle \in \mathbb{P}(\mathbb{F}^n)$. Then, the pair

$(\boldsymbol{C}, \boldsymbol{A})$ is called *perspectively observable* on $\mathbb{P}(\mathbb{F}^n)$ if for any two subspaces $\mathcal{V}_1, \mathcal{V}_2 \in \mathbb{P}(\mathbb{F}^n)$ the implication

$$\boldsymbol{C}\mathrm{e}^{\boldsymbol{A}t}\mathcal{V}_1 = \boldsymbol{C}\mathrm{e}^{\boldsymbol{A}t}\mathcal{V}_2 \text{ for almost all } t \in \mathbb{R} \implies \mathcal{V}_1 = \mathcal{V}_2 \tag{2.26}$$

holds. Note again that we do not require equality for *all* $t \in \mathbb{R}$ in the above definition.

To state sufficient conditions for perspectiveobservability we need the following concepts. A matrix $\boldsymbol{A} \in \mathbb{C}^{n\times n}$ is called *strongly cyclic* if $\mathrm{Real}\,\lambda_i \neq \mathrm{Real}\,\lambda_j$ for all different eigenvalues of $\boldsymbol{A}$. Thus there is only one Jordan block of $\boldsymbol{A}$ corresponding to eigenvalues with a fixed real part. Let $\lambda_1, \ldots, \lambda_r \in \mathbb{C}$ denote the distinct real or complex eigenvalues of $\boldsymbol{A} \in \mathbb{R}^{n\times n}$ and let $l \leq 2r$ be the number of non-real, complex-conjugate pairs of eigenvalues. Then, $\boldsymbol{A} \in \mathbb{R}^{n\times n}$ is called *strongly regular* if $r = n$ and $\mathrm{Real}\,\lambda_i \neq \mathrm{Real}\,\lambda_j$ for $i \neq j$ and $\lambda_i \neq \overline{\lambda_j}$. Moreover, $\boldsymbol{A} \in \mathbb{R}^{n\times n}$ is said to satisfy the *irrationality condition* if two eigenvalues $i \neq j$, $\lambda_i \neq \overline{\lambda_j}$ and $\lambda_i, \lambda_j \notin \mathbb{R}$ a ratnionally independent.

Theorem 2.10. [101]
1) Assume that $\boldsymbol{A} \in \mathbb{C}^{n\times n}$ is strongly cyclic. Then the following statements are equivalent:
(a) The pair $(\boldsymbol{C}, \boldsymbol{A})$ is perspectively observable on $\mathbb{P}(\mathbb{C}^n)$.
(b) $\dim \boldsymbol{C}\mathcal{V} = \dim \mathcal{V}$ holds for each complex 2-dimensional $\boldsymbol{A}$-invariant subspaces $\mathcal{V}$.
(c) For all $\alpha, \beta \in \mathbb{C}$ one has

$$\mathrm{rank}\begin{pmatrix} \boldsymbol{A}^2 + \alpha\boldsymbol{A} + \beta\boldsymbol{I}_n \\ \boldsymbol{C} \end{pmatrix} = n.$$

2) Assume that $\boldsymbol{A} \in \mathbb{R}^{n\times n}$ is strongly regular and satisfies the irrationality condition. Then the following statements are equivalent:
(a) The pair $(\boldsymbol{C}, \boldsymbol{A})$ is perspectively observable on $\mathbb{P}(\mathbb{R}^n)$.
(b) For each 1- and 2-dimensional $\boldsymbol{A}$-invariant subspaces $\mathcal{V}$ one has the equality $\dim \boldsymbol{C}\mathcal{V} = \dim \mathcal{V}$.
(c) For all $\alpha, \beta \in \mathbb{R}$ one has

$$\mathrm{rank}\begin{pmatrix} \boldsymbol{A}^2 + \alpha\boldsymbol{A} + \beta\boldsymbol{I}_n \\ \boldsymbol{C} \end{pmatrix} = n.$$

Note that the complex case in Theorem 2.10 covers earlier re-derives results from [134]. For recent results an efficient tests for perspective observability we refer to [264].

2.2.3 Sensitivity Analysis of Networks

In this section the focus is on the following question: How to measure the sensitivity of the behavior of the overall network with respect to variations in the network structure? This problem has been investigated in a subproject that has led to the publications [176], [177]. For simplicity we stick to the sensitivity of networks with identical SISO node systems. Although this theory can be extended to heterogenous networks [177]. In the sequel, we consider the input-to-state and the state-to-output interconnections since this leads to an interesting connection to model reduction based on balanced truncation.

Let the N identical SISO node systems be described by the matrices $\boldsymbol{\alpha} \in \mathbb{F}^{n\times n}$, $\boldsymbol{\beta} \in \mathbb{F}^{n}$ and $\boldsymbol{\gamma} \in \mathbb{F}^{1\times n}$. Further, the network structure is described the matrices $\boldsymbol{L} \in \mathbb{F}^{N\times N}$ and $\boldsymbol{B} \in \mathbb{F}^{N\times m}$ and $\boldsymbol{C} \in \mathbb{F}^{p\times N}$. Recall that the network state-space representation

$$\begin{aligned} \boldsymbol{x}(k+1) =& \bar{\boldsymbol{A}}\boldsymbol{x}(k) + \bar{\boldsymbol{B}}\boldsymbol{u}(k) \\ \boldsymbol{y}(k) =& \bar{\boldsymbol{C}}\boldsymbol{x}(k) \end{aligned} \tag{2.27}$$

is given by

$$\bar{\boldsymbol{A}} = \boldsymbol{I}_N \otimes \boldsymbol{\alpha} + \boldsymbol{L} \otimes \boldsymbol{\beta}\boldsymbol{\gamma}, \quad \bar{\boldsymbol{B}} = \boldsymbol{B} \otimes \boldsymbol{\beta} \quad \text{and} \quad \bar{\boldsymbol{C}} = \boldsymbol{C} \otimes \boldsymbol{\gamma}.$$

For homogenous networks the node transfer function is given by

$$g(z) = \boldsymbol{\gamma}(z\boldsymbol{I}_n - \boldsymbol{\alpha})^{-1}\boldsymbol{\beta}. \tag{2.28}$$

Throughout this section g is supposed to be bounded real and hence, $h(z) = 1/g(z)$ is real rational satisfying $h(\infty) = \infty$ and $|h(z)| \geq 1$ for all $|z| \geq 1$. In addition, it is assumed that the network transfer function

$$\phi(z) = \boldsymbol{C}(z\boldsymbol{I}_N - \boldsymbol{L})^{-1}\boldsymbol{B}$$

has all its poles in the unit open disc $\mathbb{D} = \{z \in \mathbb{C} : |z| < 1\}$. Consequently, the global network transfer function of (2.27) can be written as

$$\boldsymbol{\Phi}(z) = \bar{\boldsymbol{C}}(z\boldsymbol{I}_{nN} - \bar{\boldsymbol{A}})^{-1}\bar{\boldsymbol{B}} = \boldsymbol{C}(h(z)\boldsymbol{I}_N - \boldsymbol{L})^{-1}\boldsymbol{B} \tag{2.29}$$

and $\boldsymbol{\Phi}$ is strictly proper.

Let $p = m$, then the Jacobians of $\boldsymbol{\Phi} = \phi \circ h$ with respect to the input/output coupling parameters $\boldsymbol{B}, \boldsymbol{C}$ are given by the stable transfer functions

$$\begin{aligned} \frac{\partial \boldsymbol{\Phi}}{\partial \boldsymbol{B}}(z) &= (h(z)\boldsymbol{I}_N - \boldsymbol{L}^{\mathrm{T}})^{-1}\boldsymbol{C}^{\mathrm{T}}, \\ \frac{\partial \boldsymbol{\Phi}}{\partial \boldsymbol{C}}(z) &= (h(z)\boldsymbol{I}_N - \boldsymbol{L})^{-1}\boldsymbol{B}. \end{aligned} \tag{2.30}$$

Then we call

$$\mathcal{S}_g(\boldsymbol{L},\boldsymbol{B},\boldsymbol{C}) := \frac{1}{2\pi i}\int_{|z|=1} \left(\|(h(z)\boldsymbol{I}_N - \boldsymbol{L})^{-1}\boldsymbol{B}\,\|^2 + \|\boldsymbol{C}(h(z)\boldsymbol{I}_N - \boldsymbol{L})^{-1}\,\|^2\right)\tfrac{\mathrm{d}z}{z} \tag{2.31}$$

the *network sensitivity* of $\boldsymbol{\Phi}$ with respect to $\boldsymbol{B}$ and $\boldsymbol{C}$. It assigns to each realization $(\boldsymbol{L},\boldsymbol{B},\boldsymbol{C})$ of $\boldsymbol{\phi}$ the sum of the L_2-norms of the Jacobians (2.30). Note that the network sensitivity depends on the realization $(\boldsymbol{L},\boldsymbol{B},\boldsymbol{C})$ of the network transfer function. It is thus of interest to find realizations so that the network sensitivity is minimal. A realization $(\boldsymbol{L}_*,\boldsymbol{B}_*,\boldsymbol{C}_*) = (\boldsymbol{T}\boldsymbol{L}\boldsymbol{T}^{-1},\boldsymbol{T}\boldsymbol{B},\boldsymbol{C}\boldsymbol{T}^{-1})$, or the associated state space coordinate transformation $\boldsymbol{T}$, is said to have *minimal sensitivity*, if among all realizations $(\boldsymbol{L},\boldsymbol{B},\boldsymbol{C})$ of $\boldsymbol{\phi}$, it holds that

$$\mathcal{S}_g(\boldsymbol{L}_*,\boldsymbol{B}_*,\boldsymbol{C}_*) \le \mathcal{S}_g(\boldsymbol{L},\boldsymbol{B},\boldsymbol{C}).$$

In order to investigate the existence and the derivation of realizations having minimal sensitivity we define the *network sensitivity Gramians*

$$\boldsymbol{W}_c^g := \frac{1}{2\pi i}\int_{|z|=1} (h(z)\boldsymbol{I}_N - \boldsymbol{L})^{-1}\boldsymbol{B}\,\boldsymbol{B}^{\mathrm{T}}(\overline{h(z)}\boldsymbol{I}_N - \boldsymbol{L}^{\mathrm{T}})^{-1}\tfrac{\mathrm{d}z}{z} \tag{2.32}$$

$$\boldsymbol{W}_o^g := \frac{1}{2\pi i}\int_{|z|=1} (\overline{h(z)}\boldsymbol{I}_N - \boldsymbol{L}^{\mathrm{T}})^{-1}\boldsymbol{C}^{\mathrm{T}}\,\boldsymbol{C}\,(h(z)\boldsymbol{I}_N - \boldsymbol{L})^{-1}\tfrac{\mathrm{d}z}{z}. \tag{2.33}$$

Using these Gramians the sensitivity of a network can be expressed as the sum of the traces of $\boldsymbol{W}_c^g$ and $\boldsymbol{W}_o^g$, i.e.

$$\mathcal{S}_g(\boldsymbol{L},\boldsymbol{B},\boldsymbol{C}) = \mathrm{tr}(\boldsymbol{W}_c^g + \boldsymbol{W}_o^g), \tag{2.34}$$

where $\mathrm{tr}(\boldsymbol{X})$ denotes the trace of the matrix $\boldsymbol{X}$.

Lemma 2.1. *Suppose that the node transfer function g is strictly proper with minimal realizations. Let the network transfer function $\boldsymbol{\phi}$ be stable. Then, the sensitivity Gramians are positive definite if and only if the realization $(\boldsymbol{L},\boldsymbol{B},\boldsymbol{C})$ is controllable and observable.*

As we are looking for state space coordinate transformations that minimize sensitivity it is of interest how the sensitivity Gramians change under similarity transformations. Let $\boldsymbol{T}$ describe a state space transformation $(\boldsymbol{L},\boldsymbol{B},\boldsymbol{C}) \mapsto (\boldsymbol{T}\boldsymbol{A}\boldsymbol{T}^{-1},\boldsymbol{T}\boldsymbol{B},\boldsymbol{C}\boldsymbol{T}^{-1})$. It is easy to see that the Gramians transform as

$$(\boldsymbol{W}_c^g,\boldsymbol{W}_o^g) \mapsto (\boldsymbol{T}\,\boldsymbol{W}_c^g\,\boldsymbol{T}^{\mathrm{T}},(\boldsymbol{T}^{\mathrm{T}})^{-1}\,\boldsymbol{W}_o^g\,\boldsymbol{T}^{-1}).$$

So, we assign to each state space transformation $\boldsymbol{T}$ (or $\boldsymbol{P} := \boldsymbol{T}^{\mathrm{T}}\boldsymbol{T}$, respectively) the network sensitivity cost as

$$\begin{aligned}\mathcal{S}_g(\boldsymbol{T}\boldsymbol{L}\boldsymbol{T}^{-1},\boldsymbol{T}\boldsymbol{B},\boldsymbol{C}\boldsymbol{T}^{-1}) &:= \mathrm{tr}(\boldsymbol{W}_c^g\boldsymbol{T}^{\mathrm{T}}\boldsymbol{T} + \boldsymbol{W}_o^g\boldsymbol{T}^{-1}(\boldsymbol{T}^{\mathrm{T}})^{-1})\\ &= \mathrm{tr}(\boldsymbol{W}_c^g\,\boldsymbol{P} + \boldsymbol{W}_o^g\,\boldsymbol{P}^{-1}).\end{aligned}$$

To characterize sensitivity optimal realizations we consider the optimization problem for the convex cost function

$$\Phi_g : \mathcal{P} \to \mathbb{R}, \quad \Phi_g(\boldsymbol{P}) := \operatorname{tr}(\boldsymbol{W}_c^g \boldsymbol{P} + \boldsymbol{W}_o^g \boldsymbol{P}^{-1}),$$

where $\mathcal{P}$ denotes the convex space of all real $N \times N$ positive definite matrices $\boldsymbol{P}$. The subsequent statement provides a characterization of sensitivity optimal realizations. Let $\boldsymbol{R}, \boldsymbol{O}, \boldsymbol{H}$ denote the reachability, observability and the Hankel matrix of $(\boldsymbol{L}, \boldsymbol{B}, \boldsymbol{C})$, respectively.

Theorem 2.11. [176] *Let $(\boldsymbol{L}, \boldsymbol{B}, \boldsymbol{C})$ be an arbitrary, not necessarily minimal, realization of the stable strictly proper transfer function $\boldsymbol{\phi}(z)$. Equivalent are:*

1. *There exists an invertible coordinate transformation $\boldsymbol{T}_*$ that minimizes the sensitivity function $\boldsymbol{T} \mapsto \mathcal{S}_g(\boldsymbol{T}\boldsymbol{L}\boldsymbol{T}^{-1}, \boldsymbol{T}\boldsymbol{B}, \boldsymbol{C}\boldsymbol{T}^{-1})$.*
2. *There exists $\boldsymbol{T}_*$ such that $(\boldsymbol{L}_*, \boldsymbol{B}_*, \boldsymbol{C}_*) = (\boldsymbol{T}_*\boldsymbol{L}\boldsymbol{T}_*^{-1}, \boldsymbol{T}_*\boldsymbol{B}, \boldsymbol{C}\boldsymbol{T}_*^{-1})$ has equal Gramians $\boldsymbol{W}_c^g(\boldsymbol{L}_*, \boldsymbol{B}_*) = \boldsymbol{W}_o^g(\boldsymbol{C}_*, \boldsymbol{L}_*)$.*
3. *There exists a unique positive definite matrix $\boldsymbol{P}_* = \boldsymbol{T}_*^{\mathrm{T}}\boldsymbol{T}_*$ that minimizes $\Phi_g(\boldsymbol{P}) = \operatorname{tr}(\boldsymbol{W}_c^g \boldsymbol{P} + \boldsymbol{W}_o^g \boldsymbol{P}^{-1})$ on $\mathcal{P}$.*
4. *There exists $\boldsymbol{P}_* \in \mathcal{P}$ with $\boldsymbol{P}_* \boldsymbol{W}_c^g \boldsymbol{P}_* = \boldsymbol{W}_o^g$.*
5. $\operatorname{rk} \boldsymbol{R} = \operatorname{rk} \boldsymbol{O} = \operatorname{rk} \boldsymbol{H}$.

Thus, if $\operatorname{rk} \boldsymbol{R} = \operatorname{rk} \boldsymbol{O} = \operatorname{rk} \boldsymbol{H}$ holds, then a realization $(\boldsymbol{L}, \boldsymbol{B}, \boldsymbol{C})$ of $\boldsymbol{\phi}$ is sensitivity optimal if and only if $\boldsymbol{W}_c^g = \boldsymbol{W}_o^g$. Moreover, if $(\boldsymbol{L}, \boldsymbol{B}, \boldsymbol{C})$ is controllable and observable, then any two sensitivity optimal realizations $(\boldsymbol{L}_*, \boldsymbol{B}_*, \boldsymbol{C}_*), (\boldsymbol{L}_{**}, \boldsymbol{B}_{**}, \boldsymbol{C}_{**})$ are similar via a unique orthogonal coordinate transformation $\boldsymbol{T} \in O_N(\mathbb{R})$.

In particular, this theorem shows that any network implementation of identical systems described by a bounded real transfer function so that input/output weights of the network are equal ($\boldsymbol{C}^{\mathrm{T}} = \boldsymbol{B}$) has minimal sensitivity.

In [176] we propose some numerical results for the computation of sensitivity optimal realizations using a simple Newton algorithm which is fast (locally quadratically convergent) and globally convergent.

For any positive definite symmetric matrices $\boldsymbol{W}_c^g, \boldsymbol{W}_o^g$ and any $\boldsymbol{P} \in \mathcal{P}$ we denote by $\boldsymbol{Z} = \boldsymbol{Z}(\boldsymbol{P})$ the unique positive definite solution of the Lyapunov equation

$$\boldsymbol{Z}\boldsymbol{W}_c^g\boldsymbol{P} + \boldsymbol{P}\boldsymbol{W}_c^g\boldsymbol{Z} = \boldsymbol{P}\boldsymbol{W}_c^g\boldsymbol{P} - \boldsymbol{W}_o^g. \tag{2.35}$$

Furthermore, we consider the iteration

$$\boldsymbol{P}_{t+1} = \boldsymbol{P}_t - \boldsymbol{Z}(\boldsymbol{P}_t), \quad t = 0, 1, \ldots, \tag{2.36}$$

where $\boldsymbol{P}_0 \in \mathcal{P}$ is so that $\boldsymbol{P}_0\boldsymbol{W}_c^g\boldsymbol{P}_0 - \boldsymbol{W}_o^g > 0$. Note that this is exactly the Riemannian Newton algorithm for optimizing Φ_g, with respect to a suitable

Riemannian metric on $\mathcal{P}$. Hence, the sequence $(\boldsymbol{P}_t)_{t\in\mathbb{N}}$ monotone decreasing in the sense that for any $t \in \mathbb{N}$ it holds that $\boldsymbol{P}_t > \boldsymbol{P}_{t+1}$. In addition, $(\boldsymbol{P}_t)_{t\in\mathbb{N}}$ converges locally quadratically to the sensitivity optimal $\boldsymbol{P}_*$.

In the sequel we compare the sensitivity Gramians to the classical discrete-time controllability and observability Gramians associated with a realization $(\boldsymbol{L}, \boldsymbol{B}, \boldsymbol{C})$. We denote the classical controllability and observability Gramians by $\boldsymbol{W}_c(\boldsymbol{L}, \boldsymbol{B})$ and $\boldsymbol{W}_o(\boldsymbol{C}, \boldsymbol{L})$, respectively, which are given by the unique solutions of the following Lyapunov equations

$$\boldsymbol{L}\boldsymbol{W}_c(\boldsymbol{L}, \boldsymbol{B})\boldsymbol{L}^{\mathrm{T}} - \boldsymbol{W}_c(\boldsymbol{L}, \boldsymbol{B}) + \boldsymbol{B}\,\boldsymbol{B}^{\mathrm{T}} = 0, \tag{2.37}$$

$$\boldsymbol{L}^{\mathrm{T}}\boldsymbol{W}_o(\boldsymbol{C}, \boldsymbol{L})\boldsymbol{L} - \boldsymbol{W}_o(\boldsymbol{C}, \boldsymbol{L}) + \boldsymbol{C}^{\mathrm{T}}\,\boldsymbol{C} = 0. \tag{2.38}$$

For model reduction based on balanced truncation the classical Gramians are of vital interest. To gain insight into sensitivity optimal model reduction and its limitations we show that the sensitivity Gramians satisfy the following Lyapunov-like equations. For brevity we use the notations $\boldsymbol{X}(z) := (h(z)\boldsymbol{I}_N - \boldsymbol{L})^{-1}\boldsymbol{B}$ and $\boldsymbol{Y}(z) := \boldsymbol{C}(h(z)\boldsymbol{I}_N - \boldsymbol{L})^{-1}$. Also we denote by $|\lambda_{\max}(\boldsymbol{A})|$ the spectral radius of the matrix $\boldsymbol{A}$.

Lemma 2.2. *Suppose that g is stable and satisfies $\max_{|z|=1} |g(z)| < \frac{1}{|\lambda_{\max}(\boldsymbol{L})|}$. Then,*

$$\boldsymbol{L}\,\boldsymbol{W}_c^g\,\boldsymbol{L}^{\mathrm{T}} - \boldsymbol{W}_c^g + \boldsymbol{B}\,\boldsymbol{B}^{\mathrm{T}} = \frac{1}{2\pi i}\int_{|z|=1} (|h(z)|^2 - 1)\,\boldsymbol{X}(z)\,\boldsymbol{X}(z)^*\tfrac{\mathrm{d}z}{z},$$

$$\boldsymbol{L}^{\mathrm{T}}\,\boldsymbol{W}_o^g\,\boldsymbol{L} - \boldsymbol{W}_o^g + \boldsymbol{C}^{\mathrm{T}}\,\boldsymbol{C} = \frac{1}{2\pi i}\int_{|z|=1} (|h(z)|^2 - 1)\,\boldsymbol{Y}(z)^*\,\boldsymbol{Y}(z)\tfrac{\mathrm{d}z}{z}.$$

The subsequent statement draws a comparison between the sensitivity Gramians $\boldsymbol{W}_c^g, \boldsymbol{W}_o^g$ and the classical Gramians $\boldsymbol{W}_c(\boldsymbol{L}, \boldsymbol{B}), \boldsymbol{W}_o(\boldsymbol{C}, \boldsymbol{L})$, respectively. To this end we write $\boldsymbol{X} < (\leq)\boldsymbol{Y}$ if the matrix $\boldsymbol{Y} - \boldsymbol{X}$ is positive (semi-)definite. Recall that g is called lossless if it is bounded real and satisfies $|g(z)| = 1$ for all $|z| = 1$.

Theorem 2.12. [176]
(a) Suppose that g is bounded real with $|g(z)| \not\equiv 1$ for $|z| = 1$. Then,

$$\boldsymbol{W}_c^g < \boldsymbol{W}_c(\boldsymbol{L}, \boldsymbol{B}) \quad \textit{and} \quad \boldsymbol{W}_o^g < \boldsymbol{W}_o(\boldsymbol{C}, \boldsymbol{L}).$$

(b) The transfer function g is lossless if and only if

$$\boldsymbol{W}_c^g = \boldsymbol{W}_c(\boldsymbol{L}, \boldsymbol{B}) \quad \textit{and} \quad \boldsymbol{W}_o^g = \boldsymbol{W}_o(\boldsymbol{C}, \boldsymbol{L}).$$

(c) Suppose that $1 \leq |g(z)| < \frac{1}{|\lambda_{\max}(\boldsymbol{L})|}$ for all $|z| = 1$. Then,

$$\boldsymbol{W}_c^g \geq \boldsymbol{W}_c(\boldsymbol{L}, \boldsymbol{B}) \quad \textit{and} \quad \boldsymbol{W}_o^g \geq \boldsymbol{W}_o(\boldsymbol{C}, \boldsymbol{L}).$$

On the other hand, the relation of the sensitivity Gramians to the classical Gramians associated with the realization $(\bar{\boldsymbol{A}}, \bar{\boldsymbol{B}}, \bar{\boldsymbol{C}})$ is also of interest. In [268] it is shown that

$$(h(z)\boldsymbol{I}_N - \boldsymbol{L})^{-1}\boldsymbol{B} = (I \otimes \gamma)\,(z\boldsymbol{I}_{nN} - \bar{\boldsymbol{A}})^{-1}\bar{\boldsymbol{B}},$$
$$\boldsymbol{C}(h(z)\boldsymbol{I}_N - \boldsymbol{L})^{-1} = \bar{\boldsymbol{C}}\,(z\boldsymbol{I}_{nN} - \bar{\boldsymbol{A}})^{-1}(I \otimes \boldsymbol{\beta}).$$

Therefore, the sensitivity Gramians can be expressed via the classical Gramians associated with $(\bar{\boldsymbol{A}}, \bar{\boldsymbol{B}}, \bar{\boldsymbol{C}})$ by

$$\boldsymbol{W}_c^g = (\boldsymbol{I} \otimes \gamma)\,\boldsymbol{W}_c(\bar{\boldsymbol{A}}, \bar{\boldsymbol{B}})\,(\boldsymbol{I} \otimes \gamma^{\mathrm{T}}),$$
$$\boldsymbol{W}_o^g = (\boldsymbol{I} \otimes \boldsymbol{\beta}^{\mathrm{T}})\,\boldsymbol{W}_o(\bar{\boldsymbol{C}}, \bar{\boldsymbol{A}})\,(\boldsymbol{I} \otimes \boldsymbol{\beta}).$$

This implies that the sensitivity Gramians can be computed as

$$\boldsymbol{W}_c^g = (\boldsymbol{I} \otimes \gamma)\,\boldsymbol{X}\,(\boldsymbol{I} \otimes \gamma^{\mathrm{T}})$$

via the linear matrix inequality

$$\bar{\boldsymbol{A}}\boldsymbol{X}\bar{\boldsymbol{A}}^{\mathrm{T}} - \boldsymbol{X} + \bar{\boldsymbol{B}}\bar{\boldsymbol{B}}^{\mathrm{T}} = 0\,, \quad \boldsymbol{X} > 0.$$

Moreover, in [211] it is shown that, if the transfer function g is bounded real the classical controllability and observability Gramians for $(\bar{\boldsymbol{A}}, \bar{\boldsymbol{B}}, \bar{\boldsymbol{C}})$ and $(\boldsymbol{L}, \boldsymbol{B}, \boldsymbol{C})$ are related via

$$\boldsymbol{W}_c(\bar{\boldsymbol{A}}, \bar{\boldsymbol{B}}) \leq \boldsymbol{W}_c(\boldsymbol{L}, \boldsymbol{B}) \otimes \boldsymbol{P}^{-1} \quad \text{and} \quad \boldsymbol{W}_o(\bar{\boldsymbol{C}}, \bar{\boldsymbol{A}}) \leq \boldsymbol{W}_o(\boldsymbol{C}, \boldsymbol{L}) \otimes \boldsymbol{P},$$

where $\boldsymbol{P}$ is the unique positive definite solution to the Riccati equation

$$\boldsymbol{\alpha}\boldsymbol{P}\boldsymbol{\alpha}^{\mathrm{T}} - \boldsymbol{P} + \frac{\boldsymbol{\alpha}^{\mathrm{T}}\boldsymbol{P}\boldsymbol{\beta}\boldsymbol{\beta}^{\mathrm{T}}\boldsymbol{P}\boldsymbol{\alpha}}{1 - \boldsymbol{\beta}^{\mathrm{T}}\boldsymbol{P}\boldsymbol{\beta}} + \gamma^{\mathrm{T}}\gamma = 0$$

in terms of the matrices $(\boldsymbol{\alpha}, \boldsymbol{\beta}, \gamma)$ describing the node systems. Since it holds that $\boldsymbol{\beta}^{\mathrm{T}}\boldsymbol{P}\boldsymbol{\beta} \leq 1$ and $\gamma^{\mathrm{T}}\boldsymbol{P}^{-1}\gamma \leq 1$ the sensitivity Gramians can be estimated by the classical Gramians corresponding to the network realization $(\boldsymbol{L}, \boldsymbol{B}, \boldsymbol{C})$ and realization $(\boldsymbol{\alpha}, \boldsymbol{\beta}, \boldsymbol{C})$ of the node transfer function.

Corollary 2.4. *Suppose that the node transfer function g is bounded real and $\boldsymbol{P}$ is as above. Then,*

$$\boldsymbol{W}_c^g \leq \gamma^{\mathrm{T}}\boldsymbol{P}^{-1}\gamma \cdot \boldsymbol{W}_c(\boldsymbol{L}, \boldsymbol{B}), \quad \boldsymbol{W}_o^g \leq \boldsymbol{\beta}^{\mathrm{T}}\boldsymbol{P}\boldsymbol{\beta} \cdot \boldsymbol{W}_o(\boldsymbol{C}, \boldsymbol{L}).$$

At the end of this subsubsection we examine the issue of balancing with respect to the sensitivity Gramians. A network realization $(\boldsymbol{L}, \boldsymbol{B}, \boldsymbol{C})$ is said to be *g-sensitivity balanced* if the sensitivity Gramians coincide and are diagonal, i.e.

$$\boldsymbol{W}_c^g = \boldsymbol{W}_o^g = \boldsymbol{\Sigma}^g = \mathrm{diag}(\sigma_1^g, ..., \sigma_N^g).$$

The entries $\sigma_1^g \geq ... \geq \sigma_N^g$ are called the *g-singular values* of the network. By Theorem 2.12 (b) it turns out that if the node transfer function is lossless then sensitivity balancing of the overall network is equivalent to balancing of the network structure.

Theorem 2.13. [176] *Suppose that the node transfer function g is lossless. Then the network is g-sensitivity balanced if and only if $(\boldsymbol{L}, \boldsymbol{B}, \boldsymbol{C})$ is balanced, i.e. $\boldsymbol{W}_c(\boldsymbol{L}, \boldsymbol{B}) = \boldsymbol{W}_o(\boldsymbol{C}, \boldsymbol{L}) = \boldsymbol{\Sigma}$. In particular, the g-singular values coincide with the singular values of $(\boldsymbol{L}, \boldsymbol{B}, \boldsymbol{C})$.*

2.3 Minimal Bit Rates and Entropy for Control Tasks

2.3.1 Motivation and Earlier Results

A prerequisite for the analysis of the communication demands within networked systems is a thorough analysis of the information needed for performing control tasks. This problem has been considered for more than twenty years, cf. the survey [271], however with limited success. Early landmarks are the paper [92] which considered quantized information for stabilization and proposed to use statistical methods from ergodic theory, and [390] discussing stabilization of linear systems via coding. From the wealth of literature on this topic we also cite [93, 114, 254, 270, 357, 396]. We propose an approach that is different from quantization of the state space. In particular, no information pattern is specified describing how the information on the states of the system is available for the controller.

In non-technical terms, the basic idea for the approach is the following. Consider a control task on the time interval $[0, \infty)$. For example, this may be the problem to make a subset of the state space invariant or the problem to stabilize the system about an equilibrium. Usually, a controller device is constructed which performs the control task based on measurements of the output of the system. If successful, the controller will generate control actions on the system such that the desired behavior is achieved for all initial values in a given set $\mathcal{K}$ in the state space.

If continual measurement of the output is not possible due to data rate constraints (in a noiseless communication channel), the controller only has a finite amount of information available on any finite interval $[0, \tau]$. Hence, it may appear reasonable that the controller can only generate a finite number of time-dependent control functions $\boldsymbol{u}(t)$, $t \in [0, \tau]$, which are to guarantee the desired behavior on $[0, \tau]$ for all initial states in $\mathcal{K}$. If time increases, the amount of information for the controller increases, and hence it may generate more controls. Looking at this from the other side, the number of controls, which are necessary for accomplishing the control task on $[0, \tau]$, determines the minimal data rate. Thus, the growth rate of the minimal number of controls as time tends to infinity is a measure for the minimal data rate necessary to accomplish the control task on $[0, \infty)$ for all initial values in $\mathcal{K}$.

More specifically, we see that the combination of the aspects of control and information, which traditionally are separated, requires a modification of the familiar feedback paradigm. If the communication channel between the system and the controller only admits a finite bit rate, then it is not justified that the controller generates infinitely many different control functions on a finite time interval, since only a finite amount of information is available. This excludes state-dependent controls $\boldsymbol{u}(\boldsymbol{x}(t, \boldsymbol{x}_0))$, where depending on the initial state $\boldsymbol{x}_0$ one obtains infinitely many control functions. Instead, the controller might use information on the system to compute open-loop controls on some time interval $[t_i, t_{i+1}]$; then, using updated information, a new control function might be computed and used on the next time interval $[t_{i+1}, t_{i+2}]$ (the t_i might depend on the initial value $\boldsymbol{x}_0$.) The computations may be based on quantization of the state space (cf. Section 5.4), symbolic controllers or may be done via devices like model predictive control (MPC) (cf. Chapter 4). In any case, this results in time-dependent control functions $\boldsymbol{u}(t)$ defined on $[0, \infty)$.

Instead of concentrating on the algorithmic question how to generate these controls, we discuss the minimal bit rate needed to discern the time-dependent control functions $\boldsymbol{u}(t)$ on any time interval $[0, \tau]$, $\tau > 0$. This, in fact, is the point of view taken in [357], p. 1057, that estimates the minimal bit rate for stabilization of discrete-time linear systems from below (see Proposition 3.2 in [357]).

This basic idea can be formalized in different ways. We will discuss in some detail invariance entropy for the problem to keep the system within a subset of the state space, and an entropy for the stabilization problem.

In this problem formulation, a close analogy to the notion of topological entropy in the theory of dynamic systems becomes apparent; the monographs [110, 197, 376] contain expositions of this theory. Here one counts, how fast trajectories of a dynamic system move apart, and hence one has to look at initial points. In control, the decisive parameter which determines the behavior of trajectories is the control function. Hence, we will count control functions and then we use rather analogous mathematical constructions.

In Section 2.3.2 the problem to render a compact subset of the state space invariant will be discussed in detail; cf. [66, 70, 71, 73, 198–201]. Section 2.3.3 presents results on exponential stabilization.

Notation. The closure of a set $\mathcal{A}$ is $\mathrm{cl}\,\mathcal{A}$ and the cardinality of a set $\mathcal{A}$ is $\#\mathcal{A}$; thus if $\mathcal{A}$ is a finite set, $\#\mathcal{A}$ is the number of its elements and otherwise $\#\mathcal{A} = \infty$.

2.3.2 Invariance Entropy

Overview of New Results. The concept of *invariance entropy* has its origins in the problem to determine data rates needed in order to achieve desired control objectives. Nair, Evans, Mareels and Moran introduced ideas from topological dynamics into this field. They defined and studied *topological feedback entropy* for the problem of stabilizing a discrete-time system into a compact subset of the state space and *local topological feedback entropy* for stabilization at an equilibrium. We modify this approach by introducing the notion of *invariance entropy* for continuous-time systems which measures how fast the number of open-loop control functions grows which are needed to keep the system inside a compact set for longer and longer times. More precisely, for a compact and controlled invariant subset $\mathcal{Q}$ of the state space, and a compact set $\mathcal{K} \subset \mathcal{Q}$ of allowed initial states, the minimal number of open-loop control functions necessary to keep the system in $\mathcal{Q}$ up to time $\tau > 0$, from an arbitrary initial state in $\mathcal{K}$, is denoted by $r^*_{\text{inv}}(\tau, \mathcal{K}, \mathcal{Q})$. Then the exponential growth rate

$$h^*_{\text{inv}}(\mathcal{K}, \mathcal{Q}) := \limsup_{\tau \to \infty} \frac{1}{\tau} \ln r^*_{\text{inv}}(\tau, \mathcal{K}, \mathcal{Q})$$

is called the *strict invariance entropy of* $(\mathcal{K}, \mathcal{Q})$. Requiring that the solutions only stay in an ε-neighborhood of $\mathcal{Q}$, another quantity $h_{\text{inv}}(\mathcal{K}, \mathcal{Q})$, simply called the *invariance entropy of* $(\mathcal{K}, \mathcal{Q})$, is defined similarly, now also taking the limit for ε going to zero. We started a thorough investigation of these notions with an emphasis on the derivation of upper and lower bounds.

In this section, we give an overview of the main results obtained so far. After introducing the basic notions and explaining the elementary properties of invariance entropy, we show that the strict invariance entropy $h^*_{\text{inv}}(\mathcal{Q})$ can be characterized as the minimal data rate for the control objective of rendering the set $\mathcal{Q}$ invariant by a coder-controller device. Also relations between invariance entropy and Lyapunov exponents are addressed. In particular, the following two results are highlighted:

> For linear systems, the invariance entropy equals the sum of the unstable eigenvalues counted according to their multiplicities. For nonlinear systems, under additional controllability assumptions on the set $\mathcal{Q}$, the strict invariance entropy $h^*_{\text{inv}}(\mathcal{K}, \mathcal{Q})$ is upper bounded by the sum of the unstable Lyapunov exponents of any regular periodic solution in the interior of $\mathcal{Q}$.

The latter result can be used to obtain a formula for the invariance entropy of a control set for a scalar control-affine system with a single control vector field. An application to a linearized inverted pendulum is given.

Definition of Invariance Entropy. We consider a continuous-time control system

$$\dot{\boldsymbol{x}}(t) = \boldsymbol{f}(\boldsymbol{x}(t), \boldsymbol{u}(t)), \quad \boldsymbol{u} \in \underline{\mathcal{U}} \tag{2.39}$$

on a smooth manifold $\mathcal{M}$, endowed with a metric d (not necessarily a Riemannian distance). The set of admissible control functions is given by

$$\underline{\mathcal{U}} = \{\boldsymbol{u} : \mathbb{R} \to \mathbb{R}^m \ : \ \boldsymbol{u} \text{ measurable with } \boldsymbol{u}(t) \in \mathcal{U} \text{ a.e.}\}$$

with a compact control range $\mathcal{U} \subset \mathbb{R}^m$, and the right-hand side $\boldsymbol{f} : \mathcal{M} \times \mathbb{R}^m \to T\mathcal{M}$ is a continuous function which is continuously differentiable in the first argument. Then, for each initial value $\boldsymbol{x} \in \mathcal{M}$ and for each $\boldsymbol{u} \in \underline{\mathcal{U}}$, there exists a unique solution (in the sense of Carathéodory), which we denote by $\varphi(t, \boldsymbol{x}, \boldsymbol{u})$. Let $\mathcal{Q} \subset \mathcal{M}$ be a compact and *controlled invariant* set, that is, for every $\boldsymbol{x} \in \mathcal{Q}$ there is $\boldsymbol{u} \in \underline{\mathcal{U}}$ with $\varphi(t, \boldsymbol{x}, \boldsymbol{u}) \in \mathcal{Q}$ for all $t \geq 0$. Furthermore, let $\mathcal{K} \subset \mathcal{Q}$ be another compact set.

Definition 2.3. *For $\tau > 0$, a set $\mathcal{S} \subset \underline{\mathcal{U}}$ is called $(\tau, \mathcal{K}, \mathcal{Q})$-spanning if for every $\boldsymbol{x} \in \mathcal{K}$ there is $\boldsymbol{u} \in \mathcal{S}$ with*

$$\varphi(t, \boldsymbol{x}, \boldsymbol{u}) \in \mathcal{Q} \text{ for all } t \in [0, \tau].$$

Let $r^*_{\text{inv}}(\tau, \mathcal{K}, \mathcal{Q})$ be the minimal cardinality of such a set. The strict invariance entropy *of $(\mathcal{K}, \mathcal{Q})$ is defined as the exponential growth rate of $r^*_{\text{inv}}(\tau, \mathcal{K}, \mathcal{Q})$ for $\tau \to \infty$, that is,*

$$h^*_{\text{inv}}(\mathcal{K}, \mathcal{Q}) := \limsup_{\tau \to \infty} \frac{1}{\tau} \ln r^*_{\text{inv}}(\tau, \mathcal{K}, \mathcal{Q}).$$

If $\mathcal{K} = \mathcal{Q}$, we simply write $r^*_{\text{inv}}(\tau, \mathcal{Q})$ and $h^*_{\text{inv}}(\mathcal{Q})$.

One problem with this definition is that the numbers $r^*_{\text{inv}}(\tau, \mathcal{K}, \mathcal{Q})$ need not be finite. To overcome this problem, we also work with another version of invariance entropy.

Definition 2.4. *For $\tau, \varepsilon > 0$, a set $\mathcal{S} \subset \underline{\mathcal{U}}$ is called $(\tau, \varepsilon, \mathcal{K}, \mathcal{Q})$-spanning if for every $\boldsymbol{x} \in \mathcal{K}$ there is $\boldsymbol{u} \in \mathcal{S}$ with*

$$\operatorname{dist}(\varphi(t, \boldsymbol{x}, \boldsymbol{u}), \mathcal{Q}) < \varepsilon \text{ for all } t \in [0, \tau].$$

Let $r_{\text{inv}}(\tau, \varepsilon, \mathcal{K}, \mathcal{Q})$ be the minimal cardinality of such a set. The invariance entropy of $(\mathcal{K}, \mathcal{Q})$ is defined as

$$h_{\text{inv}}(\mathcal{K}, \mathcal{Q}) := \lim_{\varepsilon \searrow 0} \limsup_{\tau \to \infty} \frac{1}{\tau} \ln r_{\text{inv}}(\tau, \varepsilon, \mathcal{K}, \mathcal{Q}).$$

If $\mathcal{K} = \mathcal{Q}$, we simply write $r_{\text{inv}}(\tau, \varepsilon, \mathcal{Q})$ and $h_{\text{inv}}(\mathcal{Q})$.

In the preceding definition, we always have $r_{\text{inv}}(\tau, \varepsilon, \mathcal{K}, \mathcal{Q}) < \infty$, which easily follows from compactness of $\mathcal{K}$ and continuous dependence on the initial value. Existence of the limit in the definition of $h_{\text{inv}}(\mathcal{K}, \mathcal{Q})$ follows from monotonicity of $r_{\text{inv}}(\tau, \varepsilon, \mathcal{K}, \mathcal{Q})$ in ε. Some elementary properties of $h_{\text{inv}}(\mathcal{K}, \mathcal{Q})$ and $h^*_{\text{inv}}(\mathcal{K}, \mathcal{Q})$ are the following.

1. $0 \leq h_{\text{inv}}(\mathcal{K}, \mathcal{Q}) \leq h^*_{\text{inv}}(\mathcal{K}, \mathcal{Q})$ and $h_{\text{inv}}(\mathcal{K}, \mathcal{Q}) < \infty$.
2. $h_{\text{inv}}(\mathcal{K}, \mathcal{Q})$ is independent of the choice of the metric on $\mathcal{M}$.
3. $h_{\text{inv}}(\mathcal{K}, \mathcal{Q})$ and $h^*_{\text{inv}}(\mathcal{K}, \mathcal{Q})$ are invariant under C^0-state equivalence.
4. If $\mathcal{K} = \mathcal{K}_1 \cup \ldots \cup \mathcal{K}_n$ ($\mathcal{K}_i$ compact), then

$$h^{(*)}_{\text{inv}}(\mathcal{K}, \mathcal{Q}) = \max_{i=1}^{n} h^{(*)}_{\text{inv}}(\mathcal{K}_i, \mathcal{Q}).$$

5. For the system with right-hand side $s \cdot F$ $(s > 0)$ instead of F, the set $\mathcal{Q}$ is also controlled invariant and

$$h_{\text{inv}}^{(*)}(\mathcal{K}, \mathcal{Q}; s \cdot F) = s \cdot h_{\text{inv}}^{(*)}(\mathcal{K}, \mathcal{Q}; F).$$

In general, the equality $h_{\text{inv}}(\mathcal{K}, \mathcal{Q}) = h_{\text{inv}}^*(\mathcal{K}, \mathcal{Q})$ does not hold, since there are cases when $h_{\text{inv}}^*(\mathcal{K}, \mathcal{Q}) = \infty$. However, in some special situations equality can be shown.

Minimum Transmission Data Rate. We relate the strict invariance entropy $h_{\text{inv}}^*(\mathcal{Q})$ to minimal transmission data rates in feedback loops with a digital channel between coder and controller. Consider system (2.39) and suppose that a sensor, which is connected to a controller via a digital noiseless channel, measures its states at sampling times $k\tau$, $k \geq 0$, for some fixed time step $\tau > 0$. The state at time $k\tau$ is coded using a finite alphabet $\mathcal{S}_k$ of (time-varying) size μ_k (Fig. 2.1). The coder transmits a symbol $s_k \in \mathcal{S}_k$, which may depend on the present state and on all past states. The corresponding coder mapping is denoted by $\gamma_k : \mathcal{M}^{k+1} \to \mathcal{S}_k$. At time $k\tau$ the controller has $k+1$ symbols $s_0, s_1, \ldots, s_k$ available and generates a finite-time control function $\boldsymbol{u}(k) : [0, \tau] \to \mathcal{U}$. We denote the corresponding controller mapping by $\delta_k : \mathcal{S}_0 \times \mathcal{S}_1 \times \cdots \times \mathcal{S}_k \to \underline{\mathcal{U}}_\tau := \{\boldsymbol{u}_{|[0,\tau]} : \boldsymbol{u} \in \underline{\mathcal{U}}\}$.

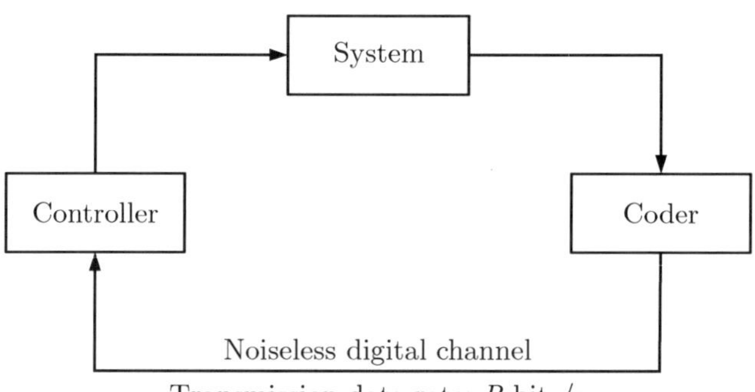

Fig. 2.1 Feedback system with coder and controller

We call the quadruple $\mathcal{H} := (\mathcal{S}, \gamma, \delta, \tau)$, where $\mathcal{S} = (\mathcal{S}_k)_{k \geq 0}$, $\gamma = (\gamma_k)_{k \geq 0}$, and $\delta = (\delta_k)_{k \geq 0}$, a *coder-controller*, and we define the (asymptotic, average) *transmission data rate* of $\mathcal{H}$ by

$$R(\mathcal{H}) := \liminf_{k \to \infty} \frac{1}{k\tau} \sum_{j=0}^{k-1} \log_2 \mu_j.$$

Additionally, we require that this number is finite. (This definition is essentially taken from [270].) We say that $\mathcal{H}$ *renders* $\mathcal{Q}$ *invariant* if for all $\boldsymbol{x}(0) \in \mathcal{Q}$the sequence $(\boldsymbol{x}(k))_{k \geq 0}$, defined recursively by

$$\boldsymbol{x}(k) := \varphi\left(\tau, \boldsymbol{x}(k-1), \boldsymbol{u}(k-1)\right)$$

with

$$\boldsymbol{u}(k-1) = \delta_{k-1}(\gamma_0(\boldsymbol{x}(0)), \gamma_1(\boldsymbol{x}(0), \boldsymbol{x}(1)), \ldots, \gamma_{k-1}(\boldsymbol{x}(0), \ldots, \boldsymbol{x}(k-1)))$$

satisfies $\varphi(t, \boldsymbol{x}(k), \boldsymbol{u}(k)) \in \mathcal{Q}$ for all $k \geq 0$ and $t \in [0, \tau]$. Then we have the following result, cf. [198, 202].

Theorem 2.14. *Let* $\mathcal{Q}$ *be a compact and controlled invariant set for system* (2.39). *Then*

$$h^*_{\text{inv}}(\mathcal{Q}) = \inf_{\mathcal{H}} \frac{R(\mathcal{H})}{\log_2 \mathrm{e}}, \qquad (2.40)$$

where the infimum is taken over all coder-controllers $\mathcal{H}$ *that render* $\mathcal{Q}$ *invariant.*

To prove this theorem, one first shows that every (finite) $(\tau, \mathcal{Q})$-spanning set $\{u_1, \ldots, u_n\}$ defines coding regions $\mathcal{C}_1, \ldots, \mathcal{C}_n$ with $\mathcal{Q} = \bigcup_{i=1}^{n} \mathcal{C}_i$, such that a coder-controller $\mathcal{H}$ can be defined as follows: The coder sends the symbol i to the controller if the measured state is in $\mathcal{C}_i$. Upon receiving the symbol i, the controller generates the control function $\boldsymbol{u}_i$. Then it is easy to see that the corresponding transmission data rate is $\log_2(n)/\tau$. This proves the existence of a sequence $\mathcal{H}_k$ of coder-controllers whose transmission data rates come arbitrarily close to $h^*_{\text{inv}}(\mathcal{Q})$. On the other hand, every coder-controller $\mathcal{H} = (\mathcal{S}, \gamma, \delta, \tau)$ gives $(k\tau, \mathcal{Q})$-spanning sets for all $k \geq 1$ by concatenation of the control functions generated by the controller at times $i\tau$, $(i = 0, 1, \ldots, k)$. The cardinalities of these sets are $\prod_{i=0}^{k-1} \mu_i$, which implies the inequality "$\leq$" in (2.40).

Relation between Invariance Entropy and Lyapunov Exponents. Next, we present two results which show that the invariance entropy is related to the Lyapunov exponents of the given control system. For linear systems, where the Lyapunov exponents coincide with the real parts of the eigenvalues of the dynamic matrix of the system, we have the following result.

Theorem 2.15. *For the linear control system*

$$\dot{\boldsymbol{x}}(t) = \boldsymbol{A}\boldsymbol{x}(t) + \boldsymbol{B}\boldsymbol{u}(t), \quad \boldsymbol{u} \in \underline{\mathcal{U}},$$

with $\boldsymbol{A} \in \mathbb{R}^{n \times n}$ *and* $\boldsymbol{B} \in \mathbb{R}^{n \times m}$ *it holds that*

$$h_{\text{inv}}(\mathcal{K}, \mathcal{Q}) = \sum_{\lambda \in \text{spec}(A)} \max\{0, n_\lambda \text{Real}(\lambda)\}, \qquad (2.41)$$

if $\mathcal{K}$ *has positive Lebesgue measure. Here,* n_λ *denotes the algebraic multiplicity of the eigenvalue* λ.

One may wonder why the value of the invariance entropy does only depend on the matrix $\boldsymbol{A}$, but not on $\boldsymbol{B}$. As one sees in the proof of Theorem 2.15 in [70], this is basically a consequence of the specific structure of the solutions of a linear system. In particular, the distance of two solutions $\varphi(t, \boldsymbol{x}, \boldsymbol{u})$ and $\varphi(t, \boldsymbol{y}, \boldsymbol{u})$ is independent of the matrix $\boldsymbol{B}$, and the same is true for the volume growth factor under application of $\varphi(t, \cdot, \boldsymbol{u})$. Both of these quantities are strongly related to the numbers $r_{\text{inv}}(\tau, \varepsilon, \mathcal{K}, \mathcal{Q})$. However, we should remark that the assumptions of Theorem 2.15 cannot be satisfied for arbitrary pairs of matrices $(\boldsymbol{A}, \boldsymbol{B})$. The existence of a compact controlled invariant set $\mathcal{Q}$ with nonempty interior (and hence positive Lebesgue measure) can be guaranteed if the pair $(\boldsymbol{A}, \boldsymbol{B})$ is controllable, the matrix $\boldsymbol{A}$ is hyperbolic, and the control range $\mathcal{U}$ is compact and convex with nonempty interior. Then there exists a unique control set $\mathcal{D}$ whose closure $\mathcal{Q} = \operatorname{cl} \mathcal{D}$ has the desired properties, cf. [70].

Remark 2.1. For inhomogeneous bilinear systems on $\mathbb{R}^n$ a similar result holds in form of an estimate from below. Here, the real parts of the eigenvalues are replaced by minimal Lyapunov exponents of the associated homogeneous system on invariant subbundles of $\underline{\mathcal{U}} \times \mathbb{R}^n$, cf. [198, 202].

For nonlinear systems we have a finite set of Lyapunov exponents for each solution, that is, for each pair $(\boldsymbol{x}, \boldsymbol{u}) \in \mathcal{M} \times \underline{\mathcal{U}}$. Hence, we cannot expect a formula for the invariance entropy as simple as the one in Theorem 2.15. In fact, it seems improbable that a general formula can be given here. However, under additional controllability assumptions on the system, an upper bound, similar to the expression in (2.41), can be derived.

Recall that a *control set* $\mathcal{D} \subset \mathcal{M}$ for system (2.39) is a maximal set of controlled invariance and *approximate controllability* (for each two states $\boldsymbol{x}, \boldsymbol{y} \in \mathcal{D}$ and $\varepsilon > 0$ there are $\boldsymbol{u} \in \underline{\mathcal{U}}$ and $t \geq 0$ with $d(\varphi(t, \boldsymbol{x}, \boldsymbol{u}), \boldsymbol{y}) < \varepsilon$.) For $\mathcal{Q}$ being the closure of a control set, the following result holds, cf. [198, 199, 202].

Theorem 2.16. *Assume that the right-hand side $\boldsymbol{f}$ of system (2.39) is continuously differentiable, and let $\mathcal{D}$ be a control set with compact closure $\mathcal{Q} = \operatorname{cl} \mathcal{D}$ for this system. Let $(\varphi(\cdot, \boldsymbol{x}_0, \boldsymbol{u}_0), \boldsymbol{u}_0(\cdot))$ be a controlled periodic solution in $\operatorname{int} \mathcal{D} \times \operatorname{int} \mathcal{U}$ which is regular, that is, the linearization along this solution is controllable. Then for each compact set $\mathcal{K} \subset \mathcal{D}$ it holds that*

$$h^*_{\text{inv}}(\mathcal{K}, \mathcal{Q}) \leq \sum_{\lambda} \max\{0, n_\lambda \lambda\},$$

where the sum is taken over the different Lyapunov exponents of $\varphi(\cdot, \boldsymbol{x}_0, \boldsymbol{u}_0)$ and n_λ denotes the multiplicity of λ.

Remark 2.2. Assuming local accessibility for system (2.39) on the interior of a control set $\mathcal{D}$, there is a periodic solution in $\operatorname{int}\mathcal{D}$ with initial value $\boldsymbol{x}$ for every $\boldsymbol{x} \in \operatorname{int}\mathcal{D}$ (cf. Proposition 4.3.3 in [72]). In general, it is not clear if such periodic solutions are regular. However, for real-analytic systems, existence of regular periodic solutions through every point $\boldsymbol{x} \in \operatorname{int}\mathcal{D}$ can be guaranteed if the system is strongly accessible, which follows from results of [344, 345].

For scalar control-affine systems of the form

$$\dot{x}(t) = f(x(t)) + u(t)g(x(t)), \quad u \in \underline{\mathcal{U}}, \tag{2.42}$$

with C^1-vector fields $f, g : \mathbb{R} \to \mathbb{R}$ the estimate of Theorem 2.16 can be used together with other results to prove the following result, cf. [198, 199, 202].

Theorem 2.17. *Let $\mathcal{D}$ be a bounded control set of system* (2.42). *Further assume that local accessibility holds on $\mathcal{Q} := \operatorname{cl}\mathcal{D}$. Then for every compact set $\mathcal{K} \subset \mathcal{D}$ with nonempty interior it holds that*

$$h_{\text{inv}}(\mathcal{K}, \mathcal{Q}) = h^*_{\text{inv}}(\mathcal{K}, \mathcal{Q}) = \max\left\{0, \min_{x \in \mathcal{Q}}\left(f'(x) - \frac{f(x)}{g(x)}g'(x)\right)\right\}. \tag{2.43}$$

The expression to be minimized in (2.43) is the derivative of the right-hand side of (2.42) at the equilibrium pair (x, u_x) with $u_x = -f(x)/g(x)$, which coincides with the Lyapunov exponent for this pair. Since in dimension one (under appropriate regularity assumptions), the full Lyapunov spectrum is already attained by the Lyapunov exponents of equilibria, Eq. (2.43) can also be written as

$$h_{\text{inv}}(\mathcal{K}, \mathcal{Q}) = h^*_{\text{inv}}(\mathcal{K}, \mathcal{Q}) = \max\left\{0, \min \Sigma_{\text{Ly}}(\mathcal{Q})\right\},$$

where $\Sigma_{\text{Ly}}(\mathcal{Q})$ denotes the Lyapunov spectrum over $\mathcal{Q}$, that is, the set of all limits

$$\limsup_{\tau \to \infty} \frac{1}{\tau} \ln\left|\frac{\partial \varphi}{\partial x}(\tau, x, u)\right|,$$

such that $\varphi(t, x, u) \in \mathcal{Q}$ for all $t \geq 0$.

Example 2.1 *The inverted pendulum*

Theorem 2.17 can be applied to a controlled linearized inverted pendulum. The pendulum is represented by

$$\ddot{y}(t) + 2b\dot{y}(t) - (1 + u(t))y(t) = 0, \quad u \in \underline{\mathcal{U}}$$

with $b > 0$ and control range $\mathcal{U} = [-\rho, \rho]$, where $0 < \rho < b^2 + 1$. This equation describes the linearization of a controlled damped mathematical pendulum at the unstable position. The corresponding state-space representation is:

$$\dot{\boldsymbol{x}}(t) = \underbrace{\begin{pmatrix} 0 & 1 \\ 1 & -2b \end{pmatrix}}_{=:\,\boldsymbol{A}_0} \boldsymbol{x}(t) + u(t) \underbrace{\begin{pmatrix} 0 & 0 \\ 1 & 0 \end{pmatrix}}_{=:\,\boldsymbol{A}_1} \boldsymbol{x}(t), \quad u \in \underline{\mathcal{U}}.$$

The eigenvalues of the matrix $\boldsymbol{A}_0$ are given by

$$\lambda_{\pm} = -b \pm \sqrt{b^2 + 1}.$$

Since $b > 0$, λ_- is negative and λ_+ is positive. Hence, the uncontrolled system has one stable and one unstable direction. We can project the bilinear system to the unit circle $\mathcal{S}^1$. An easy computation yields the equations for the projected system:

$$\dot{\varphi} = \left(-2b\sin(\varphi)\cos(\varphi) - \sin^2(\varphi) + \cos^2(\varphi)\right) + u(t)\cos^2(\varphi), \quad u \in \underline{\mathcal{U}}.$$

From Theorem 8.1.1 in [72] it follows that the control sets on $\mathcal{S}^1$ consist of equilibria. Hence, in order to determine these sets, we have to find the zeros of the right-hand side. Therefore, we divide by $\cos^2(\varphi)$ (possible for $2\varphi \notin \{\pi, 3\pi\}$) to get

$$\tan^2(\varphi) + 2b\tan(\varphi) - (1 + u) = 0 \quad \Leftrightarrow \quad \tan(\varphi) = -b \pm \sqrt{b^2 + 1 + u}.$$

Hence, we obtain the solutions

$$\varphi_{\pm} = \arctan\left(-b \pm \sqrt{b^2 + 1 + u}\right) \in \left(-\frac{\pi}{2}, \frac{\pi}{2}\right)$$

and two other solutions in $(\pi/2, (3\pi)/2)$. The solutions $\varphi_{\pm}$ are real numbers, since

$$b^2 + 1 + u \in [b^2 + 1 - \rho, b^2 + 1 + \rho] \subset (0, 2(b^2 + 1)).$$

Hence, in $(-\pi/2, \pi/2)$ we obtain the following two intervals of equilibria (cf. Fig. 2.2), which are the closures of control sets:

$$\begin{aligned} \mathcal{Q}_- &= \left[\arctan\left(-b - \sqrt{b^2 + 1 + \rho}\right), \arctan\left(-b - \sqrt{b^2 + 1 - \rho}\right)\right], \\ \mathcal{Q}_+ &= \left[\arctan\left(-b + \sqrt{b^2 + 1 - \rho}\right), \arctan\left(-b + \sqrt{b^2 + 1 + \rho}\right)\right]. \end{aligned}$$

Using the formula of Theorem 2.17, we find

$$h_{\text{inv}}(\mathcal{K}, \mathcal{Q}_{\pm}) = \max\left\{0, \min_{\varphi \in \mathcal{Q}_{\pm}} \left(-2b - 2\tan(\varphi)\right)\right\}.$$

Hence, we obtain

$$\begin{aligned} h_{\text{inv}}(\mathcal{K}, \mathcal{Q}_-) &= \max\left\{0, \min_{u \in [-\rho, \rho]} \left(2\sqrt{b^2 + 1 - u}\right)\right\} = 2\sqrt{b^2 + 1 - \rho}, \\ h_{\text{inv}}(\mathcal{K}, \mathcal{Q}_+) &= 0. \end{aligned}$$

We can interpret this result as follows: The control set $\mathcal{D} = \operatorname{int} \mathcal{Q}_-$ contains $\varphi_0 := \arctan(-b - \sqrt{b^2 + 1})$, which is an equilibrium for the control $u = 0$, that is, the vector $(\cos(\varphi_0), \sin(\varphi_0))^{\mathrm{T}}$ is an eigenvector of the matrix $\boldsymbol{A}_0$ corresponding to the stable eigenvalue $\lambda_- = -b - \sqrt{b^2 + 1}$. On $\mathcal{D}$, the projected system is controllable. This implies that the cone $\pi^{-1}(\mathcal{D}) \subset \mathbb{R}^2$ over $\mathcal{D}$ is the maximal subset of $\mathbb{R}^2$ where it is possible to steer to the stable axis (that is, to the one-dimensional eigenspace corresponding to λ_-) with the bilinear system. Thus, we have computed the invariance entropy of the maximal subset of $\mathbb{R}^2$, where the system can be stabilized to the equilibrium $(0, 0)$. The control set $\mathcal{Q}_+$ is easily seen to be invariant and hence its invariance entropy is zero. □

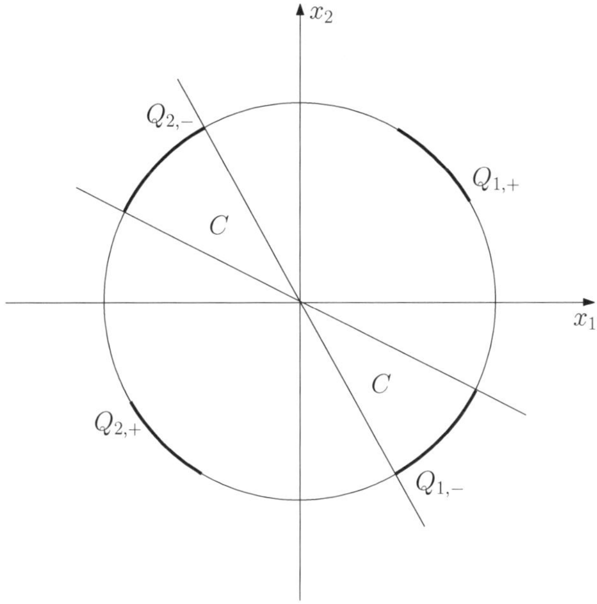

Fig. 2.2 Control sets on the unit circle

Open Problems and Extensions. The most important open problem in the theory of invariance entropy concerns the relation between the two quantities $h_{\mathrm{inv}}(\mathcal{K},\mathcal{Q})$ and $h^*_{\mathrm{inv}}(\mathcal{K},\mathcal{Q})$. We always have $h_{\mathrm{inv}}(\mathcal{K},\mathcal{Q}) \leq h^*_{\mathrm{inv}}(\mathcal{K},\mathcal{Q})$ and we know that $h_{\mathrm{inv}}(\mathcal{K},\mathcal{Q}) < \infty$, while $h^*_{\mathrm{inv}}(\mathcal{K},\mathcal{Q}) = \infty$ is possible. In the one-dimensional control-affine case, Theorem 2.16 shows that under appropriate controllability assumptions, the two quantities coincide. An analogous result holds for linear control systems in arbitrary dimensions, which is shown in [198]. In general, however, this problem remains completely open.

There are other open problems which are somewhat similar. For example, if $\mathcal{Q}$ is the closure of a control set $\mathcal{D}$, we can only give an upper bound for $h^*_{\mathrm{inv}}(\mathcal{K},\mathcal{Q})$ when $\mathcal{K} \subset \mathcal{D}$, but not for $h^*_{\mathrm{inv}}(\mathcal{Q})$. While $h^*_{\mathrm{inv}}(\mathcal{K},\mathcal{Q})$ in this case is independent of $\mathcal{K}$ as long as $\mathcal{K}$ has nonempty interior and $\mathcal{K} \subset \mathcal{D}$, it is not clear whether $h^*_{\mathrm{inv}}(\mathcal{K},\mathcal{Q}) = h^*_{\mathrm{inv}}(\mathcal{Q})$. This question is of importance, since $h^*_{\mathrm{inv}}(\mathcal{Q})$, as explained above, is a measure for the minimal data rate for rendering $\mathcal{Q}$ invariant, while we cannot show something similar for $h^*_{\mathrm{inv}}(\mathcal{K},\mathcal{Q})$ with $\mathcal{K} \neq \mathcal{Q}$.

The upper estimate of Theorem 2.16 can be improved by imposing stronger regularity assumptions on the given control system. Indeed, assuming real analyticity and strong accessibility plus some weak hyperbolicity condition, one can show that the assumptions of regularity and periodicity of the controlled solution can be omitted. This result can be found in the forthcoming monograph [202].

The results for invariance entropy presented above are restricted to compact controlled invariant sets $\mathcal{Q}$ for finite-dimensional systems. They can be extended in several directions. In [68, 69] the set $\mathcal{Q}$ is replaced by a controlled invariant subspace and techniques from geometric control are used to derive estimates for a corresponding notion of invariance entropy. In [183], invariance entropy for certain linear infinite-dimensional systems (including parabolic partial differential equations and delay equations) is characterized.

2.3.3 Entropy for Exponential Stabilization

Section Overview. Using an approach which is similar to invariance entropy, we consider the time-dependent controls which have to be generated by a controller in order to achieve exponential stability, cf. [67]. Again, we consider continuous-time systems.

There are two ways to measure the information needed for exponential stabilization for a given set $\mathcal{Q}$ of initial states: Either one can look at sets of control functions on $[0, \infty)$ which admit exponential stabilization and at the required bits to discern them on any finite interval $[0, \tau]$ and then analyze the bit rate for $\tau \to \infty$; finally, one can minimize the bit rate over such sets of controls. Alternatively, one may look at sets of control functions defined on a bounded interval $[0, \tau]$ admitting exponential decay on $[0, \tau]$ and then look at the minimal number of bits for such a set of control functions. Then, again, one can analyze what happens for $\tau \to \infty$. In the first case we speak about minimal bit rates and in the second case, we speak about entropy, since this approach is close to entropy in the theory of dynamic systems.

A discouraging example (Example 2.2) shows that finitely many controls are not sufficient for an exponential estimate on a finite interval. Hence, we impose somewhat weaker conditions. Then the stabilization entropy can be shown to be finite. In the linear case, a formula in terms of eigenvalues holds. It is worth to emphasize that this formula pertains to a fixed exponential decay rate (Theorem 2.19). Thus, it also applies to cases, where the control goal is to increase the exponential decay rate for a stable system, a situation where finite communication channels might more easily be tolerated. On the other hand, the stabilization entropy provides an upper bound for the minimal bit rate. In particular, the minimal bit rate is finite (Theorem 2.18), and in the linear case, an explicit estimate is available.

Stabilization with Limited Bit Rates. Consider a nonlinear control system of the form

$$\dot{\boldsymbol{x}}(t) = \boldsymbol{f}(\boldsymbol{x}(t), \boldsymbol{u}(t)), \quad \boldsymbol{u} \in \underline{\mathcal{U}}, \tag{2.44}$$

where $\boldsymbol{f} : \mathbb{R}^d \times \mathbb{R}^m \to \mathbb{R}^d$ is continuous and Lipschitz continuous with respect to the first argument. The admissible controls are given by

$$\underline{\mathcal{U}} = \{\boldsymbol{u} : [0, \infty) \to \mathbb{R}^m \ : \ \boldsymbol{u} \text{ measurable with } \boldsymbol{u}(t) \in \mathcal{U} \text{ a.e.}\},$$

where we assume that the controls $\boldsymbol{u}$ are integrable on every bounded interval and the control range $\mathcal{U}$ is a subset of $\mathbb{R}^m$. We assume that (i) unique global solutions $\varphi(t, \boldsymbol{x}_0, \boldsymbol{u})$, $(t \geq 0)$ of the differential equation with initial condition $\boldsymbol{x}(0) = \boldsymbol{x}_0 \in \mathbb{R}^d$ and control $\boldsymbol{u} \in \underline{\mathcal{U}}$ exist and (ii) on compact intervals, the solutions depend continuously on the initial value.

We consider the bit rates for the problem to stabilize an equilibrium $\boldsymbol{x}^* \in \mathbb{R}^d$ corresponding to an admissible control value $\boldsymbol{u}^* \in \mathcal{U}$, that is, $\boldsymbol{0} = \boldsymbol{f}(\boldsymbol{x}^*, \boldsymbol{u}^*)$. Without loss of generality, the equilibrium is given by $\boldsymbol{x}^* = \boldsymbol{0}$ corresponding to the control value $\boldsymbol{u}^* = \boldsymbol{0} \in \mathcal{U}$. Suppose that the system is exponentially controllable to the equilibrium for all $\boldsymbol{x}_0$ in a neighborhood of $\boldsymbol{x}^* = \boldsymbol{0}$, that is, there are constants $M, \alpha > 0$ such that for all $\boldsymbol{x}_0 \neq \boldsymbol{0}$ there is $\boldsymbol{u} \in \underline{\mathcal{U}}$ with

$$\|\varphi(t, \boldsymbol{x}_0, \boldsymbol{u})\| < M\mathrm{e}^{-\alpha t} \|\boldsymbol{x}_0\| \text{ for all } t \geq 0. \tag{2.45}$$

This holds, for example, if the linearized system is stabilizable by a linear feedback; (see, for instance, [346] (Section 5.8) for asymptotic stability and [151] (Satz 9.8) for exponential stability). For linear control systems, an estimate of the form (2.45) holds iff the unstable part is controllable.

Since we are interested in problems where the controller does not have continual access to the present state $\boldsymbol{x}(t)$ (or to observed values $\boldsymbol{y}(t) = \boldsymbol{g}(\boldsymbol{x}(t))$), we discuss the minimal bit rate needed to discern the required time-dependent control functions $\boldsymbol{u}$ on any time interval $[0, \tau]$, $\tau > 0$.

The following example (the simplest controllable system) shows that a direct approach is not possible. Here, on a finite time interval $[0, \tau]$ one cannot find finitely many controls such that for every point in a neighborhood of the origin the exponential estimate (2.45) is satisfied.

Example 2.2 *A discouraging example*

Consider the following scalar system:

$$\dot{x}(t) = u(t), \quad u(t) \in \mathcal{U} = \mathbb{R}. \tag{2.46}$$

Let $\alpha > 0$, $M > 1$ and fix $\tau > 0$. We claim that for τ large enough there is no finite set $\mathcal{S}$ of control functions on $[0, \tau]$, such that for every $0 \neq x_0 \in \mathcal{K} := [-1, 1]$ there is $u \in \mathcal{S}$ such that the exponential estimate

$$|\varphi(t, x_0, u)| = \left| x_0 + \int_0^t u(s)\mathrm{d}s \right| < M\mathrm{e}^{-\alpha t} |x_0| \text{ for all } t \in [0, \tau] \tag{2.47}$$

holds. We proceed by contradiction: Suppose that a finite set $\mathcal{S} = \{u_1, \ldots, u_n\}$ with the stated properties exists and define

$$\mathcal{K}_j := \left\{ x_0 \in \mathcal{K} \;:\; |\varphi(t, x_0, u_j)| < M\mathrm{e}^{-\alpha t} |x_0| \text{ for all } t \in [0, \tau] \right\}.$$

Observe that $0 \notin \mathcal{K}_j$. For the control $u_0(t) \equiv 0$, every point x_0 is an equilibrium. Thus, for $x_0 \neq 0$ the control u_0 does not satisfy (2.47) if τ is large enough such that $Me^{-\alpha\tau} < 1$ and we may assume that $u_0 \notin \mathcal{S}$. Hence, for every j one finds $t_j \in [0, \tau]$ with

$$c_j := \max_{t\in[0,\tau]} \left| \int_0^t u_j(s)\mathrm{d}s \right| = \left| \int_0^{t_j} u_j(s)\mathrm{d}s \right| > 0.$$

Consider an initial point $x_0 \in \mathcal{K}$ with $|x_0| < \min_j(c_j/(2M))$. Observe that then $|x_0| < M|x_0| < (1/2)\min_j c_j$, since $M > 1$. We claim that there is no control $u_j \in \mathcal{S}$ satisfying (2.47). In fact, one computes

$$\begin{aligned} \left| x_0 + \int_0^{t_j} u_j(s)\mathrm{d}s \right| &\geq \left| \int_0^{t_j} u_j(s)\mathrm{d}s \right| - |x_0| \geq c_j - \frac{c_j}{2} \\ &= \frac{c_j}{2} > \mathrm{e}^{-\alpha t_j}\frac{c_j}{2} > M\mathrm{e}^{-\alpha t_j}|x_0|, \end{aligned}$$

which proves the claim. □

In contrast to the linear example (2.46), the scalar bilinear system

$$\dot{x}(t) = (1 + u(t))x(t), \quad u(t) \in \mathcal{U} = \mathbb{R}$$

can be stabilized by the single constant control $u(t) \equiv -2$. Thus, no transfer of bits is necessary. Reference [83] describes other situations where finitely many bits are sufficient. While it might be worthwhile to study bilinear control systems in this context, we follow a different path and relax the exponential stability property (2.45) by introducing a small additive term. The following simple, but basic lemma shows that then only finitely many bits are required on a finite interval. It follows by using continuous dependence on initial values and compactness.

Lemma 2.3. *Consider control system* (2.44) *and let $\mathcal{K}$ be a compact subset of $\mathbb{R}^n$. Assume that there are constants $M > 1$ and $\alpha > 0$ such that for all $0 \neq \boldsymbol{x} \in \mathcal{K}$ there is a $\boldsymbol{u} \in \underline{\mathcal{U}}$ with*

$$\|\varphi(t, \boldsymbol{x}, \boldsymbol{u})\| < M\mathrm{e}^{-\alpha t}\|\boldsymbol{x}\| \text{ for all } t \geq 0. \tag{2.48}$$

Let $\varepsilon > 0$. Then for every $\tau > 0$ there is a finite set $\mathcal{S} = \{\boldsymbol{u}_1, \ldots, \boldsymbol{u}_n\} \subset \underline{\mathcal{U}}$ such that for every $\boldsymbol{x} \in \mathcal{K}$ there is a $\boldsymbol{u}_j \in \mathcal{S}$ with

$$\|\varphi(t, \boldsymbol{x}, \boldsymbol{u}_j)\| < \mathrm{e}^{-\alpha t}(\varepsilon + M\|\boldsymbol{x}\|) \text{ for all } t \in [0, \tau].$$

Stabilization Entropy. In view of the preceding lemma, we will consider weakened versions of the estimate (2.48). There are the two ways to measure the information needed for stabilization as indicated above and we begin with an entropy-like notion. Consider a compact set $\mathcal{K} \subset \mathbb{R}^n$ of initial states, and let $\alpha > 0$, $M > 1$ and $\varepsilon > 0$. For a time $\tau > 0$ we call a subset $\mathcal{S} \subset \underline{\mathcal{U}}$ a (τ, ε)-spanning set of controls if for every $\boldsymbol{x} \in \mathcal{K}$ there is $\boldsymbol{u} \in \mathcal{S}$ with

$$\|\varphi(t, \boldsymbol{x}, \boldsymbol{u})\| < \mathrm{e}^{-\alpha t}(\varepsilon + M\|\boldsymbol{x}\|) \text{ for all } t \in [0, \tau]. \tag{2.49}$$

The minimal cardinality of all these sets is

$$s_{\text{stab}}(\tau, \varepsilon) := \min\{\#\mathcal{S} \ : \ \mathcal{S} \text{ is } (\tau, \varepsilon)\text{-spanning}\}.$$

If there is no finite set $\mathcal{S}$ of controls with this property (in particular, if there is no such set at all), we let $s_{\text{stab}}(\tau, \varepsilon) := \infty$. Lemma 2.3 shows that the exponential controllability condition (2.48) implies the existence of finite (τ, ε)-spanning sets.

We want to determine which information has to be transmitted through a digital communication channel in order to identify a control function in such a finite set $\mathcal{S}$. The elements can be encoded by symbols given by finite sequences of 0's and 1's in the set

$$\Sigma_k := \{(s_0 s_1 s_2 \dots s_{k-1}) \ : \ s_i \in \{0, 1\} \text{ for } i = 0, 1, \dots, k-1\},$$

where $k \in \mathbb{N}$ is the least integer greater than or equal to $\log_2 \#\mathcal{S}$. Thus, $\#\mathcal{S}$ is bounded above by 2^k. Equivalently, the number of bits determining an element of $\mathcal{S}$ is $\log_2(2^k) = k$. It will be convenient to use here the natural logarithm instead of the logarithm with base 2. Now we consider what happens for time tending to infinity and then ε tending to 0 to obtain the following notion describing an entropy property for exponential stabilization.

Definition 2.5. *Let $\mathcal{K}$ be a compact set in $\mathbb{R}^n$ and $\alpha > 0$, $M > 1$. Then the (α, M)-stabilization entropy $h_{\text{stab}}(\alpha, M, \mathcal{K})$ is defined by*

$$h_{\text{stab}}(\alpha, M, \mathcal{K}) := \lim_{\varepsilon \searrow 0} \limsup_{\tau \to \infty} \frac{1}{\tau} \ln s_{\text{stab}}(\tau, \varepsilon).$$

In the following, we drop the argument $\mathcal{K}$ in this and in similar notions if the choice of $\mathcal{K}$ is clear or if its specification is irrelevant in the corresponding context. The existence of the limit for $\varepsilon \searrow 0$ is obvious, since it equals the supremum over all $\varepsilon > 0$. (The value $+\infty$ is allowed.) Furthermore, the inequality $h_{\text{stab}}(\alpha', M') \leq h_{\text{stab}}(\alpha, M)$ holds for $\alpha \geq \alpha' > 0$ and $M' \geq M > 1$.

Remark 2.3. If one would consider $\alpha = 0$, condition (2.49) implies that every trajectory starting in $\mathcal{K}$ remains in the ball around the origin with radius $\varepsilon + M \max_{x \in \mathcal{K}} \|\boldsymbol{x}\|$.

Minimal Bit Rate. A second way of counting bits is the following. Consider a set of control functions defined on $[0, \infty)$ which allow us to steer the system asymptotically to the equilibrium $\boldsymbol{x}^* = \boldsymbol{0}$ satisfying the following conditions. Let $M > 1$, $\alpha > 0$, $\varepsilon > 0$ and let γ be a decreasing function on $[0, \infty)$ with $\gamma(0) = \varepsilon$ and $\lim_{t \to \infty} \gamma(t) = 0$. For brevity, we call γ an $\mathcal{L}_\varepsilon$-function (note that continuity of γ is not required.)

Let $\mathcal{R}(\gamma, \varepsilon) \subset \underline{\mathcal{U}}$ be a set of control functions such that for every $\boldsymbol{x} \in \mathcal{K}$ there is $\boldsymbol{u} \in \mathcal{R}(\gamma, \varepsilon)$ with

$$\|\varphi(t, \boldsymbol{x}, \boldsymbol{u})\| < \gamma(t) + M \mathrm{e}^{-\alpha t} \|\boldsymbol{x}\| \ \text{ for all } t \geq 0. \tag{2.50}$$

Then $\mathcal{R}(\gamma,\varepsilon)$ is called (γ,ε)-stabilizing for $\mathcal{K}$. Thus, in the ε-neighborhood of the equilibrium, the decay given by the exponential rate α may slow down, but still convergence holds for $t\to\infty$. Let

$$\mathcal{R}_\tau(\gamma,\varepsilon) := \left\{\boldsymbol{u}_{|[0,\tau]} \ : \ \boldsymbol{u}\in\mathcal{R}(\gamma,\varepsilon)\right\}$$

be the corresponding restrictions of the controls in $\mathcal{R}(\gamma,\varepsilon)$. Suppose that the cardinality $\#\mathcal{R}_\tau(\gamma,\varepsilon)$ is finite. The bit rate on the time interval $[0,\tau]$ is defined as $(1/\tau)\log_2\#\mathcal{R}_\tau(\gamma,\varepsilon)$. If there is no set $\mathcal{R}(\gamma,\varepsilon)$ with (2.50) or if $\mathcal{R}_\tau(\gamma,\varepsilon)$ contains infinitely many elements, we set $\#\mathcal{R}_\tau(\gamma,\varepsilon):=\infty$. The required bit rate for stabilization using controls in $\mathcal{R}(\gamma,\varepsilon)$ is

$$b(\mathcal{R}(\gamma,\varepsilon)) := \liminf_{\tau\to\infty}\frac{1}{\tau}\log_2\#\mathcal{R}_\tau(\gamma,\varepsilon).$$

Definition 2.6. *With the notions introduced above, the minimal bit rate for* (α,M)*-stabilization at* $\boldsymbol{x}^*=\boldsymbol{0}$ *of a compact set* $\mathcal{K}\subset\mathbb{R}^n$ *is*

$$b_{\mathrm{stab}}(\alpha,\mathcal{K},M) := \lim_{\varepsilon\searrow 0}\inf_{\gamma\in\mathcal{L}_\varepsilon}\inf_{\mathcal{R}(\gamma,\varepsilon)} b(\mathcal{R}(\gamma,\varepsilon)),$$

where the inner infimum is taken over all (γ,ε)*-stabilizing sets* $\mathcal{R}(\gamma,\varepsilon)\subset\underline{\mathcal{U}}$ *of controls and the outer infimum is taken over all* $\mathcal{L}_\varepsilon$*-functions* γ*.*

The limit for $\varepsilon\searrow 0$ is the supremum for $\varepsilon>0$. Note also the inequality $b_{\mathrm{stab}}(\alpha',M')\le b_{\mathrm{stab}}(\alpha,M)$ for $\alpha\ge\alpha'>0$ and $M'\ge M>1$.

An example for an allowed $\mathcal{L}_\varepsilon$-function is $\gamma(t)=\varepsilon e^{-\alpha t}$, $t\ge 0$. However, for this function, we cannot prove our main result for the stabilization bit rate (Theorem 2.20) showing that the stabilization entropy provides an upper bound. Also, this theorem will only give a result for the limit inferior for $\tau\to\infty$ (not for the limit superior). Consideration of the limit inferior may be justified by the fact that we are interested in the minimal bit rate for $\tau\to\infty$, hence the times may be chosen appropriately.

Comparison of Stabilization Entropy and Minimal Bit Rate. The stabilization entropy indicates how much the number of required control functions increases, when time increases. Here minimization is performed on each interval $[0,\tau]$ separately. If one wants to enlarge the time interval where the exponential decay holds, one may have to consider controls which, when restricted to the smaller interval, are different from the earlier ones. This is in contrast to minimal bit rates, where restrictions to $[0,\tau]$ are considered for control functions defined on $[0,\infty)$. Thus, while stabilization entropy certainly merits its own interest, the minimal bit rate might appear more appealing from this point of view.

The difference between these two concepts can also be seen by looking at them from a quantization point of view. Let $\mathcal{R}(\gamma,\varepsilon)$ be a (γ,ε)-stabilizing set such that for every $\tau>0$ the set $\mathcal{R}_\tau(\gamma,\varepsilon)$ of restrictions to $[0,\tau]$ is finite. Then define for every $\boldsymbol{u}\in\mathcal{R}_T(\gamma,\varepsilon)$

$$\mathcal{K}(\boldsymbol{u},\tau) := \left\{\boldsymbol{x} \in \mathcal{K} \;:\; \|\varphi(t,\boldsymbol{x},\boldsymbol{u})\| < \gamma(t) + M\mathrm{e}^{-\alpha t}\,\|\boldsymbol{x}\| \text{ for all } t \in [0,\tau]\right\}.$$

The sets $\mathcal{K}(\boldsymbol{u},\tau)$ form an open cover of $\mathcal{K}$ which may be viewed as a finite quantization. For $\tau' > \tau$, the same construction for $\mathcal{R}_{\tau'}(\gamma,\varepsilon)$ again yields a finite quantization of $\mathcal{K}$ which is obtained by refining the quantization at time τ, since both are obtained by restrictions of controls in $\mathcal{R}(\gamma,\varepsilon)$. In contrast, the quantization for $\tau' > \tau$ obtained by a (τ',ε)-spanning set of controls used for defining the entropy is not related to the quantization associated with a (τ,ε)-spanning set.

Results on the Stabilization Entropy. The next theorem gives upper and lower bounds for the stabilization entropy, which is easier to analyze than the minimal bit rate.

Theorem 2.18. *Consider control system* (2.44) *and let* $\boldsymbol{x}^* = \boldsymbol{0}$ *be an equilibrium with* $\boldsymbol{0} = \boldsymbol{f}(\boldsymbol{0},\boldsymbol{0})$, $\boldsymbol{0} \in \mathcal{U}$. *Suppose that the derivative* $\boldsymbol{f}_x(\boldsymbol{x},\boldsymbol{u})$ *of* $\boldsymbol{f}$ *with respect to* $\boldsymbol{x}$ *exists and is continuous in* $(\boldsymbol{x},\boldsymbol{u})$. *Let* $\mathcal{K} \subset \mathrm{I\!R}^n$ *be a compact neighborhood of the origin and denote* $\kappa := \max_{\boldsymbol{x}\in\mathcal{K}} \|\boldsymbol{x}\|$.

1. *Assume further that* $\mathrm{div}_x \boldsymbol{f}(\boldsymbol{x},\boldsymbol{u})$ *is bounded below for* $\|\boldsymbol{x}\| \leq \kappa+1$ *and* $\boldsymbol{u} \in \mathcal{U}$. *Then for* $\alpha > 0$, $M > 1$ *the stabilization entropy of* $\mathcal{K}$ *satisfies the estimate*

$$h_{\mathrm{stab}}(\alpha, M) \geq \alpha d + \min_{u\in\mathcal{U}} \mathrm{div}_x \boldsymbol{f}(\boldsymbol{0},\boldsymbol{u}).$$

2. *Assume that for every* $\boldsymbol{0} \neq \boldsymbol{x} \in \mathcal{K}$ *there is a control* $\boldsymbol{u} \in \underline{\mathcal{U}}$ *with*

$$\|\varphi(t,\boldsymbol{x},\boldsymbol{u})\| < M\mathrm{e}^{-\alpha t}\,\|\boldsymbol{x}\| \text{ for all } t \geq 0$$

and that there is a Lipschitz constant $L > 0$ *such that for all* $\boldsymbol{x}_1, \boldsymbol{x}_2$ *in an open set containing* $\mathcal{K}$

$$\|\boldsymbol{f}(\boldsymbol{x}_1,\boldsymbol{u}) - \boldsymbol{f}(\boldsymbol{x}_2,\boldsymbol{u})\| \leq L\,\|\boldsymbol{x}_1 - \boldsymbol{x}_2\| \text{ for all } \boldsymbol{u} \in \mathcal{U}.$$

Then the stabilization entropy of $\mathcal{K}$ *satisfies*

$$h_{\mathrm{stab}}(\alpha, M) \leq (L+\alpha)d.$$

Next we consider the stabilization entropy for linear control systems in $\mathrm{I\!R}^n$ of the form

$$\dot{\boldsymbol{x}}(t) = \boldsymbol{A}\boldsymbol{x}(t) + \boldsymbol{B}\boldsymbol{u}(t), \quad \boldsymbol{u} \in \underline{\mathcal{U}}, \tag{2.51}$$

with matrices $\boldsymbol{A} \in \mathrm{I\!R}^{n\times n}$ and $\boldsymbol{B} \in \mathrm{I\!R}^{n\times m}$ and control range $\mathcal{U} \subset \mathrm{I\!R}^m$ containing the origin.

Theorem 2.19. *Consider a linear control system of the form* (2.51) *with* $\boldsymbol{0} \in \mathcal{U}$. *Assume that there are* $\alpha > 0$, $M > 1$ *such that for every initial value* $\boldsymbol{0} \neq \boldsymbol{x} \in \mathrm{I\!R}^n$ *there is a control* $\boldsymbol{u} \in \underline{\mathcal{U}}$ *with*

$$\|\varphi(t, \boldsymbol{x}, \boldsymbol{u})\| < M\mathrm{e}^{-\alpha t}\|\boldsymbol{x}\| \text{ for all } t \geq 0.$$

For every compact neighborhood $\mathcal{K}$ of the origin in $\mathbb{R}^n$ the (α, M)-stabilization entropy of system (2.51) satisfies

$$h_{\text{stab}}(\alpha, M, \mathcal{K}) = \sum_{\text{Real}(\lambda) > -\alpha} (\alpha + \text{Real}(\lambda)); \tag{2.52}$$

here summation is over all eigenvalues λ of $\boldsymbol{A}$, counted according to their algebraic multiplicities, with $\text{Real}(\lambda_i) > -\alpha$. *In particular, $h_{\text{stab}}(\alpha, M, \mathcal{K})$ is independent of $\mathcal{K}$.*

The next theorem establishes the announced relation between the stabilization bit rate and the stabilization entropy.

Theorem 2.20. *Consider a control system of the form* (2.44) *and suppose that $\mathcal{K} \subset \mathbb{R}^d$ is a compact neighborhood of the equilibrium $\boldsymbol{x}^* = \boldsymbol{0}$. Assume that there are constants $M > 1$ and $\alpha^* > 0$ such that for all $\boldsymbol{0} \neq \boldsymbol{x} \in \mathcal{K}$ there is $\boldsymbol{u} \in \underline{\mathcal{U}}$ with*

$$\|\varphi(t, \boldsymbol{x}, \boldsymbol{u})\| < M\mathrm{e}^{-\alpha^* t}\|\boldsymbol{x}\| \text{ for all } t \geq 0. \tag{2.53}$$

Then for $\alpha \in (0, \alpha^*)$ the stabilization bit rate and the stabilization entropy satisfy

$$b_{\text{stab}}(\alpha, M) \leq \log_2 \mathrm{e} \cdot h_{\text{stab}}(\alpha^*, M) < \infty.$$

For linear control systems, one gets an easy corollary.

Corollary 2.5. *Suppose that the linear control system* (2.51) *satisfies the assumptions of Theorem 2.20. Then for all $\alpha \in (0, \alpha^*)$ the stabilization bit rate satisfies*

$$\begin{aligned}(\log_2 \mathrm{e})^{-1} \cdot b_{\text{stab}}(\alpha, M) &\leq h_{\text{stab}}(\alpha, M) \\ &= \sum_{\text{Real}(\lambda) > -\alpha} (\alpha + \text{Real}(\lambda))\end{aligned}$$

and

$$\begin{aligned}(\log_2 \mathrm{e})^{-1} \cdot \inf_{\alpha > 0} b_{\text{stab}}(\alpha, M) &= \inf_{\alpha > 0} h_{\text{stab}}(\alpha, M) \\ &= \sum_{\text{Real}(\lambda) > 0} \text{Real}(\lambda).\end{aligned}$$

For smooth nonlinear control systems and arbitrarily small control range, we conjecture that a formula analogous to (2.52) holds, now for the Jacobian at the equilibrium (cf. [270], Theorem 3, for such a claim in the context of local uniform asymptotic stabilization).

2.4 Dynamic Quantization for Feedback Stabilization with Delayed Data-Rate Limited Communication

2.4.1 Problem Statement

In this section we consider networked control systems over a digital channel with limited data rate. An inherent property of such channels is that data needs to be quantized. Here we employ dynamic quantization for stabilization. The term dynamic quantization refers to an approach where the size of the quantization region as well as the center of the region change in time according to the input received from system measurements. This information from the system allows to estimate the region in which quantization is required. With this approach limited bandwidth can be employed to communicate increasingly accurate information, which will eventually be sufficient to achieve stabilization.

The class of systems under consideration is

$$\dot{\boldsymbol{x}}(t) = \boldsymbol{f}(\boldsymbol{x}(t), \boldsymbol{k}(\boldsymbol{x}(t) + \boldsymbol{e}(t))), \tag{2.54}$$

where $\boldsymbol{x} \in \mathbb{R}^n$ and $\boldsymbol{f} : \mathbb{R}^n \times \mathbb{R}^m \to \mathbb{R}^n$. The aim is to stabilize the system at the equilibrium $(\boldsymbol{x}^*, \boldsymbol{u}^*) = (\boldsymbol{0}, \boldsymbol{0})$. We assume that a controller $\boldsymbol{k}$ has been designed which renders system (2.54) input-to-state stable (ISS) with respect to the quantization error $\boldsymbol{e}$. For a discussion of the ISS property, see Section 5.5. We assume that $\boldsymbol{f}$ is sufficiently smooth to guarantee existence and uniqueness of solutions. In particular $\boldsymbol{f}$ is Lipschitz in the first component with Lipschitz constant L.

We will assume that the state of the system can be measured and that this measured information needs to be transmitted to the controller via a digital channel. We will present a encoder/decoder scheme that will result in a stable system, given the information on $\boldsymbol{f}$ and $\boldsymbol{k}$ in (2.54). A particular problem in networked control systems are delays and packet loss. Here we describe how dynamic quantization can be implemented to cope with these phenomena.

2.4.2 Dynamic Quantization

In a first pass we ignore delay and packet loss and explain the idea behind dynamic quantization with the help of Fig. 2.3. In our setup communication attempts occur at the time instances $t_0 < t_1 < t_2 < \ldots$. At every communication instant the quantization region is a hypercube that is determined by its center $\boldsymbol{C}$ and the edge length ℓ. This hypercube is partitioned into N^n smaller hypercubes of equal size; recall that n is the dimension of the state space. The smaller hypercubes are called subregions. At the communication event t_k the encoder sends the number of the subregion in which the state $\boldsymbol{x}(t_k)$ of the system lies. At the time of transmission the estimate of the state $\hat{\boldsymbol{x}}$ is the center of the subregion. In this case $\boldsymbol{e}$ represents the error between the actual value of the system and the midpoint of the subregion, which is used as an approximation thereof.

The problem that has to be solved by the scheme lies in the fact that encoder and decoder have to agree on center and length of the subregion that is determined by the transmitted code word, so that the transmitted information can be used to compute the control value $\boldsymbol{k}(\boldsymbol{x}+\boldsymbol{e})$.

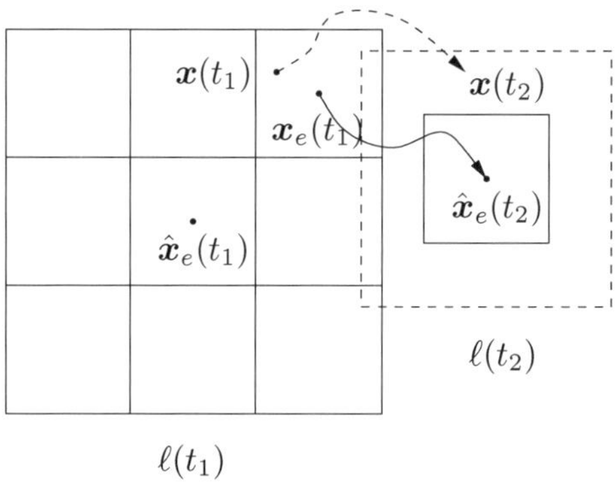

Fig. 2.3 Dynamic quantization

At a time t_k a certain quantization region, i.e. its center and edge length, is determined by the previous dynamics and it has to be ensured that the state $\boldsymbol{x}(t_k)$ is within the quantization region at time t_k. Given that encoder and decoder agree on the value $\hat{\boldsymbol{x}}(t_k)$, they perform identical numerical estimates of the evolution of the system

$$\dot{\boldsymbol{x}}(t) = \boldsymbol{f}(\boldsymbol{x}(t), \boldsymbol{k}(\boldsymbol{x}(t))), \quad \boldsymbol{x}(t_k) = \hat{\boldsymbol{x}}(t_k),$$

and the solution $\hat{\boldsymbol{x}}(t_{k+1}) := \boldsymbol{x}(t_{k+1}; t_k, \boldsymbol{x}(t_k))$ of this initial value problem at time t_{k+1} is the center of the quantization region at time t_{k+1}.

In other words, the sensor determines the actual subregion in which the state lies and calculates the center $\boldsymbol{x}_e(t_k)$. We will refer to this process as encoding. The decoder on the other end of the channel has a copy of these values, i.e. the decoder knows the center and length of the region and the values n and N. Thus if we transmit the number of the subregion in which the state lies, the decoder is able to reconstruct the value $\boldsymbol{x}_e(t_k)$.

If both encoder and decoder let the center $\boldsymbol{x}_e$ of the subregion follow the closed-loop dynamics until time t_2, the error between the estimate $\boldsymbol{x}_e$ and the state $\boldsymbol{x}$ can grow by a certain factor. If we let the subregion grow by the same amount (the augmented region is the dashed box in Fig. 2.3), we are sure that the state at time t_2 is still within the subregion. This subregion becomes the new quantization region with the new center $\hat{\boldsymbol{x}}_e(t_2) = \boldsymbol{x}_e(t_2^-)$. Now we are in the same situation as previously, namely, we know a hypercube containing the state and we can now repeat the same steps.

If the quotient between the growth of the quantization region and the reduction of the error due to N is smaller than 1, the quantization error converges to 0 as the number of communication events increases.

During continuous flow of the system, the growth of the uncertainty of the state (i.e., the quantization error) may be bounded with the help of the Lipschitz constant L of the system (depicted by the dotted box in Fig. 2.3). This bound ensures that the state is at any time in some region with known size and center.

The growth of the error is countered by a reduction of the size of the quantization region by a defined number N (in Fig. 2.3: $N = 3$), which is the number of smaller regions (per dimension) within the quantization range (depicted in Fig. 2.3 as the jump from the center of the large region to the center of the smaller subregion in which the actual state lies).

Delays and Packet Loss. As sketched above it is important that encoder and decoder agree on certain values. This is easily achieved, if there is no delay in the channel, because the decoder can copy the behavior of the encoder exactly.

If delays and packet loss are present then the information available at the decoder and encoder side may be different. To account for this fact, we will in the sequel use variables available at the encoder by $\boldsymbol{x}_e, \hat{\boldsymbol{x}}_e$, etc., and those available at the decoder by $\boldsymbol{x}_d, \hat{\boldsymbol{x}}_d$, etc. The scheme described in the following relies crucially on time stamps and on acknowledgments (acks). The time stamps provide the information at which time certain information was available at the other side and acks provide the encoder with the information that data has been received by the decoder. Note that we assume acks to be time-stamped as well.

For the communication channel we assume the following properties:

Assumption 2.1

1. *All packets are time stamped with the current time they are sent.*
2. *Only packets sent from encoder to decoder are lost.*
3. *There exists a minimal delay from encoder to decoder, given by $\tau_{\min}$, i.e., $t_k^* - t_k \geq \tau_{\min}$ and $t_k - t_{k-1}^* \geq \tau_{\min}$.*
4. *The channel is able to transmit packets containing a value from a set of N^n (N odd) discrete values within $\tau_{\min}$ units of time.*
5. *If $\tau_{\max}$ time elapses without receiving an ack, the packet sent last time is considered lost and a new packet will be sent.*

These assumptions have the following consequences:

- According to Part 1 we have to send the actual time together with the encoded state information. It is not reasonable to be able to transmit the state information quantized and the time information not. For the sake of simplicity we omit details on time quantization but refer to [326] for a discussion.

- Part 2 is a major restriction on the channel used. As the acks are much smaller than the state information the decoder could send many acks to ensure that at least one arrives at the encoder. Without this assumption we could not guarantee that the encoder and the decoder agree on their states.

- Part 3 is in general not a restrictive one. In every real communication channel such a minimal delay exists.

- Part 4 states that the bandwidth of the channel B must be large enough to transmit the state information within $\tau_{\min}$ units of time. For instance, if binary encoding is used we require

$$B \geq \frac{n \log_2 N}{\tau_{\min}}. \tag{2.55}$$

  If this condition is not met, the decoder could introduce an artificial delay by waiting to ensure that $\tau_{\min}$ is large enough to fulfill the bandwidth constraint.
  The choice for N to be odd guarantees that the center of the quantization region lies in the interior of one of the subregions. Although this is not needed in general, it eases the presentation.

- By Part 5, the values $\tau_{\min}$ and $\tau_{\max}$ may be regarded as design parameters. While choosing $\tau_{\min}$ too small can violate the bandwidth constraint, a larger value can degrade the performance of the overall system. Similar statements hold for $\tau_{\max}$. If $\tau_{\max}$ is chosen too small, no ack will arrive at the encoder before a new packet will be sent and again choosing $\tau_{\max}$ too large may have a negative effect on the performance.

Further assumptions that will be of importance are the following:

> **Assumption 2.2.** *The clocks of encoder and decoder are synchronized and the time $t_1 = 0$ when the encoder sends the first packet is known by the encoder and the decoder.*

> **Assumption 2.3.** *Both the encoder and the decoder know the same bound of the initial state of the system (i.e., encoder and decoder agree on a constant $R > 0$ such that $|x(0)| \leq R$).*

We consider TCP-like packet-based transmissions over a noiseless, error-free channel with delay and packet loss.

The encoder encodes the state and sends a symbol from a finite alphabet to the decoder together with the time when the state was encoded (time stamping). As soon as a packet arrives, the decoder reconstructs the encoded state and sends an acknowledgment back to the encoder. If this ack arrives at the encoder or a predefined time elapses without receiving one, it repeats the encoding. Denote by t_k the k-th time instance the encoder received an ack.

The time when the k-th information sent by the encoder is received by the decoder is denoted by t_k^*. Note that we assume that there is no time delay between the arrival of an information and the sending of the next packet, i.e. t_k and t_k^* are also the time instances when the encoder sends information and the decoder sends an ack, respectively.

We now detail the complete description of encoder and decoder. A fundamental element of the encoder is the map φ, which yields the subregion in which the state $\boldsymbol{x}$ lies as a function of the center of the quantization region $\boldsymbol{x}_e$ and the edge length ℓ. It is given by

$$\varphi(\boldsymbol{x}_e, \boldsymbol{x}, \ell) = \left\lfloor \frac{N}{\ell}(\boldsymbol{x} - \boldsymbol{x}_e) + \frac{1}{2}\left[1 \dots 1\right]^\top \right\rfloor , \tag{2.56}$$

where the floor function $\lfloor \cdot \rfloor$ is applied in each component.

Encoder and the decoder are initialized trivially by

$$\begin{aligned} &k = 1,\ t_0 = 0,\ t_0^* = 0,\ t_1 = 0,\ t_1^* = 0 \text{ and } \hat{\boldsymbol{x}}_d(0) = \hat{\boldsymbol{x}}_e(0) = 0 \\ &\boldsymbol{x}_e(0^-) = \boldsymbol{x}_d(0^-) = 0 \text{ and } \ell_e(0^-) = \ell_d(0^-) = 2R. \end{aligned}$$

Encoder Dynamics. For the encoder three cases have to be treated: (i) times at which an acknowledgment of a previous packet is received by the encoder, (ii) times at which the maximal admissible time span $\tau_{\max}$ passes without the receipt of an acknowledgment, at which time the previous data will be considered lost, and (iii) intermediate times. The required actions for each of these cases are given as follows. In both cases ((i) and (ii)) t_s and $s(t_s)$ are auxiliary variables, describing the data payload of the packets sent from encoder to decoder.

(i) Every time an ack arrives at the encoder ($t = t_k$) the following operations are invoked:

$$t_s := t \tag{2.57}$$

$$\ell_e(t_k) := \ell_e(t_{k-1}) e^{L(t_k - t_{k-1})}/N \tag{2.58}$$

$$\begin{aligned} \boldsymbol{x}_e(t_k^-) := \boldsymbol{x}_e(t_{k-1}) &+ \int_{t_{k-1}}^{t_{k-1}^*} \boldsymbol{f}(\boldsymbol{x}_e(\tau), \boldsymbol{k}(\hat{\boldsymbol{x}}_e(\tau)))\, \mathrm{d}\tau \\ &+ \int_{t_{k-1}^*}^{t_k} f(\boldsymbol{x}_e(\tau), \boldsymbol{k}(\boldsymbol{x}_e(\tau)))\, \mathrm{d}\tau \end{aligned} \tag{2.59}$$

$$\boldsymbol{s}(t_s) := \varphi(\boldsymbol{x}_e(t_k^-), \boldsymbol{x}(t_k), \ell_e(t_k)) \tag{2.60}$$

$$\boldsymbol{x}_e(t_k) := \boldsymbol{x}_e(t_k^-) + \boldsymbol{s}(t_s)\frac{\ell_e(t_k)}{N} \tag{2.61}$$

$$\hat{\boldsymbol{x}}_e(t_k) = \boldsymbol{x}_e(t_k^-). \tag{2.62}$$

Accordingly, the encoder updates the length of the quantization region according to the growth of the error on the last interval (2.58). The center of the quantization region is updated via (2.59). Both integrals are needed to account for the change in the control action on the decoder side at time t^*_{k-1} which is known to the encoder by the time stamp of the ack. The subregion in which the state lies is determined by (2.60). This information will be sent to the decoder together with the actual time (2.57). The jump from the center to the subregion is done by Eq. (2.61). The value of the old quantization region is copied by (2.62) in order to have access to this information in case that the current communication event fails.

(ii) If $\tau_{\max}$ time instants elapse without receipt of an ack the following simplified actions need to be taken:

$$t_s := t \tag{2.63}$$

$$\ell_e(t) := \ell_e(t_k)e^{L(t-t_k)} \tag{2.64}$$

$$\boldsymbol{x}_e(t^-) := \hat{\boldsymbol{x}}_e(t) \tag{2.65}$$

$$\boldsymbol{s}(t_s) := \varphi(\boldsymbol{x}_e(t^-), \boldsymbol{x}(t), \ell_e(t)) \tag{2.66}$$

$$\boldsymbol{x}_e(t) := \boldsymbol{x}_e(t^-) + \boldsymbol{s}(t_s)\frac{\ell_e(t)}{N}. \tag{2.67}$$

Accordingly, the packet sent last time is considered lost and a new one is sent. Similarly to the case of no loss, the encoder updates the length of the quantization region (2.64). Note that there is no division by N, as the reduction effect provided by the transmitted information is not present. Equation (2.65) cancels the jump from the center to the subregion performed in the last encoding step. Equations (2.66) and (2.67) follow the same reasoning as in the case of no loss.

(iii) In between transmission events only the center of the quantization region has to be updated by

$$\dot{\hat{\boldsymbol{x}}}_e(t) = \boldsymbol{f}(\hat{\boldsymbol{x}}_e(t), k(\hat{\boldsymbol{x}}_e(t))). \tag{2.68}$$

We need (2.68) to know the trajectory which will be used to close the loop on the decoder side. It is also needed to treat the case of packet loss in (2.65).

Decoder Dynamics. For the decoder only two cases have to be distinguished: (i) the arrival of information at times t^*_k and (ii) the times in between arrivals.

(i) Every time a packet arrives at the decoder ($t = t^*_k$) the following operations are carried out:

$$\ell_d(t_s) := \ell_d(t_{k-1})e^{L(t_s-t_{k-1})}/N \tag{2.69}$$

$$\boldsymbol{x}_d(t_s^-) := \hat{\boldsymbol{x}}_d(t^*_{k-1}) + \int_{t^*_{k-1}}^{t_s} \boldsymbol{f}(\boldsymbol{x}_d(\tau), \boldsymbol{k}(\boldsymbol{x}_d(\tau)))\,\mathrm{d}\tau \tag{2.70}$$

$$\boldsymbol{x}_d(t_s) := \boldsymbol{x}_d(t_s^-) + s(t_s)\frac{\ell_d(t_s)}{N} \tag{2.71}$$

$$\hat{\boldsymbol{x}}_d(t_k^*) := \boldsymbol{x}_d(t_s) + \int_{t_s}^{t_k^*} \boldsymbol{f}(\boldsymbol{x}_d(\tau), \boldsymbol{k}(\hat{\boldsymbol{x}}_d(\tau)))\,\mathrm{d}\tau. \tag{2.72}$$

(ii) Otherwise

$$\dot{\hat{\boldsymbol{x}}}_d(t) = \boldsymbol{f}(\hat{\boldsymbol{x}}_d(t), \boldsymbol{k}(\hat{\boldsymbol{x}}_d(t))). \tag{2.73}$$

The decoder copies the behavior of the encoder with the help of (2.69)-(2.71). Equation (2.72) compensates for the delay between encoder and decoder.

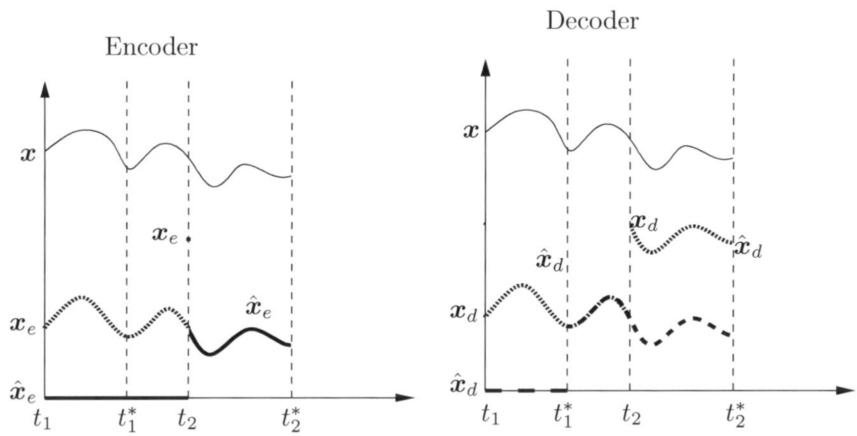

Fig. 2.4 Snapshot of the different trajectories at time t_2^*

A sketch of the evolution of the different trajectories is depicted in Fig. 2.4. The overall effect of this encoder/decoder scheme is that if at time t the state lies within the quantization region, the error between the state and the estimate shrinks by N because of the jump from the center of the region to the center of a subregion. Hence we guarantee that

$$|\boldsymbol{x}(t) - \boldsymbol{x}_e(t^-)| \leq \frac{\ell}{2} \Rightarrow |\boldsymbol{x}(t) - \boldsymbol{x}_e(t)| \leq \frac{\ell}{2N} \tag{2.74}$$

holds, which can be seen from (2.56) and (2.61). Overall, the evolution of the closed-loop system is given by

$$\dot{\boldsymbol{x}}(t) = \boldsymbol{f}(\boldsymbol{x}(t), \boldsymbol{k}(\hat{\boldsymbol{x}}_d(t)))\,. \tag{2.75}$$

For this system it is possible to show the following stabilization result.

Theorem 2.21. *Consider a control system with a time-stamped dynamic quantization encoder/decoder scheme. If*

$$N > e^{L\tau^*},$$

where τ^* is the long time average between successful transmissions, then the closed-loop system is asymptotically stable.

It is worth mentioning that the quantity N is directly related to the bandwidth of the communication channel.

2.4.3 Markovian Communication Models

One shortcoming of the results presented so far lies in the deterministic nature of the model of the communication channel. For many practical examples the appropriate models for the dynamic behavior are of stochastic nature. This can be seen for example for TCP and for the wireless case, where packet loss frequently is due to the influence of an environment which can be suitably modeled as random.

In this section we derive conditions for the stabilization of system (2.54) under the condition that the communication channel can be described by a (time-homogeneous) Markov process.

In the following we assume as given a communication channel in which external perturbations such as average load of other users is stationary. We consider a Markov chain $\{X(k)\}_{k\in\mathbb{N}}$, where the state X lies in a state space of the chain S. This state space has to be specified with the concrete situation in mind. As we only send information at discrete time instances, a Markov process which is discrete in time is justified.

We assume as given two continuous maps describing the communication, namely

$$T : S \to [T_{\min}, \infty), \quad g : S \to \mathbb{N}, \tag{2.76}$$

where $T(X)$ denotes the length of the next communication interval depending on the state X of the channel and $g(X)$ denotes the number of bits that can be sent in that interval. Thus if communication starts at a time $t_1 \in \mathbb{R}$ and $l := n \log_2 N + b$ bits have to be sent, where b is the acknowledgment size in bits, we define a stopping time of the Markov chain by

$$k_1^* = \min\left\{ k \;\middle|\; \sum_{j=0}^{k} g(X(j)) \geq l \text{ or } \sum_{j=0}^{k} T(X(j)) \geq \tau_{\max} \right\} \tag{2.77}$$

and until k_1^*

$$\sum_{j=0}^{k_1^*} T(X(j)) =: \tau(1)$$

units of time elapse. If $\tau(1) \geq \tau_{\max}$ (with $\tau_{\max} > 0$ as a design parameter), we consider the information to be lost. To ensure the Markovian property of our description, we assume that $X(k)$ has the strong Markov property, i.e. the evolution of the process only depends on the state of the chain at the stopping time k^*. The next time we want to send information, i.e. at time $t_1 + \tau(1)$ we define the next stopping time k_2^* by

$$k_2^* = \min \left\{ k \;\middle|\; \sum_{j=k_1^*}^{k} g(X(j)) \geq l \text{ or } \sum_{j=k_1^*}^{k} T(X(j)) \geq \tau_{\max} \right\}$$

and the duration from k_1^* until k_2^* by $\sum_{j=k_1^*}^{k_2^*} T(X(j)) =: \tau(2)$. If we repeat this procedure a sequence of time instances is given by $\{\tau(j)\}_{j \in \mathbb{N}}$.

We now assume ergodicity of the Markov chain, which ensures that almost surely

$$\tau_M^* := \lim_{k \to \infty} \frac{1}{k} \sum_{j=0}^{k} \tau(j) \mathbb{1}\{\tau(j) < \tau_{\max}\} \tag{2.78}$$

exists, where $\mathbb{1}$ denotes the indicator function.

With the above considerations we are able to state a stochastic version of Theorem 2.21, where the long time average is given by τ_M^* instead of τ^*.

Theorem 2.22. *Consider a control system with a time-stamped dynamic quantization encoder/decoder scheme. Assume that the communication channel is given by a strongly ergodic Markov process X, and maps T and g given in (2.76) and with stopping times as in (2.77). If*

$$N > e^{L\tau_M^*}, \tag{2.79}$$

where τ_M^* is given by (2.78), then the closed loop system is asymptotically stable with probability 1.

Again this result relates the error growth due to the systems dynamics given by the Lipschitz constant L with the information that can be sent on average using the communication channel and the refinement of the quantization.

As this point we see the relation to the results in Section 2.3: In general, with the information provided by entropy considerations finer bounds on the necessary information can be obtained. However, information on the entropy of the process is frequently hard to obtain, so that in the results presented in this section we have settled for a coarse estimate as given by the Lipschitz constant.

Example 2.3 *Dynamic quantization for the inverted pendulum*

We discuss the results of the previous section for the example of a pendulum on a cart that is stabilized in the upright position. The dynamic equations are given by

$$\frac{\mathrm{d}}{\mathrm{dt}}\begin{bmatrix}x_1\\x_2\\x_3\\x_4\end{bmatrix}=\begin{bmatrix}0&0&1&0\\0&0&0&1\\0&\frac{mg}{M}&0&0\\0&-\frac{g(m+M)}{lM}&0&0\end{bmatrix}\begin{bmatrix}x_1\\x_2\\x_3\\x_4\end{bmatrix}+\begin{bmatrix}0\\0\\\frac{1}{M}\\-\frac{1}{lM}\end{bmatrix}u \tag{2.80}$$

where x_1, x_3 are angle and angular velocity of the pendulum and x_2 resp. x_4 denote the position and velocity of the cart.

In the simulations the values $m = 0.329, M = 3.2, l = 0.44, g = 9.81$ have been used. The shaded grey area in Figs. 2.5 and 2.6 represents the quantization region as a function of time. Between communication events this region grows according to the Lipschitz constant of the system. It shrinks if new information is received.

In the simulation a stochastic model has been used for the generation of delays, but so that the conditions of Theorem 2.22 are satisfied. It can be seen that the quantization region quickly approximates the trajectory, so that the information at the controller becomes increasingly accurate. The discontinuous lines in both figures represent the value that is used for controlling the system. Because of the delay in the system the discontinuities of these control values do not occur at the communication times; the latter are indicated by dots on the horizontal axis.

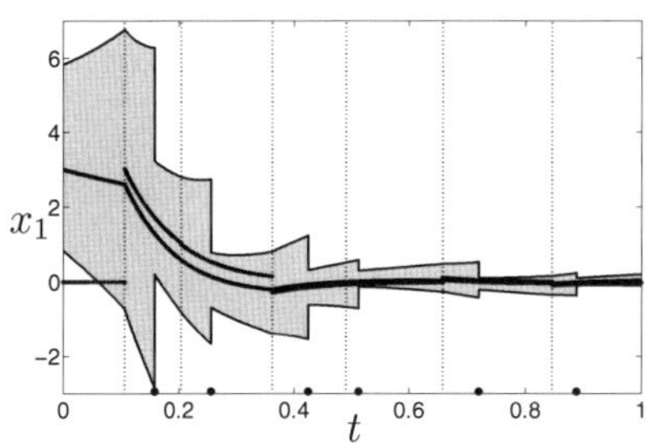

Fig. 2.5 Trajectory of angle (x_1)

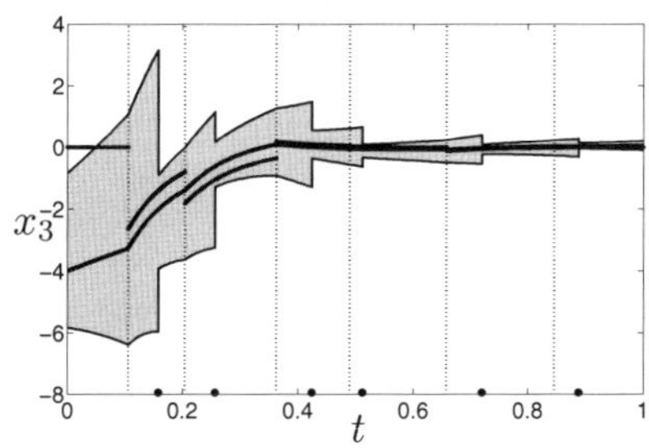

Fig. 2.6 Trajectory of angular velocity (x_3)

The error between system state and estimate used by the controller is shown in Figs. 2.7 and 2.8. □

Bibliographical Notes

The approach of dynamic quantization we are using here, was introduced by [41] and has been extended to the nonlinear case in [231]. However, the notion of non-static quantization is not new. It was first mentioned within the control community in [359] or even earlier within the communication community (c.f. [137]).

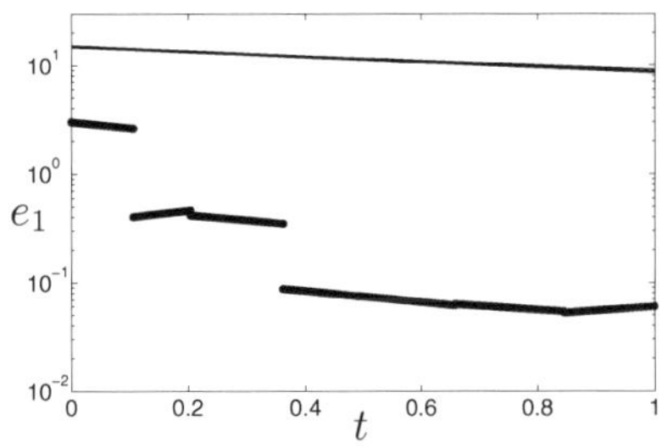

Fig. 2.7 Error e_1

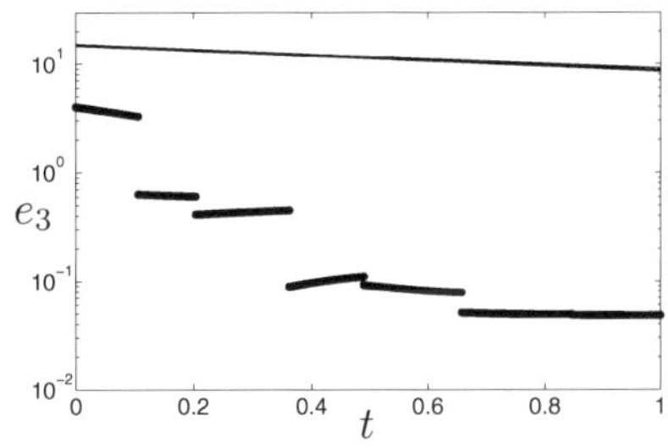

Fig. 2.8 Error e_3

One of the main problems in this area is that the amount of information that can be sent from the sensing device to the controller is limited and can be corrupted in several ways. This type of problems has received considerable attention, see [84–86, 163, 178, 276, 326] as a starting point. First steps in this area considered communication constraints such as limited bandwidth or data rate, but issues as delays and packet loss were not treated. Also the communication channel was treated as static [86, 178, 276], whereas many realizations of communication channels use protocols which define internal dynamics of the channel. Examples of this are given by TCP and certain wireless protocols.

The results presented here are amenable to more dynamic communication channels. They were first published in [325, 326].

3

Distributed Estimation and Control

B. Lohmann, F. Deroo, S. Hirche, J. Lunze, M. Meinel,
P. Philipp, and M. Ulbrich

For interconnected systems, the state estimation and control algorithms associated with the subsystems can communicate over the network to improve their performance. This chapter describes methods for structuring the overall control system including the communication network and shows how to use the communicated information for estimation and control if the network induces transmission delays and packet loss.

Chapter Contents

3.1 Estimation and Control Architectures

3.1.1 Distribution of Control Tasks

The availability of powerful communication networks leads to new architectures for the control of interconnected systems. The current hierarchical structure shown in the left part of Fig. 1.14 on p. 23 is replaced by distributed structures depicted in the right part of the figure. The grey boxes represent physical systems and the white boxes estimation and control algorithms. This idea is explained in more detail in this chapter.

The architecture of a control system describes which components of the plant, the state estimator and the controller work together and which information is exchanged among these components. Which architecture should be used depends upon different practical aspects like the availability of computer resources, the properties of the plant and the software used for the implementation of the estimation and control algorithms.

Classical control theory has concentrated on an *embedded systems approach* where all components of the controller are placed on a single computer board and connected to the overall plant. Hence, all information delivered by sensors of the plant are communicated to a single entity that is responsible for the estimation of the overall state and the generation of the overall system input. In control theory, this structure is called a centralized control loop and the controller a *centralized controller*.

There are several reasons why centralized control should be extended to more involved architectures:

- The plant may be large and geographically distributed. Then the centralized control architecture is no longer applicable because it is impossible to send all sensor information to a common controller, to provide the necessary computing power in one centralized entity, or to set up the overall plant model. Then it is necessary to distribute the estimation and control tasks over the subsystems and, possibly, a coordinator.

- The plant consists of independent subsystems that have their own control equipment and have to fulfill a common control goal. Then the natural way is to associate the control algorithms with the local computing resources.

- If the structure of the plant changes, for example due to the appearance and the disappearance of subsystems during operation, it is reasonable to simultaneously change the structure of the controller. Architectures have to be used that provide the corresponding flexibility.

Figure 3.1 illustrates several options of how to extend the centralized control architecture. The following explanations refer to the control problem but similar investigations can be made for state estimation and other control

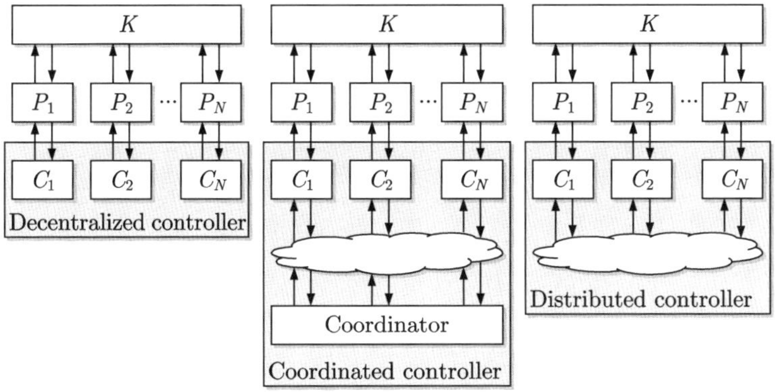

Fig. 3.1 Three basic control architectures

tasks. In all cases the plant is considered as an interconnection of several subsystems P_i, $(i = 1, 2, ..., N)$, which have local control stations C_i, $(i = 1, 2, ..., N)$ that have access only to the local sensor information $\boldsymbol{y}_i(t)$ and the local actuator inputs $\boldsymbol{u}_i(t)$. The architectures differ with respect to the information exchange among the local controllers:

- **Decentralized control:** The subsystems are controlled by independent control stations, which do not have any information exchange and, thus, have to determine the current control input $\boldsymbol{u}_i(t)$ of the i-th subsystem in dependence upon the output $\boldsymbol{y}_i(t)$ of the same subsystem only:

$$\boldsymbol{u}_i(t) = \boldsymbol{k}_i(\boldsymbol{y}_i(t)).$$

  The local controllers C_i are represented by the functions $\boldsymbol{k}_i$, $(i = 1, 2, ..., N)$. This structure has been studied in the classical literature on large-scale systems more than two decades ago, when the communication networks did not allow extensive on-line information exchange [245, 339].

- **Coordinated control:** If the couplings K among the plant subsystems are strong, decentralized control may not produce sufficient control performance of the overall system, because the interactions cannot be compensated by corresponding interactions among the control stations. Then a coordinator can be used to provide the local controllers C_i with the necessary information about the performance of the neighboring subsystems, for example, with some information $\boldsymbol{s}_i(t)$ that describes the influence of the neighboring subsystem on the subsystem P_i. The control law is represented by some function $\boldsymbol{k}_i$:

$$\boldsymbol{u}_i(t) = \boldsymbol{k}_i(\boldsymbol{y}_i(t), \boldsymbol{s}_i(t)).$$

- **Distributed Control:** In distributed control the local controllers decide themselves which information should be sent to which other control station and no separate centralized authority like a coordinator is available for this task. The control input $\boldsymbol{u}_i(t)$ of the i-th subsystem depends upon some but not all outputs $\boldsymbol{y}_j(t)$ of other subsystems:

$$\boldsymbol{u}_i(t) = \boldsymbol{k}_i(\boldsymbol{y}_i(t), \underbrace{\boldsymbol{y}_1(t), ..., \boldsymbol{y}_N(t)}_{\text{communicated outputs}}).$$

If subsystems disappear or new subsystems are introduced into the plant, the distributed control architecture is flexible enough to change the controller structure accordingly.

Clearly, the best possible performance of the overall system can be obtained by coordinated or distributed control, because in these structures the local controllers of the subsystem can exchange, directly or indirectly, information that is necessary to cope with the influence of the physical interconnections on the subsystem behavior. The decentralized structure, however, has the advantage that it does not necessitate any communication infrastructure and that structural changes of the plant are directly associated with the corresponding changes of the overall controller.

3.1.2 Distributed Control Design

What has been said so far for on-line control can be applied analogously for the design process of the controller. Each design problem is described by a plant model M and a set S of specifications to be satisfied. A control law $\boldsymbol{k}(\boldsymbol{y})$ is to be found such that the closed-loop systems satisfies the specifications.

Classical control theory has mainly dealt with *centralized design* where the model M represents the overall plant and a centralized controller has to be found such that the overall closed-loop system satisfies the specifications. The entity D that solves the design problem is also called a decision-making unit (or decision maker) or a control agent.

In all distributed design schemes, several decision makers D_i, $(i = 1, ..., N)$ are involved each of which does not have a model M of the complete plant, but dedicated models M_i are used together with dedicated specification sets S_i to find the local controllers C_i. The model M_i may refer to the subsystem P_i or to the subsystem together with some partial knowledge about the neighboring subsystems. Accordingly, the performance specifications S_i refer to the i-th subsystem only or to the subsystem and some directly interconnected parts of the overall plant.

Hence, the distribution of the design concerns the information structure with respect to the models and specifications used in the design tasks. In the following explanations it is assumed that the control law C_i of the i-th subsystem is designed by a decision maker D_i:

- **Decentralized design** means that the entities D_i, $(i = 1, 2, ..., N)$ involved in the design of the local controllers C_i have different plant models and different performance criteria each of which reflect a part of the overall plant or a part of the overall performance specifications. No interaction among the activities of the decision makers D_i occur.

- **Coordinated design** includes two layers. In the lower layer separate design problems are solved with respect to the subsystems P_i. The model M_i and the specification set S_i used for the design of the control station C_i are modified by a coordinator such that the set of controllers C_i, $(i = 1, 2, ..., N)$ obtained by all decision makers together represent a controller of the overall plant that takes into account the subsystem interactions.

- **Distributed design** represents a design scheme in which the local entities D_i solve separate design problems, which refer in part to the models M_j and design specifications S_j of the neighboring subsystems. Communication among the entities D_i, $(i = 1, 2, ..., N)$ ensure that the separate design problems include requirements on the overall system behavior so that the control stations C_i, $(i = 1, 2, ...N)$ designed independently represent an overall controller that satisfies the overall system requirements.

The information structures of the design process mentioned above can be combined with all the controllers introduced in Section 3.1.1 as will be discussed in more detail in Section 3.3 (see also [245]).

3.1.3 Contributions of this Book to Distributed Control

Although the communication network appears in both the coordinated control and the distributed control structures in Fig. 3.1, the main aim of the theory of networked control systems is to create new methods for distributed control, where the intelligence about state estimation and control is put on the subsystem level and, hence, may be implemented in smart sensors or smart actuators. The main questions to be solved concern systematic methods for deciding which information should be sent to which other subsystem and methods for making the overall system tolerant to the imperfections of the communication medium.

The aim of distributing the controller among the subsystems is tackled in the remainder of this chapter and in several further sections of this book:

- **State estimation in networked control systems:** An architecture for distributed state estimation is described in Section 3.2 with the aim to make the estimation algorithm tolerant to imperfections of the communication network.

- **Two-layer optimal control of distributed systems:** Section 3.3 describes in more detail the distribution of the on-line control algorithms and the off-line control design tasks among the subsystems to find optimal or suboptimal solutions without having the overall plant model M or the overall measurement information available in a single component on the subsystem layer.

- **Distributed event-based control:** Sections 5.3, 5.5 and 5.6 extend the idea of event-based control to interconnected systems where the local controllers are implemented independently of each other as event-triggered feedback.
- **Distributed model predictive control:** Chapter 4 describes several algorithms for model predictive control, where the optimization problem is decomposed and solved in a distributed way by the subsystems with partial information about the overall system state.
- **Distributed control with situation-dependent communication:** Section 6.5 describes how the local control stations can decide under what conditions they should exchange information about the current state of their associated subsystem.

3.2 Centralized and Decentralized Moving-Horizon Estimation for Networked Control Systems

3.2.1 Centralized and Decentralized Networked Control Architecture

Networked control systems (NCS) are characterized by the fact that sensor and/or control data are packet-based transmitted over an imperfect communication network. This situation induces several new aspects, such as variable sampling/transmission intervals, random packet delays, packet reorderings, packet drops and unsynchronized clocks which aggravate the control task.

In this section, we consider the centralized and decentralized NCS architectures depicted in Fig. 3.2 (a) and Fig. 3.2 (b), respectively. Both architectures offer the advantage of applying the unaltered control inputs instantaneously to the plant by avoiding its transmission over an imperfect communication network. Information about the state of the system can be gathered by stationary as well as mobile sensors which transmit their measurements wired or wireless to the estimator site over a network. The objective of the estimators is on the one hand to reconstruct the full state vector of the system and on the other hand to compensate the above mentioned network induced imperfections. Although, at a first glance, this objective may differ for the two architectures, both problems can be tackled within a common framework, where the estimation problem is formulated as a suitable optimization

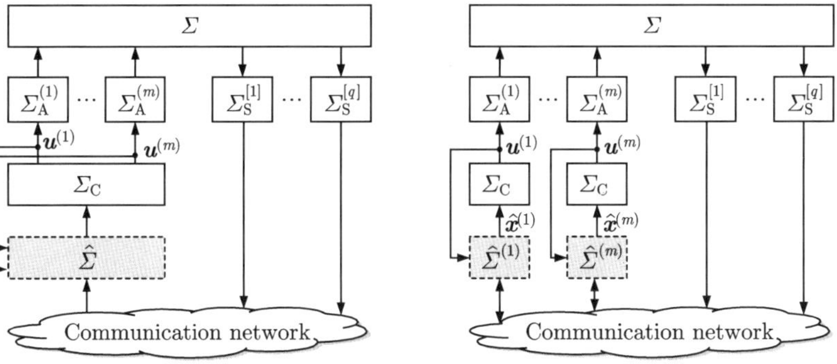

(a) Centralized NCS architecture (b) Decentralized NCS architecture

Fig. 3.2 Networked control system architectures with system Σ, sensors $\Sigma_{\mathrm{S}}^{[j]}$, communication network Σ_{N}, estimators $\hat{\Sigma}^{(i)}$, controller Σ_{C} and actuators $\Sigma_{\mathrm{A}}^{(i)}$

problem. This common formulation results for the two architectures in the centralized and the decentralized moving-horizon estimator concept.

In the first part of this section, we present the centralized moving-horizon estimator for the NCS architecture depicted in Fig. 3.2 (a). Each sensor is equipped with an unsynchronized clock and transmits its time-stamped and non-uniformly sampled measurements via packets to the estimator site over a digital communication network. These packets are subject to random delay or might even be completely lost. The estimation problem is formulated as a nonlinear program which can be efficiently implemented in real time. The key features of the proposed method include the independence of the underlying communication protocols as well as of clock synchronization. Furthermore, the estimator concept allows the controller to be designed for the nominal case, i.e. without considering the network.

The centralized moving-horizon estimator is validated on a networked pendulum test-rig. The swing-up and the stabilization of the pendulum represent a challenging benchmark for the estimator due to the unstable and non-minimum phase system dynamics along with the network-induced non-negligible time delays and packet drops. A two-degree-of-freedom control scheme is used in the experiments. Due to the good performance of the proposed estimator, a simple linear LQR controller is sufficient for the feedback part.

In the second part, we present the decentralized moving-horizon estimator for the NCS architecture depicted in Fig. 3.2 (b). The starting point is the previous formulation of the estimation problem as a centralized optimization problem. The cost function is defined on a sliding window involving a finite number of past measurements. In order to derive a decentralized procedure, the measurements are distributed among the estimators. Thereby, local observability of the full state vector is not required for the estimators.

The communication topology of the estimators is modeled as a directed graph and is reflected in the optimization problem as additional consensus constraints. Applying a dual decomposition technique reveals a separable dual problem which is solved by a subgradient method to achieve a decentralized algorithm. The resulting decentralized moving-horizon estimator shares the same benefits as the centralized counterpart including e.g. the possibility of designing a centralized controller for the nominal case which, however, is implemented fully decentralized. Simulation results of a closed-loop weir system illustrate the main features of the proposed method.

Notation. For a vector $\boldsymbol{x} \in \mathbb{R}^n$, $\hat{\boldsymbol{x}}$ denotes its estimated value and $\|\boldsymbol{x}\|_{\boldsymbol{S}} = \sqrt{\boldsymbol{x}^{\mathrm{T}}\boldsymbol{S}\boldsymbol{x}}$ is the weighted Euclidean norm where $\boldsymbol{S} \in \mathbb{R}^{n\times n}$ is a positive definite matrix. For the vectors $\boldsymbol{x}^{(i)} \in \mathbb{R}^n$ with $i \in \mathcal{I} = \{1, 2, \ldots, I\}$, the column operator is defined as $\mathrm{col}(\boldsymbol{x}^{(i)}, i \in \mathcal{I}) = \left(\boldsymbol{x}^{(1)^{\mathrm{T}}}, \boldsymbol{x}^{(2)^{\mathrm{T}}}, \ldots, \boldsymbol{x}^{(I)^{\mathrm{T}}}\right)^{\mathrm{T}}$ and stacks all the vectors $\boldsymbol{x}^{(i)}$ whose index i belongs to the index set $\mathcal{I}$ in an ordered sequence in a single column vector. The matrix $\mathrm{diag}(\boldsymbol{A}_1, \ldots, \boldsymbol{A}_n)$ is block-diagonal with blocks $\boldsymbol{A}_i$.

3.2.2 Problem Formulation

Consider the centralized and decentralized architecture of the networked control system depicted in Fig. 3.2. The plant Σ is a MIMO continuous-time nonlinear time-invariant system

$$\dot{\boldsymbol{x}}(t) = \boldsymbol{f}(\boldsymbol{x}(t), \boldsymbol{u}(t)) + \boldsymbol{w}(t), \tag{3.1}$$

where $\boldsymbol{x}(t) \in \mathbb{R}^n$ is the state vector with the initial value $\boldsymbol{x}(0) \in \mathbb{R}^n$, $\boldsymbol{u}(t) \in \mathbb{R}^m$ is the control input and $\boldsymbol{w}(t) \in \mathbb{R}^n$ is an additive disturbance affecting the system dynamics. We assume that the right-hand side of (3.1) is sufficiently smooth to guarantee existence and uniqueness of solutions.

Each of the q sensors $\Sigma_{\mathrm{S}}^{[j]}$ in the set $\mathcal{S} = \{1, 2, \ldots q\}$ is equipped with an individual clock which possesses the local time $\bar{t}^{[j]}$. Moreover, each sensor generates a sequence of packets $\bar{P}^{[j]}(i) = \{\boldsymbol{y}^{[j]}(\bar{t}_i^{[j]}), \bar{t}_i^{[j]}\}$ consisting of measurements $\boldsymbol{y}^{[j]}(\bar{t}_i^{[j]}) \in \mathbb{R}^{n_{y,j}}$ and corresponding sensor time stamps $\bar{t}_i^{[j]}$. These measurements are derived by non-uniformly sampling the sensing model at the sensor times $\bar{t}_i^{[j]}$, i.e.

$$\boldsymbol{y}^{[j]}(\bar{t}_i^{[j]}) = \boldsymbol{h}^{[j]}(\boldsymbol{x}(\bar{t}_i^{[j]})) + \boldsymbol{v}^{[j]}(\bar{t}_i^{[j]}), \quad j \in \mathcal{S}, \quad i = 0, 1, 2, \ldots, \tag{3.2}$$

where $\boldsymbol{v}^{[j]}(\bar{t}_i^{[j]}) \in \mathbb{R}^{n_{y,j}}$ are the measurement disturbances. It is important to note that in the case of unsynchronized clocks, the sensor times $\bar{t}_i^{[j]}$ are unknown in the global time t.

A feature of both NCS architectures is the possibility of designing a centralized controller Σ_C for the nominal case, i.e. without considering the network. This means that for both cases, each of the m actuators in the set $\mathcal{A} = \{1, 2, \dots m\}$ is controlled by the centralized controller

$$\dot{\boldsymbol{\xi}}(t) = \boldsymbol{g}(\boldsymbol{\xi}(t), \boldsymbol{x}(t), \boldsymbol{r}(t)) \tag{3.3}$$

$$\underbrace{\begin{pmatrix} \boldsymbol{u}^{(1)}(t) \\ \vdots \\ \boldsymbol{u}^{(m)}(t) \end{pmatrix}}_{\boldsymbol{u}(t)} = \underbrace{\begin{pmatrix} \boldsymbol{c}^{(1)}(\boldsymbol{\xi}(t), \boldsymbol{x}(t), \boldsymbol{r}(t)) \\ \vdots \\ \boldsymbol{c}^{(m)}(\boldsymbol{\xi}(t), \boldsymbol{x}(t), \boldsymbol{r}(t)) \end{pmatrix}}_{\boldsymbol{c}(\boldsymbol{\xi}(t), \boldsymbol{x}(t), \boldsymbol{r}(t))}, \tag{3.4}$$

where $\boldsymbol{\xi}(t) \in \mathrm{I\!R}^{n_\xi}$ is the control state vector with the initial value $\boldsymbol{\xi}(0) \in \mathrm{I\!R}^{n_\xi}$ and $\boldsymbol{r}(t) \in \mathrm{I\!R}^{n_r}$ is the reference input.

The successful implementation of the controller Σ_C requires knowledge about the full state vector. To this end, the packets $\bar{P}^{[j]}(i)$ are transmitted over a packet-delaying and packet-dropping network Σ_N to the controller site. The objective for both architectures depicted in Fig. 3.2 is to develop an estimator which generates an estimate of the full state vector based on the information in the packets $\bar{P}^{[j]}(i)$ and thus to compensate the network induced imperfections.

3.2.3 Centralized Moving-Horizon Estimation

In this section, we propose a centralized moving-horizon estimator (CMHE) for the architecture depicted in Fig. 3.2 (a). For simplicity of presentation, we consider only one sensor and drop the superscript in (3.2) indicating the number of the sensor. Thus, the transmitted packets can be written as $\bar{P}(i) = \{\boldsymbol{y}(\bar{t}_i), \bar{t}_i\}$ with i being the packet number. Each successful received packet is extended by the arrival time stamp t_j to yield the augmented packet $P(j, i) = \{\boldsymbol{y}(\bar{t}_i), \bar{t}_i, t_j\}$. Note that the packets may arrive out-of-order. The estimator possesses a buffer (moving horizon) which stores an ordered sequence $B(k)$ consisting of $N + 1$ sorted (in ascending order by the sensor time stamps $\bar{t}_i$) augmented packets which contain the latest sensor time stamps.

Example 3.1 *Transmitted and received packet and resulting buffer sequence*

Consider the sequence of transmitted and received packets depicted in Fig. 3.3. Then the resulting buffer sequences are for $k \in \{2, 3, 4\}$ and $N + 1 = 2$ as follows: $B(2) = \{P(2, 1), P(1, 2)\}$, $B(3) = \{P(1, 2), P(3, 4)\}$ and $B(4) = \{P(3, 4), P(4, 5)\}$. □

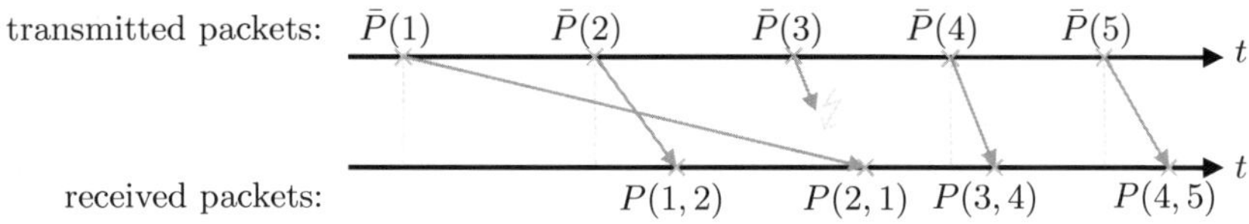

Fig. 3.3 Transmission of packets through the network

The following time relation is assumed:

Assumption 3.1. *The relation between the global time t and the sensor time $\bar{t}$ in a sufficiently small time interval is given by the affine clock model $t = s\,\bar{t} + t_o$, where s and t_o represent the unknown clock drift rate and the unknown clock offset, respectively.*

Based on this assumption, the sensing model (3.2) is recast in global time

$$\boldsymbol{y}(s\,\bar{t}_i + t_o) = \boldsymbol{h}(\boldsymbol{x}(s\,\bar{t}_i + t_o)) + \boldsymbol{v}(s\,\bar{t}_i + t_o), \qquad i = 0, 1, 2, \ldots, \tag{3.5}$$

and abbreviated by

$$\boldsymbol{y}(i) = \boldsymbol{h}(\boldsymbol{x}(i)) + \boldsymbol{v}(i), \qquad i = 0, 1, 2, \ldots. \tag{3.6}$$

The main two parts of the centralized moving-horizon estimator are the update step and the prediction step. The former is responsible for synthesizing information in the arrived packets, which is then used in the latter as an update to perform the actual state prediction.

Whenever the buffer sequence $B(k)$ changes, the update step estimates optimal clock parameters $\hat{s}$ and $\hat{t}_o$ as well as an optimal initial state $\hat{\boldsymbol{x}}(k-N)$ and optimal state disturbances $\hat{\boldsymbol{w}}(k-N), \ldots, \hat{\boldsymbol{w}}(k-1)$. More precisely, the following nonlinear program is solved

$$\min_{\substack{\hat{s}, \hat{t}_o, \hat{\boldsymbol{x}}(k-N), \\ \hat{\boldsymbol{w}}(k-N), \ldots, \hat{\boldsymbol{w}}(k-1)}} J(\hat{s}, \hat{t}_o, \hat{\boldsymbol{x}}(k-N), \hat{\boldsymbol{w}}(k-N), \ldots, \hat{\boldsymbol{w}}(k-1)) \tag{3.7a}$$

subject to

$$\hat{\boldsymbol{x}}(i+1) = \hat{\boldsymbol{x}}(i) + \int_{\hat{s}\,\bar{t}_i + \hat{t}_o}^{\hat{s}\,\bar{t}_{i+1} + \hat{t}_o} \boldsymbol{f}(\hat{\boldsymbol{x}}(t), \boldsymbol{u}(t))\,\mathrm{d}t + \hat{\boldsymbol{w}}(i), \quad i \in \mathcal{I}(k) \backslash \{k\} \tag{3.7b}$$

$$\boldsymbol{y}(i) = \boldsymbol{h}(\hat{\boldsymbol{x}}(i)) + \hat{\boldsymbol{v}}(i), \qquad i \in \mathcal{I}(k) \tag{3.7c}$$

where the set $\mathcal{I}(k) = \{k-N, k-N+1, \ldots, k\}$ denotes the set of indices corresponding to the sorted elements in the buffer sequence $B(k)$. The cost function is given by

$$\begin{aligned} J(\hat{s}, \hat{t}_o, \hat{\boldsymbol{x}}(k-N), \hat{\boldsymbol{w}}(k-N), \ldots, \hat{\boldsymbol{w}}(k-1)) &= \Theta(\hat{\boldsymbol{s}}, \hat{\boldsymbol{t}}_o, \bar{\boldsymbol{s}}, \bar{\boldsymbol{t}}_o) \\ + \Gamma(\hat{\boldsymbol{x}}(k-N), \bar{\boldsymbol{x}}(k-N)) + \frac{1}{2} \sum_{i=k-N}^{k} \|\hat{\boldsymbol{v}}(i)\|^2_{\boldsymbol{R}^{-1}} &+ \frac{1}{2} \sum_{i=k-N}^{k-1} \|\hat{\boldsymbol{w}}(i)\|^2_{\boldsymbol{Q}^{-1}} \end{aligned} \tag{3.8}$$

and consists of four terms. The first two terms are known as the arrival costs and penalize the distance from the clock parameters $\hat{s}$, $\hat{t}_o$ and the estimate $\hat{\boldsymbol{x}}(k-N)$ to some predictions $\bar{s}$, $\bar{t}_o$ and $\bar{\boldsymbol{x}}(k-N)$, respectively. These predictions are obtained by incorporating past information, which is not explicitly accounted for in the objective function. The third and fourth term are a penalizations of the measurement disturbances and the state disturbances, respectively. From the solution of this problem, the current state is propagated by a forward simulation of (3.1).

Example 3.2 *Swing-up and stabilization of a networked pendulum [289]*

The pendulum test rig used for the swing-up maneuver and the subsequent stabilization experiment is shown in Fig. 3.4 (a). The cart is driven by a toothed belt connected to a synchronous motor and the position of the cart x and the angle of the rod φ are measured by incremental angle encoders. Both measurements are combined to emulate the sensor Σ_S depicted in Fig. 3.2 (a) with one local sensor time $\bar{t}$ and the clock parameters $s = 1.2$ and $t_o = -3$. A packet $\bar{P}(i) = \{x(\bar{t}_i), \varphi(\bar{t}_i), \bar{t}_i\}$ is transmitted over the network, if either x, φ or $\bar{t}$ changes by 0.01 m, 0.0175 rad or 10 ms based on the information in the latest transmitted packet.

The model of the pendulum with the state vector $\boldsymbol{x} = (x_1, x_2, x_3, x_4)^{\mathrm{T}} = (x, \dot{x}, \varphi, \dot{\varphi})^{\mathrm{T}}$ is derived via the Lagrange method and can be written as

$$\dot{x}_1 = x_2$$

$$\dot{x}_2 = \frac{(ml\cos x_3)(mlg\sin x_3 - d_\varphi x_4) + (\Theta + ml^2)(bu - d_x x_2 - mlx_4^2\sin x_3)}{(\Theta + ml^2)(M+m) - m^2l^2\cos^2 x_3}$$

$$\dot{x}_3 = x_4$$

$$\dot{x}_4 = \frac{(ml\cos x_3)(bu - d_x x_2 - d_\varphi x_4 + mlgx_4^2\sin x_3) - (M+m)(d_\varphi x_4 + mlg\sin x_3)}{(\Theta + ml^2)(M+m) - m^2l^2\cos^2 x_3}$$

The pendulum is stabilized along the nominal trajectory by a time-varying LQR-feedback controller. The upper plots of Fig. 3.5 show the performance of the centralized moving-horizon estimator for a packet drop rate of 20% and a packet delay characterized by the histogram depicted in the lower plot of Fig. 3.5. The estimation error for the clock parameters is within 2%. □

(a) Pendulum test rig (b) Pendulum schematics

Fig. 3.4 Pendulum test rig and its schematics

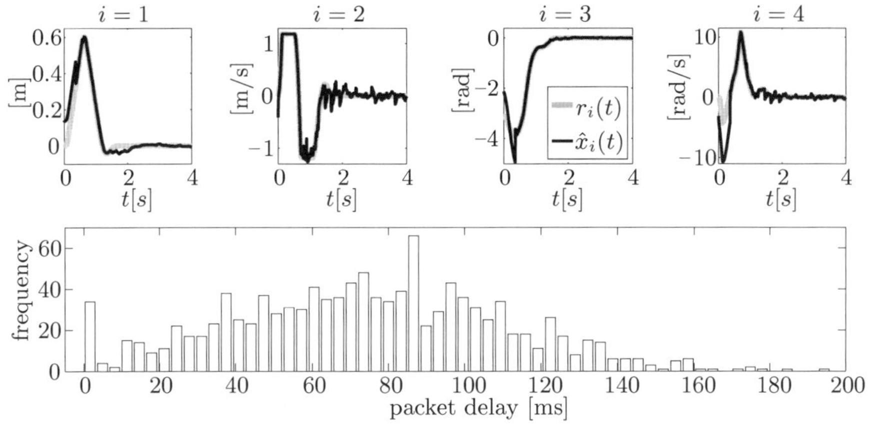

Fig. 3.5 Upper plots: Comparison between the nominal trajectories $r_i(t)$ (grey) and the estimated states of the centralized moving-horizon estimator $\hat{x}_i(t)$ (black). Lower plot: Histogram of the packet delays of the successfully transmitted packets

3.2.4 Decentralized Moving-Horizon Estimation

In this section, we address the decentralized moving-horizon estimator (DMHE) design for the architecture depicted in Fig. 3.2 (b). We consider the general case of q sensors $\Sigma_{\mathrm{S}}^{[j]}$, however, with synchronized clocks, i.e. each sensor posses the global time t. Moreover, we assume that only random packet-delays but no packet-drops occur in the network Σ_{N}. For ease of presentation, we restrict ourself to the case where each sensor samples uniformly and synchronized a continuous-time linear time-invariant system. Then the nonlinear program (3.7) which has to be solved in the update step in the centralized case can be recast as the quadratic program

$$\min_{\hat{\boldsymbol{x}}(k-N),\hat{\boldsymbol{w}}(k-N),\ldots,\hat{\boldsymbol{w}}(k-1)} J(\hat{\boldsymbol{x}}(k-N),\hat{\boldsymbol{w}}(k-N),\ldots,\hat{\boldsymbol{w}}(k-1)) \tag{3.9a}$$

subject to

$$\hat{\boldsymbol{x}}(i+1) = \boldsymbol{A}\,\hat{\boldsymbol{x}}(i) + \boldsymbol{B}\,\boldsymbol{u}(i) + \hat{\boldsymbol{w}}(i), \qquad i \in \mathcal{I}(k)\backslash\{k\} \tag{3.9b}$$

$$\boldsymbol{y}^{[j]}(i) = \boldsymbol{C}^{[j]}\,\hat{\boldsymbol{x}}(i) + \hat{\boldsymbol{v}}^{[j]}(i), \qquad i \in \mathcal{I}(k), \quad j \in \mathcal{S} \tag{3.9c}$$

where the cost function is given by

$$\begin{aligned} J(\hat{\boldsymbol{x}}(k-N),\hat{\boldsymbol{w}}(k-N),\ldots,\hat{\boldsymbol{w}}(k-1)) &= \Gamma(\hat{\boldsymbol{x}}(k-N),\bar{\boldsymbol{x}}(k-N)) \\ &+ \frac{1}{2}\sum_{i=k-N}^{k}\sum_{j=1}^{q}\|\hat{\boldsymbol{v}}^{[j]}(i)\|^2_{\boldsymbol{R}^{[j]-1}} + \frac{1}{2}\sum_{i=k-N}^{k-1}\|\hat{\boldsymbol{w}}(i)\|^2_{\boldsymbol{Q}^{-1}}. \end{aligned} \tag{3.10}$$

From the solution of this problem, the current state is propagated by

$$\begin{aligned}\hat{\boldsymbol{x}}(k-N+i) &= \boldsymbol{A}^{i}\hat{\boldsymbol{x}}(k-N) + \sum_{j=1}^{i} \boldsymbol{A}^{i-j}\boldsymbol{B}\boldsymbol{u}(k-N-1+j)\\ &+ \sum_{j=1}^{\min(i,N)} \boldsymbol{A}^{i-j}\hat{\boldsymbol{w}}(k-N-1+j).\end{aligned} \tag{3.11}$$

This expression computes a filtered state estimate if $i = N$, and a predicted state estimate if $i > N$. The choice of the prediction time i depends on the random and variable packet delays and thus may vary for each update step.

Before we tackle the problem of decentralized estimation, we state the equivalence between the centralized moving-horizon estimator and the Kalman filter. To this end, we denote

$$\begin{aligned}\boldsymbol{y}(i) &= \mathrm{col}(\boldsymbol{y}^{[j]}(i), j \in \mathcal{S})\\ \boldsymbol{v}(i) &= \mathrm{col}(\boldsymbol{v}^{[j]}(i), j \in \mathcal{S})\\ \boldsymbol{C} &= \mathrm{col}(\boldsymbol{C}^{[j]}, j \in \mathcal{S})\\ \boldsymbol{R} &= \mathrm{diag}(\boldsymbol{R}^{[1]}, \ldots, \boldsymbol{R}^{[q]}).\end{aligned}$$

Theorem 3.1. *Suppose that $\boldsymbol{w}(i)$ and $\boldsymbol{v}(i)$ are white noise with covariance equal to $\boldsymbol{Q}$ and $\boldsymbol{R}$, respectively. If the arrival cost is chosen as*

$$\Gamma(\hat{\boldsymbol{x}}(k-N), \bar{\boldsymbol{x}}(k-N)) = \frac{1}{2}\|\hat{\boldsymbol{x}}(k-N) - \bar{\boldsymbol{x}}(k-N)\|^2_{\boldsymbol{P}(k-N)^{-1}}, \tag{3.12}$$

where $\bar{\boldsymbol{x}}(k-N)$ denotes the optimal predicted estimate at time $k-N$ given all measurements up to time $k-N-1$ and where the covariance matrix $\boldsymbol{P}(k-N)$ is updated by the Riccati equation

$$\begin{aligned}\boldsymbol{P}(k+1) &= \boldsymbol{Q} + \boldsymbol{A}\,\boldsymbol{P}(k)\,\boldsymbol{A}^{\mathrm{T}}\\ &\quad -\boldsymbol{A}\,\boldsymbol{P}(k)\,\boldsymbol{C}^{\mathrm{T}}(\boldsymbol{C}\,\boldsymbol{P}(k)\,\boldsymbol{C}^{\mathrm{T}} + \boldsymbol{R})^{-1}\boldsymbol{C}\,\boldsymbol{P}(k)\,\boldsymbol{A}^{\mathrm{T}},\end{aligned} \tag{3.13}$$

then the state estimate obtained by (3.11) is equivalent to the state estimate derived by a Kalman filter.

Positive definiteness (and thus invertibility) of $\boldsymbol{P}(k)$ is guaranteed by Theorem 3.2 which can be proven by using the results of [32, 91].

Theorem 3.2. *If $(\boldsymbol{C}, \boldsymbol{A})$ is detectable and $(\boldsymbol{A}, \boldsymbol{Q}^{\frac{1}{2}})$ is controllable, then $\lim_{k\to\infty} \boldsymbol{P}(k) = \boldsymbol{P}(\infty)$ where $\boldsymbol{P}(\infty) > 0$ is the unique steady sate solution to the Riccati equation (3.13). Furthermore, if $\boldsymbol{P}(0) \geq \boldsymbol{P}(\infty)$ (meaning that $\boldsymbol{P}(0) - \boldsymbol{P}(\infty)$ is nonnegative definite), then $\boldsymbol{P}(k)$ is positive definite for all $k \geq 0$.*

The objective is now to distribute the centralized moving-horizon estimation problem (3.9) among the m estimators $\hat{\Sigma}^{(i)}$ associated to the m actuators $\hat{\Sigma}_{\mathrm{A}}^{(i)}$ depicted in Fig. 3.2 (b) such that the centralized controller Σ_{C} can be implemented in a decentralized manner. To this end, each estimator $\hat{\Sigma}^{(i)}$ has to converge to the optimal solution of (3.9) without knowing all of the most recent $N+1$ measurements, which are transmitted to different estimators to reduce the local computational load and to lower the bandwidth consumption of the network Σ_{N}.

By denoting

$$\begin{aligned}
\underline{\boldsymbol{y}}(k) &= \mathrm{col}(\boldsymbol{y}(i), i \in \mathcal{I}(k)) \\
\underline{\boldsymbol{v}}(k) &= \mathrm{col}(\boldsymbol{v}(i), i \in \mathcal{I}(k)) \\
\underline{\boldsymbol{w}}(k) &= \mathrm{col}(\boldsymbol{w}(i), i \in \mathcal{I}(k)\backslash\{k\}) \\
\underline{\boldsymbol{u}}(k) &= \mathrm{col}(\boldsymbol{u}(i), i \in \mathcal{I}(k)\backslash\{k\}) \\
\underline{\boldsymbol{R}} &= \mathrm{diag}(\boldsymbol{R}, \ldots, \boldsymbol{R}) \\
\underline{\boldsymbol{Q}} &= \mathrm{diag}(\boldsymbol{Q}, \ldots, \boldsymbol{Q})
\end{aligned}$$

$$\boldsymbol{F} = \begin{pmatrix} \boldsymbol{C} \\ \boldsymbol{CA} \\ \vdots \\ \boldsymbol{CA}^N \end{pmatrix}, \quad \boldsymbol{G} = \begin{pmatrix} \boldsymbol{0} & \boldsymbol{0} & \ldots & \boldsymbol{0} \\ \boldsymbol{CB} & \boldsymbol{0} & \ddots & \vdots \\ \boldsymbol{CAB} & \boldsymbol{CB} & \ddots & \boldsymbol{0} \\ \vdots & \vdots & \ddots & \boldsymbol{0} \\ \boldsymbol{CA}^{N-1}\boldsymbol{B} & \boldsymbol{CA}^{N-2}\boldsymbol{B} & \ldots & \boldsymbol{CB} \end{pmatrix}$$

$$\boldsymbol{H} = \begin{pmatrix} \boldsymbol{0} & \boldsymbol{0} & \ldots & \boldsymbol{0} \\ \boldsymbol{C} & \boldsymbol{0} & \ddots & \vdots \\ \boldsymbol{CA} & \boldsymbol{C} & \ddots & \boldsymbol{0} \\ \vdots & \vdots & \ddots & \boldsymbol{0} \\ \boldsymbol{CA}^{N-1} & \boldsymbol{CA}^{N-2} & \ldots & \boldsymbol{C} \end{pmatrix}$$

we can express the constrained problem (3.9) in terms of vectors and matrices as the unconstrained problem

$$p^* = \min_{\hat{\boldsymbol{x}}(k-N), \underline{\hat{\boldsymbol{w}}}(k)} J(\hat{\boldsymbol{x}}(k-N), \underline{\hat{\boldsymbol{w}}}(k)), \tag{3.14}$$

where the cost function is

$$\begin{aligned}
J(\hat{\boldsymbol{x}}(k-N), \underline{\hat{\boldsymbol{w}}}(k)) = &\frac{1}{2} \left\| \underline{\boldsymbol{y}}(k) - \boldsymbol{F}\hat{\boldsymbol{x}}(k-N) - \boldsymbol{G}\underline{\boldsymbol{u}}(k) - \boldsymbol{H}\underline{\hat{\boldsymbol{w}}}(k) \right\|^2_{\underline{\boldsymbol{R}}^{-1}} \\
&+ \frac{1}{2} \left\| \underline{\hat{\boldsymbol{w}}}(k) \right\|^2_{\underline{\boldsymbol{Q}}^{-1}} + \frac{1}{2} \left\| \hat{\boldsymbol{x}}(k-N) - \bar{\boldsymbol{x}}(k-N) \right\|^2_{\boldsymbol{P}^{-1}_{k-N}} .
\end{aligned} \tag{3.15}$$

The first step in deriving a distributed algorithm is to allocate the measurements $\underline{\boldsymbol{y}}(k)$, previously stored in the buffer $B(k)$, to the m buffers $B^{(i)}(k)$ of the m estimators $\hat{\Sigma}^{(i)}$. Consequently, the index set $\mathcal{Y}(k) = \mathcal{I}(k) \times \mathcal{S}$ associated to $B(k)$ is separated in m index sets $\mathcal{Y}^{(i)}(k)$ associated to the m buffers $B^{(i)}(k)$ such that $\bigcup_{i=1}^{m} \mathcal{Y}^{(i)}(k) = \mathcal{Y}(k)$ and $\bigcap_{i=1}^{m} \mathcal{Y}^{(i)}(k) = \{\emptyset\}$. This ensures that each measurement $\boldsymbol{y}^{[j]}(i)$ with $(i,j) \in \mathcal{Y}(k)$ is assigned precisely to one buffer. The measurements stored in the i-th buffer at the time k are denoted by $\underline{\boldsymbol{y}}^{(i)}(k) = \mathrm{col}(\boldsymbol{y}^{[j]}(i), (i,j) \in \mathcal{Y}_k^{(i)})$ and its dynamics is written as

$$\underline{\boldsymbol{y}}^{(i)}(k) = \boldsymbol{F}^{(i)}(k)\,\hat{\boldsymbol{x}}(k-N) + \boldsymbol{G}^{(i)}(k)\,\underline{\boldsymbol{u}}(k) + \boldsymbol{H}^{(i)}(k)\,\hat{\underline{\boldsymbol{w}}}(k) + \hat{\underline{\boldsymbol{v}}}^{(i)}(k). \quad (3.16)$$

Next, we introduce local solutions $\hat{\boldsymbol{x}}^{(i)}(k-N), \hat{\underline{\boldsymbol{w}}}^{(i)}(k)$ for each estimator in order to divide the cost function (3.15) into m decoupled parts

$$\begin{aligned} J(\hat{\boldsymbol{x}}^{(1)}(k-N), \hat{\underline{\boldsymbol{w}}}^{(1)}(k), \ldots, \hat{\boldsymbol{x}}^{(m)}(k-N), \hat{\underline{\boldsymbol{w}}}^{(m)}(k)) \\ = \sum_{i=1}^{m} J^{(i)}(\hat{\boldsymbol{x}}^{(i)}(k-N), \hat{\underline{\boldsymbol{w}}}^{(i)}(k)), \end{aligned} \quad (3.17a)$$

where the addend $J^{(i)}$ is

$$\begin{aligned} J^{(i)} = & \frac{1}{2}\left\| \underline{\boldsymbol{y}}^{(i)}(k) - \boldsymbol{F}^{(i)}(k)\,\hat{\boldsymbol{x}}^{(i)}(k-N) - \boldsymbol{G}^{(i)}(k)\,\underline{\boldsymbol{u}}(k) - \boldsymbol{H}^{(i)}(k)\,\hat{\underline{\boldsymbol{w}}}^{(i)}(k) \right\|^2_{\underline{\boldsymbol{R}}^{(i)\,-1}} \\ & + \frac{1}{2m}\left\| \hat{\underline{\boldsymbol{w}}}^{(i)}(k) \right\|^2_{\underline{\boldsymbol{Q}}^{-1}} + \frac{1}{2m}\left\| \hat{\boldsymbol{x}}^{(i)}(k-N) - \bar{\boldsymbol{x}}(k-N) \right\|^2_{\boldsymbol{P}(k-N)^{-1}}. \end{aligned} \quad (3.17b)$$

The reason for this choice of $J^{(i)}$ will become clear later. The solution to the problem stemming from (3.17) is identical to the solution of (3.9) if and only if each local solution $\hat{\boldsymbol{x}}^{(i)}(k-N), \hat{\underline{\boldsymbol{w}}}^{(i)}(k)$ coincides with $\hat{\boldsymbol{x}}(k-N), \hat{\underline{\boldsymbol{w}}}(k)$. In order to reach this consensus, the estimators have to exchange information. The information flow of the primal variables $\boldsymbol{p}^{(i)} = \left(\hat{\boldsymbol{x}}^{(i)}(k-N)^{\mathrm{T}}, \hat{\underline{\boldsymbol{w}}}^{(i)}(k)^{\mathrm{T}}\right)^{\mathrm{T}} \in \mathbb{R}^{(N+1)n}$ among the estimators is described by the directed graph $\mathcal{G} = \{\mathcal{V}, \mathcal{E}\}$, where the nodes in $\mathcal{V} = \{1, 2, \ldots, m\}$ represent the estimators and the edge (i,j) in the set $\mathcal{E} \subseteq \mathcal{V} \times \mathcal{V}$ models that estimator i can transmit information about $\boldsymbol{p}^{(i)}$ to estimator j. We assume that the graph is weakly connected, i.e. there is an undirected path between any pair of vertices, and that there are no selfloops, i.e. $(i,i) \notin \mathcal{E}, \forall i \in \mathcal{V}$. Moreover, we associate to the graph $\mathcal{G}$ the adjacency matrix $\boldsymbol{\Omega} = [\omega_{ij}] \in \mathbb{R}^{m \times m}$, with entries

$$\omega_{ij} = \begin{cases} 1 & (i,j) \in \mathcal{E} \\ 0 & (i,j) \notin \mathcal{E}. \end{cases} \quad (3.18)$$

Then, (3.9) can be expressed as

$$p^* = \min_{\boldsymbol{p}^{(1)}} J^{(1)}(\boldsymbol{p}^{(1)}) + \ldots + \min_{\boldsymbol{p}^{(m)}} J^{(m)}(\boldsymbol{p}^{(m)}) \quad (3.19a)$$

$$\text{subject to} \qquad \boldsymbol{p}^{(i)} = \boldsymbol{p}^{(j)}, \quad \forall (i,j) \in \mathcal{E}, \quad (3.19b)$$

where the consensus constraints (3.19b) guarantee that each local solution $\boldsymbol{p}^{(i)}$ is identical to the global solution $\boldsymbol{p}$.

Although the cost function in (3.19a) is decoupled, the primal problem (3.19) cannot be solved in a distributed manner because of the coupling through the consensus constraints (3.19b). Decoupling can be achieved by the dual problem of (3.19). To this end, the dual variables $\boldsymbol{\lambda}^{(i,j)} \in \mathbb{R}^{(N+1)n}$ are introduced which are transmitted from estimator i to estimator j. The information flow of $\boldsymbol{\lambda}^{(i,j)}$ among the estimators is described by the reverse graph $\bar{\mathcal{G}} = \{\mathcal{V}, \bar{\mathcal{E}}\}$, which is obtained by reversing the order of the nodes of all the pairs in $\mathcal{E}$. That is, if $\mathcal{E}$ contains an edge (i, j), then $\bar{\mathcal{E}}$ contains an edge (j, i) and vice versa. The adjacency matrix $\bar{\boldsymbol{\Omega}} = [\bar{\omega}_{ij}] \in \mathbb{R}^{m \times m}$ associated to the graph $\bar{\mathcal{G}}$ is $\bar{\boldsymbol{\Omega}} = \boldsymbol{\Omega}^{\mathrm{T}}$. The resulting dual problem can be reformulated as

$$d^* = \max_{\mathrm{col}(\boldsymbol{\lambda}^{(i,j)}, (i,j) \in \bar{\mathcal{E}})} \sum_{i=1}^{m} \left(\min_{\boldsymbol{p}^{(i)}} \left(J^{(i)}(\boldsymbol{p}^{(i)}) - \sum_{j=1}^{m} \left(\bar{\omega}_{ij} \boldsymbol{\lambda}^{(i,j)} - \bar{\omega}_{ji} \boldsymbol{\lambda}^{(j,i)} \right)^{\mathrm{T}} \boldsymbol{p}^{(i)} \right) \right) \tag{3.20}$$

and is decoupled regarding the primal variables. Moreover, strong duality holds, i.e. the duality gap $p^* - d^*$ reduces to zero. In other words, the dual problem (3.20) can be used to solve the primal problem (3.9) in a distributed manner, e.g. by a suitable tailored subgradient method. The resulting algorithm that is followed by the i-th estimator to calculate $\boldsymbol{p}^{(i)}$ is presented in Algorithm 3.1.

The primal and dual variables are initialized by zero in line 1. Several termination conditions can be used in line 2, like the number of iterations or the convergence tolerance of the dual variables [207]. There exist several step size rules for which convergence of the algorithm can be proven, like a constant step size [338].

The update step of the primal variables in line 6 has some interesting properties. First of all, due to the special choice of the local cost functions $J^{(i)}$ in (3.17b), the inverse of $\boldsymbol{M}^{(i)}$ always exists, regardless of the measurement allocation. This means that there is no need for the measurement allocation to imply local observability of the states. Second, there is no need for the i-th estimator to exchange information in order to calculate the predicted estimate $\bar{\boldsymbol{x}}(k-N)$, the covariance matrix $\boldsymbol{P}(k-N)$ or the control input $\underline{\boldsymbol{u}}(k)$. All these values can be calculated and stored fully decentralized due to the facts that all estimators are initialized identically by $\bar{\boldsymbol{x}}(0)$ and $\boldsymbol{P}(0)$ and that all estimators converge to the identical optimal primal variable $\boldsymbol{p}$. Third, all matrices including the inverse of $\boldsymbol{M}^{(i)}$ can be calculated offline such that only simple matrix vector multiplications have to be performed online in each iteration step.

Algorithm 3.1. Decentralized moving-horizon estimator for calculating $\boldsymbol{p}^{(i)}$

Initialization: $l = 0$ (iteration index), $\boldsymbol{\lambda}^{(i,j)}(0) = \mathbf{0}$, $\boldsymbol{p}^{(i)}(0) = \mathbf{0}$

while termination condition = false **do**

Increase iteration index: $l = l + 1$

Update dual variable $\boldsymbol{\lambda}^{(i,j)}(l)$:

$$\boldsymbol{\lambda}^{(i,j)}(l) = \boldsymbol{\lambda}^{(i,j)}(l-1) - \alpha(l-1)\left(\boldsymbol{p}^{(i)}(l-1) - \boldsymbol{p}^{(j)}(l-1)\right)$$

Communicate $\boldsymbol{\lambda}^{(i,j)}(l)$ according to the graph $\bar{\mathcal{G}}$.

Update primal variable $\boldsymbol{p}^{(i)}(l)$:

$$\boldsymbol{p}^{(i)}(l) = \boldsymbol{M}^{(i)^{-1}}\left(\boldsymbol{o}^{(i)} + \sum_{j=1}^{m}\left(\bar{\omega}_{ij}\boldsymbol{\lambda}^{(i,j)}(l) - \bar{\omega}_{ji}\boldsymbol{\lambda}^{(j,i)}(l)\right)\right)$$

where

$$\begin{aligned}
\boldsymbol{M}^{(i)} &= \frac{1}{m}\mathrm{diag}\left(\boldsymbol{P}(k-N), \underline{\boldsymbol{Q}}\right)^{-1} + \boldsymbol{N}^{(i)^{\mathrm{T}}}\boldsymbol{N}^{(i)} \\
\boldsymbol{N}^{(i)} &= \frac{1}{\sqrt{2}}\begin{pmatrix} \underline{\boldsymbol{R}}^{(i)^{-1/2}} & \mathbf{0} \\ \mathbf{0} & \underline{\boldsymbol{R}}^{(i)^{-1/2}} \end{pmatrix}\begin{pmatrix} \boldsymbol{F}^{(i)}(k) & \boldsymbol{H}^{(i)}(k) \\ \boldsymbol{F}^{(i)}(k) & \boldsymbol{H}^{(i)}(k) \end{pmatrix} \\
\boldsymbol{o}^{(i)} &= \begin{pmatrix} \boldsymbol{F}^{(i)}(k)^{\mathrm{T}}\underline{\boldsymbol{R}}^{(i)^{-1}} \\ \boldsymbol{H}^{(i)}(k)^{\mathrm{T}}\underline{\boldsymbol{R}}^{(i)^{-1}} \end{pmatrix}\left(\underline{\boldsymbol{y}}^{(i)}(k) - \boldsymbol{G}^{(i)}(k)\,\underline{\boldsymbol{u}}(k)\right) \\
&\quad + \begin{pmatrix} \frac{1}{m}\boldsymbol{P}(k-N)^{-1} \\ \mathbf{0} \end{pmatrix}\bar{\boldsymbol{x}}(k-N)
\end{aligned}$$

Communicate $\boldsymbol{p}^{(i)}(l)$ according to the graph $\mathcal{G}$.

end while

Example 3.3 *Weir system [292]*

Consider the artificial weir system depicted in Fig. 3.6 which can be described by the third-order discrete-time linear system

$$\boldsymbol{x}(k+1) = \begin{pmatrix} 0.90 & 0 & 0 \\ 0.43 & 0.80 & 0 \\ 0.15 & 0.35 & 0.75 \end{pmatrix}\boldsymbol{x}(k) + \begin{pmatrix} -0.3 & 0 & 0 \\ 0.3 & -0.4 & 0 \\ 0 & 0.4 & -0.2 \end{pmatrix}\boldsymbol{u}(k) + \boldsymbol{w}(k) \tag{3.21a}$$

$$y^{[1]}(k) = (1\ \ 0\ \ 0)\,\boldsymbol{x}(k) + v^{[1]}(k) \tag{3.21b}$$

$$y^{[2]}(k) = (0\ \ 0\ \ 1)\,\boldsymbol{x}(k) + v^{[2]}(k). \tag{3.21c}$$

The system involves three states $\boldsymbol{x}(k) = (x_1(k), x_2(k), x_3(k))^{\mathrm{T}}$ corresponding to the fill levels of the water reservoirs with the initial condition $\boldsymbol{x}(0) = (60, 40, 70)^{\mathrm{T}}$. The covariance of the state noise $\boldsymbol{w}(k)$ is $\boldsymbol{Q} = \mathrm{diag}(1,1,1)$. The control inputs $\boldsymbol{u}(k) = (u^{(1)}(k), u^{(2)}(k), u^{(3)}(k))^{\mathrm{T}}$ are the flow rates of the three valves. Sensor 1 and 2 measure the first and the third state where the covariance of $v^{[1]}(k)$ and $v^{[2]}(k)$ is $R^{[1]} = 5$ and $R^{[2]} = 5$, respectively. Note that the system is not observable by sensor 1 alone.

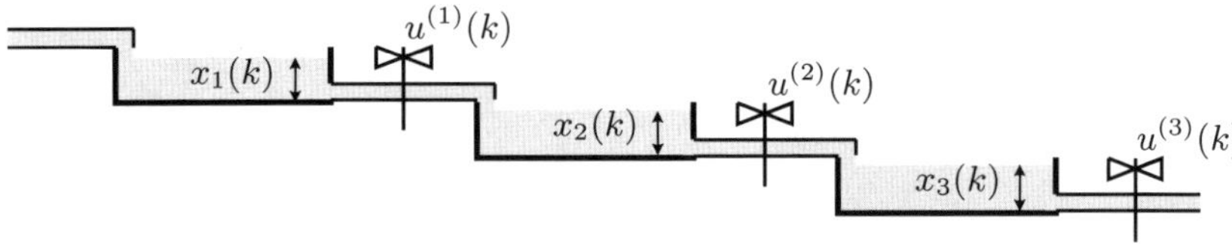

Fig. 3.6 Schematic diagram of a weir system

We design a combined linear static feedforward and LQR-feedback controller

$$\boldsymbol{u}(k) = -\boldsymbol{K}\boldsymbol{x}(k) + (\boldsymbol{K} - \boldsymbol{B}^{-1}(\boldsymbol{A} - \boldsymbol{I}))\boldsymbol{r}(k)$$

with the reference input $\boldsymbol{r}(k) = (r_1(k), r_2(k), r_3(k))^{\mathrm{T}} \in \mathbb{R}^3$ such that the closed-loop system becomes

$$\boldsymbol{x}(k+1) = (\boldsymbol{A} - \boldsymbol{B}\boldsymbol{K})(\boldsymbol{x}(k) - \boldsymbol{r}(k)) + \boldsymbol{r}(k) + \boldsymbol{w}(k).$$

The LQR feedback gain $\boldsymbol{K} \in \mathbb{R}^{3\times 3}$ is determined for the state weighting matrix $\boldsymbol{Q}_{\mathrm{LQR}} = \mathrm{diag}(80, 30, 80)$ and the control input weighting matrix $\boldsymbol{R}_{\mathrm{LQR}} = \mathrm{diag}(1, 1, 1)$.

The three estimators are placed next to the different valves and are identically initialized by $\bar{\boldsymbol{x}}(0) = (0, 0, 0)^{\mathrm{T}}$ and $\boldsymbol{P}(0) = \mathrm{diag}(1, 1, 1)$. The latter is updated according to the Ricatti equation (3.13). The length of the moving-horizon is $N + 1 = 4$ and the associated sets are $\mathcal{I}(k) = \{k-3, k-2, k-1, k\}$ and $\mathcal{S} = \{1, 2\}$. The measurements from sensor 1 are allocated alternately to the estimators 1 and 2 while the measurements from sensor 2 are always assigned to the estimator 3. This results in the alternating index sets for estimators 1 and 2:

$$\begin{aligned}
\mathcal{Y}^{([k+1 \bmod 2]+1)}(k) &= \{(k-3, 1), (k-1, 1)\} \\
\mathcal{Y}^{([k \bmod 2]+1)}(k) &= \{(k-2, 1), (k, 1)\} \\
\mathcal{Y}^{(3)}(k) &= \{(k-3, 2), (k-2, 2), (k-1, 2), (k, 2)\}.
\end{aligned}$$

The communication topology between the estimators is represented by the graphs $\mathcal{G}$ and $\bar{\mathcal{G}}$ shown in Fig. 3.7. Due to the fact that the sampling time of the sensors is much higher than the packet delays, the prediction time i in (3.11) is chosen to $N + 1$. Moreover, 20 iterations per optimization are performed.

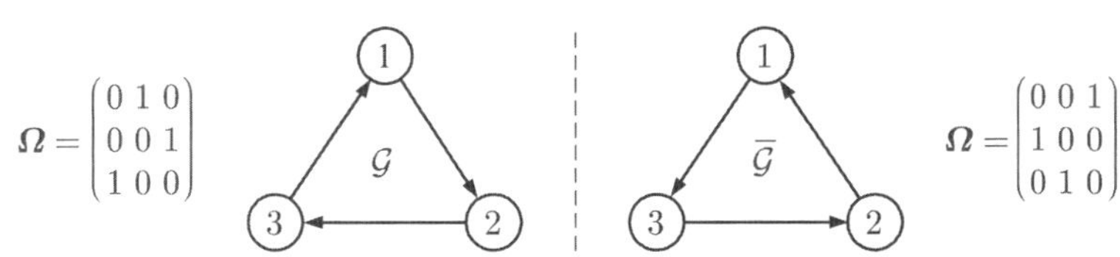

Fig. 3.7 Information flow of the primal and dual variables for a ring communication topology represented by the graphs $\mathcal{G}$ and $\bar{\mathcal{G}}$ and the associated adjacency matrix $\boldsymbol{\Omega}$ and $\bar{\boldsymbol{\Omega}}$, respectively

Figure 3.8 depicts the closed-loop performance for the ring communication topology resulting from $\mathcal{G}$. As long as $k < 4$, the feedback part is turned off and the state is predicted according to (3.11) with $N = 0$. After the buffers are filled, i.e. $k > 3$, the estimators reconstruct the state of the system from the noisy measurements. Note that the performance depicted in Fig. 3.8 is identical to that of a centralized Kalman filter. □

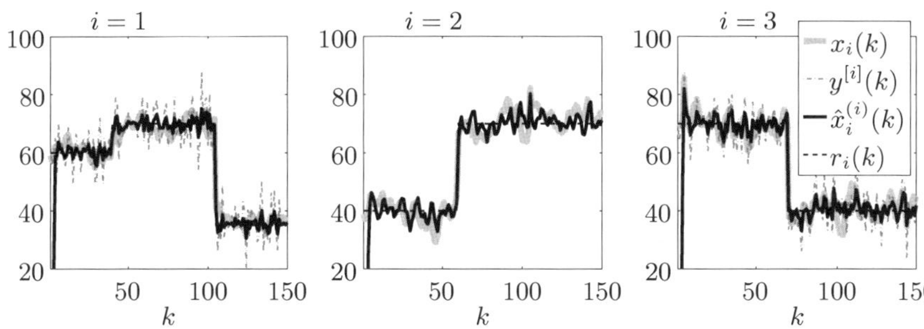

Fig. 3.8 Comparison between the noisy measurements $y^{[i]}(k)$, nominal trajectories $r_i(k)$, true states $x_i(k)$ and estimated states $\hat{x}_i^{(i)}(k)$ for the i-th water reservoir

Bibliographical Notes

Additional information about the presented centralized moving-horizon estimator concept can be found in [291]. A discussion on centralized moving-horizon observers for networked control systems can be found in [285, 290]. The possibility of estimating the clock parameters is of fundamental importance. An extensive discussion about this topic in the context of observability of NCS can be found in [288]. Another important aspect is the efficient solution of the nonlinear program representing the update step. The arising derivatives can be efficiently calculated by exploiting the structure of the moving-horizon estimation formulation in general [286], and especially of the extended networked structure [287]. The practical implementation in a real-time environment with application to a networked pendulum as shown in example 3.2 is discussed in [289].

Additional information about the presented decentralized moving horizon estimator concept can be found in [292]. For a discussion on decentralized moving horizon observers for network control systems, see [293]. A detailed theory of subgradient methods containing their convergence properties for different step size sequences can be found in [207, 338].

3.3 Towards a Two-Layer Optimal Control Architecture for Distributed Systems

New application fields like smart power grids, water distribution systems, traffic systems, or large arrays of micro-electromechanical systems (MEMS) require novel optimization and control algorithms that operate in a distributed, event-triggered fashion and are able to react to dynamic changes within the overall system on different time scales. The research presented in this section is targeted at a two-layer architecture for the optimal control of distributed systems. It consists of two interacting, distributed layers: an upper optimization layer and a lower feedback control layer. The general idea is that the feedback controller adapts itself to the changing physical structure and is able to deal with fast dynamics, while the distributed optimization layer can be adjusted to the communication topology and addresses global aspects on a slower time scale. The overall control scheme uses an event-triggered, local information exchange.

We start by giving an overview on a selection of important techniques for both distributed optimization and distributed control. We then focus on two particular methods, one from optimization in the context of DC optimal power flow, the other from controller design. The approaches have in common that they both are based on dual decomposition techniques. Furthermore, the controller design uses a distributed optimization approach.

3.3.1 Distributed Optimization

Distributed optimization algorithms are gaining significant importance due to the ubiquitous parallel processing of new computer generations and the growing need for distributed approaches to optimized decision making and control in large scale networked systems. In this section we review several central ideas that are at the core of distributed optimization algorithms.

Optimization problems in a multi-agent system have the following structure: every agent $i \in \{1, \dots, N\}$ has control over a subblock $\boldsymbol{x}_i \in \mathbb{R}^{n_i}$ of the decision variables $\boldsymbol{x} = (\boldsymbol{x}_1^{\mathrm{T}}, \dots, \boldsymbol{x}_N^{\mathrm{T}})^{\mathrm{T}} \in \mathbb{R}^n$ and the aim is to solve the following constrained optimization problem:

$$\min_{\boldsymbol{x} \in \mathcal{X} \subseteq \mathbb{R}^n} f(\boldsymbol{x}), \tag{3.22}$$

where $\mathcal{X} \subseteq \mathbb{R}^n$ is the feasible set and $f : \mathcal{U} \to \mathbb{R}$ is the cost function, defined on an open set $\mathcal{U} \supset \mathcal{X}$. The structure of f and $\mathcal{X}$ determines to what extent the agents have to interact in order to jointly solve (3.22). Favorable properties for decomposing the problem are separability of the cost function, i.e., $f(\boldsymbol{x}) = \sum_{i=1}^{N} f_i(\boldsymbol{x}_i)$ or separability of the constraint set, i.e., $\mathcal{X} = \mathcal{X}_1 \times \cdots \times \mathcal{X}_N$. If both are separable then problem (3.22) decomposes into N subproblems $\min_{\boldsymbol{x}_i \in \mathcal{X}_i} f_i(\boldsymbol{x}_i)$ that can be solved independently and in parallel by the agent i, $(i = 1, 2, ..., N)$. Of course, this full separability is usually not given and depending on the type of non-separability different approaches can be followed to obtain distributed solution methods.

For the case of a separable feasible set $\mathcal{X} = \mathcal{X}_1 \times \cdots \times \mathcal{X}_N$ and a non-separable objective function $f(\boldsymbol{x})$, variants of block coordinate descent methods are appropriate. *Parallel variable distribution* (PVD) [118, 324, 341, 342] is a general class of methods that follow this paradigm. In each iteration, the PVD algorithm performs a parallelization step and a synchronization step to compute the new iterate $\boldsymbol{x}^{k+1}$ from the current iterate $\boldsymbol{x}^k$. In the parallelization step, the main responsibility of agent i is to find an improvement for the i-th block of $\boldsymbol{x}$ by solving a suitable subproblem. This is done in parallel by all agents. In addition, the search space of agent i can be extended by one dimension for each of the other $N-1$ subblocks $\boldsymbol{x}_j$, $(j \neq i)$. This "forget-me-not approach" improves the robustness and accelerates the convergence compared to earlier methods such as block Jacobi method [31] and coordinate descent method [361]. The subproblem of the i-th agent looks as follows:

$$\min_{\boldsymbol{y}_i \in \mathbb{R}^{n_i}, (\lambda_{ij})_{j \neq i} \in \mathbb{R}^{N-1}} f_i\left(\boldsymbol{y}_i, (\boldsymbol{x}_j + \lambda_{ij}\boldsymbol{d}_j^k)_{j \neq i}\right)$$
$$\text{s.t. } \boldsymbol{y}_i \in \mathcal{X}_i, \quad \boldsymbol{x}_j^k + \lambda_{ij}\boldsymbol{d}_j^k \in \mathcal{X}_j, \quad j \neq i.$$

Here, $\boldsymbol{x}_{\bar{i}} = (\boldsymbol{x}_1, \ldots, \boldsymbol{x}_{i-1}, \boldsymbol{x}_{i+1}, \ldots, \boldsymbol{x}_N)$ denoting $\boldsymbol{x}$ after removal of the i-th block $\boldsymbol{x}_i$, the function $f_i(\boldsymbol{x}_i, \boldsymbol{x}_{\bar{i}}) := f(\boldsymbol{x}_1, \ldots, \boldsymbol{x}_N)$ is just f with conveniently reordered variables $\boldsymbol{x}_j$. This problem is solved to obtain (approximate) solutions $\boldsymbol{y}_i^k$ and λ_{ij}^k, $(j \neq i)$. Convergence analyses exist for different flavors of accuracy requirements to which these subproblems are solved: exact, ϵ-solutions, or sufficient decrease conditions [324, 341, 342]. The synchronization step then collects the agents' new candidate iterates $\boldsymbol{x}^{k,i}$ with subblocks $\boldsymbol{x}_i^{k,i} = \boldsymbol{y}_i^k$ and $\boldsymbol{x}_j^{k,i} = \boldsymbol{x}_j^k + \lambda_{ij}^k \boldsymbol{d}_j^k$, $(j \neq i)$. Here, the vectors $\boldsymbol{d}_j^k$ are subblocks of a suitable direction $\boldsymbol{d}^k \in \mathbb{R}^n$ (e.g. steepest descent). The new iterate $\boldsymbol{x}^{k+1} \in \mathcal{X}$ is determined such that $f(\boldsymbol{x}^{k+1}) \leq \min_{1 \leq i \leq m} f(\boldsymbol{x}^{k,i})$.

Another class of distributed optimization methods, *bilevel decomposition algorithms* (BDA) [37, 257, 258, 355], is suitable for partially separable optimization problems. These are separable with respect to N local variables $\boldsymbol{x}_i \in \mathbb{R}^{n_i}$, $(i = 1, \ldots, N)$, and coupled via global variables $\boldsymbol{y} \in \mathbb{R}^p$:

$$\min_{\boldsymbol{y}, \boldsymbol{x}_1, \ldots, \boldsymbol{x}_N} \sum_{i=1}^{N} f_i(\boldsymbol{y}, \boldsymbol{x}_i) \quad \text{s.t.} \quad \boldsymbol{g}_j(\boldsymbol{y}, \boldsymbol{x}_j) \leq \boldsymbol{0}, \quad 1 \leq j \leq N$$

with $\boldsymbol{g}_i : \mathbb{R}^p \times \mathbb{R}^{n_i} \to \mathbb{R}^{l_i}$. The main idea of BDA to solve this problem is to break it up into a master problem and a set of N subproblems each depending only on the local variable block $\boldsymbol{x}_i$. A more formal description of the algorithm is the following: solve the master problem

$$\min_{\boldsymbol{y}} \sum_{i=1}^{N} F_i^*(\boldsymbol{y}),$$

where $F_i^*(\boldsymbol{y})$ is the optimal value function of the i-th subproblem:

$$F_i^*(\boldsymbol{y}) = \min\{f_i(\boldsymbol{y},\boldsymbol{x}_i) \mid g_i(\boldsymbol{y},\boldsymbol{x}_i) \leq 0,\ \boldsymbol{x}_i \in \mathbb{R}^{n_i}\}$$

for $i = 1,\ldots,N$. The latter problems can be solved in parallel.

Both, PVD methods and BDA require to some extent a global data exchange. PVD distributes the current iterate to all agents before the synchronization step and collects the results in the parallelization step. In BDA, each evaluation of the master problem's cost function requires to distribute the global variable $\boldsymbol{y}$ to the agents and to collect the function values $F_i^*(\boldsymbol{y})$ afterwards. Some optimization procedures for the master problem require additional information like gradients of F_i^*, which have to be collected as well.

Alternatives to BDA were proposed that tackle the whole coupled problem by interior point methods and perform a BDA-like decomposition on the linear algebra level by a Schur complement approach [258].

Many optimization problems arising in the context of large-scale networks such as utility maximization problems (NUM) [282], distributed estimation [327], and the DC optimal power flow problem [300] have the following common structure: The cost function is convex and separable and the coupling occurs only via linear constraints:

$$\min_{\boldsymbol{x}\in\mathcal{X}} \sum_{i=1}^{N} f_i(\boldsymbol{x}_i) \quad \text{s.t.} \quad \boldsymbol{A}\boldsymbol{x} \leq \boldsymbol{c}\ , \quad \boldsymbol{B}\boldsymbol{x} = \boldsymbol{d}$$

with $\mathcal{X} = \mathcal{X}_1 \times \cdots \times \mathcal{X}_N$, $\mathcal{X}_i \subset \mathbb{R}^{n_i}$ convex and closed, $\boldsymbol{A} = (\boldsymbol{A}_1,\ldots,\boldsymbol{A}_N) \in \mathbb{R}^{l\times n}$, $\boldsymbol{A}_i \in \mathbb{R}^{l\times n_i}$, $\boldsymbol{c} \in \mathbb{R}^l$, $\boldsymbol{B} = (\boldsymbol{B}_1,\ldots,\boldsymbol{B}_N) \in \mathbb{R}^{n\times n}$, $\boldsymbol{B}_i \in \mathbb{R}^{n\times n_i}$, and $\boldsymbol{d} \in \mathbb{R}^n$. Due to this structure *dual decomposition algorithms* [56, 82, 273, 377] find broad application, where the constraints are decoupled using the Lagrangian $\mathcal{L}(\boldsymbol{x},\lambda,\mu) = \sum_{i=1}^{N}[f_i(\boldsymbol{x}_i) + \lambda^{\mathrm{T}}\boldsymbol{A}_i\boldsymbol{x}_i + \mu^{\mathrm{T}}\boldsymbol{B}_i\boldsymbol{x}_i] - \lambda^{\mathrm{T}}\boldsymbol{c} - \mu^{\mathrm{T}}\boldsymbol{d}$. Then the dual problem is considered:

$$\max_{\lambda\in\mathbb{R}_+^l,\ \mu\in\mathbb{R}^n} \boldsymbol{d}(\lambda,\mu),$$

where $\boldsymbol{d}(\lambda,\mu) = \inf_{\boldsymbol{x}\in\mathcal{X}} \mathcal{L}(\boldsymbol{x},\lambda,\mu)$. Obviously, the evaluation of the dual function can be done in parallel and a classical approach to solve the dual problem is the application of the projected gradient method [31] (or the projected subgradient method [31], if the dual function is not differentiable) which is easily parallelizable too. Often, the structure of the network optimization problems mentioned above in combination with dual decomposition requires only local communication between the agents, i.e., only neighboring agents in the network need to communicate with each other (see Section 3.3.3). By incorporating event-triggered communication [256] into the dual decomposition approach, additional flexibility with respect to changes of the physical and the communication network topology can be achieved. This is sketched in Section 3.3.3.

We close this section by mentioning that many aspects of (distributed) optimization can be interpreted from a control perspective [378]. The iteration performed in optimization methods can often be interpreted as a time discretization of a dynamic system (e.g., gradient flow). The aim, however, is not to accurately trace the trajectory, but rather to efficiently approach an equilibrium point.

3.3.2 Control of Distributed Systems

The size of large-scale systems is the reason why classical, centralized control methods cannot be applied. Traditionally, completely decentralized control methods are used which means that the individual subsystems do not exchange information and only use their own measurements for the control task. However, the wide-spread use of modern communication networks has sparked ideas and interest to employ these new technologies in the context of control to find an intermediate between centralized and decentralized control. The basic idea is to allow communication between some – but not all – subsystems, in order to improve performance when compared to decentralized control, while maintaining the low complexity compared to centralized (full information) control. The structure of the allowed communication connections – sometimes referred to as the *communication topology* – imposes structural constraints on the controller.

In this section, we will restrict ourselves to structured linear-time invariant systems with N subsystems. We assume that the subsystems can influence each other via the state, but that the input of each agent is restricted to the one respective agent. The dynamics of each subsystem i is given by

$$\dot{\boldsymbol{x}}_i(t) = \boldsymbol{A}_{ii}\boldsymbol{x}_i(t) + \boldsymbol{B}_i\boldsymbol{u}_i(t) + \sum_{\substack{j=1\\ j\neq i}}^{N} \boldsymbol{A}_{ij}\boldsymbol{x}_j(t), \quad i = 1, ..., N,$$

where $\boldsymbol{x}_i \in \mathbb{R}^{n_i}$, $\boldsymbol{u}_i \in \mathbb{R}^{m_i}$, $\boldsymbol{A}_{ii} \in \mathbb{R}^{n_i \times n_i}$, $\boldsymbol{A}_{ij} \in \mathbb{R}^{n_i \times n_j}$ and $\boldsymbol{B}_i \in \mathbb{R}^{n_i \times m_i}$.

The coupling structure of the dynamic system can be represented by a directed graph $\mathcal{G} = (\mathcal{V}, \mathcal{E})$ where $\mathcal{V}$ is the collection of nodes (i.e. subsystems, agents) and $\mathcal{E}$ is the collection of edges (i.e. interconnections). The edge set describes which subsystems influence each other. If edge $(j, i) \in \mathcal{E}$, subsystem i is influenced directly by the state of agent j. Hence, the relation $\boldsymbol{A}_{ij} \neq \boldsymbol{O}$ holds only if $(j, i) \in \mathcal{E}$. The set of neighboring agents of agent i is defined by

$$\mathcal{N}_i = \{j \,|\, (i, j) \in \mathcal{E} \text{ or } (j, i) \in \mathcal{E}\}\,.$$

By concatenation of the subsystems' states, the overall interconnected system can be written compactly as the continuous-time time-invariant linear system

$$\dot{\boldsymbol{x}}(t) = \boldsymbol{A}\boldsymbol{x}(t) + \boldsymbol{B}\boldsymbol{u}(t), \quad \boldsymbol{x}(0) = \boldsymbol{x}_0, \tag{3.23}$$

where $\boldsymbol{x} \in \mathbb{R}^n$ is the state, $\boldsymbol{u} \in \mathbb{R}^m$ is the input and $\boldsymbol{x}_0 \in \mathbb{R}^n$ is the initial condition. The matrix $\boldsymbol{A}$ consists of the blocks $\boldsymbol{A}_{ij}$, and $\boldsymbol{B}$ is block diagonal: $\boldsymbol{B} = \mathrm{diag}(\boldsymbol{B}_i)$.

The goal of distributed optimal control is to design a controller which also has a specific sparsity structure, and which is optimal in some sense. For example, we want to design a linear static state feedback control where communication is allowed among neighbors:

$$\boldsymbol{K}_{ij} \neq \boldsymbol{O} \text{ only if } j \in \mathcal{N}_i. \tag{3.24}$$

The literature offers a variety of performance specifications which will be explored in more detail below.

Literature Survey. As already mentioned, the earliest approaches dealing with large-scale systems are decentralized control [339] where each subsystem uses only its own information for the control task. While stability may be guaranteed, the performance is usually degraded when compared to centralized (full information) control. So distributed control can be regarded as a compromise between classical centralized control and decentralized control. A big distinction in the approaches can be made which is especially interesting when large-scale systems are considered. This divides all approaches into those which assume that complete, centralized model information is available while others consider the control problem with only limited or locally available model information. We begin our overview with the complete-model approaches.

An early result [387] in distributed control already illustrates the difficulty of the problem by showing that the optimal controller for a two-agent linear quadratic cost problem is nonlinear when a structural constraint is imposed. A more recent result is given in [322] where conditions on the communication topology are given based on the plant transfer function such that the controller design problem can be formulated as a convex optimization problem. The paper uses the Youla parametrization to parameterize all stabilizing controllers and if the set of admissible controllers satisfies the so-called property of quadratic invariance, the optimal controller can be found via a convex problem. Also, a computational test for quadratic invariance based on the structure of the controller matrix $\boldsymbol{K}$ and on the structure of the plant transfer function $\boldsymbol{G}$ is given. However, the resulting optimization problems are in general infinite-dimensional and thus, though convex, still difficult to solve.

In some cases, it may be of interest to have a minimal communication topology while maintaining good performance levels in comparison to the centralized case and this is achieved in [332]. This problem is first formulated as a minimization of the l_0-norm, i.e. the direct minimization of the number of entries in the feedback matrix. Since this problem is hard to solve, a convex relaxation is introduced in form of a weighted l_1-minimization. The minimization is subject to a given performance degradation in an $\mathcal{H}_\infty$-sense. This means that the performance of the decentralized controller is compared to the performance of the optimal centralized controller and the discrepancy cannot become too large.

Interesting approaches for distributed controller synthesis based on state-space formulations are for example [215, 365] which are based on LMIs and where the controller topology is identical to the plant topology, i.e. if subsystems are coupled dynamically, they are allowed to exchange information. This assumption on the structure is fairly common and is based on the idea that such a communication topology would be relatively easy to implement.

In [336], a state-space method for distributed $\mathcal{H}_2$ optimal controller design is presented which is applicable when the subsystems are partially ordered, e.g. chains, trees or nested structures. It is shown that the problem can be decomposed into N subproblems, where N is the number of subsystems in the whole system. Each subproblem then consists of the solution of an uncoupled Riccati equation for which efficient techniques are known.

If one is interested in the largest eigenvalue as a performance measure, the approach in [164] is applicable. The minimization of the real part of the largest eigenvalue aims at maximizing the convergence rate. In order to do so, a decentralized controller is designed in a first step. This controller is simply used to stabilize the system and no performance is optimized. The advantage of this is that the system remains stable even if communication links fail. In a second step, a limited number of communication links between the subsystems are added to get optimal performance. This constitutes a mixed integer problem which is NP-hard. For a special case of scalar subsystems and several conditions on the dynamics matrix, a more efficient solution is also derived.

Controller Design with Partial Model Information. All the methods presented above have in common that the controller is implemented in a distributed way but it is designed in a centralized way, i.e. with the assumption that a complete model is available. So in the following, we want to present some of the very few approaches that try to satisfy their control task with only limited model information. One of these papers is [116]. The authors show for discrete-time LTI systems that when each agent knows only its own system model, deadbeat control cannot be outperformed by any other control strategy when an LQR cost functional is considered. For a specific structure of the dynamics matrix $\boldsymbol{A}$ (tridiagonal), which is present in vehicle following problems, the authors in [115] formulate a distributed optimization problem using dual decomposition to find the optimal linear quadratic distributed controller. In order to facilitate convergence of the problem, they use a relaxed augmented Lagrangian technique. Another approach based on distributed optimization is presented in [135]. The authors use dual decomposition in a distributed model predictive method which involves repeated minimization of a cost function to determine an optimal control action profile.

In [253] it is assumed that subsystems know their own dynamic model and the model of their direct neighbors (neighborhood is defined through direct physical coupling). The optimal distributed controller for an LQR cost functional is then computed iteratively by introducing an adjoint (dual) state and using simulated trajectories of the system and adjoint states to compute a gradient descent direction for the state feedback matrix.

3.3.3 Dual Decomposition of the DC Optimal Power Flow Problem

We now consider the static DC optimal power flow problem, where a convex and separable energy production cost function in an electrical power network has to be minimized subject to coupling constraints that enforce the power balance within the network as well as bounds on power generation and power line flow. The power system is modeled by a directed graph $\mathcal{G} = (\mathcal{V}, \mathcal{E})$ with $\mathcal{V} = \{1, \ldots, n\}$ representing the set of buses, where all buses contain a load and for simplicity of notation are directly connected to a generator. The generalization of only $p < n$ buses directly connected to a generator is straightforward. The edge $e_{ij} \in \mathcal{E} \subseteq \mathcal{V} \times \mathcal{V}$ with $|\mathcal{E}| = m$ represents the transmission line from bus i to bus j. Let I be the $m \times n$ incidence matrix of the graph $\mathcal{G}$ and define a diagonal matrix $\boldsymbol{D} \in \mathbb{R}^{m \times m}$ with $d_{ll} = 1/x_l$ where x_l is the reactance of the lth transmission line.

Denoting the generated power by P_i^g, the load by P_i^d and the voltage phase angle at bus i by θ_i, the DC optimal power flow problem can be stated as follows:

$$
\begin{aligned}
\min_{P_i^g \in \mathcal{P}_i,\ \theta_i \in \Theta_i} \; & \sum_{i=1}^{n} C_i(P_i^g) \\
\text{s.t. } & \boldsymbol{B}\theta = P^g - P^d && (3.25) \\
& F^{\min} \leq \boldsymbol{D}\boldsymbol{I}\theta \leq F^{\max}, && (3.26)
\end{aligned}
$$

where $\mathcal{P}_i$, Θ_i are compact and convex sets, $C_i(x) = a_{i2}x^2 + a_{i1}x + a_{i0}$ is the quadratic cost of power production at bus i with nonnegative coefficients a_{i2}, a_{i1}, a_{i1} for $i = 1, \ldots, n$. Matrix $\boldsymbol{B} \in \mathbb{R}^{n \times n}$ is defined as

$$
\boldsymbol{B} = \boldsymbol{I}^{\mathrm{T}}\boldsymbol{D}\boldsymbol{I} \text{ with elements } b_{ij} = \begin{cases} \sum_{k \in \mathcal{N}_i} \frac{1}{x_{ik}}, & \text{if } i = j \\ -\frac{1}{x_{ij}}, & \text{if } j \in \mathcal{N}_i \\ 0, & \text{else,} \end{cases}
$$

where $\mathcal{N}_i = \{j \mid (i,j) \in \mathcal{E} \vee (j,i) \in \mathcal{E}\}$ denotes the set of indices of bus i's neighbors and x_{ij} denotes the reactance of the transmission line connecting bus i and bus j. The constraint (3.26) limits the power flow on each transmission line l to $[F_l^{\min}, F_l^{\max}]$, whereas constraint (3.25) expresses the power flow equations.

We used the partially separable structure of this problem in combination with a dual decomposition approach [272] to derive a dual problem that allows for a parallel evaluation of the dual cost function:

$$f(\lambda,\mu) = \min_{P_i^g\in\mathcal{P}_i,\ \theta_i\in\Theta_i} \sum_{i=1}^{n}[C_i(P_i^g)+\mu_i P_i^g] + \sum_{i=1}^{n}\Big[\sum_{l\in\mathcal{L}_i}(\lambda_l-\lambda_{l+m})(DI)_{li} - \sum_{j\in\mathcal{N}_i\cup i}\mu_j b_{ji}\Big]\theta_i + \sum_{l=1}^{m}[\lambda_{l+m}F_l^{\min} - \lambda_l F_l^{\max}] - \sum_{i=1}^{n}\mu_i P_i^d,$$

where $\mathcal{L}_i = \{l \mid v_i \in e_l \wedge e_l \in \mathcal{E}\}$ denotes the set of indices of lines connected to bus i. Nesterov's smoothing method [275] is utilized by adding strongly convex functions $d_i(x) = \frac{\sigma_i}{2}\|x\|_2^2$, $\sigma_i > 0$ for $i = 1,\dots,2n$ scaled with a smoothing parameter $c > 0$ to obtain a Lipschitz continuously differentiable smoothed dual cost function that inherits the favorable structure of the original dual function:

$$\begin{aligned} f_c(\lambda,\mu) = {} & \min_{P_i^g\in\mathcal{P}_i,\ \theta_i\in\Theta_i} \sum_{i=1}^{n}[C_i(P_i^g)+\mu_i P_i^g + c\,d_i(P_i^g)] \\ & + \sum_{i=1}^{n}\Big\{\Big[\sum_{l\in\mathcal{L}_i}(\lambda_l-\lambda_{l+m})(DI)_{li} - \sum_{j\in\mathcal{N}_i\cup i}\mu_j b_{ji}\Big]\theta_i + c\,d_{i+n}(\theta_i)\Big\} \\ & + \sum_{l=1}^{m}[\lambda_{l+m}F_l^{\min} - \lambda_l F_l^{\max}] - \sum_{i=1}^{n}\mu_i P_i^d. \end{aligned}$$

The considered augmented dual problem has the form:

$$\max_{\lambda\in\mathbb{R}_+^{2m},\ \mu\in\mathbb{R}^n} f_c(\lambda,\mu).$$

In [256] we adjusted Nesterov's first order method [272, 275] to solve this problem in parallel and achieved ϵ-optimality with optimal iteration complexity $O(\sqrt{L_c/\epsilon})$, where L_c is the Lipschitz constant of the gradient of the smoothed dual function f_c. The error introduced by the smoothing is of the order $O(c)$, where c is the smoothing parameter. Since there holds $L_c = O(1/c)$, the choice $c = O(\epsilon)$ achieves an ϵ-optimal solution in $O(1/\epsilon)$ iterations. We further showed in [256] that this method can be applied in such a way that each agent needs to acquire only data that are maintained by direct neighbors, i.e., by those agents to which it is linked by a direct edge (line) of the power grid graph.

In order to further reduce communication traffic, we introduced event-triggered communication in the developed method. Here, an agent transmits data only when a certain threshold is exceeded. Convergence was proved in [256] while maintaining the favorable features of the original scheme and numerical tests show that communication between the agents could be reduced by up to 60%. Moreover, we observed that an incautious choice of the convexity parameters $\sigma_i > 0$ of the d_i can lead to a quite large Lipschitz constant L_c due to the dependence

$$L_c(\sigma) = \frac{1}{c}\sum_i \frac{\boldsymbol{u}_i}{\sigma_i},$$

where the vectors $\boldsymbol{u}_i$ are problem-depending constants. We showed that the resulting optimization problem to minimize the Lipschitz constant subject to nonnegative convexity parameters $\sigma_i > 0$ can be solved analytically and in a distributed, parallel way. Thus, it is well suited also for the case of dynamically changing network topologies. Finally, we developed a scaling technique to further reduce the number of iterations necessary to achieve ϵ-optimality.

3.3.4 Distributed Controller Design with Local Model Information

While the majority of the research on distributed control considers a centralized design and a distributed implementation, for large-scale distributed systems, distributed design methods are necessary. The reasons for this are manifold. A first issue is privacy, i.e. not every subsystem might be willing to share model information with every other subsystem they are cooperating with. Second, a dynamic model might not be readily available for a large system. Furthermore, a system model might be too large to be handled by a centralized design method. A last point is that large-scale systems are likely to change at some point (nodes are added or removed), and a centralized approach would require a lot of effort to react to these changes, while a distributed approach could react locally in the affected nodes only.

Motivated by this reasoning, and inspired by the results in [253], an accelerated iterative method to determine a distributed controller for continuous-time LTI systems given in Eq. (3.23) is presented in [99]. The finite horizon LQR cost functional

$$J(\boldsymbol{x}, \boldsymbol{u}) = \int_0^T \boldsymbol{x}^{\mathrm{T}}(t)\boldsymbol{Q}\boldsymbol{x}(t) + \boldsymbol{u}^{\mathrm{T}}(t)\boldsymbol{R}\boldsymbol{u}(t)\,\mathrm{d}t \tag{3.27}$$

is minimized in the following way. The feedback matrix $\boldsymbol{K}$ is restricted to the structure described by the constraint given in (3.24). The gradient of the cost functional (3.27) with respect to the (i, j)-th entry of the feedback matrix is given by

$$(\nabla_{\boldsymbol{K}} J)_{ij} = -2\boldsymbol{R}_i \int_0^T \boldsymbol{u}_i(t)\boldsymbol{x}_j^{\mathrm{T}}(t)\,\mathrm{d}t - \boldsymbol{B}_i^{\mathrm{T}} \int_0^T \boldsymbol{\lambda}_i(t)\boldsymbol{x}_j^{\mathrm{T}}(t)\,\mathrm{d}t,$$

where

$$\dot{\boldsymbol{\lambda}}(t) = -(\boldsymbol{A} - \boldsymbol{B}\boldsymbol{K})^{\mathrm{T}}\boldsymbol{\lambda}(t) - 2(\boldsymbol{Q} + \boldsymbol{K}^{\mathrm{T}}\boldsymbol{R}\boldsymbol{K})\boldsymbol{x}(t)$$

with $\boldsymbol{\lambda}(T) = \mathbf{0}$. Thus, the gradient is computed using simulated trajectories of the system. The (locally) optimal feedback matrix can then be determined iteratively using a gradient descent method. Thus, in order to compute their respective entries of the feedback matrix, the agents only need information from their direct neighbors.

The main problem concerns the determination of a good step size that leads to relatively fast convergence. Step size selection is a difficult problem for distributed optimization because no centralized or collected knowledge is available. In the paper, the step size is determined according to a Barzilai-Borwein scheme. Applied to the presented problem, the Barzilai-Borwein method gives the step size with

$$\gamma_k = \frac{\langle \Delta \mathrm{vec}(\boldsymbol{K}), \Delta \mathrm{vec}(\boldsymbol{K}) \rangle}{\langle \Delta \mathrm{vec}(\boldsymbol{K}), \Delta \mathrm{vec}(\nabla_{\boldsymbol{K}} J) \rangle}, \tag{3.28}$$

where Δ denotes the difference between iteration k and $(k-1)$, $\langle \cdot, \cdot \rangle$ is the scalar product, superscript (k) denotes the k-th iteration and vec represents the column-wise vectorization of a matrix. While this formula cannot be computed distributedly directly, the computation can be completely distributed using a consensus algorithm in the following way.

First, the individual agents use their own entries of both feedback and gradient matrix to determine an estimate of the BB-step size γ_k. Afterwards, the value of (3.28) is given by a distributed consensus algorithm based on the previous estimates. Specifically, as a first step, each node i, $(i = 1, ..., n)$ initializes the two scalar values

$$\rho_i(k(0)) = \left\langle \Delta \mathrm{vec}(\boldsymbol{K}_i^{\mathrm{T}}), \Delta \mathrm{vec}(\boldsymbol{K}_i^{\mathrm{T}}) \right\rangle$$

and

$$\psi_i(k(0)) = \left\langle \Delta \mathrm{vec}(\boldsymbol{K}_i^{\mathrm{T}}), \Delta \mathrm{vec}((\nabla_{\boldsymbol{K}} J)_i^{\mathrm{T}}) \right\rangle .$$

In other words, in every iteration k, the agents will calculate the completely local values $\rho_i(k(0))$ and $\psi_i(k(0))$ using only their own respective row(s) of the feedback and gradient matrix. After this, consensus iterations start which require information exchange:

$$\begin{aligned} \rho_i(k(t+1)) &= \boldsymbol{W}_{ii}\rho_i(k(t)) + \sum_{j \in \mathcal{N}_i} \boldsymbol{W}_{ij}\rho_j(k(t)) \\ \psi_i(k(t+1)) &= \boldsymbol{W}_{ii}\psi_i(k(t)) + \sum_{j \in \mathcal{N}_i} \boldsymbol{W}_{ij}\psi_j(k(t)), \end{aligned}$$

where $\boldsymbol{W}$ is a symmetric, non-negative with strictly positive diagonal entries, doubly stochastic matrix, compatible with graph $\mathcal{G}$, and it requires only information exchange between neighbors. It is then shown that the consensus algorithm converges to γ_k of the BB-formula (3.28). Additionally, using the presented step size method, convergence of the gradient method to a stationary point is shown using the so-called Armijo rule.

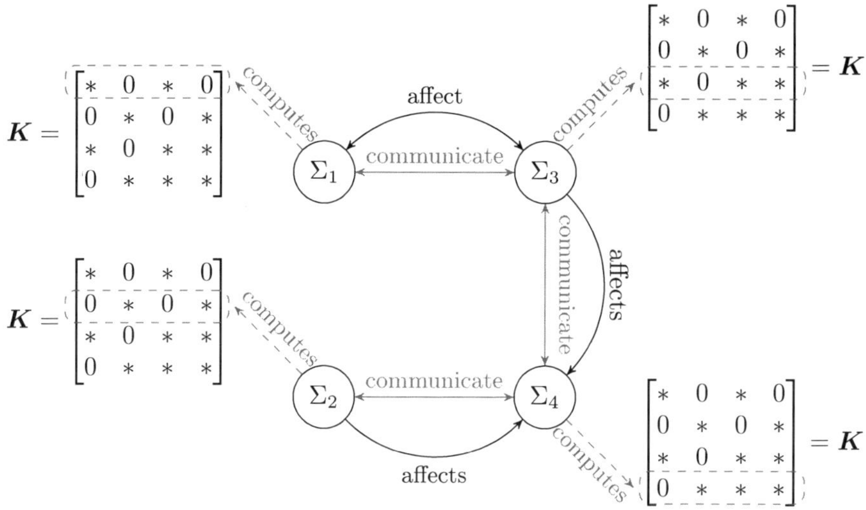

Fig. 3.9 Representation of the general approach for an example of four subsystems: Σ_i denotes subsystem i, black arrows represent physical coupling, grey arrows represent communication, grey dashed arrows computation

As a numerical demonstration of the effectiveness of the new step size method, the approach is applied to 500 randomly created stable systems of size $n = 10$ with $m = 10$ inputs. An average of 44 entries of the state matrices are non-zero. The time horizon is 10 s, all threshold values of the algorithm are 10^{-3}. A comparison of the BB step size is done with a constant step size of $\gamma = 1$. The result is that on average, the BB step size needs only 18% of the iterations that are needed with the constant step size.

Furthermore, the problem setup allows reference tracking of the state. This offers the possibility to link this control level with an upper optimization level which computes optimal setpoints through distributed optimization.

A visualization of the distributed controller computation is given in Fig. 3.9 for an example system with four subsystems.

4

Distributed and Networked Model Predictive Control

L. Grüne, F. Allgöwer, R. Findeisen, J. Fischer, D. Groß,
U. D. Hanebeck, B. Kern, M. A. Müller, J. Pannek, M. Reble,
O. Stursberg, P. Varutti, and K. Worthmann

This chapter introduces several concepts of model predictive control that use the capabilities of digital networks. Centralized and decentralized structures are considered. The control methods are illustrated by examples.

Chapter Contents

4.1 Model Predictive Control of Networked Systems

4.1.1 Chapter Overview

In this chapter, we consider the problem of controlling networked and distributed systems by means of model predictive control (MPC). The basic idea behind MPC is to repeatedly solve an optimal control problem based on a model of the system to be controlled. Every time a new measurement is available, the optimization problem is solved and the corresponding input sequence is applied until a new measurement arrives. As explained in the sequel, the advantages of MPC over other control strategies for networked systems are due to the fact that a model of the system is available at the controller side, which can be used to compensate for random bounded delays. At the same time, for each iteration of the optimization problem an optimal input sequence is calculated. In case of packet dropouts, one can reuse this information to maintain closed-loop stability and performance.

In the following, we first give an overall description of the problems we are considering. Then, in Section 4.1.3 a review on the basic ideas of MPC is provided. The following sections are concerned with distributed and networked versions of MPC:

- **Compensation of delays and packet loss:** In Section 4.2, we focus on networked control systems. In particular, we show how a model-based compensation in combination with predictive control can be used to ensure closed-loop stability.
- **Decentralized and distributed MPC:** Section 4.3 describes new methods for distributed and decentralized solutions for both the deterministic and the stochastic case.
- **Distributed MPC of communicating decentralized systems** (Section 4.4): Decentralized systems should be coupled by communication among the control stations if an information exchange can improve the performance of the overall system. Section 4.4 investigates this idea for MPC systems.
- **MPC based on virtual control inputs** (Section 4.5): In order to avoid communication, the control inputs can be replaced by expected values, which are called virtual control inputs. This idea is developed for different information structures.

4.1.2 Distributed and Decentralized Systems

In this chapter, we consider the problem of controlling single networked control systems as well as distributed ones (Figs. 4.1 and 4.2). We consider the network on a high abstraction level where it can be represented in terms of random delays and unpredictable packet dropouts. Problems such as jitter, quantization, or packet fragmentation, which typically appear in low level protocols, are not explicitly taken into account. Our objective is thus to develop an application level solution which can be used in combination with MPC to compensate for both delays and packet dropouts.

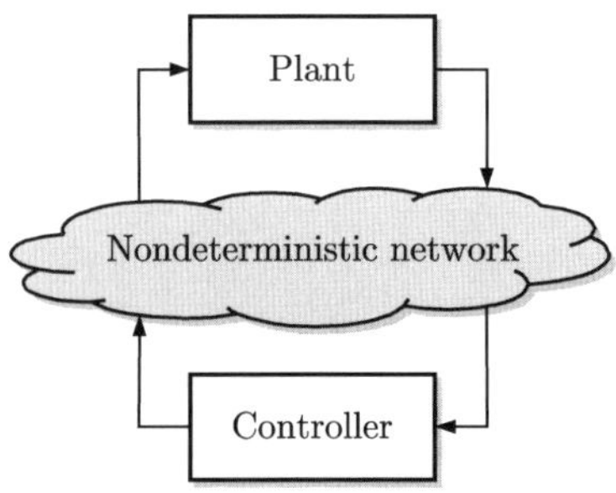

Fig. 4.1 Single networked control system

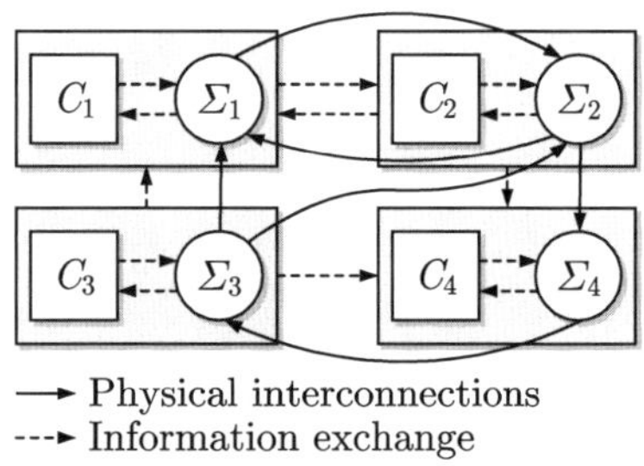

Fig. 4.2 Example of a distributed system

In the context of distributed control systems we distinguish between distributed and decentralized methods. With *distributed* we refer to those control solutions which make use of information exchange between the local controllers to achieve stability; with *decentralized* we denote problems which are solved without information exchange [329]. We consider both the case of physically coupled systems as well as systems without explicit interactions. Both deterministic and stochastic methods are presented. Although most of the results are described for discrete-time systems, these can be directly carried over to the continuous-time setting.

4.1.3 Brief Review of Model Predictive Control (MPC)

In this section we briefly review the basics of MPC for discrete and continuous-time systems. We consider continuous-time systems governed by the time-invariant ordinary differential equation

$$\dot{\boldsymbol{x}}(t) = \boldsymbol{f}(\boldsymbol{x}(t), \boldsymbol{u}(t)), \quad \boldsymbol{x}(0) = \boldsymbol{x}_0, \tag{4.1}$$

with $\boldsymbol{f} : \mathbb{R}^n \times \mathbb{R}^m \to \mathbb{R}^n$. $\boldsymbol{x}(t) = \boldsymbol{x}(t; \boldsymbol{x}_0, \boldsymbol{u})$ represents the solution of (4.1) at time t resulting from the control function $\boldsymbol{u} : \mathbb{R} \to \mathbb{R}^m$ and satisfying the initial condition $\boldsymbol{x}(0;\boldsymbol{x}_0,\boldsymbol{u})=\boldsymbol{x}_0$. The sets of feasible states and control values are denoted by $\mathbb{X}\subseteq\mathbb{R}^n$ or $\mathbb{U}\subseteq\mathbb{R}^m$, respectively. If the conditions

$$\boldsymbol{x}(t;\boldsymbol{x}_0,\boldsymbol{u})\in\mathbb{X},\ t\in[0,T]\quad\text{and}\quad\boldsymbol{u}(t)\in\mathbb{U},\ t\in[0,T)\tag{4.2}$$

are satisfied, the control function $\boldsymbol{u}$ is called *admissible* on $[0,T)$ for $\boldsymbol{x}_0$, which is denoted by $\boldsymbol{u}\in\mathcal{U}(\boldsymbol{x}_0,T)$. Here we assume existence and uniqueness of the solution $\boldsymbol{x}(\cdot\,;\boldsymbol{x}_0,\boldsymbol{u})$ until time T for all $\boldsymbol{u}\in\mathcal{U}(\boldsymbol{x}_0,T)$. Furthermore, $\boldsymbol{u}$ is said to be admissible for all times with respect to $\boldsymbol{x}_0$, i.e., $\boldsymbol{u}\in\mathcal{U}(\boldsymbol{x}_0,\infty)$, if $\boldsymbol{u}\in\mathcal{U}(\boldsymbol{x}_0,T)$ holds for each $T>0$.

Discrete-time systems can result from sampling of continuous-time ones, for instance due to the use of digital components with zero-order hold. If we refer to $\delta\in(0,T]$ as the sampling rate, given the state $\boldsymbol{x}\in\mathbb{R}^n$ and the control input $\boldsymbol{u}\in\mathbb{R}^m$, the discrete-time dynamics can be defined by

$$\boldsymbol{x}^+=\boldsymbol{f}(\boldsymbol{x},\boldsymbol{u}):=\boldsymbol{x}(\delta;\boldsymbol{x},\tilde{\boldsymbol{u}})\quad\text{with}\quad\tilde{\boldsymbol{u}}(t)=\boldsymbol{u},\ t\in[0,\delta),\tag{4.3}$$

i.e., the following state $\boldsymbol{x}^+$ is the solution of the differential equation (4.1) obtained by applying the constant control input $\boldsymbol{u}$ for a time δ starting from the current state $\boldsymbol{x}$. The discrete-time k then corresponds to the continuous-time $k\delta$. In this way, we can deal with a continuous-time system in a discrete-time setting. In general, we can describe the discrete-time dynamics by means of the difference equation

$$\boldsymbol{x}(k+1)=\boldsymbol{f}(\boldsymbol{x}(k),\boldsymbol{u}(k)),\quad \boldsymbol{x}(0)=\boldsymbol{x}_0,\tag{4.4}$$

with a nonlinear map $\boldsymbol{f}:\mathbb{R}^n\times\mathbb{R}^m\to\mathbb{R}^n$. The solution for an input sequence $(\boldsymbol{u}(k))_{k\in\mathbb{N}_0}$ and initial condition $\boldsymbol{x}_0$ at time k is denoted by $\boldsymbol{x}_{\boldsymbol{u}}(k;\boldsymbol{x}_0)$. An input sequence $\boldsymbol{u}=(\boldsymbol{u}(0),\boldsymbol{u}(1),\ldots,\boldsymbol{u}(N-1))$ of length N is called admissible for $\boldsymbol{x}_0\in\mathbb{X}$ if the conditions

$$\boldsymbol{f}(\boldsymbol{x}_{\boldsymbol{u}}(k;\boldsymbol{x}_0),\boldsymbol{u}(k))\in\mathbb{X}\quad\text{and}\quad\boldsymbol{u}(k)\in\mathbb{U}\tag{4.5}$$

hold for each $k\in\{0,1,\ldots,N-1\}$, which is denoted by $\boldsymbol{u}\in\mathcal{U}(\boldsymbol{x}_0,N)$. Analogously to the continuous-time setting, an input sequence $\boldsymbol{u}=(\boldsymbol{u}(k))_{k\in\mathbb{N}_0}$ is called admissible for all times with respect to $\boldsymbol{x}_0$, i.e., $\boldsymbol{u}\in\mathcal{U}(\boldsymbol{x}_0,\infty)$, if $\boldsymbol{u}\in\mathcal{U}(\boldsymbol{x}_0,N)$ is satisfied for each $N\in\mathbb{N}$.

In order to present the basic concept of MPC, we consider the discrete-time system (4.4). Our goal is to stabilize a steady state $\boldsymbol{x}^\star$ of (4.4) by a static state feedback $\mu:\mathbb{R}^n\to\mathbb{R}^m$ minimizing the cost

$$J_\infty(\boldsymbol{x}_0,\boldsymbol{u})=\sum_{k=0}^{\infty}\ell(\boldsymbol{x}_{\boldsymbol{u}}(k;\boldsymbol{x}_0),\boldsymbol{u}(k))\tag{4.6}$$

such that the state and control constraints are satisfied, i.e., $\boldsymbol{u}\in\mathcal{U}(\boldsymbol{x}_0,\infty)$. The stage cost $\ell:\mathbb{R}^n\times\mathbb{R}^m\to\mathbb{R}_{\geq 0}$ should be constructed according to some design specifications and be positive definite with respect to $\boldsymbol{x}^\star$. However, since nonlinear optimal control on an infinite time horizon is, in general, computationally hard, we use MPC to compute an approximately optimal feedback. MPC is a methodology which iteratively solves a sequence of optimal control problems on a truncated and, thus, finite prediction horizon.

MPC Algorithm. Let an initial state $\boldsymbol{x}_0 \in \mathbb{X}$ be given and set $\hat{\boldsymbol{x}} := \boldsymbol{x}_0$. Then, an MPC algorithm consists of the three steps described in Algorithm 4.1 (see also Fig. 4.3).

Algorithm 4.1. Basic MPC algorithm

Given: Model of the plant. Set $k := 0$.

1: Solve the optimal control problem on a truncated horizon based on the most recent state measurement $\hat{\boldsymbol{x}}$, i.e., find an admissible control input $\boldsymbol{u}^\star \in \mathcal{U}(\hat{\boldsymbol{x}}, N)$ minimizing the cost functional

$$J_N(\hat{\boldsymbol{x}}, \boldsymbol{u}) := \sum_{\hat{k}=0}^{N-1} \ell(\boldsymbol{x}_{\boldsymbol{u}}(\hat{k}; \hat{\boldsymbol{x}}), \boldsymbol{u}(\hat{k})) + E(\boldsymbol{x}_{\boldsymbol{u}}(N; \hat{\boldsymbol{x}})), \quad \boldsymbol{x}_{\boldsymbol{u}}(0; \hat{\boldsymbol{x}}) = \hat{\boldsymbol{x}}, \tag{4.7}$$

with a terminal cost $E : \mathbb{R}^n \to \mathbb{R}_{\geq 0}$ and possibly subject to an additional terminal constraint $\boldsymbol{x}_{\boldsymbol{u}}(N; \hat{\boldsymbol{x}}) \in \mathbb{X}_f \subseteq \mathbb{X}$. In order to avoid technical difficulties we assume the existence of a minimizing control, i.e., the existence of $\boldsymbol{u}^\star \in \mathcal{U}(\hat{\boldsymbol{x}}, N)$ with

$$V_N(\hat{\boldsymbol{x}}) := \inf_{\boldsymbol{u} \in \mathcal{U}(\hat{\boldsymbol{x}}, N)} J_N(\hat{\boldsymbol{x}}, \boldsymbol{u}) = J_N(\hat{\boldsymbol{x}}, \boldsymbol{u}^\star), \tag{4.8}$$

which in particular implies the existence of an admissible control sequence. More details on the terminal cost E and the terminal constraint set $\mathbb{X}_f$ will be explained later.

2: Define the static state feedback $\mu_N : \mathbb{R}^n \to \mathbb{R}^m$ by $\mu_N(\hat{\boldsymbol{x}}) := \boldsymbol{u}^\star(0)$, i.e., the MPC feedback map μ_N is given by the first element of the computed optimal input sequence $\boldsymbol{u}^\star \in \mathcal{U}(\hat{\boldsymbol{x}}, N)$.

3: Shift the prediction horizon forward in time, i.e., set $k := k + 1$, and repeat this procedure beginning with step 1.

Iterative application of this algorithm defines a feedback control law on the infinite time horizon despite using only a finite prediction horizon at each time instant. We emphasize that MPC requires a plant model in order to predict the future behavior of the plant and, thus, to determine a suitable control sequence. If neither disturbances nor measurement errors occur, i.e., in so called nominal MPC, Algorithm 4.1 yields the closed loop

$$\boldsymbol{x}_{\mu_N}(k+1) = \boldsymbol{f}(\boldsymbol{x}_{\mu_N}(k), \mu_N(\boldsymbol{x}_{\mu_N}(k))), \quad \boldsymbol{x}_{\mu_N}(0) = \boldsymbol{x}_0.$$

Then, the "measurement" $\hat{\boldsymbol{x}}$ at time instant $k+1$ coincides with $\boldsymbol{x}_{\mu_N}(k+1)$.

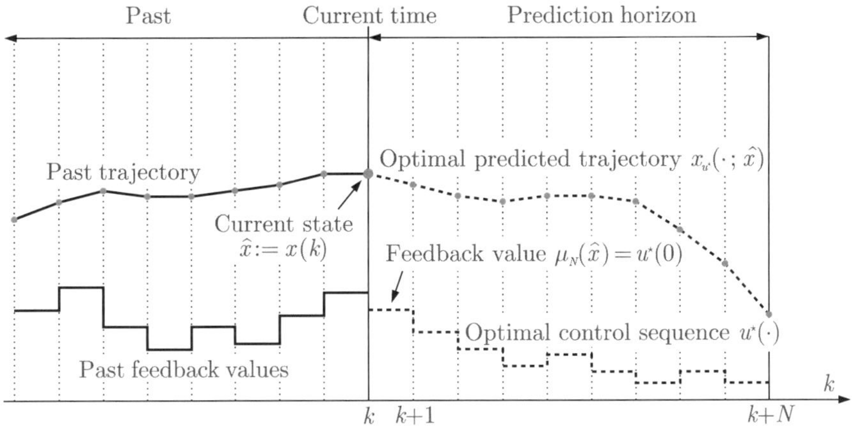

Fig. 4.3 Schematic illustration of MPC

In continuous time, when considering the continuous-time system (4.1), the counterpart to the discrete-time MPC cost functional (4.7) is given by

$$J_T(\hat{\boldsymbol{x}}, \boldsymbol{u}) := \int_0^T \ell(\boldsymbol{x}(t; \hat{\boldsymbol{x}}, \boldsymbol{u}), \boldsymbol{u}(t))\, \mathrm{d}t + E(\boldsymbol{x}(T; \hat{\boldsymbol{x}}, \boldsymbol{u})), \quad \boldsymbol{x}(0; \hat{\boldsymbol{x}}, \boldsymbol{u}) = \hat{\boldsymbol{x}}. \tag{4.9}$$

Here, the parameter T denotes the prediction horizon. Instead of the first element of a sequence of control values, the first portion $\boldsymbol{u}^\star(t)|_{t\in[0,\delta)}$ of the computed optimal control function $\boldsymbol{u}^\star \in \mathcal{U}(\hat{\boldsymbol{x}}, T)$ is implemented at the plant. Hence, the MPC feedback $\mu_T : [0, \delta) \times \mathbb{R}^n \to \mathbb{R}^m$ given by $\mu_T(t, \hat{\boldsymbol{x}}) = \boldsymbol{u}^\star(t)$ now assigns a piece of a control function to each initial value. The horizon is then shifted by δ time units. The other steps of the algorithm from above remain unchanged.

Stability Properties of the MPC Loop. The underlying idea of MPC is rather simple and intuitive. However, particular care is required in order to guarantee asymptotic stability of the closed loop when using a finite prediction horizon. This matter has been subject to extensive research in the last decade. We only give a very brief overview, restricting ourselves to the discrete-time setting and omitting technical details for the sake of a concise presentation.

Essentially, most stability proofs in MPC have in common that the optimal cost function V_N is used as a Lyapunov function. If the inequality

$$V_N(\boldsymbol{f}(\boldsymbol{x}, \mu_N(\boldsymbol{x}))) - V_N(\boldsymbol{x}) \leq -\alpha\, \ell(\boldsymbol{x}, \mu_N(\boldsymbol{x})), \tag{4.10}$$

known as a relaxed dynamic programming inequality, is satisfied for all $\boldsymbol{x} \in \mathbb{X}$ and some constant $\alpha \in \mathbb{R}_{>0}$, then asymptotic stability of the closed loop is guaranteed. The approaches available in the literature in order to guarantee (4.10) can be distinguished into two classes: while the first class of schemes uses additional (stabilizing) terminal cost terms E and/or terminal constraint sets $\mathbb{X}_f$, the so-called unconstrained MPC schemes guarantee stability without these ingredients.

In order to prove stability in the first class of schemes, usually an auxiliary local control law is employed. It is assumed that there exists a local terminal control law $\bar{\mu} : \mathbb{R}^n \to \mathbb{R}^m$ such that the following conditions hold inside the terminal region $\mathbb{X}_f \subseteq \mathbb{X}$:

- the terminal region $\mathbb{X}_f$ is control-invariant under $\bar{\mu}$, i.e., the conditions $\bar{\mu}(\boldsymbol{x}) \in \mathbb{U}$ and $f(\boldsymbol{x}, \bar{\mu}(\boldsymbol{x})) \in \mathbb{X}_f$ are satisfied for each $\boldsymbol{x} \in \mathbb{X}_f$, and
- the terminal cost E is an ℓ-conform control Lyapunov function, i.e.,

$$E(f(\boldsymbol{x}, \bar{\mu}(\boldsymbol{x}))) - E(\boldsymbol{x}) \leq -\ell(\boldsymbol{x}, \bar{\mu}(\boldsymbol{x})) \quad \forall \boldsymbol{x} \in \mathbb{X}_f. \tag{4.11}$$

Under these assumptions and initial feasibility of the finite-horizon optimal control problem (in particular satisfaction of the terminal constraint $\boldsymbol{x}_{\boldsymbol{u}}(N; \hat{\boldsymbol{x}}) \in \mathbb{X}_f$), recursive feasibility of the optimization problem and asymptotic stability of the closed loop can be proven [47, 251, 307].

In the second class of schemes, the terminal cost term E and terminal constraint set $\mathbb{X}_f$ are omitted. Instead existence of a bounded sequence $(\gamma_i)_{i \in \mathbb{N}} \subset \mathbb{R}_{\geq 1}$ is assumed such that the condition

$$V_i(\boldsymbol{x}) \leq \gamma_i V_1(\boldsymbol{x}) \quad \forall \boldsymbol{x} \in \mathbb{X} \tag{4.12}$$

holds for each $i \in \mathbb{N}$. Then, a finite prediction horizon N can be determined for which stability is guaranteed: based on (4.12), the principle of optimality, and the solution of an associated "abstract" linear program, a value of α can be found such that (4.10) holds. By choosing N large enough, it is then always possible to ensure $\alpha > 0$ and, consequently, asymptotic stability. For more details, see [154].

4.2 Compensation of Delays and Packet Loss

In networked control systems the measurement information obtained by the sensor and the input signal computed by the controller are transmitted via a communication network. As a consequence, packet loss, non-negligible delays, jitters, quantization, packet fragmentation and other problems may occur. In this section, we consider the problem of controlling nonlinear systems via non-deterministic communication networks, which can be represented in terms of random delays and packet loss (Fig. 4.4). In this situation, MPC can be used for compensating both delays and packet loss. For instance, by transmitting not only the first value of the control sequence $\boldsymbol{u}^\star$ but larger portions of this sequence, we can create a backup input in case the following transmission fails due to a packet loss. Since MPC provides a whole sequence of input values, the needed sequence is readily available without any additional computational effort. Likewise, since MPC is a prediction-based control methodology, network delays can be compensated by using the readily available prediction algorithm in the MPC scheme.

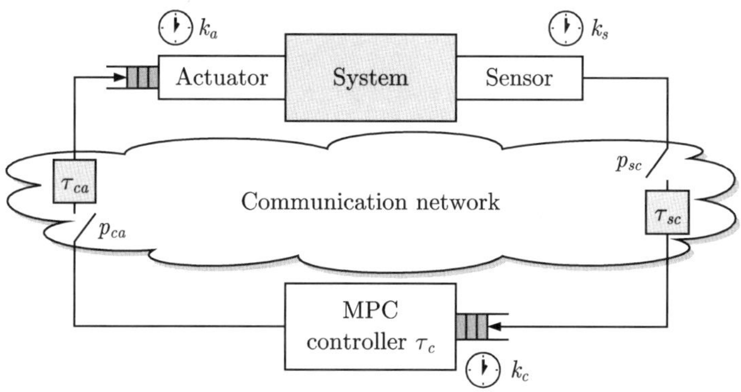

Fig. 4.4 Scheme of the networked control system

In the sequel, we first provide a precise description of the problem under investigation and then present the details of the compensation algorithm in Section 4.2.2. Moreover, we sketch how a communication protocol can be designed in order to ensure consistency of the control sequences used for predictions and the control sequences applied to the plant. We then outline how stability results for classical MPC carry over to the networked setting and present alternative stochastic and time-delay approaches to the problem.

4.2.1 Problem Setup

In the following, we consider discrete-time nonlinear systems (4.4). For continuous-time versions of analogous results we refer to [119, 161, 295, 368, 371]. In order to formalize the idea of transmitting and applying whole control sequences instead of single control values, we use the concept of multistep feedback laws. A multistep feedback law is a map $\mu_{m^\star} : \mathbb{R}^n \times \{0, 1, \ldots, m^\star - 1\} \to \mathbb{R}^m$, $(m^\star \in \mathbb{N})$ which generates an admissible control sequence of length $m^\star$, i.e.

$$\mu_{m^\star} = (\mu_{m^\star}(\boldsymbol{x}, 0), \mu_{m^\star}(\boldsymbol{x}, 1), \ldots, \mu_{m^\star}(\boldsymbol{x}, m^\star - 1)) \in \mathcal{U}(\boldsymbol{x}, m^\star)$$

depending on the state $\boldsymbol{x}$ according to (4.5).

The system is controlled through a communication network, which may be subject to both random delays and packet dropouts. The values τ_{sc}, τ_c, τ_{ca} represent the sensor-to-controller, the computational and the controller-to-actuator delay, respectively, and can be modelled deterministically by assuming knowledge of worst-case upper bounds for the delays or stochastically by assuming that they follow a known distribution. Packet dropouts can be modelled deterministically by assuming that the number of subsequent dropouts is bounded by a known number, or stochastically in terms of loss probabilities p_{sc}, p_{ca}, from the down- and up-link. Within the network, we suppose the clocks at sensor, controller and actuator to be synchronized. The clock counters are denoted by k_s, k_c and k_a, respectively (Fig. 4.4). In order to compensate for the network nondeterminism, we utilize a communication protocol which transmits time-stamped information between the network components: measurements $\boldsymbol{x}(k_s)$ are time-stamped with the instant k_s at which they are collected from the plant, whereas control sequences $\mu_{m^\star,\sigma}$ are time-stamped with $\sigma \in \mathbb{N}$ — the time at which the control inputs are supposed to be applied. For sake of simplicity, we assume that the controller computes the input sequences $\mu_{m^\star,\sigma}$ at a predefined sampling rate (typically at each sampling instant) based on the most recent measurement available in the buffer. Nevertheless, this can be extended to the event-based case, as shown for example in [368, 371].

4.2.2 Model-Based Compensation of Network Effects

The idea we propose is to exploit the plant model available at the controller side in order to compensate for uncertainties induced by the network. In other words, we use the internal model of the MPC controller to neutralize the destabilizing effects induced by delays and dropouts. The use of time stamps and the synchronization assumptions are important to establish a *common time-frame* between the controller and the actuator, as briefly explained in the following. Although restrictive, these are commonly used assumptions on networked control systems. In [370], a way to avoid the synchronization problem is presented.

Prediction Consistency. The approach we are going to sketch now is a deterministic approach, meaning that we assume knowledge of (worst-case) upper bounds for the possible delays and the maximal number of consecutive packet dropouts, as described in detail below.

For the moment, let us consider the case in which only measurement delays τ_{sc} are present. Since a model of the system is available, $\boldsymbol{x}(k_s)$ is time-stamped with k_s, and since all clocks are synchronized, we can use the system dynamics in order to compute a prediction $\tilde{\boldsymbol{x}}(k_s + \tau_{sc})$ of the current state $\boldsymbol{x}(k_c)$ based on the (now) old measurement $\boldsymbol{x}(k_s)$. The predicted state can then be used to solve the optimal control problem in the MPC algorithm introduced in Section 4.1.

However, to predict the current state, the input sequence needs to be known. Thus, a buffer at the controller side is required to store this information. As long as we suppose that no computation and actuation delays are present, the controller knows that the calculated input sequence is immediately applied. In the presence of computation and actuation delays, however, we do not know in advance when the input sequence is applied. Hence, we need to ensure that the input used for prediction coincides with the input applied at the plant. Formally, this requirement is stated in the following definition.

Definition 4.1 (Prediction consistency).
(i) A feedback control $\mu_{m^\star,\sigma}(\cdot, \tilde{\boldsymbol{x}}(\sigma))$ *is called consistently predicted if the control sequence* $\tilde{\boldsymbol{u}}(k_s), \tilde{\boldsymbol{u}}(k_s+1), \ldots, \tilde{\boldsymbol{u}}(\sigma-1)$ *used for the prediction of* $\tilde{\boldsymbol{x}}(\sigma)$ *in the controller is identical to the control sequence* $\boldsymbol{u}(k_a), \boldsymbol{u}(k_a+1), \ldots, \boldsymbol{u}(\sigma-1)$ *applied by the actuator.*
(ii) A networked control scheme is called prediction consistent if at each time instant $k \in \mathbb{N}_0$ *the applied control* $\mu_{m^\star,\sigma}$ *in the actuator is well defined, i.e.* $k \leq \sigma + m^\star - 1$, *and if* $\mu_{m^\star,\sigma}$ *is consistently predicted.*

The concept of prediction consistency allows to separate the *analysis of the feedback law*, e.g. in terms of stability and performance, from the *influence of the network* on the closed loop. More precisely, assume that a controller which generates admissible input sequences $\mu_{m^\star,\sigma}(\cdot, \tilde{\boldsymbol{x}}(\sigma))$ is given. This may be an MPC controller but the concept in fact applies to general types of controllers. Assume, furthermore, that the controller stabilizes the system regardless of how many elements of each sequence $\mu_{m^\star,\sigma}(\cdot, \tilde{\boldsymbol{x}}(\sigma))$ are applied (this property will be made precise in Section 4.2.3). Then its implementation via a prediction-consistent network (assuming that no prediction errors occur) will result in a stable networked closed-loop system. The reason for this is that in the absence of prediction errors, prediction consistency guarantees that the predicted state coincides with the exact state, hence the feedback sequences are computed based on the exact information about the system state. The influence of prediction errors and external disturbances can then be analyzed independently as, e.g., in [120].

Overview of Compensation Schemes. Examples of prediction-consistent network architectures can be found, for example, in [25, 156, 295, 297]. In [119, 156, 368], prediction consistency under unknown delays is obtained in terms of *worst-case compensation*. This means that the time stamp σ in the computation of $\mu_{m^\star,\sigma}$ is chosen such that the control sequence $\mu_{m^\star,\sigma}$ arrives at the actuator buffer at a time $k_a \leq \sigma$. Since the delays τ_c and τ_{ca} are not know in advance, bounds ${\tau_c}^{\max}$ and ${\tau_{ca}}^{\max}$ are imposed, after which the input packet is considered as lost. Consequently, $\sigma = k_s + \tau_{sc} + {\tau_c}^{\max} + {\tau_{ca}}^{\max}$ holds (Fig. 4.5).

As presented for instance in [25, 119, 156, 295, 297, 356, 368], we can use the plant model available at the controller side to predict the state of the system at time σ. This value is then used to solve the optimization problem. In order to compensate for packet loss, the complete input $\mu_{m^\star,\sigma}$ is dispatched to the actuator and used in case that subsequent control sequences do not arrive at the actuator due to packet loss [156, 368, 371]. To this end, a buffer at the actuator side is required to store this information and to use it at the correct

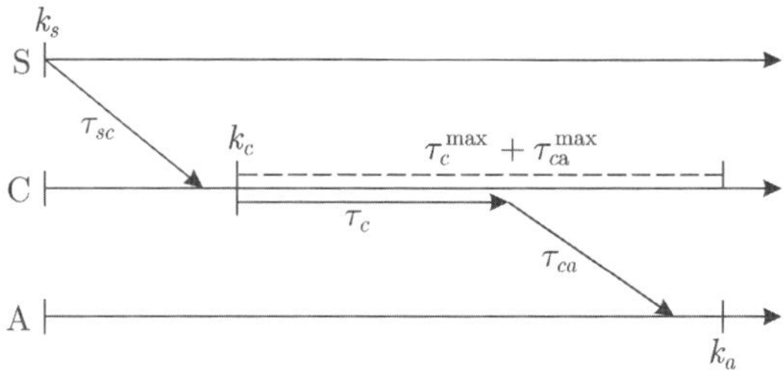

Fig. 4.5 Timeline connections within the scheme: S – sensor, C – controller, A – actuator

moment. Ensuring prediction consistency requires some additional assumptions or mechanisms in this case. In [368, 371], it is assumed that acknowledgments are available at the actuator side. The use of acknow-ledgments makes the applied input deterministically known at the controller side, thus ensuring prediction consistency. Alternatively, this can also be guaranteed by sending error messages in case of packet dropouts which allows for updating the control sequences more frequently if the network works properly [156].

Another strategy which does also not make use of acknowledgments has been introduced in [369]. Note that the maximum number of consecutive packet losses which can be compensated by such mechanisms depends on various parameters, including the delay of the acknowledgement or error message (cf. [156] for a detailed analysis).

Notice that MPC and model-based compensation fit perfectly together. In fact, a model of the plant is already available, i.e. the prediction of the state measurement $\tilde{\boldsymbol{x}}(\sigma)$ can be done easily by evaluating the system dynamics (4.4) using the control values stored in the controller buffer. Additionally, the outcome of the MPC algorithm is already an optimal control sequence defined on a control horizon of length N. Hence, for $m^\star \leq N$ a control signal of length $m^\star$ is readily available for transmission and it can be used to compensate for input packet dropouts.

4.2.3 Stability Properties of Model-Based Compensation

In case that one of the transmitted control sequences $\mu_{m^\star,\sigma}$ does not arrive or arrives too late at the actuator, the control values in the control sequence transmitted earlier are used to provide the actuator with current information. Since delays and packet loss vary depending on the current network load, this means that the length $m \in \{1, 2, \ldots, m^\star\}$ of the portion of each $\mu_{m^\star,\sigma}$ used at the actuator may vary with time. In principle, one could make this value independent of time by always using $m = m^\star$ values, however, for robustness reasons it is desirable to always use the most recent control input, i.e. to keep m as small as possible [120]. Hence, time-varying control horizons m

should be considered and thus our goal is to find a condition which ensures asymptotic stability of the MPC closed loop in this situation. To this end, we aim at employing the value function $V_N(\cdot)$ as a common Lyapunov function and show the relaxed Lyapunov inequality

$$V_N(\boldsymbol{x}_{\mu_{N,m^\star}}(m;\boldsymbol{x})) \leq V_N(\boldsymbol{x}) - \alpha \sum_{k=0}^{m-1} \ell(\boldsymbol{x}_{\mu_{N,m^\star}}(k;\boldsymbol{x}), \mu_{N,m^\star}(k,\boldsymbol{x})) \quad (4.13)$$

with $\alpha \in (0,1]$ for each feasible state $\boldsymbol{x} \in \mathbb{X}$ and all $m \in \{1,2,\ldots,m^\star\}$ [146, 233]. Here, $\mu_{N,m^\star}$ denotes the sequence consisting of the first $m^\star \in \{1,2,\ldots,N\}$ elements of the computed control $\boldsymbol{u}^\star \in \mathcal{U}(\boldsymbol{x},N)$ in step 1 of Algorithm 4.1. Assume that in addition to (4.13) the inequalities

$$\eta_1(\|\boldsymbol{x}-\boldsymbol{x}^\star\|) \leq V_1(\boldsymbol{x}) \quad \text{and} \quad V_N(\boldsymbol{x}) \leq \eta_2(\|\boldsymbol{x}-\boldsymbol{x}^\star\|) \quad (4.14)$$

hold for all $\boldsymbol{x} \in \mathbb{X}$ with $\mathcal{K}_\infty$-functions η_1, η_2. Then asymptotic stability of the MPC loop can be ensured [157, 392].

To this end, we assume that, for each feasible state $\boldsymbol{x} \in \mathbb{X}$, a control sequence $\boldsymbol{u} \in \mathcal{U}(\boldsymbol{x},\infty)$ exists such that the stage costs along the respective trajectory are exponentially decaying, i.e. a pair (C,ζ) consisting of the overshoot bound $C \in \mathbb{R}_{\geq 1}$ and the decay rate $\zeta \in (0,1)$ exists satisfying

$$\ell(\boldsymbol{x}_{\boldsymbol{u}}(k;\boldsymbol{x}), \boldsymbol{u}(k)) \leq C\zeta^k \min_{\boldsymbol{u} \in \mathcal{U}(\boldsymbol{x},1)} \ell(\boldsymbol{x},\boldsymbol{u}) =: C\zeta^k \ell^\star(\boldsymbol{x}) \quad \forall k \in \mathbb{N}_0. \quad (4.15)$$

Note that this condition implies (4.12) with $\gamma_i := C\sum_{n=0}^{i-1}\zeta^n$ [364]. Then, proceeding analogously to the indicated proof based on (4.12) yields (4.13) with suboptimality degree

$$\alpha_{N,m} := 1 - \frac{\prod_{i=m+1}^{N}(\gamma_i - 1)\prod_{i=N-m+1}^{N}(\gamma_i - 1)}{\left(\prod\limits_{i=m+1}^{N}\gamma_i - \prod\limits_{i=m+1}^{N}(\gamma_i - 1)\right)\left(\prod\limits_{i=N-m+1}^{N}\gamma_i - \prod\limits_{i=N-m+1}^{N}(\gamma_i - 1)\right)}, \quad (4.16)$$

cf. [160, 392] for a proof. Hence, $\alpha_{N,m} > 0$, $(m = 1,2,\ldots,N-1)$ for a given horizon N is a sufficient condition for stability of the networked closed loop. One can show that this condition is always ensured for sufficiently large prediction horizon N [160]. Additionally, one can prove that the infinite horizon costs along the closed–loop trajectory are bounded by $\alpha_{N,m}^{-1}V_\infty(\boldsymbol{x})$, where $V_\infty(\boldsymbol{x})$ is the optimal infinite horizon value for initial value $\boldsymbol{x}$. Hence, $\alpha_{N,m}$ can be interpreted as a performance bound. A comparative study carried out in [392, 393] shows that this bound is tighter than other bounds derived earlier in [142, 364]. Furthermore, the values $\alpha_{N,m}$ satisfy the inequality $\alpha_{N,1} \leq \alpha_{N,m}$, $(m \in \{1,2,\ldots,N-1\})$ which shows that $\alpha_{N,1} > 0$ implies $\alpha_{N,m} > 0$ for all $m \in \{1,2,\ldots,N-1\}$ and thus asymptotic stability of the MPC loop despite time-varying control horizons m. Hence, the stability criterion for $m = 1$ implies the stability criterion for arbitrary time-varying horizons m. As a conclusion, packet dropouts or non-negligible delays which typically occur in networked control systems can be dealt with rigorously [157].

We point out that these results can be transferred to a continuous-time setting. Here, the suboptimality degree was introduced in [309] and the required additional properties were rigorously proven in [161].

Stochastic Stability of the MPC Loop. All schemes discussed so far are deterministic in the sense that they require the knowledge of worst-case upper bounds for the unknown delays and the unknown number of consecutive packet losses in the network. In many realistic scenarios, however, it is not possible to ensure such hard deterministic upper bounds. In this case stochastic stability properties and stochastic network models can be employed. Frequently used network models include independent and identically distributed (i.i.d.) packet dropouts and the Gilbert-Elliot channel model, which takes the temporal correlation of the network congestion level into account and is described by a two-state Markov chain model $d(k)$

$$d(k) = \begin{cases} 1 & \text{packet dropout occurs at time } k \\ 0 & \text{packet successfully transmitted at time } k \end{cases}$$

with transition probabilities

$$\begin{aligned} &\text{Prob}(d(k+1) = 0 \,|\, d(k) = 0) = q\,, \qquad &&\text{Prob}(d(k+1) = 1 \,|\, d(k) = 0) = 1 - q\,, \\ &\text{Prob}(d(k+1) = 1 \,|\, d(k) = 1) = p\,, \qquad &&\text{Prob}(d(k+1) = 0 \,|\, d(k) = 1) = 1 - p \end{aligned}$$

(Fig. 4.6).

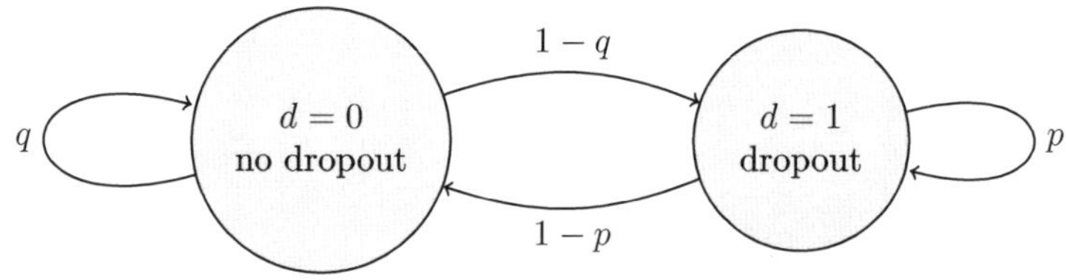

Fig. 4.6 Schematic of the Markov packet dropout model

The simpler model of a constant dropout rate $0 < p < 1$, i.e., the random variables $d(k)$ are considered as i.i.d. Bernoulli variables with

$$\text{Prob}(d(k) = 1) = p\,, \quad \text{Prob}(d(k) = 0) = 1 - p$$

is included in the two-state Markov chain model as a special case when $q = 1 - p$.

For these models, the assumption of a sufficiently long prediction horizon such that an admissible input is always available in the buffer/actuator (which is needed to ensure the existence of a well-defined control in the actuator buffer as required in Def. 4.1(ii)) cannot be satisfied. Consequently, deterministic stability cannot be guaranteed because the time between two successful transmission times has unbounded support. However, stochastic stability guarantees can be derived for stochastic stability and asymptotic mean-square stability, i.e.,

$$\mathrm{E}\left[\sum_{k=0}^{\infty} |\boldsymbol{x}(k)|^2 \Big| \boldsymbol{x}_0\right] < \infty \quad \text{and} \quad \lim_{k\to\infty} \mathrm{E}\left[|\boldsymbol{x}(k)|^2 \big| \boldsymbol{x}_0\right] = 0\,.$$

An MPC scheme using a global control Lyapunov function as terminal cost has been previously presented in [301]. A significantly less restrictive scheme has been derived in [316, 317] without terminal cost terms. It relies on a controllability assumption similar to (4.12) instead of a global control Lyapunov function, and is, therefore, readily satisfied if a global control Lyapunov function is available. Based on the suboptimality degree $\alpha_{N,1}$ from (4.16) for systems without packet dropouts and the dropout parameters p and q, explicit conditions can be derived for stochastic stability guarantees and stochastic performance estimates in terms of the expected performance. These performance estimates rigorously underpin two (intuitively expected) results: first, model-based compensation and buffering strategies improve stability and performance of the closed loop; and second, performance arbitrarily close to the infinite horizon optimal one can be achieved if the prediction horizon as well as the buffer size are chosen sufficiently large.

4.2.4 Outlook to Time-Delay Systems

In some situations, an alternative approach for delay compensation is to directly consider time-delay systems described by the functional differential equation

$$\dot{\boldsymbol{x}}(t) = \boldsymbol{f}(\boldsymbol{x}(t), \boldsymbol{x}(t-\tau), \boldsymbol{u}(t)), \quad \boldsymbol{x}(\theta) = \boldsymbol{x}_0(\theta) \;\; \forall \theta \in [-\tau, 0]$$

with $\boldsymbol{f} : \mathbb{R}^n \times \mathbb{R}^n \times \mathbb{R}^m \to \mathbb{R}^n$. Such models arise, for instance, when considering a hierarchial structure in which the underlying local control loops are subject to computation or transmission delays.

The main difficulty in the analysis of MPC schemes for this class of systems stems from the infinite-dimensional nature of these systems. However, it is possible to extend the approaches for finite-dimensional systems, (cf. Section 4.1), to this class of systems using mild additional assumptions. Stability conditions for MPC schemes with terminal constraints have been presented in [252, 314] and MPC schemes without terminal constraints are developed in [313, 315].

Another difficulty is encountered when calculating the stabilizing design parameters for these MPC schemes, in particular a positively invariant terminal region and a terminal cost functional. Several approaches exist, taking Lyapunov-Krasovskii and Lyapunov-Razumikhin arguments into account [310–313]. All of these schemes naturally recover well-known results for finite-dimensional systems, but have different properties, advantages and, drawbacks in the infinite-dimensional case [308].

4.3 Decentralized and Distributed Control Algorithms

In this section, we develop several algorithms for decentralized and distributed MPC. The interest in such kind of algorithms is due to the fact that an overall controller may be hard to design for large-scale systems (cf. [245, 329, 339] for an overview). Also, physical couplings between the subsystems in general cannot be simply neglected since they can lead to instability of the overall system. Furthermore, in the setting of several independent systems pursuing a cooperative control task, such as a team of mobile robots or satellites, each system can only communicate with some neighboring systems and compute a local control input, while still some overall cooperative goal shall be achieved.

In the following, we develop three different approaches to this kind of problems. In Section 4.3.1, dissipativity-based distributed MPC is developed for systems which are physically coupled. Sections 4.3.2 and 4.3.3 treat the case of several independent systems coupled via constraints or a common objective. Hereby, Section 4.3.2 focuses on achieving stability in distributed MPC without stabilizing terminal constraints, while in Section 4.3.3 more general control tasks such as consensus and synchronization problems are considered.

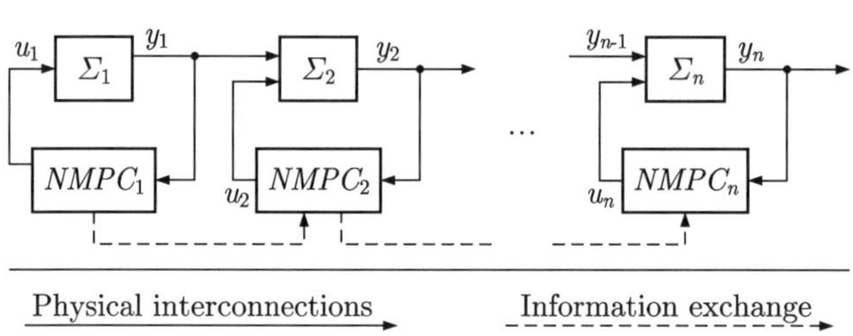

Fig. 4.7 Cascade of physically interconnected systems

4.3.1 Dissipativity-Based Distributed MPC

Consider a cascade of P physically interconnected systems (Fig. 4.7), e.g., cascade of batteries or hydroelectric plants. The physical interconnections are treated as external disturbances, which are compensated by means of predictive control. In particular, we focus on nonlinear large-scale systems representable as the cascade of $P \geq 2$ nonlinear continuous-time input affine subsystems of the form

$$\Sigma_i : \begin{cases} \dot{\boldsymbol{x}}_i(t) = \boldsymbol{f}_i(\boldsymbol{x}_i(t)) + \boldsymbol{h}_i(\boldsymbol{x}_i(t))\boldsymbol{u}_i(t) + \boldsymbol{y}_{i-1}(t) \\ \boldsymbol{y}_i(t) = \boldsymbol{g}_i(\boldsymbol{x}_i(t), \boldsymbol{u}_i(t)) \end{cases} \tag{4.17}$$

$(i = 1, 2, ..., N)$, where $\boldsymbol{x}_i(t) \in \mathbb{R}^{n-i}$, $\boldsymbol{u}_i(t) \in \mathbb{R}^{n-i}$, $\boldsymbol{y}_i(t) \in \mathbb{R}^{n-i}$, represent respectively the state, the input and the output of the ith system Σ_i. The interactions between the subsystems are described by means of $\boldsymbol{y}_{i-1}(t)$, i.e., each subsystem has a direct effect only on the following one. We assume that $\boldsymbol{f}_i(\cdot)$, $\boldsymbol{g}_i(\cdot)$, $\boldsymbol{h}_i(\cdot)$ are Lipschitz continuous, the origins of the subsystems are equilibrium points and that they are zero-state-detectable (ZSD) i.e.,

$$\lim_{t \to \infty} \|\boldsymbol{x}_i(t)\| = 0, \text{ for } \boldsymbol{y}_i(t) = 0, \quad \forall t \geq 0.$$

Additionally, we furthermore assume that every subsystem is *input-feedforward output-feedback passive*, IF-OFP(ν_i, ρ_i) w.r.t. the input $\boldsymbol{u}_i(t)$ and the output $\boldsymbol{y}_i(t)$ (cf. [399, 400] for details). The objective is to stabilize the origin of the overall system by developing local distributed controllers, which act on the subsystems Σ_i.

The main problems are due to the physical interconnections, which can lead to poor performance or instability. Normally these couplings are considered as unknown disturbances and robust solutions, e.g., min − max, max − min, invariant sets (or tubes), are used [217, 237, 245, 373]. These methods are, however, conservative, since they typically do not make explicitly use of all available information, but only consider the worst realization of the interconnections. Additionally, they are computationally challenging even for linear systems, thus difficult to apply in practice.

What we propose is to exploit the cascade structure of the problem, and use local MPC controllers to enforce local dissipativity by means of an additional dissipativity constraint. The regulators exchange (predicted) information to improve performance and guarantee overall stability.

The results we present here are based on previous work on passivity-based nonlinear MPC described in [304, 305]. In particular, we modify the original optimization problem by introducing a dissipativity constraint, whose goal is to ensure local dissipativity w.r.t. all inputs (control variables and couplings) and the corresponding output. The modified problem is then given by

$$\min_{\boldsymbol{u}_i} \int_{t_k}^{t_k+T} \ell_i(\boldsymbol{x}_i(t; \hat{\boldsymbol{x}}_i, \boldsymbol{u}_i), \boldsymbol{u}_i(t))\, \mathrm{d}t + E_i(\boldsymbol{x}_i(t_k + T; \hat{\boldsymbol{x}}_i, \boldsymbol{u}_i)) \tag{4.18}$$

subject to the dynamics (4.17) with $\hat{\boldsymbol{x}}_i = \hat{\boldsymbol{x}}_i(t_k)$, the terminal constraint $\boldsymbol{x}_i(t_k + T; \hat{\boldsymbol{x}}_i, \boldsymbol{u}_i) \in \mathbb{X}_f$, and the dissipativity condition

$$\tilde{\boldsymbol{u}}_i(t)^{\mathrm{T}} \tilde{\boldsymbol{y}}_i(t) - \nu_i \tilde{\boldsymbol{u}}_i(t)^{\mathrm{T}} \tilde{\boldsymbol{u}}_i(t) - \rho_i \tilde{\boldsymbol{y}}_i(t)^{\mathrm{T}} \tilde{\boldsymbol{y}}_i(t) \leq 0. \tag{4.19}$$

The input and output vectors

$$\begin{aligned} \tilde{\boldsymbol{u}}_i(t) &:= (\boldsymbol{y}_{i-1}^{\mathrm{T}}(t), \boldsymbol{u}_i^{\mathrm{T}}(t))^{\mathrm{T}} \\ \tilde{\boldsymbol{y}}_i(t) &:= (\boldsymbol{y}_{i-1}^{\mathrm{T}}(t), \boldsymbol{y}_i^{\mathrm{T}}(t))^{\mathrm{T}} \end{aligned}$$

are obtained by including the external effects of the previous system into the current one. Notice that, differently from the parallel-system case, the system cascade can lose passivity and asymptotic stability, even though the subsystems are not. Thus, additional conditions need to be verified to ensure closed-loop stability.

We use a *quasi-dominant condition*, as defined in [196], to further ensure that the physical couplings are not destabilizing. Finally, in order to solve (4.18) knowledge about the physical interconnection $\boldsymbol{y}_{i-1}(t)$ is required. To cope with this, we exploit the system structure and propagate the predicted information, sequentially, to the following controllers. In other words, considering the fact that the first controller does not require any other information to solve its own optimization problem, and, at the same time, it calculates the predicted output $\boldsymbol{y}_1(t)$ for $t \in [t_k, t_k + T)$, then we can hand over this piece of information to the following controller and repeat the process throughout the complete cascade sequence. The overall procedure is summarized in Algorithm 4.2. This algorithm has to be implemented in the local controllers of all the P subsystems.

Algorithm 4.2. Distributed dissipativity-based MPC for cascade systems

---

Given: Model of the ith subsystem Σ_i
Init: New measurement $\hat{\boldsymbol{x}}_i(t_k)$
$i = 1$
while $i <= P$ **do**
  Initialize the optimal control problem (4.18) with $\hat{\boldsymbol{x}}_i(t_k)$
  Solve the optimal control problem (4.18)
  Store $\boldsymbol{y}_i(t)$
  Apply $\boldsymbol{u}_i^*(t)$, for $t \in [t_k, t_{k+1})$
  if $i < P$ **then**
    Forward $\boldsymbol{y}_i(t)$ to system Σ_{i+1}
  end if
  $i = i + 1$
end while

Under mild conditions we can then derive the following results.

Theorem 4.1 (Distributed dissipativity-based MPC). *Consider a cascade of $P \geq 2$ IF-OFP(ν_i,ρ_i) subsystems Σ_i, with ν_i, $\rho_i \in \mathbb{R}$, in a closed loop with the optimal control input $\boldsymbol{u}_i^*(t)$ obtained from the solution of the local optimization problem* (4.18). *If*

- *there exist $\hat{\nu}$, $\hat{\rho} \in \mathbb{R}$ such that the matrix*

$$\boldsymbol{A} = \begin{pmatrix} \boldsymbol{A}_{uu} & \frac{1}{2}\boldsymbol{A}_{uy} \\ \frac{1}{2}\boldsymbol{A}_{uy}^{\mathrm{T}} & \boldsymbol{A}_{yy} \end{pmatrix} \quad \textit{with} \tag{4.20}$$

$$\boldsymbol{A}_{uu} = \begin{pmatrix} \nu_1 - \hat{\nu} & 0 & \cdots & 0 \\ 0 & \nu_2 & \ddots & \vdots \\ \vdots & \ddots & \ddots & 0 \\ 0 & \cdots & 0 & \nu_P \end{pmatrix}, \quad \boldsymbol{A}_{uy} = \begin{pmatrix} -1 & 0 & \cdots & 1 \\ 0 & \ddots & \ddots & \vdots \\ \vdots & \ddots & \ddots & 0 \\ 0 & \cdots & 0 & -1 \end{pmatrix},$$

$$\boldsymbol{A}_{yy} = \begin{pmatrix} \rho_1 + \rho_2 + \nu_2 - 1 & 0 & \cdots & 0 \\ 0 & \ddots & \ddots & \vdots \\ \vdots & \ddots & \rho_{P-1} + \rho_P + \nu_P - 1 & 0 \\ 0 & \cdots & 0 & \rho_P - \hat{\rho} \end{pmatrix}$$

  is quasi-dominant
- *there exists a time $t_0 \in \mathbb{R}_{\geq 0}$ such that the local optimization problems* (4.18) *with continuously differentiable $V_i(\boldsymbol{x}_i(t))$ are feasible*

then,

i) the overall cascade system is IF-OFP($\hat{\nu}$,$\hat{\rho}$), $\hat{\nu}$, $\hat{\rho} \in \mathbb{R}$, i.e., there exists a continuously differentiable storage function V, $V \geq 0$, for the overall cascade such that

$$\dot{V} \leq \boldsymbol{u}_1^{\mathrm{T}}\boldsymbol{y}_P - \hat{\nu}\boldsymbol{u}_1^{\mathrm{T}}\boldsymbol{u}_1 - \hat{\rho}\boldsymbol{y}_P^{\mathrm{T}}\boldsymbol{y}_P \tag{4.21}$$

ii) and $\lim_{t\to\infty} \|\boldsymbol{x}_i(t)\| \to 0$ holds for all $i \in \{1, \ldots, P\}$.

Proof. Based on the use of the extended inputs/outputs, the dissipativity condition and (4.20), recursive feasibility and convergence are proved. We refer to [367, 372] for the complete proof. □

To show the effectiveness of our approach, we consider the following example:

Example 4.1 *Cascade of two tank systems Σ_1 and Σ_2*

A cascade connection of two tanks is represented by

$$\Sigma : \begin{cases} \Sigma_1 : \begin{cases} \dot{x}_1(t) = -\frac{a_1}{A_1}\sqrt{2gx_1(t)} + \frac{1}{A_1}u_1(t), \\ y_1(t) = x_1(t) + u_1(t), \end{cases} \\ \\ \Sigma_2 : \begin{cases} \dot{x}_2(t) = -\frac{a_2}{A_2}\sqrt{2gx_2(t)} + \frac{1}{A_2}u_2(t) + \frac{a_1}{A_2}y_1(t), \\ y_2(t) = x_2(t) + 2u_2(t), \end{cases} \end{cases} \tag{4.22}$$

where a_i, A_i, g, are positive parameters. The scalars $x_i \in \mathbb{R}_{\geq 0}$, $u_i \in \mathbb{R}_{\geq 0}$, $y_i \in \mathbb{R}_{\geq 0}$ are the state, the input and the output, respectively. In particular, A_i represent the tank cross sections; a_i are the sections of the outgoing pipes; and $g = 981\ \frac{cm}{s^2}$ is the gravitational constant. The interconnection between the two tanks is given by the term $\frac{a_1}{A_2}y_1(t)$.

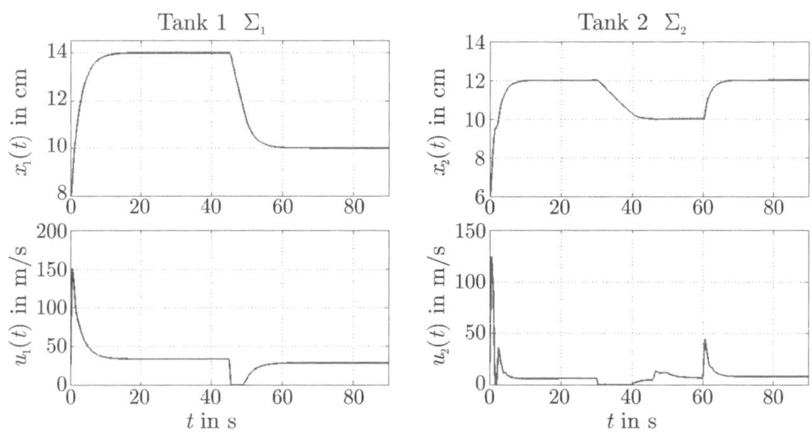

Fig. 4.8 Simulation results for the two-tank system

It is easy to show that both systems are IF-OFP(1,0) and IF-OFP(2,0), respectively, thus the overall cascade is IF-OFP($\hat{\nu}$, $\hat{\rho}$), $\hat{\nu} \leq 0$ with $\hat{\rho} \leq 1$.

In Fig. 4.8, we show the simulation results for $A_1 = 50\,\text{cm}^2$, $A_2 = 30\,\text{cm}^2$, $a_1 = 0.2\,\text{cm}^2$, $a_2 = 0.1\,\text{cm}^2$. The two tanks are required to follow precise references even in presence of the physical interconnections. The distributed controllers behave qualitatively well and easily steer the tanks to the desired set-point in a short time. □

Although for such a small cascade the benefits of distributing the computation might not be obvious, if the number of systems is large, finding a centralized solution would result in an untreatable problem. In such case, solving P small problems is preferable.

4.3.2 Decentralized MPC without Terminal Constraints

The second method deals with subsystems which are coupled via state constraints. The predicted subsystem trajectories are communicated to the other controllers in order to avoid conflicts, like collisions of vehicles. A central goal is to determine sufficient controllability conditions on the subsystems under which the stability of the proposed schemes can be guaranteed without imposing stabilizing terminal constraints.

We consider $P \in \mathbb{N}$ control systems described by the discrete-time dynamics

$$\Sigma_i : \ \boldsymbol{x}_i(k+1) = \boldsymbol{f}_i(\boldsymbol{x}_i(k), \boldsymbol{u}_i(k)) \tag{4.23}$$

with $\boldsymbol{x}_i(k) \in \mathbb{R}^n$, $\boldsymbol{u}_i(k) \in \mathbb{R}^m$, and $\boldsymbol{f}_i : \mathbb{R}^n \times \mathbb{R}^m \to \mathbb{R}^n$, $(i = 1, 2, \ldots, P)$. For a given control sequence $(\boldsymbol{u}_i(k))_{k \in \mathbb{N}_0}$ the corresponding solution emanating from the initial state $\boldsymbol{x}_i(0)$ is denoted by $\boldsymbol{x}_i^{\boldsymbol{u}}(k; \boldsymbol{x}_i(0))$, $(k \in \mathbb{N}_0)$ i.e., the subscript in $\boldsymbol{u}_i$ is omitted in order to simplify the notation. Our goal is to stabilize, for each subsystem, a desired equilibrium $\boldsymbol{x}_i^\star \in \mathbb{R}^n$. Beyond ensuring stability, we want to design the controllers such that the combined state $\boldsymbol{x}(k) = (\boldsymbol{x}_1^{\mathrm{T}}(k), \boldsymbol{x}_2^{\mathrm{T}}(k), \ldots, \boldsymbol{x}_P^{\mathrm{T}}(k))^{\mathrm{T}}$ satisfies state constraints $\boldsymbol{x}(k) \in \mathbb{X} \subseteq (\mathbb{R}^n)^P$. The coupling induced by these state constraints requires an information exchange between the otherwise independent systems since checking whether a control sequence is admissible or not requires, in general, knowledge of the states of the other subsystems. Here, we assume that this information is transmitted via a network with negligible delay.

We propose to use a sequential MPC algorithm similar to that introduced in [320, 321]. Since in MPC the future behavior is predicted for the next N time instants and, then, optimized with respect to (all) admissible control sequences, information on the predictions of the other subsystems is required. However, since we are interested in a sequential scheme in order to avoid conflicts in the computation of the individual predictions and, thus, to satisfy the imposed constraints, we may not have predictions for each subsystem for each instant on the horizon to be taken into account. Hence, the optimization has to take place based on the most recent received information. To this end, we require the concepts of partial states and partial state constraints which, roughly speaking, only check whether the state constraints can be satisfied based on the available information.

For instance, if at time k no information on the system Σ_i is available, we only check whether there exists a state of the system Σ_i such that the corresponding state satisfies the constraints. The packet transmitted to the other systems contain both a prediction and the corresponding initial time. If this initial time k_i dates back to the past, only $N - (k - k_i)$ future states can be taken into account in the MPC step at time k. As a consequence, the minimization in the first step of the MPC algorithm is performed over those trajectories which satisfy the state constraints together with the known predictions from the other systems. Denoting the available information in the subsystem Σ_i at time k by $I_i(k)$, the resulting feedback thus depends on the initial state and on the set $I_i(k)$. Combined, we obtain Algorithm 4.3.

Algorithm 4.3. Algorithm of RICHARDS and HOW

Given: Let $(\boldsymbol{x}_1(0), \ldots, \boldsymbol{x}_P(0)) \in \mathbb{X}$ be given initial values.

Initialization for $k = 0$**:**

Find admissible control sequences $\boldsymbol{u}_i$ such that the corresponding trajectories satisfy

$$(\boldsymbol{x}_1^{\boldsymbol{u}}(j; \boldsymbol{x}_1(0)), \boldsymbol{x}_2^{\boldsymbol{u}}(j; \boldsymbol{x}_2(0)), \ldots, \boldsymbol{x}_P^{\boldsymbol{u}}(j, \boldsymbol{x}_P(0))) \in \mathbb{X} \quad \text{for } j = 0, \ldots, N-1.$$

for $i = 1$ **to** P **do**

Set $k_i = 0$, $\boldsymbol{x}_i^{k_i}(j) = \boldsymbol{x}_i^{\boldsymbol{u}}(j)$ for $j = 0, 1, \ldots, N-1$ and send $(k_i, \boldsymbol{x}_i^{k_i}(\cdot))$ to all other systems.

Apply the control value $\mu_i(\boldsymbol{x}_i(0)) = \boldsymbol{u}_i(0)$ in the first step.

end for

Control loop for $k \geq 1$**:**

for $k = 1$ **to** ∞ **do**

for $i = 1$ **to** P **do**

Solve the optimal control problem based on the most recently received information, denote the resulting optimal control by $\boldsymbol{u}_i^{\star,k}$, set

$$k_i = k \quad \text{and} \quad \boldsymbol{x}_i^{k_i}(j) = \boldsymbol{x}_i^{u^{\star,k}}(j)$$

for $j = 0, \ldots, N-1$ and send $(k_i, \boldsymbol{x}_i^{k_i}(\cdot))$ to all other systems.

Apply the control value $\mu_i(\boldsymbol{x}_i(0), I_i(k)) = \boldsymbol{u}_i^{*,k}(0)$ in the k-th step.

end for

end for

System Σ_i always uses the most recently available predictions of the other systems Σ_i in order to construct the set of admissible control sequences, i.e., for Σ_q, $(q < i)$ the predictions $\boldsymbol{x}_q^k$ made at time k are used and for $q > i$ the predictions $\boldsymbol{x}_q^{k-1}$ computed at time instant $k-1$ are used in $I_i(k)$. In case of a large number P of systems this sequential optimization may cause rather long waiting times which may not be available in case of fast sampling. Possible relaxations of the strict sequential structure are subject to future research. Yet, we like to note that the scheme is not only applicable for small values of P but, as pointed out in Section 7 of [321], even for large P considerably reduces the numerical effort compared to a centralized solution of the optimization problem.

Feasibility of the closed loop is ensured if a feasible solution can be obtained in the initialization phase of the algorithm and, additionally, an admissible control exists for each system and each step of the control loop. While the first part of this assumption is tractable, the second part sounds hard to verify. Yet, it may actually be guaranteed by using terminal constraints. In this case, also stability of the corresponding closed loop can be concluded under mild technical assumptions. In turn, stabilizing terminal constraints require collision free trajectories for all systems from the initial value $\boldsymbol{x}_i(0)$ to a neighborhood $\mathcal{T}_i$ of $\boldsymbol{x}_i^\star$ even in the initialization phase. The drawback of this approach is twofold: For one, the size of the optimization horizons N directly depends on the initial conditions $\boldsymbol{x}_i(0)$, i.e., N has to be large if $\boldsymbol{x}_i(0)$ is far away from the terminal sets $\mathcal{T}_i$. And secondly, resolving all conflicts in the initialization step, that is, all possible collisions up to the point where each agent reaches its "safe" terminal constraint set $\mathcal{T}_i$, may be very hard to satisfy especially in the distributed setting.

Our approach in this context is to relax these drawbacks by dropping the requirement of stabilizing terminal constraints. Regarding feasibility in the initialization phase of Algorithm 4.3, we obtain that this requirement remains unchanged. However, without an additional terminal condition the required minimal optimization horizon N is typically smaller and hence finding an initially feasible solution is a much simpler task. This is illustrated in the numerical simulations of Example 4.2, in which the state constraints only become relevant after several steps of the algorithm. Considering the control loop, we want to use a controllability assumption similar to (4.15) in order to guarantee stability of the closed-loop. In contrast to the non-distributed setting, the set of admissible controls is time-dependent due to the fact that the information $I_i(k)$ changes with time. Hence, given the predictions of the other systems $\boldsymbol{x}_q^{\boldsymbol{u}}$, $(q \neq i)$, loosely speaking we require that there is still enough space to "squeeze in" a solution $\boldsymbol{x}_i^{\boldsymbol{u}}$. In addition, we need that starting from any point on the optimal open-loop trajectory from the last time instant, there exist solutions which approach the equilibrium $\boldsymbol{x}_i^\star$ sufficiently fast in the sense of the respective controllability assumption (cf. Section 5 of [155] for details). Note that, as the agent Σ_i selects its control, the predictions of the other agents are known. Hence, such a controllability condition can be rigorously verified at least for simple systems [155]. Note that it is not necessary to verify such a controllability condition for all initial values which may be very demanding in a distributed setting. Instead, it is sufficient to check this condition for given initial values along the closed-loop trajectory. If these assumptions hold, we can, as in the preceding sections, employ the cost function V_N^i as a Lyapunov function for the subsystem Σ_i and obtain stability of the closed loop.

In order to illustrate the proposed algorithm as well as the main advantages of using MPC without stabilizing terminal constraints, we consider the following example.

Example 4.2 *Autonomous agents in the plane*

Consider a set of $P = 4$ agents which can move in the unit square $[-1,1]^2 \subset \mathbb{R}^2$ subject to the individual dynamics

$$\boldsymbol{x}_i(k+1) = \boldsymbol{x}_i(k) + \boldsymbol{u}_i(k), \quad i \in \{1,2,3,4\},$$

where each state vector is of the form $\boldsymbol{x}_i(k) = (\boldsymbol{x}_{i1}(k), x_{i2}(k))^{\mathrm{T}}$ and all controls $\boldsymbol{u}_i \in [-\bar{u}, \bar{u}]^2 \subset \mathbb{R}^2$ are constrained by some $\bar{u} > 0$. To avoid collisions of these agents, we add the state constraints

$$\mathbb{X} := \{(\boldsymbol{x}_1^{\mathrm{T}}, \ldots, \boldsymbol{x}_P^{\mathrm{T}})^{\mathrm{T}} \in [-1,1]^{2P} \mid \|\boldsymbol{x}_{i_1} - \boldsymbol{x}_{i_2}\| \geq \delta \; \forall i_1, i_2 = 1, \ldots, P, \; i_1 \neq i_2\},$$

where $\|\cdot\|$ denotes an arbitrary norm in $\mathbb{R}^2$ and $\delta > 0$ the distance between agents we wish to guarantee with respect to this norm. Regarding the MPC setup, each agent minimizes the stage costs $\ell_i(\boldsymbol{x}_i, \boldsymbol{u}_i) = \|\boldsymbol{x}_i - \boldsymbol{x}_i^\star\|_2 + 0.1\|\boldsymbol{u}_i\|_2$ over a horizon of length $N = 3$. The targets $\boldsymbol{x}_i^\star$ of the four agents are situated in the corners of the square $[-1,1]^2$ and the initial positions of the agents are located at the corners on the opposite side of the respective target, that is, $\boldsymbol{x}_i(0) = -\boldsymbol{x}_i^\star$.

Considering Algorithm 4.3, we have to distinguish between the initialization and the control loop phase. In the control loop, the optimal control $\boldsymbol{u}_i$ of agent Σ_i is computed with respect to the state constraints which are given by the (current or old) predictions of all neighbouring agents $\boldsymbol{x}_q^{\boldsymbol{u}}$, $(q \neq i)$. During the initialization step, however, the respective data is partially not at hand, i.e., for agent Σ_i the data $\boldsymbol{x}_q^{\boldsymbol{u}}$ with $q > i$ is not yet computed. Here, we compute $\boldsymbol{u}_i$ subject to the state constraints induced by $\boldsymbol{x}_1^{\boldsymbol{u}}, \ldots, \boldsymbol{x}_{i-1}^{\boldsymbol{u}}$ only. As Fig. 4.9 illustrates, the imposed horizon length $N = 3$ is sufficient for stability.

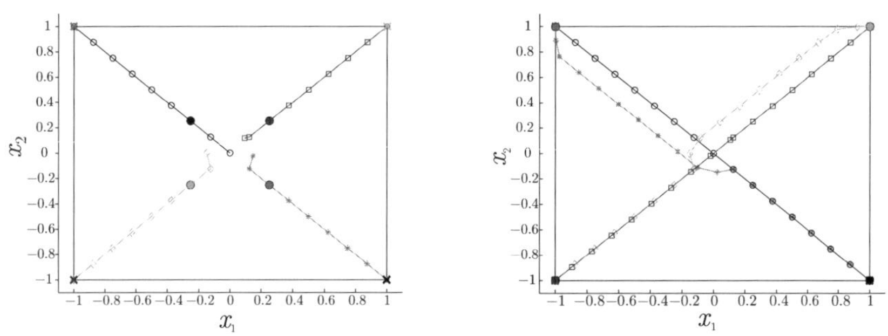

Fig. 4.9 Four agents in the plain with initial values $\boldsymbol{x}_1(0) = (-1,1)^{\mathrm{T}}$, $\boldsymbol{x}_2(0) = (1,-1)^{\mathrm{T}}$, $\boldsymbol{x}_3(0) = (1,1)^{\mathrm{T}}$, $\boldsymbol{x}_4(0) = (-1,-1)^{\mathrm{T}}$ and optimization horizon $N = 3$. Prediction for $k = 6$ (left) showing first evasive maneuver and final trajectories at $k = 16$ (right)

From Fig. 4.9 one can also observe that despite of the fact that the problem is symmetric, the solution does not necessarily inherit that property. This is due to the sequential computation of optimal controls within Algorithm 4.3. This example shows that hierarchy results in Σ_1 approaching its target directly while Σ_3 has to stop at $k = 6$ and Σ_2 and Σ_4 are even forced to take evasive actions at that particular time instant. The interested reader may download the MATLAB code for these simulations from `http://num.math.uni-bayreuth.de/en/publications/2012/gruene_worthmann_a_distrib_nmpc_scheme_2012`. □

We like to note that, in contrast to the original terminal constrained distributed MPC scheme from [320, 321], in Algorithm 4.3 most conflicts are resolved at runtime, that is in the control loop for $k \geq 1$, and not in the initialization step for $k = 0$. As the numerical simulation shows stable solutions even for larger numbers P of agents, this illustrates the ability of the scheme to resolve conflicts at runtime.

4.3.3 Distributed MPC for Cooperative Control

When controlling a network of interacting dynamic systems, besides the classical control objective of setpoint stabilization, a variety of other cooperative control tasks arises. For example, in this context consensus and synchronization problems have become important and have been subject to extensive research in recent years (cf. Section 6.1). We are interested in considering such general cooperative control tasks in a distributed model predictive control framework. In this respect, not many results have appeared in the literature so far, but the vast majority of proposed distributed MPC algorithms, including the ones presented in the previous sections, only consider the task of setpoint stabilization.

In the following, we consider cooperative control tasks which can be translated into the stabilization of some set $\mathbb{X}^0 \subseteq \mathbb{X}$. This includes typical problems like consensus and synchronization tasks and generalizes the setting described in Section 4.1, where the stabilization of the setpoint $\boldsymbol{x} = \mathbf{0}$ was considered. In order to achieve the cooperative goal, we make use of the terminal cost or terminal constraint approach introduced in Section 4.1, and we specifically show how these ingredients can be calculated in case of consensus and synchronization problems.

When considering cooperative control tasks of this kind, the typical setup is similar to the one in Section 4.3.2, i.e., $P \in \mathbb{N}$ dynamically independent systems Σ_i of the form (4.23) are present together with some input, state, and coupling constraints. In addition to the coupling via constraints as described in Section 4.3.2, the systems are now also coupled with each other via the common, cooperative control objective. This translates into coupling terms in the respective cost functional of each system.

More specifically, each system which is coupled with system Σ_i either via a common constraint or via a coupling term in the objective function is defined to be a neighbor of system Σ_i, and $\mathcal{N}_i$ is the collection of the indices of all neighbors. This means that each system Σ_i can be identified with a vertex of a graph $\mathcal{G} = (\mathcal{S}, \mathcal{E})$, where $\mathcal{S} = \{s_1, ..., s_P\}$ is the set of vertices, and the set of edges $\mathcal{E} \subseteq \{(s_i, s_j) \in \mathcal{S} \times \mathcal{S} | i \neq j\}$ describes the interconnection topology of the systems according to the couplings. Various (algebraic) properties of the graph $\mathcal{G}$ will appear later in the computation of the stabilizing terminal ingredients. In order to be able to fulfill the cooperative control objective, we assume that each system can communicate with each of its neighbors, i.e., transmit predicted trajectories.

Denote by $\boldsymbol{x}_{-i}$ the collection of the states of all neighbors of system Σ_i. Then, the cost functional J_N^i for system Σ_i is given as in (4.7) , where the stage cost ℓ_i and the terminal cost E_i can additionally depend on the neighboring system states $\boldsymbol{x}_{-i}$. As mentioned above, in order to achieve the desired cooperative control objective, we make use of the terminal cost and the terminal constraint approach as introduced in Section 4.1. Namely, we assume that there exists a (centralized) terminal region $\mathbb{X}_f \subseteq \mathbb{X}$ and for each system a local terminal control law $\bar{\mu}_i(\boldsymbol{x}_i, \boldsymbol{x}_{-i})$, such that the terminal region is controlled invariant under the local control laws, the input constraints are satisfied, and (4.11) holds with $\ell = \sum_{i=1}^P \ell_i$ and $E = \sum_{i=1}^P E_i$.

Note that while the local terminal control law is required to have a distributed structure, i.e., for each system the local control law only depends on its own and on neighboring states, we assume the existence of a centralized terminal region $\mathbb{X}_f$. We take this approach since in typical cooperative control tasks like consensus and synchronization, existence of (constant) distributed or decentralized terminal regions cannot be ensured. Of course, an important question is then how satisfaction of the centralized terminal constraint can be ensured in a distributed way. To this end, we assume that the stage and terminal cost functions ℓ_i and E_i, respectively, exhibit a certain separable structure. Namely, suppose that they can be separated into a part ℓ_{ii} (respectively, E_{ii}) consisting of the state and the input of the system Σ_i and parts ℓ_{ij} (respectively, E_{ij}) consisting of the state of Σ_i and the state of *one* of its neighbors:

$$\begin{aligned}\ell_i(\boldsymbol{x}_i, \boldsymbol{x}_{-i}, \boldsymbol{u}_i) &= \ell_{ii}(\boldsymbol{x}_i, \boldsymbol{u}_i) + \sum_{j \in \mathcal{N}_i} \ell_{ij}(\boldsymbol{x}_i, \boldsymbol{x}_j) \\ E_i(\boldsymbol{x}_i, \boldsymbol{x}_{-i}) &= E_{ii}(\boldsymbol{x}_i) + \sum_{j \in \mathcal{N}_i} E_{ij}(\boldsymbol{x}_i, \boldsymbol{x}_j). \end{aligned} \tag{4.24}$$

If the cost functions are given in such a separable structure, it turns out that the centralized terminal region $\mathbb{X}_f$, which we take as a sublevel set of the sum E of the terminal cost functions E_i, can be split up into distributed terminal regions $\mathbb{X}_f^i$. These distributed terminal regions will be time varying, (for more details cf. [266]). Furthermore, the cost functional $\bar{J}_N^i$ which system Σ_i uses when solving its optimal control problem is comprised of its own cost functional J_N^i plus those terms in the cost functionals of the neighbors Σ_j involving system Σ_i, i.e., ℓ_{ji} and E_{ji} as defined in (4.24). Similar to Algorithm 4.3, we also use the idea from [321] that the systems solve their respective optimal control problem in a sequential order. However, as we do not only consider couplings via constraints, but also via the cost functions, several modifications and additional communication between the systems are required in order to ensure fulfillment of the cooperative control task, i.e., the stabilization of the set $\mathbb{X}^0$.

Algorithm 4.4. Distributed MPC for cooperative control

Initialization: Let $k = 0$ and $\boldsymbol{x}(0) := (x_1(0), \ldots, x_P(0)) \in \mathbb{X}$ be the initial values. For all systems Σ_i, find a feasible solution $\hat{\boldsymbol{u}}_i(\cdot)$ with corresponding feasible state sequence $\hat{\boldsymbol{x}}_i(\cdot)$ such that the constraints (including the centralized terminal constraint) are satisfied. Each system transmits $\hat{\boldsymbol{x}}_i(\cdot)$ to its neighbors and applies the first control value $\hat{\boldsymbol{u}}_i(0)$.

Control loop for $k \geq 1$**:**

for $k = 1$ **to** ∞ **do**

Each system computes a feasible input sequence $\hat{\boldsymbol{u}}_i(\cdot)$ by taking the remainder of the previously optimal input sequence and adding the local terminal controller as last value.

Each system sends the corresponding feasible state sequence $\hat{\boldsymbol{x}}_i(\cdot)$ to its neighbors.

for $i = 1$ **to** P **do**

Compute the time varying terminal region $\mathbb{X}_f^i(k)$ based on the most recently received information from the neighbors.

Solve the optimal control problem with cost functional $\bar{J}_N^i$, based on the most recently received information from the neighbors.

Transmit the optimal state sequence $\boldsymbol{x}_i^{\boldsymbol{u}^*}(\cdot)$ to all neighbors.

end for

Each system applies the control value $\boldsymbol{u}_i^*(0)$.

end for

In Algorithm 4.4, the definition of the time-varying terminal regions as explained above ensures that the centralized terminal constraint is satisfied at each time step, which is needed in order to establish recursive feasibility and stability. Furthermore, the definition of the cost functional which is used in the optimal control problem for each system ensures that the sum over all optimal value functions V_N^i decreases at each time step and can hence be used as a Lyapunov function. Note that this would not be possible if each system only optimized over its own cost functional J_N^i. When defining the terminal region and solving its respective optimal control problem, each system takes into account the most recently available predictions of its neighbors. This means that system Σ_i uses the optimal predictions made at time k for all neighboring systems Σ_j with $j < i$, while it uses the feasible state sequences $\hat{\boldsymbol{x}}_j$ for all neighboring systems Σ_j with $j > i$ which were transmitted at the beginning of the time step. Furthermore, as remarked in Section 4.3.2, the sequential nature of the algorithm may be problematic in case that the number of systems P is large. However, the requirement of sequential optimization can be relaxed such that non-neighboring systems can optimize in parallel, which implies that the algorithm is scalable in case that the maximum number of neighbors of a system is limited.

As explained above, in Algorithm 4.4, we assumed a certain separable structure of the stage and terminal cost functions. While this structure is still quite general and includes various application settings, in [265, 266] we presented a second algorithm for the case where this assumption is not satisfied. However, then the centralized terminal region $\mathbb{X}_f$ cannot be split up into (time varying) distributed terminal regions as above, but the centralized terminal constraint as well as the decaying of the optimal value functions has to be ensured by means of additional communication between neighboring systems.

When applying Algorithm 4.4, we obtain the following result, the proof of which can be found in [266].

Theorem 4.2 (Distributed MPC for cooperative control). *Suppose that the terminal region $\mathbb{X}_f$ together with the local terminal controllers $\bar{\mu}_i$ is appropriately defined, and that an initial feasible solution can be found. Then, Algorithm 4.4 remains recursively feasible and the set $\mathbb{X}^0$ is asymptotically stable.*

Example 4.3 *Synchronization of five agents by distributed MPC*

The above algorithm can be used for a variety of different cooperative control tasks. Here, we show how it can be applied to consensus and synchronization problems. In this case, the target set $\mathbb{X}^0$ is given as $\mathbb{X}^0 := \{\boldsymbol{x} \in \mathbb{X} : \boldsymbol{x}_1 = \boldsymbol{x}_2 = \ldots = \boldsymbol{x}_P\}$. In [266], we showed that the terminal region $\mathbb{X}_f$ and the local terminal controllers can be computed by solving a specific linear matrix inequality (LMI). This LMI includes properties of the interconnection topology of the systems, i.e., of the underlying graph $\mathcal{G}$. For example, in case that the graph contains cycles, the matrix E_{null} containing basis vectors for the nullspace of the incidence matrix of the graph $\mathcal{G}$ plays an important role.

Figure 4.10 exemplarily shows the synchronization of five identical linear oscillators. One can see that the systems synchronize, and that systems 4 and 5 as well as 1, 2 and 3 converge faster towards each other. This behavior was specified by the choice of the cost functions for each of the systems. □

The above explained distributed MPC algorithm for cooperative control was only presented for nominal system dynamics, i.e., systems without disturbances. However, it is possible to extend these results to the case of additional unknown but bounded disturbances [267].

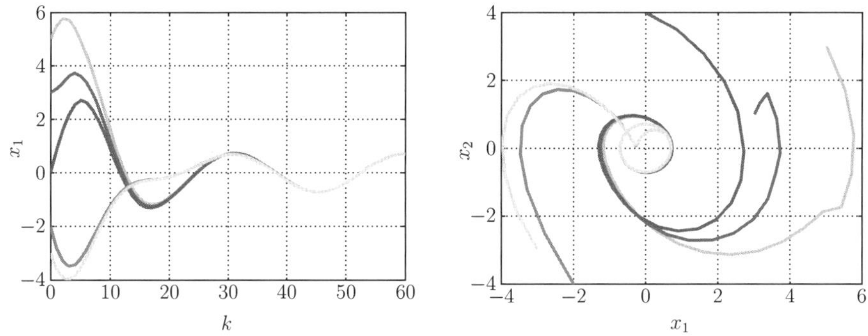

Fig. 4.10 Synchronization of five identical linear oscillators: Time domain (left) and phase plane (right)

4.4 Distributed Predictive Control of Communicating Decentralized Systems

4.4.1 Survey of Distributed MPC

This chapter considers distributed model predictive control (DMPC) of interconnected systems over digital communication networks. Usually, centralized control of such a distributed and possibly large-scale system is neither feasible nor desirable because it requires the transmission of large amounts of measured data to and from a centralized location with negligible delay, is more sensitive to faults, and leads to large and computationally complex problems for the centralized controller. An alternative option are fully decentralized control schemes consisting of local controllers C_i which do not communicate. Decentralized control is often only feasible for weakly interconnected systems. Even if it leads to feasible solutions, it often results in significantly degraded performance or robustness compared to the centralized approach, due to the lack of information about other subsystems.

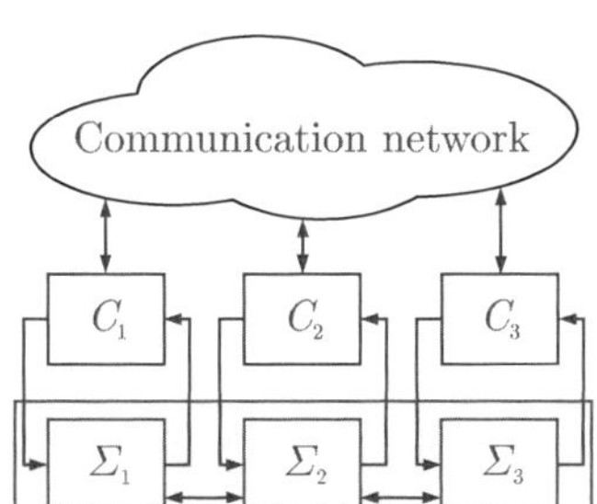

Fig. 4.11 Distributed control scheme with communication

Distributed control aims at combining the advantages of the two approaches: Using distributed local controllers C_i allows to spread the complexity of the control task among subsystems while communication between controllers can be used to increase closed-loop performance. Often a distributed control structure arises directly from the underlying system structure. If, for instance, a subsystem Σ_i is only coupled to a few other subsystems, distributed control can reduce the complexity by solving a set of smaller subproblems which are only interconnected to few other subproblems. However, the exchange of information becomes a crucial aspect (Fig. 4.11). Typically the information exchange between controllers is realized over a digital communication network. Typically two main challenges inhere in such a distributed control scheme:

- The communication network induces phenomena such as time-varying delays and packet loss, which affect the stability and performance of the distributed control scheme. Furthermore, scheduling protocols may have to be considered.

- Due to the distributed control structure, there is generally a lack of information about the state of the overall system.

MPC can be used to compensate communication uncertainties, but obtaining feasible solutions in a distributed manner may be problematic due to the lack of global state information. In addition to the well-known advantages of model predictive control, such as constraint handling, the use of MPC over communication networks is motivated by the fact that the controller predicts the future behavior of a subsystem at each time step, thus enabling a compensation of network effects by communicating a sequence of planned future states and inputs.

Classification of DMPC Algorithms. Many significantly different schemes for distributed model predictive control that utilize communication have been proposed. The assumptions with respect to the communication network as well as to the type of interactions differ largely and make a comparison difficult. However, with respect to the assumptions on the communication network, the following categorization is possible:

(a) It is assumed that the communication between controllers does not induce any uncertainties or delays. Under this assumption, all network properties are ignored in the analysis and their effect is assumed to be negligible. The vast majority of DMPC algorithms, for example [321] and Section 4.3.3, fall into this category.

(b) The communication channel is assumed to introduce a constant delay. Results for constant delays can be found in [126], [105].

(c) Time-varying communication delays and packet loss are considered in a worst-case fashion, e.g. by utilizing upper bounds on communication delays and the number of consecutive packet losses [143].

(d) The communication channel is assumed to be corrupted by additive stochastic noise. While only few investigations on DMPC consider this case, corresponding results have been obtained in the context of distributed optimization [274]. However, many of these approaches cannot model delays and packet loss adequately because of their assumptions on the stochastic noise.

To the authors' knowledge, there is no work on DMPC which considers a more general and explicit model of the properties of the communication network than the classes given above. Overall, relatively few investigations study the effect of delayed information exchange. Another aspect is the topology of the communication network, which can be classified into the following cases:

(a) Any local controller can exchange information with all other controllers at the same time, cf. [374].

(b) Each controller only exchanges information with a subset of other controllers. In such a scheme, typically only the controllers of subsystems which are directly interconnected communicate with each other. Commonly this is referred to as *neighboring communication* and considered in [143] and Section 4.3.3.

Considering that subsystems are often only interconnected with few other subsystems, an exchange of information between all controllers usually causes an unnecessary load for the communication network and is detrimental to scalability. On the other hand, the effects of only partially available information can significantly complicate the design and analysis of DMPC algorithms, further classification is possible with respect to the protocols used to exchange information between the controllers:

(a) controllers exchange information only once per sampling period (non-iterative), for example see [143] and [111],

(b) information is exchanged multiple times (i.e. iteratively) per sampling period, cf. [374].

Iterative algorithms can provide more information to the local controllers and, therefore, potentially allow for higher performance. However, they often require many iterations and are not suited to deal with communication delays and packet loss.

A related aspect is the order in which subsystems solve their local MPC problem. This leads to the following distinction:

(a) All subsystems solve their local problems and exchange information in parallel as discussed in [143] and [374].

(b) Local problems are solved in a given sequence and each local controller has to wait for the information of the preceding controller in the sequence before starting its own computations or rely on previously communicated information, cf. [321].

It is apparent that, in practice, a sequential algorithm is quite problematic for any medium to large-scale system due to computation times and communication delays.

Finally, the following types of interconnections can be considered:

(a) The models of the subsystems are coupled by common states or inputs (coupling by dynamics), this is considered in [374] and Section 4.3.1.

(b) A common control goal of the subsystems, for instance consensus or synchronization, leads to a cost function which may depend on the states and inputs of all subsystems, for details see Section 4.3.3 and [143].

(c) Interconnections can also be modeled by common constraints, which involve the states and inputs of different subsystems. For example, consider collision avoidance constraints in formation control, as discussed in [143] and Section 4.3.

In principle, constraints can be used to formulate all of these types of interconnections in an MPC framework, but this general case can result in a more complicated problem than the other formulations. For instance, reformulating a problem with a coupled cost function into a problem with coupled constraints usually complicates the analysis.

Aim of this Section. This section focuses on distributed model predictive control (DMPC) of interconnected constrained linear discrete-time systems over digital communication networks. In particular, we investigate DMPC schemes which are robust against uncertainties induced by a digital communication network such as delays and packet loss. The main idea is to model the interaction between subsystems as disturbance acting on the local subsystem. This leads to a robust MPC (RMPC) problem for each subsystem. To reduce the conservativeness of the solution, each controller communicates its predicted trajectories to other subsystems, thereby reducing the uncertainty.

First, some results on robust model predictive control for constrained linear systems with coupling over inputs in a chain structure are briefly discussed. By employing principles from set theory, robust feasibility is established for the decentralized case. The solution of the local RMPC problems is rendered less conservative and the control performance is improved by introducing communication between the controllers.

These ideas form the basis for a scheme in which subsystems communicate their predicted trajectory as well as the uncertainty of the prediction for each time step of the prediction horizon. This allows to consider a class of constrained linear systems coupled over costs and common state constraints. The resulting RMPC problems are solved by optimizing over disturbance feedback policies. By employing methods from robust optimization these problems can be solved efficiently.

The last part of this section deals with a related approach for distributed control of interconnected piecewise affine systems.

4.4.2 A Set-Invariance Approach to Distributed MPC

The first class of systems under consideration is given by a set of N_l identical linear discrete-time subsystems Σ_i that are coupled in a chain structure solely through their inputs:

$$\boldsymbol{x}_1(k+1) = \boldsymbol{A}\boldsymbol{x}_1(k) + \boldsymbol{B}\boldsymbol{u}_1(k) \tag{4.25}$$

$$\boldsymbol{x}_i(k+1) = \boldsymbol{A}\boldsymbol{x}_i(k) + \boldsymbol{B}\boldsymbol{u}_i(k) + \boldsymbol{E}\boldsymbol{w}_i(k), \quad \forall i \in \{2, \ldots, N_l\}, \tag{4.26}$$

where $\boldsymbol{x}_i(k) \in \mathbb{X} \subseteq \mathbb{R}^{n_i}$ and $\boldsymbol{u}_i(k) \in \mathbb{U}_i \subseteq \mathbb{R}^{m_i}$ are the state and the input of subsystem Σ_i. The influence of the preceding subsystem is modeled by $\boldsymbol{E} = -\boldsymbol{B}$ and $\boldsymbol{w}_i(k) = \boldsymbol{u}_{i-1}(k)$. Such a structure arises, for example, from a group of vehicles driving in a platoon. The sets $\mathbb{X}$ and $\mathbb{U}_i$ are assumed to be convex polytopes containing the origin in their interior. This structure motivates the idea to consider $\boldsymbol{w}_i(k) \in \mathbb{W}_i$ as a bounded disturbance acting on the local subsystem.

In this scheme, each subsystem solves an MPC problem with local cost function. Robustness is achieved by adding a robustness constraint $\boldsymbol{x}_i(k+1|k) \in \mathbb{X}_{i,R}$ on the first predicted state. This constraint ensures that the problem will remain feasible regardless of the disturbance $\boldsymbol{w}_i(k)$. This constraint depends on the bounds of the disturbance: in the worst case, these bounds are given by the set of inputs $\mathbb{U}_{i-1}$ of the preceding system. To ensure that the robustness constraint $\boldsymbol{x}_i(k+1|k) \in \mathbb{X}_{i,R}$ is non-empty, the input sets are artificially shrunk according to $\mathbb{U}_{i-1} := c\mathbb{U}_i$, i.e. the right-hand side of the polytopic input constraints is multiplied by an appropriate scaling factor $0 < c < 1$. Intuitively, this ensures that each system can always compensate the influence of the preceding system. Considering the worst-case disturbance will obviously lead to very conservative solutions. Thus, the first subsystem computes bounds on its future inputs from the solution of the local MPC problem. These bounds are modified and passed on along the chain of subsystems with a possible communication delay. The overall scheme is described in Algorithm 4.5 and leads to time-varying input and robustness constraints.

Here, $\mathbb{W}_{i,s}(k)$ denotes the disturbance set communicated to subsystem i at time k. The computations required in step 5 of Algorithm 4.5 usually cannot be done efficiently enough in real-time. However, if the robustness constraints

Algorithm 4.5. DMPC algorithm [105]

1: **for** all time steps $k \geq 1$ and all subsystems Σ_i, $i \neq 1$ **do**
2: Receive new bounds from subsystem $i-1$: $\mathbb{W}_i(k) := \mathbb{W}_{i,s}(k-1)$
3: Compute bounds for subsystem $i+1$ as: $\mathbb{W}_{i+1,s}(k) := \frac{1}{c}\mathbb{W}_i(k)$
4: Set the new input constraints: $\mathbb{U}_i(k) := \mathbb{W}_{i+1,s}(k)$
5: Compute for $\mathbb{U}_i(k)$ and $\mathbb{W}_i(k)$ the robustness constraint $\mathbb{X}_{i,R}(k)$ which guarantees robust strong feasibility.
6: Solve the local RMPC problem with time-dependent input and robustness constraints. If a feasible solution can be found, go to the next time step. If no feasible solution can be found, solve a relaxed optimization problem.
7: **end for**

are precomputed offline by discretizing the disturbance and input spaces the resulting online MPC problems can be solved very efficiently. Due to the time-varying constraints, feasibility of the local RMPC problems might be lost in rare cases when the control input of a vehicle drastically changes. In this case (step 6), a relaxed problem has to be solved to recover feasibility. For details refer to [105, 106].

4.4.3 Distributed MPC Based on Robust Optimization

In this section, we focus on a class of systems in which dynamically decoupled subsystems are interconnected by constraints as well as common control goals, which are modeled by a common cost function. The subsystems are controlled by local model predictive controllers which exchange information over a communication network subject to time-varying delays and packet loss. To compensate the effects induced by the communication network, uncertain predicted future behavior of a local subsystem is communicated to other subsystems. The communicated information includes nominal predicted trajectories as well corresponding uncertainties. The resulting DMPC algorithm uses neighboring communication, the local MPC problems are solved in parallel and information is exchanged only once per sample time.

In the context of dynamically decoupled systems, coupling via cost functions has for instance been addressed in [111], but communication effects are ignored. Constant communication delays in DMPC for dynamically decoupled nonlinear systems coupled by a common cost function are analyzed in an input-to-state stability framework in [126]; however, no coupling of the dynamics by constraints is considered. Coupling over constraints for linear subsystems with additive disturbances and decoupled cost functions has been considered in [321] in a sequential robust DMPC framework. Finally our approach shares some underlying ideas with the min-max MPC based DMPC proposed in [192] for systems coupled over states. For instance, the local controllers have to ensure that a subsystems' future behavior is consistent with the communicated information.

System Model. We consider N_l dynamically decoupled discrete-time linear time-invariant subsystem Σ_i:

$$\boldsymbol{x}_i(k+1) = \boldsymbol{A}_i \boldsymbol{x}_i(k) + \boldsymbol{B}_i \boldsymbol{u}_i(k) \quad \forall i \in \mathcal{N}, \tag{4.27}$$

where $\boldsymbol{x}_i(k) \in \mathbb{X}_i \subseteq \mathrm{I\!R}^{n_i}$ and $\boldsymbol{u}_i(k) \in \mathbb{U}_i \subseteq \mathrm{I\!R}^{m_i}$ are the local states and inputs of the i-th subsystem. The set $\mathcal{N} = \{1, \ldots, N_l\}$ is the index set of the subsystems. An extension to additive disturbances can be found in [143].

The states and inputs are subject to local constraints given by sets $\mathbb{X}_i$ and $\mathbb{U}_i$. In this model, the states and inputs of one subsystem do not directly influence another subsystem. Instead, the interconnection between subsystems arises from constraints on the state vector of the overall system, given by $\boldsymbol{x}(k) = (\boldsymbol{x}_1^{\mathrm{T}}(k), \ldots, \boldsymbol{x}_{N_l}^{\mathrm{T}}(k))^{\mathrm{T}}$. The corresponding input vector is given by $\boldsymbol{u}(k) = (\boldsymbol{u}_1^{\mathrm{T}}(k) \ldots, \boldsymbol{u}_{N_l}^{\mathrm{T}}(k))^{\mathrm{T}}$. The subsystems are interconnected by coupled state constraints $\boldsymbol{x}(k) \in \mathbb{X} \subseteq \mathbb{X}_1 \times \ldots \times \mathbb{X}_{N_l} \subseteq \mathrm{I\!R}^n$. Input constraints are given by $\boldsymbol{u}(k) \in \mathbb{U} = \mathbb{U}_1 \times \ldots \times \mathbb{U}_{N_l} \subseteq \mathrm{I\!R}^m$. For example, these coupled constraints can be used to model interactions such as limits on a shared resource or collision avoidance constraints.

We further assume that the state and input constraints are given by full-dimensional polytopes $\mathbb{X} = \{\boldsymbol{x}(k) \in \mathrm{I\!R}^n | \boldsymbol{C}\boldsymbol{x}(k) \leq \boldsymbol{b}_{\boldsymbol{x}}\}$ and $\mathbb{U}_i = \{\boldsymbol{u}_i(k) \in \mathrm{I\!R}^{m_i} | \boldsymbol{D}_i \boldsymbol{u}_i(k) \leq \boldsymbol{b}_{\boldsymbol{u}_i}\}$. Further coupling is induced by common control goals which are specified by a global cost function for a finite horizon of N time steps:

$$J_N = \|\boldsymbol{x}(k+N|k)\|_{\boldsymbol{P}}^2 + \sum_{l=0}^{N-1} \|\boldsymbol{x}(k+l|k)\|_{\boldsymbol{Q}}^2 + \|\boldsymbol{u}(k+l|k)\|_{\boldsymbol{R}}^2, \tag{4.28}$$

where $\boldsymbol{Q}$, $\boldsymbol{R}$, and $\boldsymbol{P}$ are given symmetric positive (semi-)definite weighting matrices, $\boldsymbol{x}(k+l|k)$ denotes the state at time $k+l$ predicted at time k, and $\|\boldsymbol{x}\|_{\boldsymbol{P}}^2 := \boldsymbol{x}^{\mathrm{T}} \boldsymbol{P} \boldsymbol{x}$ denotes a weighted norm with weighting matrix $\boldsymbol{P}$.

The interconnection of systems can be described by a directed graph $\mathcal{G} = (\mathcal{V}, \mathcal{E})$, where each node $\mathcal{V} = \{1, \ldots, N_l\}$ represents a subsystem and each edge $\mathcal{E} \subseteq \mathcal{V} \times \mathcal{V}$ represents an interconnection between two subsystems. Let the index set $\mathcal{N}_i$ denote the indices of subsystem Σ_i and of all the subsystems with which Σ_i is coupled through the cost function or constraint: $\mathcal{N}_i = \{i_1, \ldots, i_p\} \subseteq \mathcal{N}$ with $i_1 = i$ and $\{i_2, \ldots, i_p\} = \{j | (i,j) \in E\}$. Here, we require that the interconnection graph $\mathcal{G}$ has no cycles.

For example, consider the three subsystems depicted in Fig. 4.12 (b) subsystem Σ_1 is interconnected with subsystem Σ_2 and subsystem Σ_2 with subsystem Σ_3. Without the restriction that Σ_3 does not affect Σ_1, uncertainties could be amplified while propagating through the distributed controllers and could affect the subsystem they originated from. In such a case, the terminal constraint required for robust feasibility (see below) may not exist.

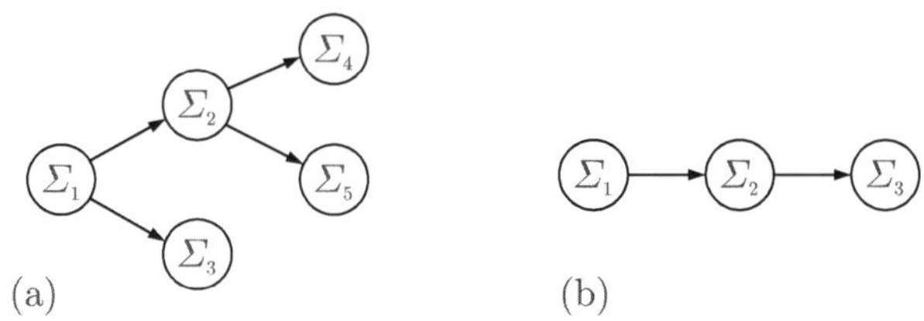

Fig. 4.12 Examples for interconnection graphs

A small-gain argument may be suitable to relax these assumptions. However, we conjecture that such an argument may be quite conservative. Methods which can handle coupling by constraints with cycles in the interconnection graph are usually of an iterative type or require other strong assumptions.

Example 4.4 *Platooning vehicles*

To illustrate the class of interconnected systems, consider the control of a platoon of vehicles in a leader-follower scenario, as specified in Fig. 4.13.

The control goal is to synchronize the trajectories of the followers with that of the lead vehicle in the sense that the leader is followed with constant spacing while avoiding collisions. Instead of following a given reference trajectory, we assume that the lead vehicle aims to achieve a control goal such as reaching a destination while reacting to external influences (e.g. traffic, speed limits) and thus can only provide uncertain predictions to the followers.

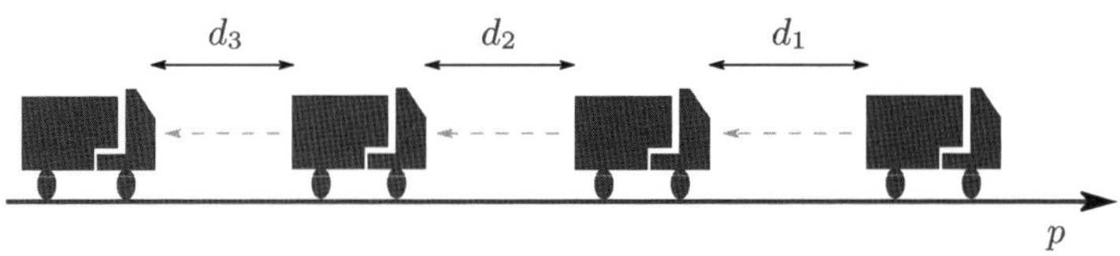

Fig. 4.13 Platooning vehicles in leader-follower configuration with distance d_i and neighboring communication (dashed)

The vehicles are modeled by double-integrator dynamics given by:

$$\boldsymbol{x}_i(k+1) = \begin{pmatrix} 1 & T \\ 0 & 1 \end{pmatrix} \boldsymbol{x}_i(k) + \begin{pmatrix} \frac{T^2}{2} \\ T \end{pmatrix} \boldsymbol{u}_i(k) \quad \forall i \in \{1, \ldots, \mathcal{N}_l\} \tag{4.29}$$

with sampling period T. The local state $x_{i,1}(k)$ represents the position and $x_{i,2}(k)$ the velocity of vehicle i. By a change of coordinates, the control goal $\boldsymbol{x}_{i+1}(k) = \boldsymbol{x}_i(k)$ corresponds to a constant spacing S_v. Here, we consider the case that each vehicle can only measure its own position and velocity. Thus, the distance can only be obtained by utilizing communication. The control goal of following the lead vehicle with a desired spacing induces coupling via the cost function. To guarantee that there are no collisions between cars, the constraints $(-S_v, -20\ \frac{\mathrm{m}}{\mathrm{s}})^{\mathrm{T}}$

$\leq \boldsymbol{x}_{i+1}(k) - \boldsymbol{x}_i(k) \leq (100\ \mathrm{m}, 20\ \frac{\mathrm{m}}{\mathrm{s}})^{\mathrm{T}}$ are imposed, i.e. the maximum distance is $100\ \mathrm{m} + \mathrm{S_v}$ and the maximum speed difference is $20\ \frac{\mathrm{m}}{\mathrm{s}}$. □

Communication. We consider the case of neighboring communication, i.e. each subsystem only communicates with directly interconnected subsystems. The communication is subject to delays and packet loss. In this framework, a packet loss can be treated as an extension of the communication delay by assuming that the time-varying delay $d(k)$ includes packet loss and is bounded by $1 \leq d(k) \leq \bar{d}$.

It is further assumed that the local clocks of the subsystems are synchronized and any remaining clock offsets are negligible compared to the sampling time. Then, communication delays can be detected by also communicating the times at which information is sent out. We assume that the prediction horizon of the MPC is chosen such that $N \gg \bar{d}$.

In the proposed scheme, each subsystem communicates its own current state, a predicted input sequence, and a time-stamp, such that the inputs and states of the connected subsystems are known at latest with a delay of $\bar{d}$ time steps. Furthermore, each subsystem communicates predicted bounded uncertainties of its own future inputs. This information on the upcoming behavior of the sender is taken into account by the connected subsystems in their planning of future control actions.

Local Prediction Model. Each local MPC requires a prediction model which adequately captures the own dynamics including the interactions with connected subsystems. To include the coupled constraints and costs, the model of the local dynamics (4.27) is augmented with the dynamics, states, and inputs of connected subsystems as follows. Let $\bar{\boldsymbol{u}}_{i,j}(k|k) \in \mathbb{R}^{m_j}$ and $\bar{\boldsymbol{x}}_{i,j}(k|k) \in \mathbb{R}^{n_j}$ denote the values that subsystem Σ_i assumes for the inputs $\boldsymbol{u}_j(k|k)$ and states $\boldsymbol{x}_j(k|k)$ of a connected subsystem Σ_j, $j \in \mathcal{N}_i \backslash i$. Due to the recalculation of the local predictions in each time step with possibly delayed communication, these values in general may differ from the true states and inputs. The corresponding additive uncertainties are considered by variables by $\boldsymbol{\delta u}_{i,j}(k|k) \in \Delta\mathcal{U}_{i,j}(k|k)$ and $\boldsymbol{\delta x}_{i,j}(k|k) \in \Delta\mathcal{X}_{i,j}(k|k)$. These estimates and uncertainties are directly obtained from the solution of the local control problems of connected subsystems (see below).

Based on this information, the local model for the i-th subsystem can be augmented to the following prediction model:

$$\tilde{\boldsymbol{x}}_i(k+1|k) = \tilde{\boldsymbol{A}}_i\tilde{\boldsymbol{x}}_i(k|k) + \tilde{\boldsymbol{B}}_i\tilde{\boldsymbol{u}}_i(k|k) + \boldsymbol{\Delta B}_i\boldsymbol{\delta}\tilde{\boldsymbol{u}}_i(k|k), \tag{4.30}$$

here the matrices are all block-diagonal and $\tilde{\boldsymbol{x}}_i(k|k) \in \mathbb{R}^{\tilde{n}_i}$, $\tilde{\boldsymbol{u}}_i(k|k) \in \mathbb{R}^{\tilde{m}_i}$ are the state and input vectors which are augmented by the communicated quantities of the connected subsystems. The initial condition of (4.30) is uncertain due to the communication delay:

$$\tilde{\boldsymbol{x}}_{i,j}(k|k) := \bar{\boldsymbol{x}}_{i,j}(k|k) + \boldsymbol{\delta x}_{i,j}(k|k), \quad \boldsymbol{\delta x}_{i,j}(k|k) \in \Delta\mathcal{X}_{i,j}(k|k), \forall j \in \mathcal{N} \setminus i.$$

The variables $\boldsymbol{\delta u}_{i,j}(k|k) \in \Delta\mathcal{U}_{i,j}(k|k) \subseteq \mathrm{I\!R}^{\tilde{m}_i}$, $\boldsymbol{\delta x}_{i,j}(k|k) \in \Delta\mathcal{X}_{i,j}(k|k) \subseteq \mathrm{I\!R}^{\tilde{n}_i - n_i}$ contain the corresponding uncertainties. Each subsystem Σ_i can accurately predict and communicate its own states $\boldsymbol{x}_i(k+1)$ at time k based on (4.27), while it knows the state $\boldsymbol{x}_i(k+p+1)$ of a connected subsystem Σ_j at latest at the time $k+p+\bar{d}$ (for $p \geq \bar{d}$)). For the example of $\bar{d} = 2$, the state $\boldsymbol{x}_j(k)$ is known to subsystem Σ_i at $k+1$. Therefore, the current states and corresponding uncertainties $\boldsymbol{\delta x}_i(k|k)$ of the connected subsystems can be computed by forward prediction based on the last accurately known state and the assumed input sequence. Thus, $\boldsymbol{\delta x}_i(k|k)$ can be eliminated from the problem.

Local Costs and Constraints. Using the augmented model, uncertain predictions for all connected subsystems Σ_j, $j \in \mathcal{N}_i$ can be obtained, enabling to compute local costs and to evaluate the constraints $\tilde{\boldsymbol{x}}_i(k|k) \in \tilde{\mathbb{X}}_i \subseteq \mathbb{X}$. The local constraints are the subset of the global constraints which involve the states of Σ_i. The local cost function is obtained by decomposition of (4.28) and may include a terminal cost:

$$\tilde{J}_{i,N} = \|\boldsymbol{\xi}_i(k+N|k)\|^2_{\boldsymbol{P}_i} + \sum_{l=0}^{N-1} \|\tilde{\boldsymbol{x}}_i(k+l|k)\|^2_{\boldsymbol{Q}_i} + \|\tilde{\boldsymbol{u}}_i(k+l|k)\|^2_{\boldsymbol{R}_i}. \quad (4.31)$$

The local terminal cost parametrized by $\boldsymbol{P}_i$ can be defined based on a delayed state-feedback controller:

$$\boldsymbol{u}_i(k) = \boldsymbol{K}_i \tilde{\boldsymbol{x}}_i(k+1-\bar{d}), \quad (4.32)$$

and an extended state vector $\boldsymbol{\xi}_i(k) = (\tilde{\boldsymbol{x}}_i^{\mathrm{T}}(k), \ldots, \tilde{\boldsymbol{x}}_i^{\mathrm{T}}(k-\bar{d}+1))^{\mathrm{T}} \in \mathrm{I\!R}^{\bar{d}\tilde{n}_i}$, which allows to explicitly included delays into a linear discrete-time system.

The cost function (4.31) can be modified to penalize deviations from previously communicated behavior. This essentially allows one to obtain a compromise between local and global convergence (cf. [143]).

RMPC Based on Robust Optimization. Based on the distributed prediction model (4.30) and the local costs and constraints we can now formulate the local MPC problem. Because the distributed model (4.30) depends on uncertain information, the local controller needs to be robust against these disturbances. In particular, the local MPC has to compute a sequence of future control inputs such that the coupled state constraints are satisfied for all possible future inputs of connected subsystems. A main idea to reduce conservativeness in such a problem is to optimize over future feedback policies, since more information is available when a predicted control input is actually applied to the system. For instance, an input $\boldsymbol{u}_i(k+l+\bar{d}|k)$ may depend on $\boldsymbol{u}_j(k+l)$ which is not known when optimizing at time k. While such problems are not tractable in general, a widely adopted solution is to reduce the number of free parameters by restricting future feedback to be affine in the states or disturbances. A similar approach is used in uncertain linear programming, where a part of the solution, referred to as “adjustable variables“, are affinely parametrized in the uncertain parameters [27]. In linear MPC, a control law which is affine in the past disturbances leads to a tractable convex optimization problem [141, 238] (see Example 4.5 for computation times).

Since the uncertain parameters in (4.30) will be known to connected subsystems with a maximal delay of $\bar{d}$ time steps, the following delayed affine feedback can be used to parametrize the future control inputs:

$$\boldsymbol{u}_i(k+l|k) = \boldsymbol{v}_i(k+l|k) + \sum_{r=-\bar{d}}^{l-\bar{d}} \boldsymbol{M}_{i,r}(k+l|k)\, \boldsymbol{\delta}\tilde{\boldsymbol{u}}_i(k+r|k). \tag{4.33}$$

This feedback law is optimized for all $l \in \{0, \ldots, N-1\}$:

$$\begin{aligned}
\min_{\boldsymbol{v}_i(k+l|k), \boldsymbol{M}_i(k+l|k)} \; & \|\hat{\boldsymbol{\xi}}_i(k+N|k)\|^2_{\boldsymbol{P}_i} + \sum_{l=0}^{N-1} \|\hat{\boldsymbol{x}}_i(k+l|k)\|^2_{\boldsymbol{Q}_i} + \|\tilde{\boldsymbol{u}}_i(k+l|k)\|^2_{\boldsymbol{R}_i} \\
\text{s.t.} \hat{\boldsymbol{x}}_i(k+l+1|k) &= \tilde{\boldsymbol{A}}_i \hat{\boldsymbol{x}}_i(k+l|k) + \hat{\boldsymbol{B}}_i \hat{\boldsymbol{u}}_i(k+l|k), \\
\hat{\boldsymbol{u}}_i(k+l|k) &= \left(\boldsymbol{v}_i(k+l|k)^{\mathrm{T}}, \bar{\boldsymbol{u}}_{i,i_2}(k+l|k)^{\mathrm{T}}, \ldots, \bar{\boldsymbol{u}}_{i,i_p}(k+l|k)^{\mathrm{T}}\right)^{\mathrm{T}}, \\
\boldsymbol{u}_i(k+l|k) &= \boldsymbol{v}_i(k+l|k) + \sum\nolimits_{r=-\bar{d}}^{l-\bar{d}} \boldsymbol{M}_{i,r}(k+l|k)\, \boldsymbol{\delta}\tilde{\boldsymbol{u}}_i(k+r|k), \\
\tilde{\boldsymbol{x}}_i(k+l+1|k) &= \tilde{\boldsymbol{A}}_i \tilde{\boldsymbol{x}}_i(k+l|k) + \tilde{\boldsymbol{B}}_i \tilde{\boldsymbol{u}}_i(k+l|k) + \boldsymbol{\Delta B}_i \boldsymbol{\delta}\tilde{\boldsymbol{u}}_i(k+l|k), \\
\tilde{\boldsymbol{u}}_i(k+l|k) &= \left(\boldsymbol{u}_i(k+l|k)^{\mathrm{T}}, \bar{\boldsymbol{u}}_{i,i_2}(k+l|k)^{\mathrm{T}}, \ldots, \bar{\boldsymbol{u}}_{i,i_p}(k+l|k)^{\mathrm{T}}\right)^{\mathrm{T}}, \\
\boldsymbol{u}_i(k+l|k) \in \mathbb{U}_i, \tilde{\boldsymbol{x}}_i(k+l|k) &\in \tilde{\mathbb{X}}_i, \tilde{\boldsymbol{x}}_i(k+N|k) \in \Xi_i, \quad \forall \delta_i(k) \in \Delta_i(k),
\end{aligned}$$

where $\boldsymbol{\delta}_i(k)$ collects the input uncertainties $\boldsymbol{\delta u}_{i,j}(k+l|k)$ of all subsystems $j \in \mathcal{N}_i \setminus i$ connected to Σ_i and for all $l \in \{-\bar{d}, \ldots, N-1\}$. The corresponding disturbance set $\Delta_i(k)$ is directly obtained from $\boldsymbol{\delta}\mathcal{U}_{i,j}(k+l|k)$. $\tilde{\boldsymbol{x}}_i(k+N|k) \in \Xi_i$ is a terminal constraint used to establish recursive feasibility and stability. While the cost is predicted based on a disturbance free model, the last constraint ensures that the state and input constraints and the terminal constraint are robustly satisfied. These conditions can be reformulated into a set of linear constraints and the uncertainties, which are subsequently communicated to connected subsystems, can be directly obtained from this problem (cf. [143]). The terminal constraint can be chosen as the maximal control invariant set of the augmented system (4.30). This choice of terminal constraint guarantees robust feasibility and stability, in the sense that the state enters the terminal set [143]. Another possibility is to construct the terminal constraint based on the terminal control law (4.32).

Distributed MPC Scheme. We can now specify the information exchange in the DMPC scheme based on the feedback control law (4.33) according to Algorithm 4.6. Each subsystem communicates its current state,

Algorithm 4.6. Distributed MPC algorithm [144]

for all subsystems $\Sigma_i \in \mathcal{N}$ **do**
  for all connected subsystems Σ_j, $(j \in \mathcal{N}_i \setminus i)$ **do**
    Receive $\boldsymbol{x}_j(k+1-d(k)|k-d(k))$, $\boldsymbol{v}_j(k+l-d(k)|k-d(k))$ and $\Delta\mathcal{U}_j(k+l-d(k)|k-d(k))$ for $l \in \{0, \dots, N-1\}$.
    Compute the estimates $\bar{\boldsymbol{x}}_{i,j}(k|k)$ and $\bar{\boldsymbol{u}}_{i,j}(k+l|k)$ of the current state and input sequence of subsystem Σ_j as well as the corresponding uncertainties.
  end for
  Measure the local state $\boldsymbol{x}_i(k|k)$, solve the local optimization problem with input consistency constraints and apply $\boldsymbol{v}_i(k|k)$.
  Compute the input consistency constraints for next time step.
  Communicate $\boldsymbol{x}_i(k+1|k)$, $\boldsymbol{v}_i(k+l|k)$ and $\Delta\mathcal{U}_i(k+l|k)$ for $l \in \{0, \dots, N-1\}$ to all subsystems Σ_j, $(j \in \mathcal{N}_i \setminus i)$.
end for

the nominal predicted input sequence $\boldsymbol{v}_i(k+l|k)$ and uncertainty bounds $\Delta\mathcal{U}_i(k+l|k)$, which arise from the feedback terms of (4.33). The connected subsystems receive this information with a delay $d(k) \le \bar{d}$ and execute the augmented prediction model (4.30) while compensating the delay $d(k)$ and using the terminal controller (4.32) to extend the received sequence. To ensure feasibility of the scheme, local control actions have to be consistent with previously communicated information. To achieve this, each subsystem shifts its previously communicated input sets one step forward to obtain new input constraints for the local RMPC problem at the next time step. Thus, replanning by the local controllers is only permitted within the previously communicated bounds. This renders the DMPC scheme recursively feasible, i.e., if the DMPC problem is feasible at time $k = 0$ the problem will remain feasible for all times. Essentially, the robustly invariant terminal constraint guarantees that a feasible solution at the next time step can always be constructed based on the previous feasible solution [143, 144].

As indicated before, the convergence properties of the DMPC scheme depend on the choice of the terminal constraint. Under suitable assumptions the DMPC scheme in Algorithm 4.6 applied in closed-loop to each augmented system renders the system input-to-state stable [144]. The interpretation of input-to-state stability is rather straightforward in this case: If the initial solutions satisfy the terminal constraint and the behavior of each system exactly matches the previously communicated information the cost function (4.28) converges to zero over time and can be used as a Lyapunov function. If a subsystem differs from its communicated behavior within the communicated bounds this leads to a remaining bounded control error, i.e. the cost does not fully converge to zero. As the following example shows, good performance can be obtained and the remaining control error is reasonably small.

Example 4.5 *Distributed control of platooning vehicles*

Consider the control of a platoon of vehicles as described in Example 4.4 with three vehicles, a sampling period of $T = 0.3$ s, an upper bound of $\bar{d} = 2$ on the communication delay, desired spacing of $S_v = 15\,\text{m}$, and a prediction horizon of $N = 10$.

Originally, the input set of all vehicles is $\mathbb{U}_i = [-3, 3]$. In order to construct a robust control invariant terminal constraint, we choose to shrink the input sets according to $\mathbb{U}_{i-1} = c\mathbb{U}_i$ with $c = 0.74$. Each vehicle is coupled to its preceding vehicle by costs and constraints on the distance and the relative velocity. Hence, the index sets are $\mathcal{N}_1 = \{1\}$, $\mathcal{N}_2 = \{2, 1\}$, $\mathcal{N}_3 = \{3, 2\}$, and the models of the followers are augmented to include the states and inputs of the preceding vehicle. Thus, e.g. the augmented state vector for the subsystem Σ_3 is $\tilde{\boldsymbol{x}}_3(k) = (\boldsymbol{x}_3^{\mathrm{T}}(k), \boldsymbol{x}_2^{\mathrm{T}}(k))^{\mathrm{T}}$. The cost function is chosen to be identical for all followers with the weight $Q = \mathrm{diag}(5, 1)$ for the distance and relative velocity, and with the input weight $R = 10^{-1}$.

The controllers are initialized to the positions of the preceding vehicle at time $k = 0$, but there is no centralized initialization of the predicted states, inputs, or disturbance sets. Until predictions are received via communication, the inputs of the preceding vehicle are assumed to be zero and consequently the disturbance set is the input set of the preceding vehicle.

Simulation results for a constant communication delay of $d(k) = 2$ are shown in Fig. 4.14. In the first row, the position, velocity, and input of the leader are shown. The subsequent rows show the distance to the preceding vehicle, the relative velocity, and the input of the followers. It can be seen that the distance between the vehicles starts to decrease at time $t = 0$ s, but due to the fact that the vehicles always deviate from previously communicated information, the control error never goes to zero.

The inputs of the first and second vehicle, as well as the predicted inputs and bounds at $t = 40\,\text{s}$, are shown in Fig. 4.15. A mismatch between the communicated and actual control actions occurs at $t = 40$ s, since the lead vehicle applies full braking but previously communicated a mild breaking maneuver. This leads to a decrease in the distance between vehicle one and two, but nonetheless collisions are robustly avoided. Similarly, the distance between the first and second vehicle temporarily increases because the lead vehicle did not communicate the acceleration at around $t = 75$ s.

It can be seen that the control error of the third vehicle with respect to the second one is smaller than the error between the first and second vehicle. The reason for this is that the input set of the third vehicle is larger than that of the second vehicle. Therefore, the third vehicle can compensate for uncertainties in the communicated information quicker and more aggressively than the second vehicle.

The computations for one vehicle and one time step as described in Algorithm 4.6 take between 70 ms and 100 ms using one core of an AMD Phenom II X4 920 with 4GB RAM using CPLEX 12.1. □

Discussion. The proposed distributed model predictive control scheme considers dynamically decoupled linear systems, in which local model predictive controllers exchange information via a possibly delayed communication in order to achieve a common control goal and satisfy coupled constraints. The problem is transformed into local RMPC problems by modeling the uncertainties due to the non-deterministic communication network as disturbances.

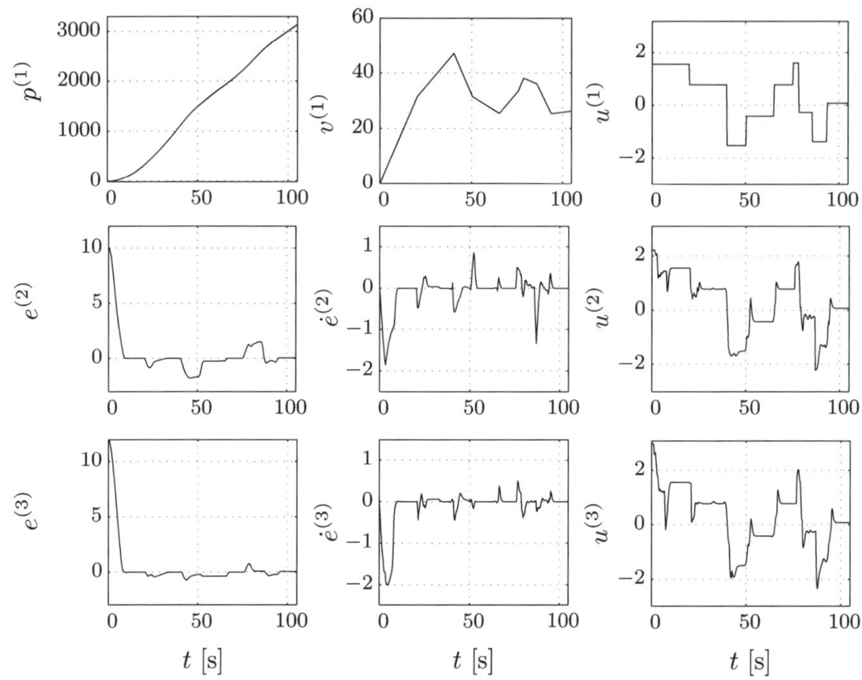

Fig. 4.14 Platoon of three vehicles controlled by the DMPC scheme

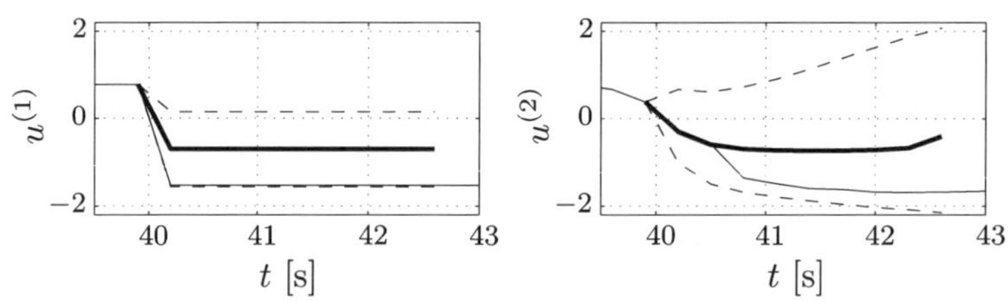

Fig. 4.15 Mismatch between actual inputs (solid line), predicted inputs (bold) and communicated bounds (dashed)

The resulting local problems are solved by an RMPC approach motivated by methods from robust optimization and feedback model predictive control.

In this DMPC scheme, we assume that information about other subsystems can only be obtained via communication, however if additional measurements are available they can be included into the scheme as described in [143]. A crucial aspect of the scheme is that the local controllers have to ensure consistency with previously communicated information. This renders the DMPC scheme recursively feasible in the presence of communication delays. The stability and convergence properties of the scheme depend on the communicated information and its corresponding uncertainty. In particular, if there is no mismatch between the actual and communicated future control actions, the scheme is asymptotically stable. It is apparent that this will almost never be the case due to replanning by the subsystems and the uncertainty induced by the communication network. In this case the mismatch between communicated and actual inputs as well as the control error are guaranteed to be bounded.

The main drawback of the proposed DMPC algorithm are the restrictions that are implicitly imposed on the interconnection topology. However, the general case with coupled constraints and without other strong assumptions on the interactions is commonly only treated by iterative schemes, which in turn offer no robustness against effects such as communication delays. Resolving this situation is a topic of future research, as well as the case of coupled dynamics. Undoubtedly, this will require more cooperation between the local controllers, in the sense that the influence on connected subsystems is adequately considered by each local controller.

4.4.4 Distributed MPC for Hybrid Systems

Hybrid systems are characterized by the combination of continuous and discrete states and inputs and can model a wide range of dynamic systems. However, properties such as discontinuities of the state derivatives or state trajectories and possibly nonlinear dynamics make distributed control of hybrid systems a difficult problem in general. Piecewise affine (PWA) systems often can be used to approximate the nonlinear continuous dynamics on a polyhedral partition of the continuous state-space $\mathbb{X}$, such that distributed control of piecewise affine systems becomes relevant.

While it may seem that results on DMPC with communication uncertainties for linear systems can easily be extended to piecewise affine systems, this is usually not the case because of the combinatorial aspect arising for PWA systems if the choice between different regions of $\mathbb{X}$ is present.

For instance, the robustness constraint used in Section 4.4.2 can, in principle, be computed for piecewise affine systems [306]. However, when applying this method to the example of platooning vehicles, for which hybrid dynamics arise from additional gear-shifts, numerical intractabilities are encountered due to the combinatorial nature of the problem. In particular, the maximal robust control invariant set of the PWA system has to be computed by an iterative procedure using on the union of convex polytopes. The number of polytopes potentially grows exponentially with the number of iterations and within each iteration any polytope has to be projected backwards for one time step, leading to a computationally expensive procedure.

The literature on RMPC of piecewise affine systems is rather sparse, but methods based on feedback min-max MPC using dynamic programming [204] and solutions based on nominal MPC with tightened constraints [216] can be found. These techniques appear, however, to be only applicable to systems of very low dimension or with only very small disturbance sets.

Therefore, modeling all interactions in distributed PWA systems as disturbance usually only works for weak or negligible interconnections, or leads to problems with far higher complexity than the original centralized problem. Considering this, a control approach which does not consider all interactions as disturbances is required. Motivated by the observation that a subsystem can directly influence the inputs of another subsystem via communication, we propose the following: Each subsystem Σ_i optimizes over the inputs of the subsystems Σ_j, $j \in \mathcal{N}_i$, where $\mathcal{N}_i$ again contains the index of subsystem Σ_i and of all subsystems connected to Σ_i. In contrast to the previous section, the solution is communicated to all subsystems Σ_j, $j \in \mathcal{N}$ with a delay $0 < d(k) < T$ that is smaller than the sampling time T.

One option to distribute the optimization is a sequential scheme, in which the subsystems optimize and communicate their input sequence in a specified order. While consistent propagation of information is ensured (compare to Section 4.3.3), a communication delay may degrade the performance significantly. Thus, the procedure to be described below makes, in contrast, use of a specific decision rule for deciding which input sequence is applied.

Consider N_l dynamically decoupled, discrete-time, piecewise affine subsystems Σ_i:

$$\boldsymbol{x}_i(k+1) = \boldsymbol{A}_{i,p}\boldsymbol{x}_i(k) + \boldsymbol{B}_{i,p}\boldsymbol{u}_i(k) + \boldsymbol{f}_{i,p}, \quad \text{if } \boldsymbol{x}_i(k) \in \mathcal{P}_{i,p}, \tag{4.34}$$

where $\boldsymbol{x}_i(k) \in \mathbb{X}_i \subseteq \mathbb{R}^{n_i}$ and $\boldsymbol{u}_i(k) \in \mathbb{U}_i \subseteq \mathbb{R}^{m_i}$ are the local states and inputs of Σ_i; $\mathcal{P}_{i,p}$ are convex polyhedral regions of the state-space associated with different affine dynamics given by $\boldsymbol{A}_{i,p}$ and $\boldsymbol{B}_{i,p}$. We assume that $\mathcal{P}_{i,0}$ contains 0 in its interior and the subsystems are subject to local input constraints $\boldsymbol{u}(k) \in \mathbb{U} := \mathbb{U}_1 \times \ldots \times \mathbb{U}_{N_l}$. Coupling is induced by global state constraints $\boldsymbol{x}(k) \in \mathbb{X}$ and a global cost function for a finite horizon of N time steps:

$$J_N = \|\boldsymbol{x}(k+N|k)\|_{\boldsymbol{P}}^2 + \sum_{l=0}^{N-1} \|\boldsymbol{x}(k+l|k)\|_{\boldsymbol{Q}}^2 + \|\boldsymbol{u}(k+l|k)\|_{\boldsymbol{R}}^2. \tag{4.35}$$

To ensure constraint satisfaction beyond the prediction Horizon N, a terminal constraint $\boldsymbol{x}(k+N|k) \in \Xi$ based on a terminal control law $\boldsymbol{u}(k+N|k) := \boldsymbol{K}\boldsymbol{x}(k+N|k)$ is employed. In many scenarios a decentralized terminal constrained may be constructed. We assume that a feasible initialization $\boldsymbol{u}_i(k+l|k)$ for $k=0$ is available to all subsystems and that information is exchanged among all controllers. Thus, the initial local input sequences and states are known to all controllers. Based on this information, each subsystem Σ_i solves at time k the following problem to compute a new input sequence

Algorithm 4.7. Distributed MPC algorithm for step k

1: **for** all subsystems $\Sigma_i \in \mathcal{N}$ **do**
2: Apply $\boldsymbol{u}_i(k|k{-}1)$, receive $\boldsymbol{x}_j(k|k{-}1)$, $\hat{\boldsymbol{u}}_j(k{+}l|k{-}1)$, $\forall j \in \mathcal{N}, \forall l \in \{0, \ldots, N-1\}$.
3: Compute the cost $J_{N,j}$ for $\hat{\boldsymbol{u}}_j(k{+}l|k-1)$ and set $\boldsymbol{u}_i(k{+}l|k) := \hat{\boldsymbol{u}}_{j^*}(k{+}l|k-1)$, $J_N := J_{N,j^*}$, with $j^* := arg\,min_j\,(J_{N,j})$.
4: Solve (4.36) and compute $\hat{\boldsymbol{u}}_i(k{+}N|k)$ using the terminal control law.
5: Communicate $\boldsymbol{x}_i(k+1|k)$, $\hat{\boldsymbol{u}}_i(k+l|k)$, $\forall l \in \{0, \ldots, N\}$ to all subsystems $\Sigma_j \in \mathcal{N}$.
6: **end for**

$\hat{\boldsymbol{u}}_i(k+l|k) := (\hat{\boldsymbol{u}}_{i,1}(k+l|k)^{\mathrm{T}}, \ldots, \hat{\boldsymbol{u}}_{i,N_l}(k+l|k)^{\mathrm{T}})^{\mathrm{T}}$ which includes inputs for all subsystems Σ_j, $j \in \mathcal{N}$.

$$\min_{\hat{\boldsymbol{u}}_i(k+l|k)} \quad \|\boldsymbol{x}(k+N|k)\|_{\boldsymbol{P}}^2 + \sum_{l=0}^{N-1} \|\boldsymbol{x}(k+l|k)\|_{\boldsymbol{Q}}^2 + \|\hat{\boldsymbol{u}}_i(k+l|k)\|_{\boldsymbol{R}}^2 \tag{4.36}$$

$$\begin{aligned}
\text{s.t. } & \text{dynamics (4.34)}, \boldsymbol{x}_j(k|k) = \boldsymbol{x}_j(k|k-1), && \forall j \neq i, i \in \mathcal{N} \\
& \boldsymbol{x}(k+l|k) \in \mathbb{X}, && \forall l \in \{1, \ldots, N-1\} \\
& \hat{\boldsymbol{u}}_i(k+l|k) \in \mathbb{U}, && \forall l \in \{1, \ldots, N-1\} \\
& \boldsymbol{x}(k+N|k) \in \mathbb{X}_{\mathbb{T}}, && \\
& \hat{\boldsymbol{u}}_{i,j}(k|k) = \boldsymbol{u}_{i,j}(k|k-1), && \forall j \in \mathcal{N} \\
& \hat{\boldsymbol{u}}_{i,j}(k+l|k) = \boldsymbol{u}_{i,j}(k+l|k-1), && \forall j \notin \mathcal{N}_i, l \in \{1, \ldots, N-1\}.
\end{aligned}$$

Due to the communication delay, the current input cannot be communicated before it has to be applied; therefore the second to last constraint restricts all inputs of the current time-step to the previously communicated values. The last constraint fixes the inputs of the subsystems $\Sigma_j \notin \mathcal{N}_i$, which are not directly interconnected to Σ_i, to previous values in order to reduce the computational load. The optimization is usually implemented as a mixed-integer program in order to account for the assignment of $\boldsymbol{A}_{i,p}$,$\boldsymbol{B}_{i,p}$, and $\boldsymbol{f}_{i,p}$ according to $\boldsymbol{x}_i(k) \in \mathcal{P}_{i,p}$. Fixing the input sequence of a subsystem, also fixes the corresponding integer variables associated with the piecewise dynamics.

Algorithm 4.7 describes the scheme and the procedure is shown over a timeline in Fig. 4.16. First, information is received from all subsystems and a previously optimized input is applied. The locally optimized inputs $\hat{\boldsymbol{u}}_i(k{+}l|k{-}1)$ are checked for the one with the lowest cost. Note that all subsystems perform the same computations in this step, thus the resulting input sequences $\boldsymbol{u}_i(k{+}l|k)$ will be identical across the subsystems. Finally the local optimization problem is solved and information is communicated to all subsystems.

Discussion. In sequential schemes for comparable problems, only one subsystem optimizes at a time step if delays are considered and subsystems may

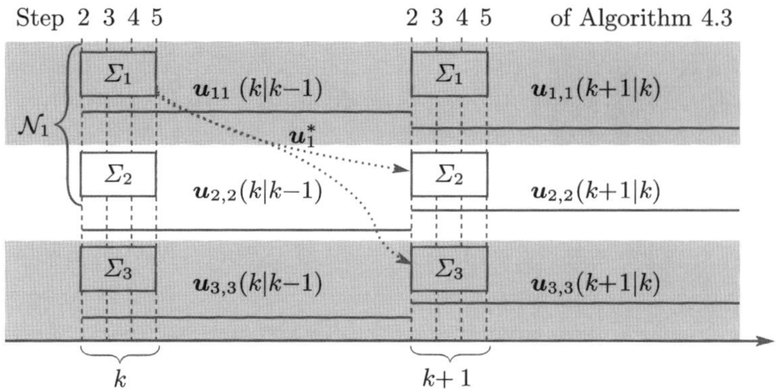

Fig. 4.16 Distributed control scheme for PWA Systems with corresponding steps from Algorithm 4.7 (dashed) and communication (dotted, only shown for Σ_1)

have to communicate more than once per local optimization with each connected subsystem, for instance, if the cost and terminal constraint are not separable (see Section 4.3.3). In the proposed scheme, each subsystem optimizes in each time step and the best local solution, which includes updated inputs for connected subsystems, is used. By combining the results of the local optimization, the costs may be improved further. However, this is beyond the scope of this section. Here, all controllers need to exchange information, what may lead to a higher communication load for large-scale systems. Due to the high computational complexity of MPC problems for PWA systems, MPC is only suitable for PWA systems with moderately fast dynamics. Thus, the assumption that the delay $d(k)$ is smaller than the sample time is not a severe restriction in most cases. Reducing the load on the communication network by exploiting the interconnection structure and using event-based communication is seen as area for future work. Also the investigation of the effects of packet loss, which have not been considered in distributed schemes for hybrid systems so far, is an interesting direction for future research.

4.5 Stochastic Model-Based Control with Virtual Control Inputs

In Section 4.2 a controller scheme is described where a controller that generates sequences of control inputs is used along with an actuator-buffer to compensate time delays and data loss in the network connection between actuator and controller. The method is particularly appealing in the context of MPC since a standard model predictive controller already calculates control input sequences as a byproduct of the receding-horizon optimization. The optimized control sequence only has to be sent to the actuator instead of being discarded. A problem of this procedure is, however, that in order to generate a control sequence, the controller needs not only to know which control inputs have been applied by the actuator, but also which ones will be applied in the future. Unfortunately, in the presence of time-varying transmission delays and stochastic data loss, in the general case, the controller only has uncertain information about the applied control inputs. One way to solve this problem, in the case of bounded time delays, is presented in Section 4.2, where different methods are proposed to ensure *prediction consistency* (see Definition 4.1). If this property holds, the controller has perfect knowledge about the content of the actuator-buffer and, thus, about the applied control inputs. However, ensuring prediction consistency comes at the price of additional time delays introduced into the system.

In this section, we choose a different design philosophy and do not try to restore the perfect information case by ensuring prediction consistency, but directly design the controller for the situation that only uncertain information about the applied control inputs is available. This requires a description of the control inputs that addresses this uncertainty. Therefore, we introduce the concept of *virtual control inputs*. Virtual control inputs characterize the potentially applied control inputs by means of probability density functions.

Based on this probabilistic description, two methods for controller design are presented. In the first method, the MPC Algorithm 4.1 is extended by the virtual control inputs for a nonlinear system with discrete-valued inputs. In the second method, we consider linear systems with continuous-valued inputs and extend a nominal feedback controller, which is designed for the system without consideration of network-induced effects, so that it generates control sequences suitable to compensate time delays and data loss in the network connection.

In the following section, we describe the system setup and its components in more detail. In Section 4.5.2, the concept of virtual control inputs is described. Based on this, Sections 4.5.3 and 4.5.4 present the two control methods.

4.5.1 System Setup and Controller-Actuator Scheme

In this section, we describe the considered system setup and introduce the notation needed in the next section. Furthermore, some aspects and assumptions regarding the controller-actuator scheme and the digital network are discussed.

We will consider three different system configurations as depicted in Fig. 4.17. In every configuration, there is a NETWORK between controller and actuator (CA-link). While in configuration (i), the state of the plant is known perfectly by the controller, in configuration (ii) the state of the plant is not directly accessible and, thus, measured by a sensor. Additionally, in configuration (iii) there is also a network present between sensor and controller (SE-link).

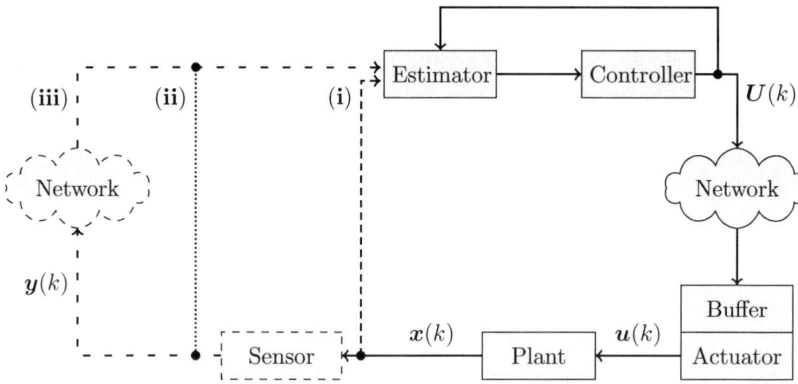

Fig. 4.17 Considered system: The controller generates a control input sequence $\boldsymbol{U}(k)$, which also contains predicted control inputs for future time steps. This sequence is transmitted in form of a data packet over a network to the actuator. The actuator holds the most recent control input sequence in a buffer and applies the time-corresponding entry of this sequence to the plant $\boldsymbol{u}(k)$. In configuration (i), the plant state $\boldsymbol{x}(k)$ is directly accessible by the estimator. In configuration (ii), the state is measured by a sensor and the measurements $\boldsymbol{y}(k)$ are available without time delay. In configuration (iii), the measurements are transmitted over a network.

To introduce the concept of virtual control inputs, we consider the following discrete-time stochastic nonlinear plant described in state-space form via

$$\boldsymbol{x}(k+1) = \boldsymbol{f}(\boldsymbol{x}(k), \boldsymbol{u}(k), \boldsymbol{w}(k), k) \tag{4.37}$$

$$\boldsymbol{y}(k) = \boldsymbol{g}\,(\boldsymbol{x}(k), \boldsymbol{v}(k), k)\ , \tag{4.38}$$

where $\boldsymbol{x}(k) \in \mathbb{R}^n$ denotes the system state at time step k, $\boldsymbol{u}(k) \in \mathbb{R}^s$ the control input applied by the actuator, and $\boldsymbol{y}(k) \in \mathbb{R}^q$ the measured output (if configuration (ii) or (iii) is considered). The terms $\boldsymbol{w}(k) \in \mathbb{R}^n$ and $\boldsymbol{v}(k) \in \mathbb{R}^q$ represent discrete-time white noise processes with probability density functions $\boldsymbol{p}^w(w(k), k)$ and $\boldsymbol{p}^v(v(k), k)$, respectively, that are independent of network-induced effects. It is assumed that the system model and the probability density functions are known by the controller. For the purpose of controller design, we consider two specializations of this setup: in Section 4.5.3, we assume that the control inputs $\boldsymbol{u}(k) \in \mathcal{U}_k$ take values from a finite discrete-valued set $\mathcal{U}(k)$ and in in Section 4.5.4 we consider a linear plant.

Controller and Actuator. As described in the introduction, the overall design idea is to use a controller that not only generates a single control input for the current control cycle, but also control inputs for future N time steps (with $N \in \mathbb{N}$). The whole control input sequence is dispatched in one data packet and sent over the network to the actuator. Attached to the actuator is a buffer, where the actuator stores the control input sequence with the most recent information received. If a data packet is received that contains older information than in the buffer, the new packet is discarded. In every time step, the actuator applies the appropriate control input of the buffered sequence to the plant, i.e., the control input of the sequence that corresponds to the current time step. By doing so, in case of time delays and data loss in the CA-link, the actuator can fall back upon former calculated control inputs. It may, however, happen that the buffer runs out of applicable control inputs. In this case, the actuator applies a default control input denoted by $\boldsymbol{u}_{\mathrm{def}}(k)$ that is known to the controller.

In the following, we will refer to a control input sequence generated by the controller at time k by $\boldsymbol{U}(k)$. Entries of that packet are denoted by $\boldsymbol{u}(k+m|k)$ with $m \in \{0, 1, ..., N\}$, where the first part of the argument $(k+m)$ gives the time step for which the control input is intended to be applied to the plant. The second part of the argument (k) specifies the time step, when the control input was generated. For a packet of length $N+1$ with $N \in \mathbb{N}$, which was generated in time step k, this gives

$$\boldsymbol{U}(k) = \{\boldsymbol{u}(k|k), \boldsymbol{u}(k+1|k), \ldots, \boldsymbol{u}(k+N|k)\} \ . \tag{4.39}$$

Example 4.6 *Controller-actuator procedure*

For example, let us assume the actual time step is $k = 5$ and the actuator receives the controller packet $\boldsymbol{U}(2)$. If the actuator has not received any of the packets $\boldsymbol{U}(3)$, $\boldsymbol{U}(4)$, and $\boldsymbol{U}(5)$, then the buffer is overwritten with $\boldsymbol{U}(2)$ and the 4-th entry of $\boldsymbol{U}(2)$, i.e., $\boldsymbol{u}(5|2)$, is applied to the plant. Otherwise, if the actuator has already received, for example, the sequence $\boldsymbol{U}(4)$, the received sequence $\boldsymbol{U}(2)$ is neglected and the 2nd entry of the buffered sequence $\boldsymbol{U}(4)$ is applied, i.e., $\boldsymbol{u}(5|4)$. □

Digital Network. The digital network between controller and actuator (CA-link) and (if present) between sensor and estimator (SE-link) are subject to time-varying delays and stochastic data loss. By interpreting lost transmissions as transmissions with infinite time delay, we unify the description of both effects by only considering time-varying but possibly *unbounded* time delays. The time delays are described by random processes $\tau^{CA}(k) \in \mathbb{N}$ and $\tau^{SE}(k) \in \mathbb{N}$ that specify how many time steps a packet will be delayed if sent at time step k. It is assumed that $\tau^{CA}(k)$ and $\tau^{SE}(k)$ are white stationary processes that are mutually independent and independent of $\boldsymbol{v}(k)$ and $\boldsymbol{w}(k)$ and that their probability density functions $\boldsymbol{p}^{CA}(\tau^{CA})$ and $\boldsymbol{p}^{SE}(\tau^{SE})$ are known.

In addition, it is assumed that the components of the control loop are time-triggered, time-synchronized, and have identical cycle times. Furthermore, the employed network is capable of transmitting large time-stamped data packets and uses a so called UDP-like protocol, i.e., the network does not provide acknowledgements for successfully transmitted data packets.

In the literature, also so called TCP-like protocols are considered, which assume that successfully transmitted data packets are instantaneously acknowledged at the sender side. In case of the CA-link, using a TCP-like protocol implies that the controller always has perfect information about the applied control inputs. It has been shown [121] that under this condition, the separation principle holds even in the sequence-based setup. Furthermore, the optimal sequence-based controller was derived in [121] and extended in [123] to the setup with multiple sensors and sequence-based information structure in the SE-link. In the following, we focus, however, on the more general case of UDP-like protocols.

4.5.2 Concept of Virtual Control Inputs

In the literature, a widely used approach for modeling time-varying time delays (and data loss) in the CA-link is to describe network and actuator as a Markov Jump Linear System (MJLS) [104]. The MJLS description consists of a set of linear systems, where at every time step, only one system of the set is active. Each of the linear systems relates controller sequences $\boldsymbol{U}(\cdot)$ with an actuator output $\boldsymbol{u}(k)$ for a specific time delay. Which one of the possible systems is active is determined by the realization of a random process that reflects the stochastic characteristics of the time delays. Therefore, the MJLS relates *realizations* of the applied control inputs $\boldsymbol{u}(k)$ with *realizations* of the time delays. This kind of model is called a generative model. It is convenient for stability analysis and for controller synthesis in the case of TCP-like protocols.

However, in the considered case of UDP-like protocols, a probabilistic description of the applied control input $\boldsymbol{u}(k)$, which relates the probability density function of the time delays with the probability density function of the applied control input $\boldsymbol{u}(k)$, is more useful. This description is advantageous as, on the one hand, the applied control inputs are in fact stochastic and, on the other hand, the state of the plant has to be estimated when it is not directly accessible as in configurations (ii) and (iii). There, the estimation of the state is strongly coupled with the estimation of the applied control inputs so that a probabilistic description is needed for the estimator design. In the following, we introduce a probabilistic description by means of so called virtual control inputs [170].

Definition 4.2 (Virtual control input). *A virtual control input $\hat{\boldsymbol{u}}(k+m|k)$ is a random variable, characterized by a probability density function $\boldsymbol{p}(\hat{\boldsymbol{u}}(k+m|k), k)$ that describes the control input $\boldsymbol{u}(k+m)$ applied to the plant at time step $k+m$ conditioned on the information available to the controller at time step k.*

The information available to the controller is denoted by the set $\mathcal{I}(k)$ and consists of the measured states $\boldsymbol{x}(0), \ldots, \boldsymbol{x}(k)$ (respectively $\boldsymbol{y}(0), \ldots, \boldsymbol{y}(k)$ in system configurations (ii) and (iii)), the last N packets $\boldsymbol{U}(k-1), \ldots, \boldsymbol{U}(k-N)$ the controller fed into the network, the model of the plant, the characteristics of the process noise and the delay distribution of the network.

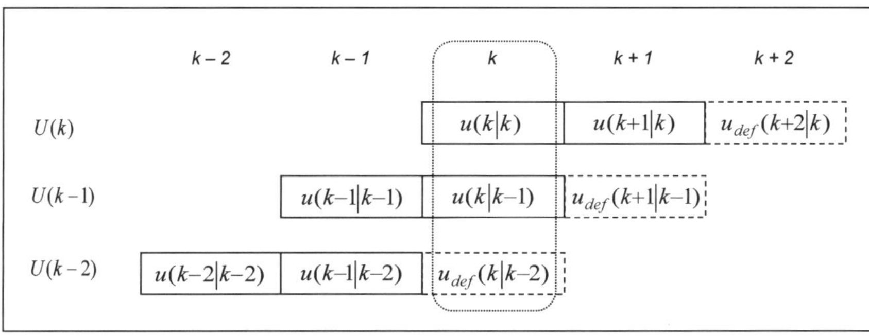

Fig. 4.18 Representation of the control input sequences $\boldsymbol{U}(k-2)$, $\boldsymbol{U}(k-1)$, and $\boldsymbol{U}(k)$, where every sequence consists of two control inputs. For the sake of completeness, the default control input $\boldsymbol{u}_{def}(\cdot)$ is added to the end of each sequence. Control inputs applicable to the same time step are vertically aligned. The rounded rectangle, e.g., marks the control inputs that could be applied at time step k

At a certain time step, there is only a finite set of deterministic control input candidates that, based on $\mathcal{I}(k)$, could be applied by the actuator. This is illustrated in Fig. 4.18. The virtual control inputs characterize these candidates via a probability density function.

Example 4.7 *Virtual control inputs*

For example, if the controller calculates at time k the current control input $\boldsymbol{u}(k|k)$ and one predicted control input for the next time step $\boldsymbol{u}(k+1|k)$, a control sequence with two elements is sent to the actuator. Let the probabilities be 0.2 that there is no delay and 0.8 that the delay is one time step. Then, the packet has to arrive at the current time step or the next time step.

From the controller's point of view, the applied control input is uncertain, since there are always two control input candidates that are possible to be applied: at time step $k+1$ these would be $\boldsymbol{u}(k+1|k+1)$ (applied with probability 0.2), which the controller is calculating in this time step, and $\boldsymbol{u}(k+1|k)$ (applied with probability

0.8) that was calculated in the last time step. A virtual control input describes this situation by means of a probability density function, which in this case is a Dirac mixture density with components at the positions $\boldsymbol{u}(k+1|k+1)$ and $\boldsymbol{u}(k+1|k)$ with weighting factors 0.2 and 0.8, respectively. □

The probability density of a set of deterministic values that are chosen with a certain probability is given by a Dirac mixture density, so that the virtual control input can be described by

$$\begin{aligned} h\left(\hat{\boldsymbol{u}}\left(k+m|k\right)\right) = \xi_{N+1}(k+m|k) \cdot \delta\left[\boldsymbol{u}^r\left(k+m|k\right) - \boldsymbol{u}_{\text{def}}\left(k\right)\right] + \\ \sum_{i=0}^{N} \xi_i(k+m|k) \cdot \delta\left[\boldsymbol{u}^r(k+m|k) - \boldsymbol{u}(k+m|k-i)\right] \end{aligned} \tag{4.40}$$

$$\text{with} \quad m \in \{0, 1, \cdots, N\} , \quad \sum_{i=0}^{N+1} \xi_i(k+m|k) = 1 ,$$

where $\delta(\cdot)$ is the Dirac delta function, $\boldsymbol{u}^r(k+m|k)$ denotes the domain of the realizations of the virtual control input $\hat{\boldsymbol{u}}(k+m|k)$, and $\boldsymbol{u}_{\text{def}}$ is the known default control input in case the buffer runs empty. The weighting factors $\xi_i(k+m|k)$ express the probability that the corresponding control input $\boldsymbol{u}(k+m|k-i)$ is applied by the actuator and can be calculated by

$$\xi_i(k+m|k) = \text{Prob}\big(\boldsymbol{u}(k+m) = \boldsymbol{u}(k+m|k-i)|\mathcal{I}(k)\big) .$$

The control input $\boldsymbol{u}(k+m|k-i)$ is applied by the actuator if the sequence buffered in the actuator at time step $k+m$ has been generated by the controller $k+m-(k-i) = m+i$ time steps ago. In other words, $\boldsymbol{u}(k+m|k-i)$ is applied by the actuator if the *age* of the buffered sequence, i.e., the difference between time step of generation and actual time step, at time step $k+m$ is equal to $m+i$. In the following, we denote the age of the buffered sequence at time step k by $\theta(k)$. With this notation, it holds that

$$\xi_i(k+m|k) = \text{Prob}\big(\theta(k+m) = i|\mathcal{I}(k)\big) .$$

Therefore, the weighting factors can be interpreted as estimates of $\theta(k+m)$. It is shown in [171] that $\theta(k)$ can be formulated as the state of a Markov chain (with state space $\{0, 1, 2, ..., N+1\}$) that is governed by the transition matrix $\boldsymbol{P}$ given by

$$\boldsymbol{P} = \begin{pmatrix} p_{0,0} & p_{0,1} & 0 & 0 & \cdots & 0 \\ p_{1,0} & p_{1,1} & p_{1,2} & 0 & \cdots & 0 \\ p_{2,0} & p_{2,1} & p_{2,2} & p_{2,3} & \cdots & 0 \\ \vdots & \vdots & \vdots & \vdots & \ddots & \vdots \\ \vdots & \vdots & \vdots & \vdots & & p_{N,N+1} \\ p_{N+1,0} & p_{N+1,1} & p_{N+1,2} & p_{N+1,3} & \cdots & p_{N+1,N+1} \end{pmatrix} \tag{4.41}$$

$$\text{with} \quad p_{i,j} = \begin{cases} 0 & \text{for } j \geq i+2 \\ 1 - \sum\limits_{r=0}^{i} q_r & \text{for } j = i+1 \\ q_j & \text{for } j \leq i < N+1 \\ 1 - \sum\limits_{r=0}^{N} q_r & \text{for } j = i = N+1\,. \end{cases}$$

The term q_i (with $i \in \mathbb{N}$) describes the probability of the event that a packet is delayed by i time steps. These probabilities can be derived as the probability density function of the time delays of the network connection are known. Arranging the weighting factors $\xi_i(k+m|k)$ in form of a vector

$$\boldsymbol{\xi}(k+m|k) = [\xi_0(k+m|k), \quad \dots, \quad \xi_{N+1}(k+m|k)]^{\mathrm{T}} \ ,$$

it holds that

$$\boldsymbol{\xi}(k+m|k) = \frac{\left[\boldsymbol{I}_{N+2-m} \ \mathbf{0}_{(N+2-m)\times m}\right] \left(\boldsymbol{P}^m\right)^{\mathrm{T}} \cdot \boldsymbol{\xi}(k|k)}{\mathbf{1}^{\mathrm{T}} \cdot \left[\boldsymbol{I}_{N+2-m} \ \mathbf{0}_{(N+2-m)\times m}\right] \left(\boldsymbol{P}^m\right)^{\mathrm{T}} \cdot \boldsymbol{\xi}(k|k)} \ , \tag{4.42}$$

where $\mathbf{1}$ denotes a column vector of appropriate dimension with all entries equal to 1. The term in the numerator represents a m-step future prediction of $\theta(k)$, where only the first $N+2-m$ entries are kept. The denominator of (4.42) normalizes the extracted subset of the predicted vector so that $\boldsymbol{\xi}(k+m|k)$ sums up to one.

The term $\boldsymbol{\xi}(k|k)$ can be interpreted as an estimate of $\theta(k)$ arranged in vector form. If $\boldsymbol{\xi}(k|k)$ is known, the other weighting factors $\boldsymbol{\xi}(k+1|k)$ up to $\boldsymbol{\xi}(k+N|k)$ can be easily derived by means of the transition matrix $\boldsymbol{P}$ according to (4.42). In the following, we briefly survey how the weighting factors $\boldsymbol{\xi}(k|k)$, i.e., the estimate of $\theta(k)$ in vector form, can be obtained. The estimation algorithm differs depending on the considered system configuration shown in Fig. 4.17. It should be noticed that in configuration (ii) and (iii), the estimation of $\theta(k)$ is strongly coupled with the estimation of the state $\boldsymbol{x}(k)$ and both have to be treated simultaneously.

(i) Perfect State Information. If the controller has perfect knowledge about the actual state of the plant, the configuration corresponds to a *hidden Markov model* with continuous-valued outputs and the continuous-valued version of the Wonham filter can be used together with (4.42). It holds

$$\boldsymbol{\xi}(k|k) = \frac{\boldsymbol{H} \cdot \boldsymbol{\xi}(k|k-1)}{\mathbf{1}^{\mathrm{T}} \cdot \boldsymbol{H} \cdot \boldsymbol{\xi}(k|k-1)} \tag{4.43}$$

$$\boldsymbol{H} = \mathrm{diag}\begin{pmatrix} \boldsymbol{p}^w\left(\boldsymbol{x}(k) - \boldsymbol{f}\left(x(k-1), \boldsymbol{u}(k-1|k-1), k\right), k\right) \\ \boldsymbol{p}^w\left(\boldsymbol{x}(k) - \boldsymbol{f}\left(x(k-1), \boldsymbol{u}(k-1|k-2), k\right), k\right) \\ \vdots \\ \boldsymbol{p}^w\left(\boldsymbol{x}(k) - \boldsymbol{f}\left(x(k-1), \boldsymbol{u}(k-1|k-N), k\right), k\right) \\ \boldsymbol{p}^w\left(\boldsymbol{x}(k) - \boldsymbol{f}\left(x(k-1), \boldsymbol{u}_{\mathrm{def}}(k), k\right), k\right) \end{pmatrix} \ , \tag{4.44}$$

where diag$(\boldsymbol{s})$ denotes a matrix with the elements of the vector $\boldsymbol{s}$ on the diagonal and zeros elsewhere.

(ii) Imperfect State Information without Time Delays. For this case the optimal estimator for $\theta(k)$ and $\boldsymbol{x}(k)$ is obtained from a bank of Kalman filters, whose memory requirement and computational load increase exponentially with time [16]. Since this is not an applicable solution, a suboptimal filter has been proposed in [122]. The algorithm is based on the *generalized pseudo Gaussian* (GPB) and *interacting multiple model* (IMM) algorithms, where a merging technique is used to reduce the increasing complexity by combining state estimates that originate from similar hypothesis about the unknown variable θ_k. Also pruning techniques are discussed [122], such as the *B-best* and *Viterbi algorithm*, which make hard decisions about the mode history and only keep the state estimates related to the most likely hypothesis about θ_k. However, it is pointed out that if the model of the system can differ significantly from the real system, a merging strategy should be preferred. A detailed description of the mentioned algorithms can also be found in [229].

(iii) Imperfect State Information with Time Delays. If a network is present in the SE-link, then also measurements can be delayed or lost. In [122], methods are surveyed how delayed measurements, measurements that arrive out of sequence, and measurements that arrive in bursts, can be efficiently incorporated in the estimator design.

In order to deal with packet loss in the SE-link, the sequence-based design philosophy can be extended to the SE-link by transmitting sequences that also contain measurements of former time steps. It has been shown [123] that, even in an extended scenario with multiple sensors, the information contained in a measurement sequence can be recursively compressed by the sensors without loss of information. Therefore, not the whole sequence has to be sent, but only a fraction of that data.

4.5.3 Model Predictive Controller Design Using Virtual Control Inputs

In this section, it is described how the concept of virtual control inputs can be used in model predictive controller design to compensate time delays and data loss in the CA-link. We consider the system (4.37) and (4.38), where, however, the control inputs $\boldsymbol{u}(k) \in \mathcal{U}_k$ take values from a finite discrete-valued set. As shown in [169], the first step to extend the basic MPC Algorithm 4.1 is to replace the deterministic cost function (4.6) by a stochastic version that is defined on the *expected* value of the cost. As a second step, the virtual control inputs have to be included in the optimization problem. This is done by replacing part 1 of Algorithm 4.1 by the following stochastic open-loop optimization problem with terminal time $T \in \mathbb{N}$:

$$\min_{\boldsymbol{U}(k)} \mathrm{E}\left\{ \sum_{i=k}^{k+T-1} l(\boldsymbol{x}(i), \hat{\boldsymbol{u}}(i|k)) + l(\boldsymbol{x}(k+T)) \middle| \mathcal{I}(k) \right\} \tag{4.45}$$

subject to

$$\begin{aligned} &\boldsymbol{x}(i+1) = \boldsymbol{f}\left(\boldsymbol{x}(i), \hat{\boldsymbol{u}}(i|k), \boldsymbol{w}(i), i\right) && \text{and} \\ &\boldsymbol{u}(i|k) \in \mathcal{U}(i) && \text{with} \\ &i = k, k+1, \ldots, k+T \ . \end{aligned}$$

Note that the cumulative cost is still minimized over *deterministic* input sequences. But, in contrast to (4.7), the control inputs not only influence the future system states $\boldsymbol{x}_u(k, x_0)$, but also the probability distributions of the virtual control inputs $\hat{\boldsymbol{u}}(k|k), \ldots, \hat{\boldsymbol{u}}(k+T-1|k)$ by adding further components to the Dirac mixture densities. Furthermore, it is worth mentioning that the expected value is not only taken over $\boldsymbol{x}(k), \ldots, \boldsymbol{x}(T)$, but also over $\hat{\boldsymbol{u}}(k|k), \ldots, \hat{\boldsymbol{u}}(k+T-1|k)$.

Due to the fact that we consider systems with discrete-valued control inputs, the optimization problem (4.45) can be represented by a decision tree. To reduce the complexity of this kind of optimization problem, techniques such as pruning, merging, and branch-and-bound can be applied.

4.5.4 Model-Based Extension of Feedback Controllers Using Virtual Control Inputs

This section is based on [170] and describes how the concept of virtual control inputs can be used to extend a given feedback-controller, so that it generates control input sequences suitable to compensate network-induced effects in the CA-link.

We consider a discrete-time *linear* plant with *continuous-valued* inputs described by

$$\boldsymbol{x}(k+1) = \boldsymbol{A}\boldsymbol{x}(k) + \boldsymbol{B}\boldsymbol{u}(k) + \boldsymbol{w}(k) \tag{4.46}$$

$$\boldsymbol{y}(k) = \boldsymbol{C}\boldsymbol{x}(k) + \boldsymbol{v}(k) \tag{4.47}$$

with $\boldsymbol{u}(k) \in \mathbb{R}^q$ and the noise processes $\boldsymbol{w}(k) \in \mathbb{R}^n$ and $\boldsymbol{v}(k) \in \mathbb{R}^q$ are supposed to be stationary and zero mean.

We assume that a linear feedback-controller of the form

$$\boldsymbol{u}(k) = \boldsymbol{K}\boldsymbol{x}(k) \tag{4.48}$$

is given, where the feedback matrix $\boldsymbol{K}$ was designed for the plant (4.46) without consideration of network effects, e.g., by pole placement, LQR, H_2 or H_∞ methods. In the following, we use the feedback matrix $\boldsymbol{K}$ to generate control input sequences based on the predicted future states of the plant, which are influenced by virtual control inputs.

Based on the estimated state $\mathrm{E}\{\boldsymbol{x}(k)\}$, which corresponds to $\boldsymbol{x}(k)$ in configuration (i), the entries of the control input sequence $\boldsymbol{U}(k)$ are calculated by

$$\boldsymbol{u}(k|k) = \boldsymbol{K} \cdot \mathrm{E}\{\boldsymbol{x}(k)\} \tag{4.49}$$

$$\boldsymbol{u}(k+1|k) = \boldsymbol{K} \cdot \mathrm{E}\{\boldsymbol{x}(k+1|k)\} \tag{4.50}$$

$$\vdots$$

$$\boldsymbol{u}(k+N|k) = \boldsymbol{K} \cdot \mathrm{E}\{\boldsymbol{x}(k+N|k)\}\ , \tag{4.51}$$

where $\boldsymbol{x}(k+m|k)$ denotes the prediction of state $\boldsymbol{x}(k+m)$ based on an estimate of $\boldsymbol{x}(k)$ and the virtual control inputs $\hat{\boldsymbol{u}}(k|k), \ldots, \hat{\boldsymbol{u}}(k+m|k)$. The state predictions are random with respect to the process noise and the virtual control inputs and can be calculated for time step $k+1$ by

$$\begin{aligned}
\mathrm{E}\{\boldsymbol{x}(k+1|k)\} &= \mathop{\mathrm{E}}_{w,u}\{\boldsymbol{A}\boldsymbol{x}(k) + \boldsymbol{B}\boldsymbol{u}(k) + \boldsymbol{w}(k)\} \\
&= \boldsymbol{A} \cdot \mathrm{E}\{\boldsymbol{x}(k)\} + \boldsymbol{B} \cdot \mathrm{E}\{\hat{\boldsymbol{u}}(k|k)\} + \mathrm{E}\{\boldsymbol{w}(k)\} \\
&= \boldsymbol{A} \cdot \mathrm{E}\{\boldsymbol{x}(k)\} + \boldsymbol{B} \cdot \left(\sum_{i=0}^{N} \xi_i(k|k) \cdot \boldsymbol{u}(k|k-i) + \xi_{N+1}(k|k) \cdot \boldsymbol{u}_{\mathrm{def}}(k) \right)
\end{aligned}$$

and, in general, for time steps $k+m$ by

$$\begin{aligned}
\mathrm{E}\{\boldsymbol{x}(k+m|k)\} &= \mathop{\mathrm{E}}_{w,u}\{\boldsymbol{A}\boldsymbol{x}(k+m-1|k) + \boldsymbol{B}\boldsymbol{u}(k+m-1) + \boldsymbol{w}(k+m-1)\} \\
&= \boldsymbol{A} \cdot \mathop{\mathrm{E}}_{w,u}\{\boldsymbol{x}(k+m-1|k)\} + \boldsymbol{B} \cdot \mathrm{E}\{\hat{\boldsymbol{u}}(k+m-1|k)\} + \mathrm{E}\{\boldsymbol{w}(k)\} \\
&= \boldsymbol{A}^m \cdot \mathrm{E}\{\boldsymbol{x}(k)\} + \sum_{j=0}^{m-1} \sum_{i=0}^{N-j} \boldsymbol{A}^{m-j-1}\boldsymbol{B} \cdot (\xi_i(k+j|k) \cdot \boldsymbol{u}(k+j|k-i) + \\
&\quad \xi_{N-j+1}(k+j|k) \cdot \boldsymbol{u}_{\mathrm{def}}(k+j|k))\,.
\end{aligned} \tag{4.52}$$

For taking the expected value, we have used the assumption that $\boldsymbol{w}(k)$ is zero mean and independent of $\boldsymbol{x}(k)$ and $\hat{\boldsymbol{u}}(k|k)$. Furthermore, we have used that given the available information to the controller at time step k, it holds for the expected value of the future control inputs

$$\begin{aligned}
&\mathrm{E}\{\hat{\boldsymbol{u}}(k+m|k)\} = \int_{-\infty}^{\infty} \boldsymbol{u}^r(k+m|k) h\left(\hat{\boldsymbol{u}}(k+m|k)\right) d\boldsymbol{u}^r(k+m|k) \\
&= \int_{-\infty}^{\infty} \boldsymbol{u}^r(k+m|k) \left(\xi_{N+1}(k+m|k) \cdot \delta\left[\boldsymbol{u}^r(k+m|k) - \boldsymbol{u}_{\mathrm{def}}(k)\right] + \right. \\
&\left. \sum_{i=0}^{N} \xi_i(k+m|k) \cdot \delta\left[\boldsymbol{u}^r(k+m|k) - \boldsymbol{u}(k+m|k-i)\right]\right) d\boldsymbol{u}^r(k+m|k)
\end{aligned}$$

$$= \sum_{i=0}^{N-m} \xi_i(k+m|k) \cdot \boldsymbol{u}(k+m|k-i) + \xi_{N-m+1}(k+m|k) \cdot \boldsymbol{u}_{\mathrm{def}} \ . \tag{4.53}$$

The extended controller given by (4.49), (4.50), (4.51), and (4.52) is nonlinear, because the weighting factors $\boldsymbol{\xi}(k|k)$ depend on the measured state as described in Section 4.5.2. To reduce the complexity of this controller, the weighting factors $\boldsymbol{\xi}(k|k)$ can be approximated by its stationary probability solution that can be computed by the equilibrium equation

$$\boldsymbol{\xi}_\infty = \boldsymbol{P}^{\mathrm{T}} \cdot \boldsymbol{\xi}_\infty \ , \tag{4.54}$$

which always has a unique solution according to Markov chain theory. With this approximation and considering configuration (i), the resulting controller is linear in the measured state $\boldsymbol{x}(k)$ and in the control input sequences $\boldsymbol{U}(k), \ldots, \boldsymbol{U}(k-N-1)$, what follows from (4.51), (4.52), and (4.54). Therefore, the approximated controller can be formulated as a linear feedback controller

$$\boldsymbol{U}(k+1) = \widetilde{\boldsymbol{K}} \cdot \boldsymbol{\eta}(k) \ ,$$

working on the augmented state

$$\boldsymbol{\eta}(k) = \left[\boldsymbol{x}^{\mathrm{T}}(k), \boldsymbol{U}^{\mathrm{T}}(k), \cdots, \boldsymbol{U}^{\mathrm{T}}(k-N-1)\right]^{\mathrm{T}} .$$

The approximated extended controller is easy to implement and, in particular, suitable for scenarios where calculation capacity is strongly limited. The stability properties of this controller have been investigated in [170], where a criterion for mean square stability of the closed-loop system was derived.

Summary. We pointed out that the controller only has uncertain information about the applied control inputs available, when NCS with time-varying transmission delays and stochastic data loss are considered. While most transmission try to reestablish the perfect information about the applied control inputs by special measures, we proposed a method to directly incorporate the uncertainties into the controller design. This method was derived by formulation the uncertain control inputs as random processes described by sets of discrete probability density functions, called virtual control inputs. Finally, we showed that the concept of the virtual control inputs cannot only be used for controller design, but also for extending a given standard feedback controller to the NCS situation.

5

Event-Based Control

L. Grüne, S. Hirche, O. Junge, P. Koltai, D. Lehmann, J. Lunze, A. Molin, R. Sailer, M. Sigurani, C. Stöcker, and F. Wirth

In event-based control, the feedback loop is closed only if an event indicates that the control error exceeds a tolerable bound and triggers a data transmission from the sensors to the controllers and the actuators. Hence, event-based control is an important method for reducing the communication load of a digital network. This chapter explains the main ideas of event-based control and proposes new loop structures and design methods.

Chapter Contents

5.1 Introduction to Event-Based Control

Event-based control is a control methodology that is currently being developed as a means to reduce the communication between the sensors, the controller and the actuators in a control loop. The sampling instants are not determined periodically by a clock, but by an event generator, which adapts the information flow in the feedback loop to the current behavior of the closed-loop system. A communication among the components is invoked only after an event has indicated that the control error exceeds a tolerable bound.

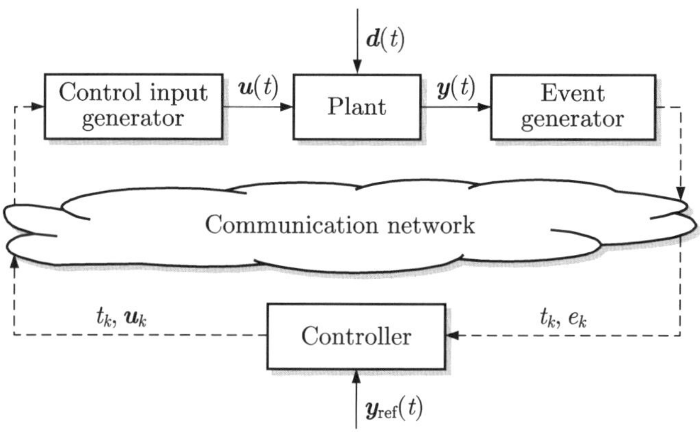

Fig. 5.1 Event-based control loop

This working principle differs fundamentally from that of sampled-data feedback loops, in which the sensor data are communicated to the controller at equidistant sampling times. In periodic sampling, a communication takes place independently of the size of the control error and, in particular, also in case of small control errors when an information feedback is not necessary to satisfy the performance requirements on the closed-loop system. In these situations, the communication and computing resources are used unnecessarily.

Figure 5.1 shows the main components of an event-based control loop. The plant has the continuous-time or discrete-time input $\boldsymbol{u}(t)$ and the state $\boldsymbol{x}(t)$ (or output $\boldsymbol{y}(t)$), which are continuously generated by the control input generator or continuously evaluated by the event generator, respectively. The communication links drawn by dashed arrows are only used after the event generator has indicated that the control error exceeds a tolerable bound at some time t_k, where an event name e_k, the current state $\boldsymbol{x}(t_k)$ or the current output $\boldsymbol{y}(t_k)$ is transmitted to the controller. The controller determines the new input $\boldsymbol{u}_k$, which is used by the control input generator to determine the continuous input $\boldsymbol{u}(t)$ in the time interval $[t_k, t_{k+1})$ until the next event occurs at time t_{k+1}.

Theoretical Challenges. In event-based control, the fundamental assumption of sampled-data control theory claiming a periodic triggering scheme invoked by a clock is violated. Hence, for event-based control the well-known discrete-time models of the plant, the controller and the closed-loop system cannot be applied but a new theory has to be developed, which takes into account the asynchronous component behavior. The main novel analysis and design aims of this new theory refer to the choice of the event generator and of the control input generator.

The **event generator** determines

- the time instants t_k, $(k = 0, 1, ...)$ at which the next communication between the event generator, the controller and the control input generator is invoked, and
- the information that is communicated from the sensor towards the controller.

The **control input generator** determines the signal $\boldsymbol{u}(t)$ continuously for the time interval $t \in [t_k, t_{k+1})$ in dependence upon the information obtained at time t_k.

The **main questions** to be answered ask

- at which time t_k a feedback loop has to be closed by using the communication links,
- which information e_k should be communicated and
- how the control input generator should determine the control input $\boldsymbol{u}(t)$ between succeeding event times.

Several different methods for event-based control have been proposed in the recent years, which distinguish with respect to the answers given to these questions. Some of them have been published under different names like *event-driven control*, *event-triggered control*, *Lebesgue sampling*, *deadband control* or *send-on-delta control* . Surveys and introductions to these techniques can be found in [241, 354].

A similar, but conceptually different methodology is *self-triggered control.* Here the plant state is not continuously supervised by the event generator, but the next event time t_{k+1} is determined by the event generator at the event time t_k [6, 379]. Then the sensors can "sleep" until the predicted next sampling instant.

Application Scenarios. There are several reasons for using event-based feedback. First, the information exchange in the feedback loop should be reduced to the minimum communication that is necessary to ensure a required system performance. If the information is transferred to and from the controller by a digital communication network, a reduced information flow decreases the risk of a network overload. For wireless nodes, reduced activity saves energy.

The second motivation occurs for systems, where the physical structure requires that measurements or control actions have to be taken at time instants prescribed by the dynamics of the plant. For example, position measuring devices of rotating components work with markings on the axis, which provide position information not at specific clock times but in certain positions.

Third, asynchronous communication protocols and real-time software do not allow to transfer and process information at specific clock times but lead to an inherently asynchronous behavior of all components of a feedback loop. Sensors and actuators likewise are triggered by events, because they work if some new information arrives. For such components the event-based working scheme is more "natural" than periodic sampling.

Fundamental Properties of Event-Based Control. The event-driven function principle implies that the plant input $\boldsymbol{u}(t)$ is determined by a combination of feedforward control and feedback control. At event times t_k, the input $\boldsymbol{u}(t_k)$ depends in a closed-loop fashion upon the current state $\boldsymbol{x}(t_k)$ (provided that the communication links do not introduce a substantial time delay), whereas between two consecutive event times t_k and t_{k+1} the input $\boldsymbol{u}(t)$ is generated as open-loop control in dependence upon "old data" $\boldsymbol{u}_k$.

Due to the aim of event-based control to sample only if a severe performance degradation has to be avoided, most event-based control schemes cannot ensure asymptotic stability of the closed-loop system. Instead, the plant state $\boldsymbol{x}(t)$ should be held in the surroundings Ω of the equilibrium state $\bar{\boldsymbol{x}}$. The property $\boldsymbol{x}(t) \in \Omega$ for all $t \geq \bar{t}$ is called *ultimate boundedness* or *practical stability* of the closed-loop system (cf. Def. 5.1). Usually the size of the set Ω depends upon the event threshold $\bar{e}$.

Comparison of Event-Based and Sampled-Data Control. Analytical results showing the difference of event-based control and sampled-data control can only be obtained for first-order systems [11, 13, 226, 303]. They show that event-based sampling can lead indeed to a considerable reduction of the communication within the control loop. Furthermore, if the system is heavily disturbed, the event-based control loop may have a better performance than the sampled-data loop, because in this situation it invokes the communication more often than the clock.

Chapter Overview. This chapter surveys methods that are applicable for high-order systems and different event-triggering and control input generation methods. Five different methods will be described in the sequel:

- **Event-based state feedback** (Section 5.2): The control input generator is constructed such that the closed-loop system mimics the behavior of a continuous state-feedback loop with adjustable precision. It is shown that the control input generator has to include a model of the continuous state-feedback loop.

- **Distributed event-based control** (Section 5.3): For interconnected systems, the idea of event-based state feedback can either be implemented as a distributed controller, which leads to the same overall system performance as the centralized feedback, or can be applied separately to the isolated subsystens resulting in a decentralized event-based control scheme.

- **Optimization-based control** (Section 5.4): The event-based controller can be obtained as the solution of an optimal control problem, if the state space is partitioned and the best possible constant input is applied as long as the state remains in the same state-space partition.

- **Event-based stabilization of large-scale systems** (Section 5.5): The stability of an interconnected system is tested by a small-gain theorem, which also evaluates the robustness of the event-based control loop against uncertainties in the communication channel.

- **Event-based control of interconnected nonlinear systems** (Section 5.6): This section extends the idea of event-based control to nonlinear interconnected systems and proves that robustly stable controllers of the subsystems lead to an input-to-state stable overall system.

- **Event-based control of stochastic systems** (Section 5.7): If formulated as a stochastic optimization problem, the event generator and the control input generator have to be designed simultaneously leading to a very complex optimization problem. If, however, both components lie in a nested information structure, where the event generator knows the plant state $\boldsymbol{x}(t)$ for all time t and the control input generator knows only the state $\boldsymbol{x}(t_k)$ at an event time t_k, the overall problem can be decomposed into two nested optimization problems that can be solved sequentially.

There is an interesting similarity concerning the structure of the event-based controllers that are developed in Sections 5.2 and 5.7 by starting from quite different viewpoints. Section 5.7 looks for optimal event-triggering and input-generating policies and ends up with an event generator and an control input generator that include a model of the closed-loop system. This structure is quite similar to the control loop elaborated in Section 5.2, which is obtained by investigating the control law of a state-feedback controller and implementing it in an event-based fashion. Likewise, the event generator and the control input generator have to include a model of the continuous control loop.

The methods developed in Sections 5.3 and 5.6 have been experimentally evaluated at a common thermofluid process, which is described in Section 5.8.

5.2 Disturbance Attenuation by Event-Based State Feedback

5.2.1 Control Aim

This section focusses on the influence of unknown disturbances on event-based control. It introduces a state-feedback approach to event-based control with the design aim to mimic the disturbance behavior of a given continuous closed-loop system, which is assumed to have the "best possible" disturbance attenuation properties.

In the sequel, the event-based control loop shown in Fig. 5.1 is simplified as depicted in Fig. 5.2. Here, the controller is incorporated in the control input generator.

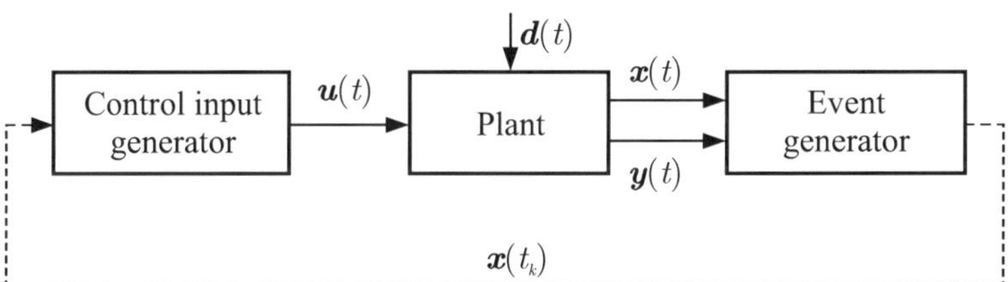

Fig. 5.2 Simplified event-based control loop

The plant is represented by the linear state-space model

$$\dot{\boldsymbol{x}}(t) = \boldsymbol{A}\boldsymbol{x}(t) + \boldsymbol{B}\boldsymbol{u}(t) + \boldsymbol{E}\boldsymbol{d}(t), \quad \boldsymbol{x}(0) = \boldsymbol{x}_0 \tag{5.1}$$

$$\boldsymbol{y}(t) = \boldsymbol{C}\boldsymbol{x}(t), \tag{5.2}$$

where $\boldsymbol{x} \in \mathbb{R}^n$ denotes the state of the system with the initial value $\boldsymbol{x}_0$, $\boldsymbol{u} \in \mathbb{R}^m$ and $\boldsymbol{y} \in \mathbb{R}^r$ are the inputs or measured outputs, respectively, and $\boldsymbol{d} \in \mathbb{R}^l$ represents exogenous disturbances.

The pair $(\boldsymbol{A}, \boldsymbol{B})$ is assumed to be controllable and the disturbance $\boldsymbol{d}(t)$ to be bounded:

$$\|\boldsymbol{d}(t)\| \leq d_{\max}. \tag{5.3}$$

The notations $\|\boldsymbol{x}\|$ and $\|\boldsymbol{A}\|$ denote an arbitrary vector norm or the induced matrix norm, and the absolute value is denoted by $|x|$. The expression $\|\boldsymbol{x}(t)\|$ denotes an arbitrary vector norm at time t. It is further assumed that

- the plant dynamics are accurately known,
- the state $\boldsymbol{x}(t)$ is measurable, and
- the information exchange between the event generator and the control input generator is instantaneous and imposes no restrictions on the information to be sent at event times.

Hence, the reason to communicate information via the dashed arrows in Fig. 5.2 is primarily given by the situation that the disturbance $\boldsymbol{d}(t)$ has caused an intolerable behavior of the control output $\boldsymbol{y}(t)$ or the plant state $\boldsymbol{x}(t)$.

Main Idea. As a main characteristic of the scheme proposed, the event generator uses a model of the continuous control loop to compare the current plant state $\boldsymbol{x}(t)$ with the desired state that occurs in the continuous closed-loop system. If the difference between both states exceeds an upper bound $\bar{e}$, an event is triggered and the current state $\boldsymbol{x}(t_k)$ is transmitted to the control input generator. As a further important fact, the control input generator incorporates the model of the continuous control loop to determine the future control input $\boldsymbol{u}(t)$, $(t \geq t_k)$. It will be shown that the event-based control loop with these characteristics has the following properties:

- The state $\boldsymbol{x}(t)$ of the event-based state-feedback loop is ultimately bounded in the sense that it remains, for all times t, in a bounded neighborhood Ω_{e} of the desired state $\boldsymbol{x}_{\mathrm{CT}}(t)$ of the continuous state-feedback loop.
- The communication over the feedback channel in the event-based control loop is bounded and depends explicitly on the disturbance $\boldsymbol{d}(t)$.
- Both the accuracy in terms of approximating the behavior of the continuous state-feedback loop and the minimum time interval between two consecutive events (*minimum inter-event time*) can be adjusted by changing the threshold $\bar{e}$ of the event generator in order to adapt the event-based state-feedback loop to the requested needs.

5.2.2 Continuous State Feedback

This section summarizes the main properties of the continuous state-feedback loop which is later used as the reference system to evaluate the behavior of the event-based state-feedback loop. Plant (5.1), (5.2) together with the state feedback

$$\boldsymbol{u}(t) = -\boldsymbol{K}\boldsymbol{x}(t) \tag{5.4}$$

yields the continuous closed-loop system

$$\dot{\boldsymbol{x}}_{\mathrm{CT}}(t) = \underbrace{(\boldsymbol{A} - \boldsymbol{B}\boldsymbol{K})}_{\bar{\boldsymbol{A}}}\boldsymbol{x}_{\mathrm{CT}}(t) + \boldsymbol{E}\boldsymbol{d}(t), \quad \boldsymbol{x}_{\mathrm{CT}}(0) = \boldsymbol{x}_0 \tag{5.5}$$

$$\boldsymbol{y}_{\mathrm{CT}}(t) = \boldsymbol{C}\boldsymbol{x}_{\mathrm{CT}}(t). \tag{5.6}$$

The index "CT" is used to distinguish the signals of this model from the corresponding signals of the event-based control loop considered later.

The state-feedback matrix $\boldsymbol{K}$ is assumed to be designed so that the matrix $\bar{\boldsymbol{A}}$ is Hurwitz and the closed-loop system has desired disturbance attenuation properties.

As $\bar{\boldsymbol{A}}$ is Hurwitz and the disturbance $\boldsymbol{d}(t)$ is assumed to be bounded according to Eq. (5.3), the state $\boldsymbol{x}_{\mathrm{CT}}(t)$ of the continuous state-feedback loop (5.5), (5.6) is *GUUB* according to the following definition:

Definition 5.1. [205] *The solution $\boldsymbol{x}(t)$ of the continuous control loop* (5.5), (5.6) *is said to be globally uniformly ultimately bounded (GUUB) if for every $\boldsymbol{x}_0 \in \mathbb{R}^n$ there exists a positive constant p and a time $\bar{t}$ such that holds:*

$$\boldsymbol{x}(t) \in \Omega_{\mathrm{t}} = \{\boldsymbol{x} : \|\boldsymbol{x}\| \leq p\}, \ \ \forall t \geq \bar{t}.$$

Then one says that the continuous control loop (5.5), (5.6) is ultimately bounded.

For the linear continuous control loop (5.5), (5.6) the state $\boldsymbol{x}(t)$ is GUUB if the matrix $\bar{\boldsymbol{A}}$ is Hurwitz and the disturbance $\boldsymbol{d}(t)$ is bounded.

Behavior of the Continuous State-Feedback Loop. The control input generated by the state-feedback controller (5.4) is given by

$$\boldsymbol{u}(t) = -\boldsymbol{K}\mathrm{e}^{\bar{\boldsymbol{A}}t}\boldsymbol{x}_0 - \int_0^t \boldsymbol{K}\mathrm{e}^{\bar{\boldsymbol{A}}(t-\tau)}\boldsymbol{E}\boldsymbol{d}(\tau)\,\mathrm{d}\tau.$$

This equation shows that the input $\boldsymbol{u}(t)$ does not only depend upon the initial state $\boldsymbol{x}_0$ but also on the disturbance input $\boldsymbol{d}(t)$. In the setting of event-based control, this aspect is important. If at time t_k the state $\boldsymbol{x}(t_k)$ is communicated to the control input generator, the control input generator is able to determine the same control input $\boldsymbol{u}(t_k) = -\boldsymbol{K}\boldsymbol{x}(t_k)$ as a continuous state-feedback controller. However, for all future times $t > t_k$, the control input generator has to know the disturbance $\boldsymbol{d}(t)$ for $t > t_k$:

$$\boldsymbol{u}(t) = -\boldsymbol{K}\mathrm{e}^{\bar{\boldsymbol{A}}(t-t_k)}\boldsymbol{x}_{\mathrm{CT}}(t_k) - \int_{t_k}^t \boldsymbol{K}\mathrm{e}^{\bar{\boldsymbol{A}}(t-\tau)}\boldsymbol{E}\boldsymbol{d}(\tau)\,\mathrm{d}\tau, \ \ t \geq t_k. \quad (5.7)$$

This analysis shows two important facts:

- Continuous state-feedback control (5.4) gets the information about the current disturbance implicitly by the continuous communication of the current state $\boldsymbol{x}_{\mathrm{CT}}(t)$.
- Any feedback without continuous communication has to make assumptions about the disturbance to be attenuated. Unless the disturbance is measurable, any discontinuous feedback cannot have the same performance as the feedback loop with continuous communication.

The main idea of the event-based state-feedback approach discussed in the following is to replace the continuous state feedback (5.4) by an event-based controller so that the state $\boldsymbol{x}(t)$ of the event-based state-feedback loop remains, for all times t, in the neighborhood $\Omega_{\mathrm{e}}(\boldsymbol{x}_{\mathrm{CT}}(t))$ of the desired state $\boldsymbol{x}_{\mathrm{CT}}(t)$ of the continuous state-feedback loop (5.5), (5.6).

5.2.3 Event-Based State Feedback

Control Input Generator. A direct consequence of the analysis in the preceding section is the fact that for the time $t \geq t_k$ the plant (5.1), (5.2) with the control input (5.7) behaves exactly like the continuous control loop (5.5), (5.6). If the control input generator uses Eq. (5.7) to determine the control input for $t \geq t_k$, then the best possible performance is obtained. To enable the control input generator to use this equation the state $\boldsymbol{x}(t_k)$ has to be measured and communicated to the control input generator, and an assumption concerning the disturbance has to be made.

In the following, the control input generator assumes that the disturbance is constant

$$\boldsymbol{d}(t) = \hat{\boldsymbol{d}}_k \quad \text{for} \quad t \geq t_k$$

with known magnitude $\hat{\boldsymbol{d}}_k$. Hence, it uses the equation

$$\boldsymbol{u}(t) = -\boldsymbol{K}\mathrm{e}^{\bar{\boldsymbol{A}}(t-t_k)}\boldsymbol{x}(t_k) - \boldsymbol{K}\bar{\boldsymbol{A}}^{-1}\left(\mathrm{e}^{\bar{\boldsymbol{A}}(t-t_k)} - \boldsymbol{I}_n\right)\boldsymbol{E}\hat{\boldsymbol{d}}_k, \quad t \geq t_k \tag{5.8}$$

which directly follows from Eq. (5.7) for constant disturbances, until it gets the next information $\boldsymbol{x}(t_{k+1})$. $\boldsymbol{I}_n$ denotes the identity matrix of size n.

The **control input generator** determines the input (5.8) by means of a model of the continuous closed-loop system (5.5)

$$\dot{\boldsymbol{x}}_{\mathrm{s}}(t) = \bar{\boldsymbol{A}}\boldsymbol{x}_{\mathrm{s}}(t) + \boldsymbol{E}\hat{\boldsymbol{d}}_k, \quad \boldsymbol{x}_{\mathrm{s}}(t_k^+) = \boldsymbol{x}(t_k),\ t \geq t_k \tag{5.9}$$

$$\boldsymbol{u}(t) = -\boldsymbol{K}\boldsymbol{x}_{\mathrm{s}}(t). \tag{5.10}$$

Here, $\boldsymbol{x}_{\mathrm{s}}$ is used to denote *the state of the control input generator.* Note that the signal $\boldsymbol{u}(t)$ obtained by Eq. (5.8) is the same as the solution of (5.9), (5.10).

The time t_k^+ indicates the update of the model state $\boldsymbol{x}_{\mathrm{s}}$ with the measured state $\boldsymbol{x}(t_k)$, which the control input generator gets from the event generator at event time t_k. Figure 5.3 shows the block diagram of the control input generator. Suitable ways for determining the event time t_k and the disturbance estimate are presented later in this section.

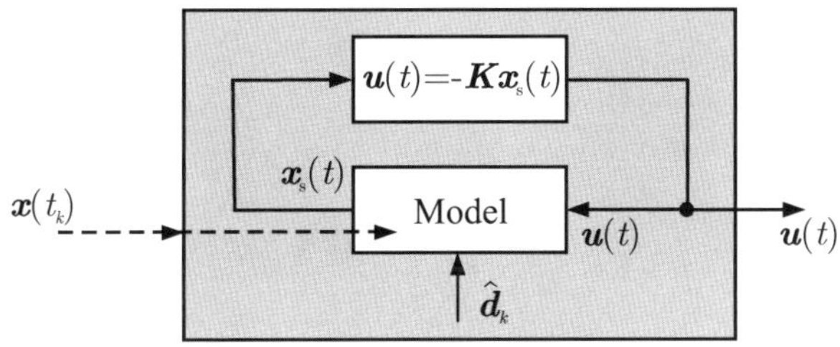

Fig. 5.3 Control input generator

Behavior of the Event-Based State-Feedback Loop. The analysis in this paragraph is valid for arbitrary event generators and arbitrary methods to estimate the disturbance magnitude $\hat{\boldsymbol{d}}_k$. It investigates the behavior of the event-based control loop in the time interval $[t_k, t_{k+1})$ between the consecutive event times t_k and t_{k+1}.

The plant (5.1), (5.2) together with the control input generator (5.9), (5.10) is described for the time period $[t_k, t_{k+1})$ by the state-space model

$$\begin{pmatrix} \dot{\boldsymbol{x}}(t) \\ \dot{\boldsymbol{x}}_{\mathrm{s}}(t) \end{pmatrix} = \begin{pmatrix} \boldsymbol{A} & -\boldsymbol{B}\boldsymbol{K} \\ \boldsymbol{O} & \bar{\boldsymbol{A}} \end{pmatrix} \begin{pmatrix} \boldsymbol{x}(t) \\ \boldsymbol{x}_{\mathrm{s}}(t) \end{pmatrix} + \begin{pmatrix} \boldsymbol{E} \\ \boldsymbol{O} \end{pmatrix} \boldsymbol{d}(t) + \begin{pmatrix} \boldsymbol{O} \\ \boldsymbol{E} \end{pmatrix} \hat{\boldsymbol{d}}_k$$

$$\begin{pmatrix} \boldsymbol{x}(t_k^+) \\ \boldsymbol{x}_{\mathrm{s}}(t_k^+) \end{pmatrix} = \begin{pmatrix} \boldsymbol{x}(t_k) \\ \boldsymbol{x}(t_k) \end{pmatrix}$$

$$\boldsymbol{y}(t) = (\boldsymbol{C} \;\; \boldsymbol{O}) \begin{pmatrix} \boldsymbol{x}(t) \\ \boldsymbol{x}_{\mathrm{s}}(t) \end{pmatrix}.$$

This model takes into account that the closed-loop system is subject to the disturbance $\boldsymbol{d}(t)$, whereas the control input generator uses the constant disturbance estimate $\hat{\boldsymbol{d}}_k$. The expression $\boldsymbol{x}_i(t_k^+) = \boldsymbol{x}_i(t_k)$ is used in the following to explicitly indicate that the respective state is not changed at the corresponding time instance.

By introducing the state transformation

$$\begin{pmatrix} \boldsymbol{x}_{\Delta}(t) \\ \boldsymbol{x}_{\mathrm{s}}(t) \end{pmatrix} = \begin{pmatrix} \boldsymbol{I}_n & -\boldsymbol{I}_n \\ \boldsymbol{O} & \boldsymbol{I}_n \end{pmatrix} \begin{pmatrix} \boldsymbol{x}(t) \\ \boldsymbol{x}_{\mathrm{s}}(t) \end{pmatrix} \tag{5.11}$$

the following result can be obtained

Lemma 5.1. [248] *The output of the event-based state-feedback loop* (5.1), (5.2), (5.9), (5.10) *subject to the disturbance* $\boldsymbol{d}(t) = \hat{\boldsymbol{d}}_k + \boldsymbol{d}_{\Delta}(t)$ *consists of two components* $\boldsymbol{y}(t) = \boldsymbol{y}_{\mathrm{s}}(t) + \boldsymbol{y}_{\Delta}(t)$ *given by*

$$\boldsymbol{y}_{\mathrm{s}}(t) = \boldsymbol{C}\mathrm{e}^{\bar{\boldsymbol{A}}(t-t_k)}\boldsymbol{x}(t_k) + \boldsymbol{C}\bar{\boldsymbol{A}}^{-1}\left(\mathrm{e}^{\bar{\boldsymbol{A}}(t-t_k)} - \boldsymbol{I}_n\right)\boldsymbol{E}\hat{\boldsymbol{d}}_k \tag{5.12}$$

$$\boldsymbol{y}_{\Delta}(t) = \int_{t_k}^{t} \boldsymbol{C}\mathrm{e}^{\boldsymbol{A}(t-\tau)}\boldsymbol{E}\boldsymbol{d}_{\Delta}(\tau)\mathrm{d}\tau. \tag{5.13}$$

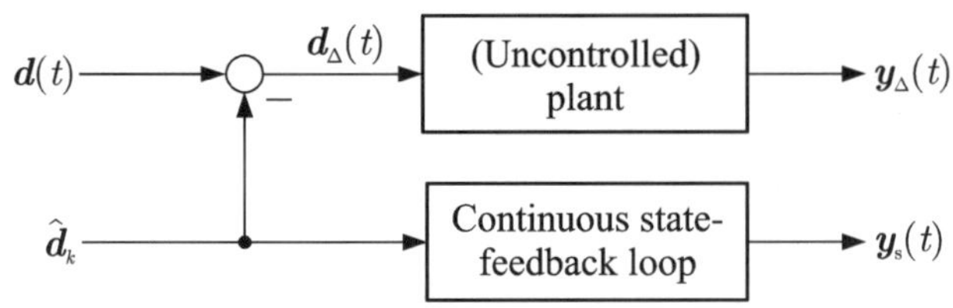

Fig. 5.4 Interpretation of Eqs. (5.12), (5.13)

The interpretation of this lemma is illustrated in Fig. 5.4. It shows three facts:

- The constant disturbance $\hat{\boldsymbol{d}}_k$ used by the control input generator, has the same effect on the event-based control as on the continuous state feedback. In the time interval $[t_k, t_{k+1})$ with the initial state $\boldsymbol{x}(t_k)$ both systems generate the output $\boldsymbol{y}_{\mathrm{s}}(t)$.
- The difference $\boldsymbol{d}_{\Delta}(t) = \boldsymbol{d}(t) - \hat{\boldsymbol{d}}_k$ between the actual disturbance $\boldsymbol{d}(t)$ and the constant disturbance estimate $\hat{\boldsymbol{d}}_k$ affects the (uncontrolled) plant and results in the output $\boldsymbol{y}_{\Delta}(t)$, which describes the difference between the outputs of the continuous state-feedback loop and the event-based state-feedback loop.
- For a good approximation $\hat{\boldsymbol{d}}_k$ of the disturbance $\boldsymbol{d}(t)$ in the time interval $[t_k, t_{k+1})$, i.e., $\boldsymbol{d}(t) - \hat{\boldsymbol{d}}_k \approx \boldsymbol{0}$, the plant subject to the open-loop control (5.8) behaves like the continuous state-feedback loop. No communication is necessary in this time interval.

According to the state transformation (5.11), the state $\boldsymbol{x}(t) = \boldsymbol{x}_{\mathrm{s}}(t) + \boldsymbol{x}_{\Delta}(t)$ of the event-based state-feedback loop can be decomposed into two components:

$$\boldsymbol{x}_{\mathrm{s}}(t) = \mathrm{e}^{\bar{\boldsymbol{A}}(t-t_k)}\boldsymbol{x}(t_k) + \bar{\boldsymbol{A}}^{-1}\left(\mathrm{e}^{\bar{\boldsymbol{A}}(t-t_k)} - \boldsymbol{I}_n\right)\boldsymbol{E}\hat{\boldsymbol{d}}_k \tag{5.14}$$

$$\boldsymbol{x}_{\Delta}(t) = \int_{t_k}^{t} \mathrm{e}^{\boldsymbol{A}(t-\tau)}\boldsymbol{E}\boldsymbol{d}_{\Delta}(\tau)\mathrm{d}\tau. \tag{5.15}$$

Like the output $\boldsymbol{y}_{\mathrm{s}}(t)$, the model state $\boldsymbol{x}_{\mathrm{s}}(t)$ is identical to the state trajectory of the continuous state-feedback system (5.5), (5.6) in the time interval $[t_k, t_{k+1})$ with initial state $\boldsymbol{x}_{\mathrm{s}}(t_k) = \boldsymbol{x}(t_k)$ and affected by the constant disturbance $\boldsymbol{d}(t) = \hat{\boldsymbol{d}}_k$.

Event Generator. Events are generated by comparing the measured state trajectory $\boldsymbol{x}(t)$ with the state trajectory $\boldsymbol{x}_{\mathrm{s}}(t)$ that would occur in the continuous state-feedback loop for the constant disturbance $\boldsymbol{d}(t) = \hat{\boldsymbol{d}}_k$. As the state $\boldsymbol{x}_{\mathrm{s}}(t)$ determined according to Eq. (5.9) represents the desired reference signal, the measured state $\boldsymbol{x}(t)$ should be kept in the surroundings

$$\Omega_{\mathrm{s}}(\boldsymbol{x}_{\mathrm{s}}(t)) = \{\boldsymbol{x} : \|\boldsymbol{x} - \boldsymbol{x}_{\mathrm{s}}(t)\| \leq \bar{e}\}$$

of this state with adjustable size $\bar{e}$.

The **event generator** triggers an event whenever the difference between the measured plant state $\boldsymbol{x}(t)$ and the reference state $\boldsymbol{x}_{\mathrm{s}}(t)$ reaches the event threshold $\bar{e}$:

$$\|\boldsymbol{x}(t) - \boldsymbol{x}_{\mathrm{s}}(t)\| = \bar{e}. \tag{5.16}$$

At this time instance t, which denotes the event time t_k, the state information $\boldsymbol{x}(t_k)$ is communicated to the control input generator.

In order to avoid a continuous transmission of the state $\boldsymbol{x}_{\mathrm{s}}(t)$ from the control input generator to the event generator, a copy of the control input generator is included in the event generator so that the event generator can determine the state $\boldsymbol{x}_{\mathrm{s}}(t)$ by means of Eq. (5.9).

As, at event time t_k, the state $\boldsymbol{x}_{\mathrm{s}}(t_k)$ is immediately updated with the measured state $\boldsymbol{x}(t_k)$, the following property holds.

Lemma 5.2. [248] *Event condition* (5.16) *ensures that the difference state* $\boldsymbol{x}_{\Delta}(t) = \boldsymbol{x}(t) - \boldsymbol{x}_{\mathrm{s}}(t)$ *is bounded and remains in the set* Ω_{Δ}*:*

$$\boldsymbol{x}_{\Delta}(t) \in \Omega_{\Delta} = \{\boldsymbol{x}_{\Delta} : \|\boldsymbol{x}_{\Delta}\| \leq \bar{e}\}, \quad \forall t \geq 0.$$

Disturbance Estimator. The following investigations show how to get an estimate $\hat{\boldsymbol{d}}_k$ of the disturbance magnitude at the event time t_k. Assume that in the preceding time interval $[t_{k-1}, t_k)$ the disturbance estimate $\hat{\boldsymbol{d}}_{k-1}$ has been used. Consider now the difference $\boldsymbol{x}_{\Delta}(t) = \boldsymbol{x}(t) - \boldsymbol{x}_{\mathrm{s}}(t)$ and assume that the disturbance $\boldsymbol{d}(t)$ has been constant in this time interval

$$\boldsymbol{d}(t) = \bar{\boldsymbol{d}} \quad \text{for} \quad t \in [t_{k-1}, t_k),$$

where $\bar{\boldsymbol{d}}$ is the actual disturbance magnitude, which usually differs from the estimate $\hat{\boldsymbol{d}}_k$. Equation (5.15) yields

$$\boldsymbol{x}(t) - \boldsymbol{x}_{\mathrm{s}}(t) = \boldsymbol{A}^{-1}\left(\mathrm{e}^{\boldsymbol{A}(t - t_{k-1})} - \boldsymbol{I}_n\right)\boldsymbol{E}(\bar{\boldsymbol{d}} - \hat{\boldsymbol{d}}_{k-1})$$

which is used to determine, at time $t = t_k$, the unknown disturbance magnitude $\bar{\boldsymbol{d}}$:

The **disturbance estimator** determines the estimate $\hat{\boldsymbol{d}}_k$ recursively:

$$\hat{\boldsymbol{d}}_0 = \boldsymbol{0} \tag{5.17}$$

$$\hat{\boldsymbol{d}}_k = \hat{\boldsymbol{d}}_{k-1} + \left(\boldsymbol{A}^{-1}\left(\mathrm{e}^{\boldsymbol{A}(t_k - t_{k-1})} - \boldsymbol{I}_n\right)\boldsymbol{E}\right)^{+}(\boldsymbol{x}(t_k) - \boldsymbol{x}_{\mathrm{s}}(t_k)). \tag{5.18}$$

The pseudoinverse $(\star)^+$ in Eq. (5.18) exists if, as usual, the number of disturbances is lower than the number of state variables ($n \geq l$) and the occurring matrices have full rank. Note that this disturbance estimation explicitly requires the existence of the inverse system matrix $\boldsymbol{A}^{-1}$ but it does not require the stability of the plant.

The disturbance estimator is included in the control input generator as well as in the event generator to provide both components with a current disturbance estimate at event times t_k, $(k = 0, 1, 2, ...)$. It has the following property.

Lemma 5.3. [248] *If at time t_1 the first event has been generated in the event-based state-feedback loop affected by a constant disturbance $\boldsymbol{d}(t) = \bar{\boldsymbol{d}}$, the disturbance estimator* (5.17), (5.18) *correctly determines the disturbance magnitude:*

$$\hat{\boldsymbol{d}}_1 = \bar{\boldsymbol{d}}.$$

If the disturbance $\boldsymbol{d}(t)$ changes, the estimate $\hat{\boldsymbol{d}}_k$ represents the "mean" value of $\boldsymbol{d}(t)$ in the time interval (t_0, t_1).

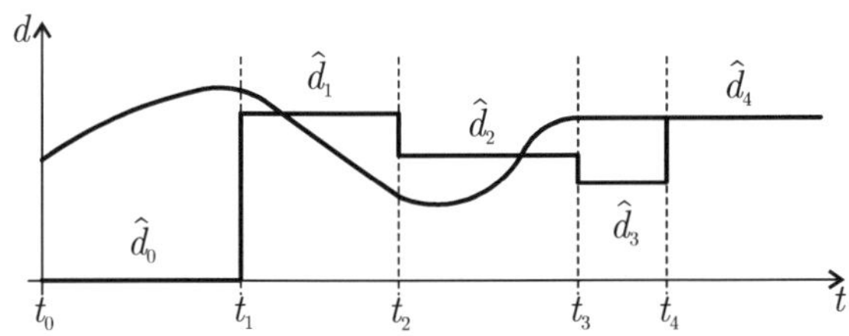

Fig. 5.5 Disturbance $d(t)$ and disturbance estimates $\hat{d}_0, \hat{d}_1, \hat{d}_2$...

For a scalar time-varying disturbance $d(t)$ the disturbance estimation is illustrated in Fig. 5.5, which shows the behavior of the disturbance $d(t)$ and the corresponding sequence of disturbance estimates. Here, $\hat{d}_1$ is the weighted average of the disturbance $d(t)$ for the time interval $[t_0, t_1)$. Similarly, $\hat{d}_2$ describes a weighted average of the actual disturbance for the time interval $[t_1, t_2)$. If the disturbance remains constant over two time intervals, then in the second time interval the estimate $\hat{d}_k$ coincides with the true magnitude of the disturbance. This happens in the example for $t \geq t_4$.

Summary of the Components. The event-based state-feedback loop has the structure depicted in Fig. 5.6. It has the following components:

- the plant (5.1), (5.2),
- the control input generator (5.9), (5.10) which also estimates the disturbance according to Eqs. (5.17), (5.18), and

- the event generator which includes a copy of the control input generator (5.9), (5.10) and the disturbance estimator (5.17), (5.18) and determines the event times t_k according to Eq. (5.16).

At event times t_k, $(k = 0, 1, 2, ...)$ the measured state information $\boldsymbol{x}(t_k)$ is sent from the event generator towards the control input generator and is used there as well as in the event generator to update the model state $\boldsymbol{x}_\mathrm{s}$ according to $\boldsymbol{x}_\mathrm{s}(t_k^+) = \boldsymbol{x}(t_k)$ and to determine the new disturbance estimate $\hat{\boldsymbol{d}}_k$. Since by assumption the data transmission is accomplished in no time, the models in the control input generator and the event generator work synchronously.

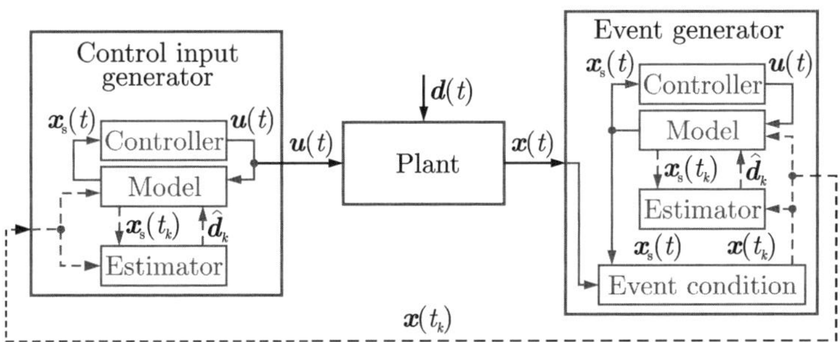

Fig. 5.6 Event-based state-feedback control loop

5.2.4 Main Properties of the Event-Based State-Feedback Loop

The central properties to be investigated when considering event-based control concern the stability and the communication over the feedback link. The main results of the subsequent analysis are the following:

- The state $\boldsymbol{x}(t)$ of the event-based control loop is *GUUB* and there exists an upper bound on its approximation error in terms of emulating the behavior of the continuous state-feedback loop (Theorem 5.1).
- There exists a lower bound on the minimum inter-event time (Theorem 5.2).
- If the disturbances are sufficiently small, no event is generated for $t > 0$ (Lemma 5.4).

Comparison between the Event-Based and the Continuous State-Feedback Loop. The following theorem compares the event-based control loop with the continuous state-feedback loop.

Theorem 5.1. [248] *The difference*

$$\boldsymbol{e}(t) = \boldsymbol{x}(t) - \boldsymbol{x}_{\mathrm{CT}}(t)$$

between the state $\boldsymbol{x}(t)$ *of the event-based state-feedback loop* (5.1), (5.2), (5.9), (5.10), (5.16) – (5.18) *and the state* $\boldsymbol{x}_{\mathrm{CT}}(t)$ *of the continuous state-feedback loop* (5.5), (5.6) *is bounded from above by*

$$\|\boldsymbol{e}(t)\| \leq e_{\max} = \bar{e} \cdot \int_0^\infty \left\| \mathrm{e}^{\bar{\boldsymbol{A}}\tau} \boldsymbol{B}\boldsymbol{K} \right\| \mathrm{d}\tau. \tag{5.19}$$

This theorem shows that the event-based controller can be made to mimic a continuous state feedback system with arbitrary precision by accordingly choosing the event threshold $\bar{e}$. It can be used to determine for every tolerable upper bound on the approximation error $\|\boldsymbol{e}(t)\|$ the event threshold $\bar{e}$. The price for a higher precision (smaller $e_{\max}$) is a more frequent communication between the event generator and the control input generator. The state $\boldsymbol{x}(t)$ remains in the set

$$\boldsymbol{x}(t) \in \Omega_{\mathrm{e}}(\boldsymbol{x}_{\mathrm{CT}}(t)) = \{\boldsymbol{x} : \|\boldsymbol{x} - \boldsymbol{x}_{\mathrm{CT}}(t)\| \leq e_{\max}\},$$

which describes a bounded neighborhood of the state $\boldsymbol{x}_{\mathrm{CT}}(t)$ of the continuous state-feedback loop for all times t as depicted in Fig. 5.7. Hence, as the state $\boldsymbol{x}_{\mathrm{CT}}(t)$ of the continuous state-feedback loop is *GUUB*, the state $\boldsymbol{x}(t)$ of the event-based state-feedback loop is *GUUB* as well.

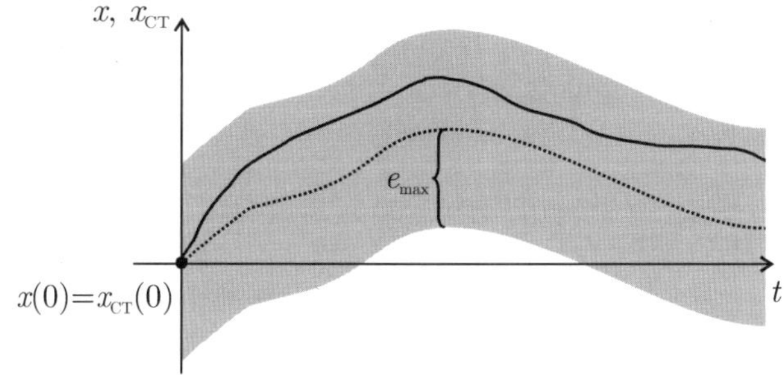

Fig. 5.7 Behavior of the event-based feedback loop: Solid line: state variable $x(t)$ of the event-based control loop; dotted line: state variable $x_{\mathrm{CT}}(t)$ of the continuous control loop

Minimum Inter-event Time. This paragraph shows that the minimum inter-event time

$$T_{\min} = \min_{k} (t_{k+1} - t_k)$$

of the event-based state-feedback loop is bounded from below and depends upon the disturbance $\boldsymbol{d}(t)$. Assume that the disturbance estimation error $\boldsymbol{d}_{\Delta}(t) = \boldsymbol{d}(t) - \hat{\boldsymbol{d}}_k$ is bounded by

$$\|\boldsymbol{d}_{\Delta}(t)\| \leq \gamma\, d_{\max} \quad \text{for} \quad t \geq 0, \tag{5.20}$$

with $0 \leq \gamma \leq 2$. The minimum inter-event time $T_{\min}$ is given by

$$T_{\min} = \arg\min_{t} \max_{\boldsymbol{d}_{\Delta}} \left\{ \left\| \int_0^t \mathrm{e}^{\boldsymbol{A}(t-\tau)} \boldsymbol{E} \boldsymbol{d}_{\Delta}(\tau)\, \mathrm{d}\tau \right\| = \bar{e} \right\}. \tag{5.21}$$

Theorem 5.2. [248] *For any bounded disturbance $\boldsymbol{d}(t)$, the minimum inter-event time $T_{\min}$ of the event-based state-feedback loop* (5.1), (5.2), (5.9), (5.10), (5.16) – (5.18) *is bounded from below by $\bar{T}$ ($T_{\min} \geq \bar{T}$) given by*

$$\bar{T} = \arg\min_{t} \left\{ \int_0^t \left\| \mathrm{e}^{\boldsymbol{A}\tau} \boldsymbol{E} \right\| \mathrm{d}\tau = \frac{\bar{e}}{\gamma\, d_{\max}} \right\}. \tag{5.22}$$

This theorem highlights how the communication depends on the disturbances. This phenomenon contrasts with sampled-data control, where the sampling frequency is chosen with respect to the plant properties (time constants) rather than the disturbance magnitude.

As the lower bound for the inter-event time decreases by decreasing the event threshold $\bar{e}$ and increases for large thresholds $\bar{e}$, Theorems 5.1 and 5.2 show that a higher precision generally leads to a more frequent communication from the event generator towards the control input generator.

Small Disturbances. The disturbance is represented in the following as $\boldsymbol{d}(t) = \bar{d}\tilde{\boldsymbol{d}}(t)$, where $\tilde{\boldsymbol{d}}(t)$ is an arbitrary finite vector function satisfying the inequality

$$\|\tilde{\boldsymbol{d}}(t)\| \leq 1 \quad \text{for} \quad t \geq 0$$

and $\bar{d}$ is the disturbance magnitude.

Lemma 5.4. [248] *Suppose that the plant* (5.1), (5.2) *is asymptotically stable. Then, for every bounded disturbance $\boldsymbol{d}(t) = \bar{d}\tilde{\boldsymbol{d}}(t)$ with magnitude $\bar{d}$ satisfying the relation*

$$|\bar{d}| < \frac{\bar{e}}{\int_0^\infty \|\mathrm{e}^{\boldsymbol{A}\tau} \boldsymbol{E}\|\, \mathrm{d}\tau} = \bar{d}_{\mathrm{UD}} \tag{5.23}$$

the event generator does not generate any event for $t > 0$.

This result gives a quantitative bound for the disturbance, for which no feedback occurs after the initial event at time $t = 0$. It shows that in event-based control the communication is adapted to the severity of the disturbance. If the disturbance is small enough, no feedback is necessary at all to meet the performance requirements.

5.2.5 Extensions

The state feedback approach to event-based control presented in this section has been extended in various ways. The main purpose of these extensions lies in the relaxation of the assumptions stated in Section 5.2.1.

In order to make the approach more attractive for practical applications, an *event-based PI controller* has been developed [222]. The extended scheme guarantees setpoint tracking for constant reference and disturbance signals while significantly reducing the communication compared to sampled-data PI control.

The effect of *model uncertainties* has been analyzed by specifying upper bounds on the uncertainties of the model parameters. The analysis shows that model uncertainties affect both the approximation accuracy and the frequency of communication but can be compensated by using more involved disturbance estimators [219].

For dealing with immeasurable state variables, an *event-based output-feedback* control has been proposed in [224]. There, a state observer is included in the event generator in order to determine an approximate state $\hat{\boldsymbol{x}}(t)$ of the plant state $\boldsymbol{x}(t)$ based on the measured output $\boldsymbol{y}(t)$. By using an adapted event condition which monitors the difference between the observer state $\hat{\boldsymbol{x}}(t)$ and the model state $\boldsymbol{x}_{\mathrm{s}}(t)$ and by sending the observer state $\hat{\boldsymbol{x}}(t_k)$ at event times to the control input generator, a bound on the approximation error and a minimum inter-event time can be guaranteed.

In networked control systems, the assumption of having an ideal communication channel is often violated. Hence, non-ideal effects like *transmission delays, packet loss or a quantization* of the transmitted information have to be taken into account. By modifying the structure of the event-based control loop mainly in terms of adapting the update mechanism, a stable behavior and a bounded communication can be preserved in all three cases if certain conditions on the delay, the number of consecutive packet losses or the resolution of the quantization are met [221, 223, 225].

In this section, the event-based state-feedback loop has been analyzed by evaluating the system behavior in between consecutive events. A different approach to the analysis has been presented in [351], based on the formulation of the event-based state-feedback loop as an impulsive system and which allows to uniformly investigate the system dynamics at and between event times. In contrast to the analysis presented in this section which only proves ultimate boundedness, the method presented in [351] also detects the asymptotic stability of the undisturbed event-based control system with stable plant dynamics.

The scheme presented in this section refers to a continuous realization of its components. As the control input generator and the event generator need to be implemented on digital hardware, a discrete-time approach becomes important. The main difference to the continuous approach is given by the fact that events cannot be triggered at any time but only at the sampling instants. The results obtained in [158, 220] are very similar to the continuous situation but, as expected, deteriorate with an increasing sampling period.

Finally, to relax the necessity of linear dynamics, an extension to nonlinear systems which are input-output linearizable has been investigated. Basically, the consideration of nonlinear dynamics requires a more involved event condition by means of which the desired properties of event-based control in terms of its stability and the boundedness of the communication can be proven [348, 349].

Example 5.1 *Event-based control of a thermofluid process*

This example illustrates the behavior of the event-based state-feedback loop in different scenarios. The plant is the thermofluid process shown in Fig. 5.8. The level $x_1(t)$ and the temperature $x_2(t)$ of the liquid in the tank TB have to be stabilized at the set-points by using the inflow $u_1(t)$ from tank T3 and the heating power $u_2(t)$ as control inputs. Hot water inflow from tank HW is the scalar disturbance $d(t)$ to be attainuated.

The linearized model of the plant is given by

$$\dot{\boldsymbol{x}}(t) = 10^{-3}\begin{pmatrix} -0.8 & 0 \\ -1\cdot 10^{-7} & -1.7 \end{pmatrix}\boldsymbol{x}(t) + 10^{-3}\begin{pmatrix} 211 & 0 \\ -108 & 20 \end{pmatrix}\boldsymbol{u}(t) + 10^{-3}\begin{pmatrix} 148 \\ -80 \end{pmatrix} d(t)$$

with $\boldsymbol{x}_0 = \boldsymbol{0}$. The controller is chosen to be

$$\boldsymbol{K} = \begin{pmatrix} 0.08 & -0.02 \\ 0.17 & 0.72 \end{pmatrix},$$

for which the continuous closed-loop system (5.5), (5.6) is stable and has desired disturbance attenuation properties. The event threshold $\bar{e}$ is set to $\bar{e} = 2$ and the event generator uses the supremum norm, for which the event condition reads as

$$\|\boldsymbol{x}_\Delta(t)\|_\infty = \|\boldsymbol{x}(t) - \boldsymbol{x}_\mathrm{s}(t)\|_\infty = 2. \tag{5.24}$$

Simulation Results. In the first investigation (left-hand side of Fig. 5.9) the plant is subject to a constant disturbance $d(t) = \bar{d}$ drawn by the solid line in the top subplot. After the initializing event at time $t_0 = 0$, an event takes place at time t_1 due to the level behavior, where the equality

$$|x_{\Delta,1}(t_1)| = |x_1(t_1) - x_{\mathrm{s},1}(t_1)| = 2 \tag{5.25}$$

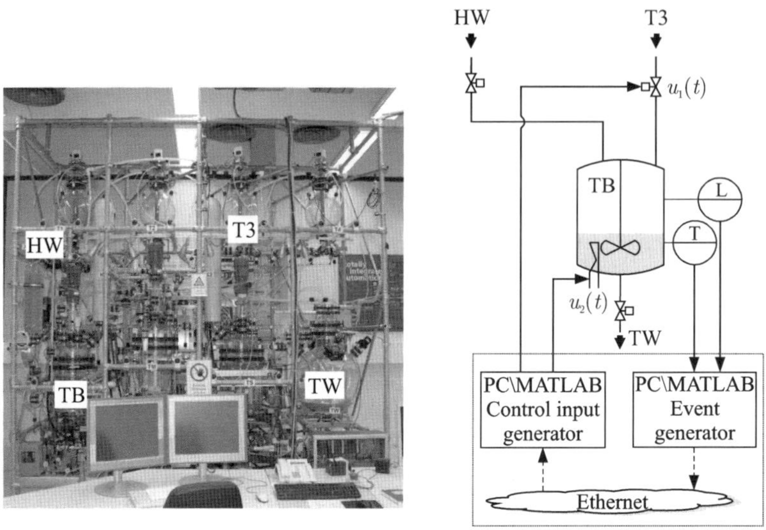

Fig. 5.8 Experimental set-up and implementation structure of the event-based controller for a thermofluid process

holds (see second subplot from top). At this event time, which is indicated in the bottom subplot, the disturbance magnitude $\bar{d}$ is correctly estimated by the disturbance estimator ($\hat{d}_1 = \bar{d}$, dotted line). Afterwards, $\boldsymbol{x}(t)$ and $\boldsymbol{x}_{\mathrm{s}}(t)$ coincide and approach the state $\boldsymbol{x}_{\mathrm{CT}}(t)$ of the continuous state-feedback loop. No further event is triggered.

The third subplot shows the temperature behavior. Note that the minus sign results from the fact that a linearized model is considered. This likewise holds for the level behavior.

In the second investigation (middle plots of Fig. 5.9), five events take place until the disturbance remains constant and its magnitude is estimated with sufficient accuracy. As the disturbance varies, the estimate $\hat{d}_1$ and the disturbance $d(t)$ differ for $t \geq t_1$ and a new event occurs at time t_2. The new disturbance estimate $\hat{d}_2$ describes a weighted average of $d(t)$ in the preceding time interval $[t_1, t_2)$, which also holds at the subsequent event times.

The disturbance estimate $\hat{d}_4$ does not coincide with the disturbance magnitude $\bar{d}$ ($d(t) - \hat{d}_4 \approx 0.003$) but the difference satisfies inequality (5.23) with $\bar{d}_{\mathrm{UD}} \approx 0.0108$ and, hence, no further event occurs.

The right plots of Fig. 5.9 show the importance of the disturbance estimation (5.17), (5.18) included in the control input generation and in the event generation. Here, the disturbance $d(t)$ is considered to be constant with the same magnitude as in the left plots but the control input generator does not apply the disturbance estimation according to Eq. (5.18). Instead, it uses the disturbance estimate $\hat{d}_k = 0$ for all $t \geq 0$.

As a consequence, after the first event at time t_1 further events are triggered because the control input generator provides an input $\boldsymbol{u}(t)$ which is generated without taking the current disturbance into account. Therefore, the plant state $\boldsymbol{x}(t)$ and the model state $\boldsymbol{x}_{\mathrm{s}}(t)$ diverge and events are generated periodically for $t > t_1$.

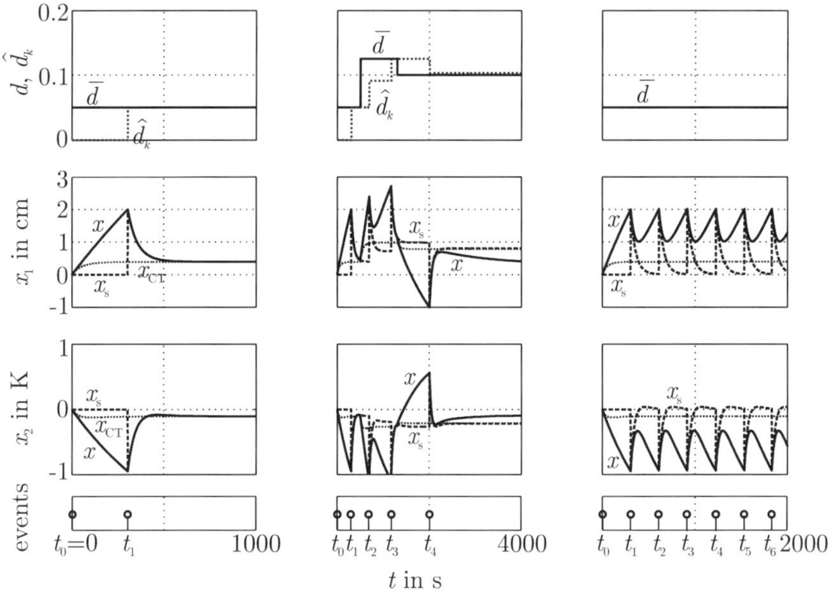

Fig. 5.9 Behavior of the event-based state-feedback loop subject to a constant disturbance (left plots), a time-varying disturbance (middle plots) and without disturbance estimation (right plots). Solid lines: plant state $\boldsymbol{x}(t)$; dashed lines: model state $\boldsymbol{x}_{\rm s}(t)$; dotted lines: state $\boldsymbol{x}_{\rm CT}(t)$ of the continuous state-feedback loop.

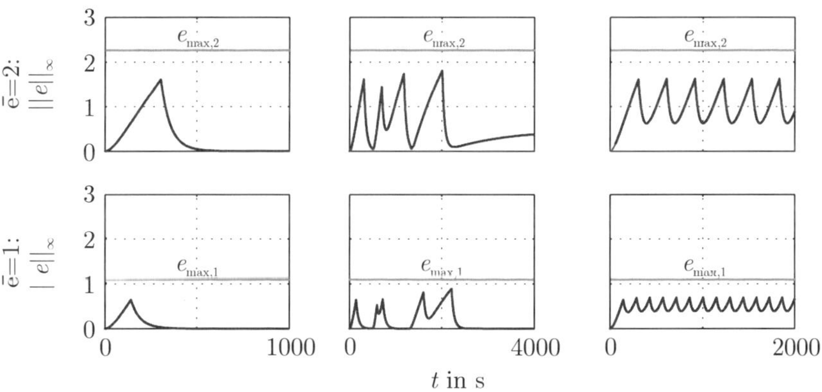

Fig. 5.10 Approximation error $\|\boldsymbol{e}(t)\|_\infty$ of the event-based state-feedback loop subject to a constant disturbance (left plots), a time-varying disturbance (middle plots) and without disturbance estimation (right plots) for $\bar{e} = 2$ and $\bar{e} = 1$

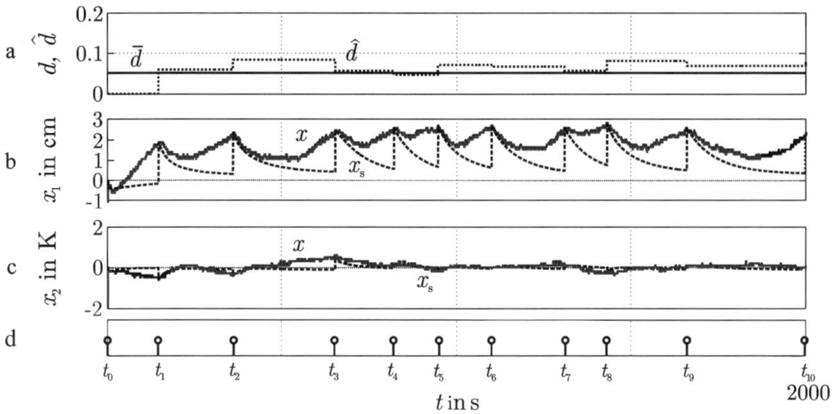

Fig. 5.11 Experimental behavior of the event-based state-feedback loop. Solid lines: plant state $\boldsymbol{x}(t)$; dashed lines: model state $\boldsymbol{x}_s(t)$.

According to Eq. (5.19), the upper approximation error bound is given by

$$\|\boldsymbol{e}(t)\|_\infty \leq e_{\max} = 2.26.$$

The actual approximation accuracy of the event-based state feedback in the situations considered in Fig. 5.9 is shown in the upper plots of Fig. 5.10. Note that the peaks in these subplots appear at the respective event times in Fig. 5.9. It can be seen that the error $\boldsymbol{e}(t)$ increases before an event occurs, immediately decreases afterwards, and never reaches the error bound $e_{\max}$.

To outline the effect of adjusting the event threshold, the lower plots show the approximation error for $\bar{e} = 1$ which yields the error bound $e_{\max} = 1.13$. As expected, the actual approximation error decreases and never reaches $e_{\max}$. However, the communication increases which can be seen in all three cases as the first event at time t_1 occurs always earlier than for $\bar{e} = 2$. Using no disturbance estimation (right subplots), this fact becomes even more evident as instead of seven events overall 14 events are triggered.

Experimental Results. Figure 5.11 shows the trajectories of the event-based state-feedback loop subject to the constant disturbance $\bar{d} = 0.05$ (Fig. 5.11a). An event takes place at time t_1, where inequality (5.25) is satisfied (Fig. 5.11b). At this time instance, the disturbance magnitude $\bar{d}$ is estimated (Fig. 5.11a). Overall 11 events occur in the time interval $[0, 2000\ \mathrm{s}]$ (Fig. 5.11d).

The experimental results differ slightly from the simulation results in Fig. 5.9 due to model uncertainties, which affect the disturbance estimation and cause the variation of the disturbance estimates $\hat{d}_k$. These uncertainties lead to the divergence between $\boldsymbol{x}(t)$ and $\boldsymbol{x}_s(t)$ after event time t_1 despite the good disturbance estimate $\hat{d}_1$.

In summary, the experiments show that the event-based state feedback is robust against severe model uncertainties, because the difference of the model and the measured output is dealt with as disturbance, for which an estimate is used as an additional input to the control input generator. This can be seen by the varying disturbance estimates and the considerable deviation between the disturbance estimate and the actual disturbance. □

5.3 Event-Based Stabilization of Interconnected Systems

5.3.1 Control of Interconnected Systems

The event-based control approach proposed in the previous section assumes the plant to have a centralized sensor unit that measures the complete state $\boldsymbol{x}(t)$ and a centralized actuator unit that applies the whole input vector $\boldsymbol{u}(t)$. However, physically interconnected systems as illustrated in Fig. 5.12 do in general not fulfill this requirement. Therefore, this section explains how the event-based state-feedback approach introduced in Section 5.2 can be extended to a distributed and a decentralized feedback structure [352].

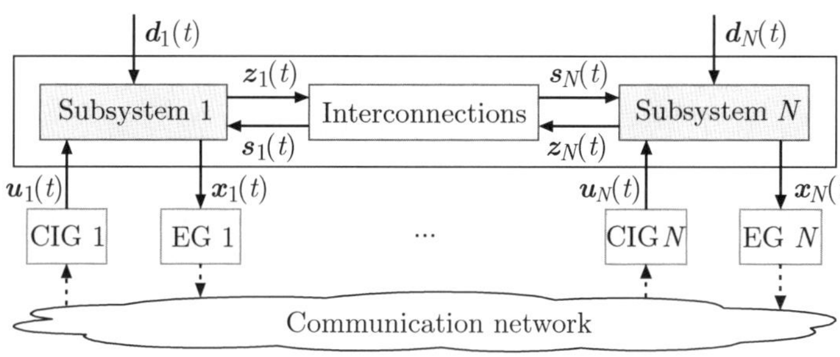

Fig. 5.12 Distributed event-based state-feedback control scheme

It is assumed that the plant (5.1), (5.2) is composed of N physically interconnected subsystems. The i-th subsystems is described by the linear state-space model

$$\Sigma_i : \begin{cases} \dot{\boldsymbol{x}}_i(t) = \boldsymbol{A}_i\boldsymbol{x}_i(t) + \boldsymbol{B}_i\boldsymbol{u}_i(t) + \boldsymbol{E}_i\boldsymbol{d}_i(t) + \boldsymbol{E}_{\mathrm{s}i}\boldsymbol{s}_i(t), \quad \boldsymbol{x}_i(0) = \boldsymbol{x}_{i0} \\ \boldsymbol{z}_i(t) = \boldsymbol{C}_{\mathrm{z}i}\boldsymbol{x}_i(t) \end{cases} \tag{5.26}$$

where $\boldsymbol{x}_i \in \mathbb{R}^{n_i}$, $\boldsymbol{u}_i \in \mathbb{R}^{m_i}$, $\boldsymbol{d}_i \in \mathbb{R}^{l_i}$, $\boldsymbol{s}_i \in \mathbb{R}^{p_i}$ and $\boldsymbol{z}_i \in \mathbb{R}^{q_i}$ denote the state, control input, disturbance, coupling input or coupling output of subsystem Σ_i, respectively. The interactions between all subsystems (5.26) are described by the algebraic relation

$$\boldsymbol{s}(t) = \boldsymbol{L}\boldsymbol{z}(t) \tag{5.27}$$

with

$$\boldsymbol{s}(t) = \left(\boldsymbol{s}_1^{\mathrm{T}}(t) \dots \boldsymbol{s}_N^{\mathrm{T}}(t)\right)^{\mathrm{T}}, \quad \boldsymbol{z}(t) = \left(\boldsymbol{z}_1^{\mathrm{T}}(t) \dots \boldsymbol{z}_N^{\mathrm{T}}(t)\right)^{\mathrm{T}}.$$

5.3.2 Distributed Realization of the Event-Based State Feedback

This section explains how the event-based state-feedback approach introduced in Section 5.2 can be realized in a distributed manner if the requirement of a centralized sensor and actuator unit is not complied.

Distributed Control Input Generation. The input generation is now distributed among N control input generators each of which produces a part of the input signal

$$\boldsymbol{u}(t) = \left(\boldsymbol{u}_1^{\mathrm{T}}(t) \ldots \boldsymbol{u}_N^{\mathrm{T}}(t)\right)^{\mathrm{T}}.$$

The control input (5.9), (5.10) for the overall system is obtained by means of a state-feedback with the feedback-gain

$$\boldsymbol{K} = \left(\boldsymbol{K}_1^{\mathrm{T}} \ldots \boldsymbol{K}_N^{\mathrm{T}}\right)^{\mathrm{T}}$$

with $\boldsymbol{K}_i \in \mathbb{R}^{m_i \times n}$, which implies that the component $\boldsymbol{u}_i(t)$ of the control input $\boldsymbol{u}(t)$ is determined according to

$$\boldsymbol{u}_i(t) = -\boldsymbol{K}_i \boldsymbol{x}_{\mathrm{s}}(t). \tag{5.28}$$

Equation (5.28) implies that the i-th control input generator needs to know the overall model state $\boldsymbol{x}_{\mathrm{s}}$. Hence, each control input generator incorporates the model of the overall continuous-time closed-loop system (5.5)

$$\dot{\boldsymbol{x}}_{\mathrm{s}}(t) = \bar{\boldsymbol{A}}\boldsymbol{x}_{\mathrm{s}}(t) + \boldsymbol{E}\hat{\boldsymbol{d}}_k, \quad \boldsymbol{x}_{\mathrm{s}j}(t_k^+) = \boldsymbol{x}_j(t_k). \tag{5.29}$$

At event time t_k, the j-th event generator triggers an event and broadcasts the state $\boldsymbol{x}_j(t_k)$ of Σ_j to all other components of the control loop. The state information $\boldsymbol{x}_j(t_k)$ is used in all generators in order to simultaneously update the j-th component $\boldsymbol{x}_{\mathrm{s}j}$ of the model state $\boldsymbol{x}_{\mathrm{s}}$.

Distributed Event Generation. The state-space model of the plant (5.1), (5.2), controlled by the input generators (5.28), (5.29) for all $i = 1, \ldots, N$ is given by

$$\dot{\boldsymbol{x}}(t) = \bar{\boldsymbol{A}}\boldsymbol{x}(t) + \boldsymbol{B}\boldsymbol{K}\left(\boldsymbol{x}(t) - \boldsymbol{x}_{\mathrm{s}}(t)\right) + \boldsymbol{E}\boldsymbol{d}(t), \quad \boldsymbol{x}(0) = \boldsymbol{x}_0. \tag{5.30}$$

Since $\bar{\boldsymbol{A}}$ is Hurwitz and the disturbance $\boldsymbol{d}(t)$ is bounded ($\|\boldsymbol{d}(t)\| \leq d_{\max}$), the closed-loop system is *GUUB* if the difference state

$$\boldsymbol{x}_\Delta(t) = \boldsymbol{x}(t) - \boldsymbol{x}_{\mathrm{s}}(t)$$

is bounded. This condition is fulfilled if the i-th event generator triggers an event whenever the equality

$$\|\boldsymbol{x}_{\Delta i}(t_k)\| = \|\boldsymbol{x}_i(t_k) - \boldsymbol{x}_{\mathrm{s}i}(t_k)\| = \bar{e}_i \tag{5.31}$$

holds, with $\bar{e}_i \in \mathbb{R}_+$ denoting the i-th event threshold. As a consequence of the event generation and the reset of the model state $\boldsymbol{x}_{\mathrm{s}i}$, the difference state $\boldsymbol{x}_{\Delta i}$ is bounded from above by

$$\|\boldsymbol{x}_{\Delta i}(t)\| \leq \bar{e}_i, \quad \forall t \geq 0. \tag{5.32}$$

Lemma 5.5. *Due to the event triggering condition* (5.31) *and the reinitialization defined in* (5.29)*, Eq.* (5.32) *holds for every subsystem* Σ_i*,* $(i \in \mathcal{N})$ *and the difference state* $\boldsymbol{x}_\Delta(t)$ *for the overall system is bounded by*

$$\|\boldsymbol{x}_\Delta(t)\| \leq \left\|\left(\bar{e}_1 \ldots \bar{e}_N\right)\right\| = \bar{\delta}. \tag{5.33}$$

Disturbance Estimation. In order to evaluate the disturbance estimation (5.22), the overall plant state $\boldsymbol{x}(t_k)$ must be measurable. However, this requirement is not complied in the i-th event generator, since there only the state $\boldsymbol{x}_i(t_k)$ of the corresponding subsystem is known. Hence, the disturbance estimation according to (5.22) in the component related to subsystem Σ_i is carried out under the assumption that $\boldsymbol{x}_j(t_k) = \boldsymbol{x}_{\mathrm{s}j}(t_k)$ holds for all $j \in \mathcal{N}, (j \neq i)$. This means that instead of the true difference state $\boldsymbol{x}_\Delta(t_k) = \boldsymbol{x}(t_k) - \boldsymbol{x}_{\mathrm{s}}(t_k)$ the estimate

$$\hat{\boldsymbol{x}}_\Delta(t_k) = \left(\boldsymbol{0}^{\mathrm{T}} \ \ldots \ \boldsymbol{0}^{\mathrm{T}} \ \ \boldsymbol{x}_{\Delta i}^{\mathrm{T}}(t_k) \ \ \boldsymbol{0}^{\mathrm{T}} \ \ldots \ \boldsymbol{0}^{\mathrm{T}}\right)^{\mathrm{T}} \tag{5.34}$$

with $\boldsymbol{x}_{\Delta i}(t_k) = \boldsymbol{x}_i(t_k) - \boldsymbol{x}_{\mathrm{s}i}(t_k)$ is applied for the estimation resulting in the recursion:

$$\hat{\boldsymbol{d}}_0 = \boldsymbol{0}, \tag{5.35}$$

$$\hat{\boldsymbol{d}}_k = \hat{\boldsymbol{d}}_{k-1} + \left(\boldsymbol{A}^{-1}\left(\mathrm{e}^{\boldsymbol{A}(t_k - t_{k-1})} - \boldsymbol{I}_n\right)\boldsymbol{E}\right)^+ \hat{\boldsymbol{x}}_\Delta(t_k). \tag{5.36}$$

The disturbance estimator (5.35), (5.36) is incorporated in every event generator. Whenever the event condition (5.31) is met, the i-th event generator broadcasts the current plant state $\boldsymbol{x}_i(t_k)$ as well as the disturbance estimate $\hat{\boldsymbol{d}}_k$ to all control input generators and event generators. Due to the simultaneous state reset and the application of the identical disturbance estimates $\hat{\boldsymbol{d}}_k$ in all control input generators and event generators, the models (5.29) used in these components are synchronized for all $t \geq 0$.

Summary of the Distributed Event-Based Controller. In summary, the event-based control loop consists of the following components:

- the plant composed of N physically interconnected subsystems (5.26) – (5.27),
- N control input generators (5.28), (5.29) that determine the control input $\boldsymbol{u}_i(t), (i \in \mathcal{N})$ and

- N event generators that also use the model (5.29) to detect the event times t_k by continuously checking the trigger condition (5.31).

When at time t_k the event generator $j \in \mathcal{N}$ triggers an event, the state information $\boldsymbol{x}_j(t_k)$ is broadcasted to all control input generators and event generators and is used there for updating the model state $\boldsymbol{x}_{\mathrm{s}j}$ according to $\boldsymbol{x}_{\mathrm{s}j}(t_k^+) = \boldsymbol{x}_j(t_k)$ and for determining the new disturbance estimate $\hat{\boldsymbol{d}}_k$ according to (5.35), (5.36).

Deviation between the Behavior of the Event-Based and the Continuous-Time State-Feedback Loop. Following the analysis presented in Section 5.2.3, the disturbance rejection behavior of the event-based control system is now investigated. The analysis yields a comparison between the state trajectories $\boldsymbol{x}(t)$ of the event-based system (5.1), (5.2), (5.28), (5.29), (5.31) and the continuous-time state-feedback loop (5.5), (5.6) the state of which is denoted by $\boldsymbol{x}_{\mathrm{CT}}(t)$. The difference $\boldsymbol{e}(t) = \boldsymbol{x}(t) - \boldsymbol{x}_{\mathrm{CT}}(t)$ is bounded by

$$\|\boldsymbol{e}(t)\| \leq \bar{\delta} \cdot \int_0^\infty \left\| \mathrm{e}^{\bar{\boldsymbol{A}}\tau} \boldsymbol{B}\boldsymbol{K} \right\| \mathrm{d}\tau$$

with $\bar{\delta}$ defined in (5.33). This result shows two important facts (cf. the discussion of Theorem 5.2):

- The event-based control system with distributed generation of the control input and event triggering can be made to mimic the continuous-time state-feedback with arbitrary accuracy.

- Since the state of continuous-time state-feedback is *GUUB* (Theorem 5.1), the state of the distributed realization of the event-based state-feedback is *GUUB* as well.

Extension. The distributed realization of the event-based state-feedback loop presented in this section requires the state of each subsystem Σ_i to be measurable. The problem where the state of only some subsystems is accessible and no feedback information is available from the remaining subsystems has been investigated in [350]. In order to ensure a maximum allowable deviation between the continuous-time state-feedback system with full state information and the event-based control system, the triggering conditions are appropriately adjusted. Therefore, a method for the design of the event thresholds has been proposed which has been formulated as a linear programming problem.

Evaluation of the Distributed Realization Scheme. A prerequisite for the distributed realization of the event-based state-feedback approach is the requirement that every control input generator and every event generator has to include a model of the overall continuous-time state-feedback system which brings along several issues:

- Interconnected systems are usually considered to be of large scale which implies that the evaluation of the behavior of the continuous-time state-feedback system in each generator necessitates extensive data storage capacity and computational resources.

- At the event times t_k a broadcast communication has to be used to simultaneously reset the state in all models and, thus, to ensure that these models are synchronized.

- In case of additional subsystems extending the overall plant, the event generators and control input generators of all subsystems must be redesigned.

These three requirements will be released in the decentralized scheme introduced in the next section.

The main merit of the distributed scheme lies in the reduction of the amount of data to be transmitted at the event times compared to the event-based single-loop system proposed in Section 5.2, which is due to asynchronous event triggering and reinitialization.

5.3.3 Decentralized Event-Based State Feedback

This section analyzes the stability of N interconnected event-based state-feedback systems shown in Fig. 5.13. The overall system consists of

- N subsystems and their physical interconnections,
- N event generators that trigger events at times t_k and
- N control input generators each of which generates the control input in a decentralized manner using the received information at time t_k.

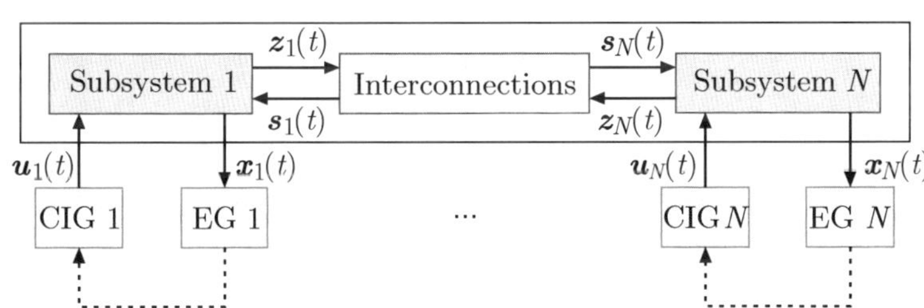

Fig. 5.13 Decentralized event-based state-feedback control scheme

The control input generator and event generator for each subsystem is designed using the method presented in Section 5.2, which is applied now for the isolated and undisturbed subsystems

$$\Sigma_i : \quad \dot{\boldsymbol{x}}_i(t) = \boldsymbol{A}_i \boldsymbol{x}_i(t) + \boldsymbol{B}_i \boldsymbol{u}_i(t), \quad \boldsymbol{x}_i(0) = \boldsymbol{x}_{i0}. \tag{5.37}$$

Hence, the decentralized event-based controllers are known to stabilize the isolated subsystems. The analysis problem to be solved concerns the question how large the interconnections between the subsystems can be such that the stability of the decoupled event-based control loops implies the stability of the overall system.

Decentralized Control Input Generation. The i-th control input generator incorporates the model

$$\dot{\boldsymbol{x}}_{\mathrm{s}i}(t) = \bar{\boldsymbol{A}}_i \boldsymbol{x}_{\mathrm{s}i}(t), \quad \boldsymbol{x}_{\mathrm{s}i}(t_k^+) = \boldsymbol{x}_i(t_k) \tag{5.38}$$

of the decoupled continuous control loop, where $\boldsymbol{x}_{\mathrm{s}i} \in \mathbb{R}^{n_i}$ denotes the model state. It is assumed that a decentralized state feedback gain

$$\boldsymbol{K} = \mathrm{diag}\left(\boldsymbol{K}_1 \dots \boldsymbol{K}_N\right)$$

with $\boldsymbol{K}_i \in \mathbb{R}^{m_i \times n_i}$ exists such that the matrices

$$\bar{\boldsymbol{A}}_i = (\boldsymbol{A}_i - \boldsymbol{B}_i \boldsymbol{K}_i), \quad i = 1, 2, ..., N$$

are asymptotically stable. As before, the model state $\boldsymbol{x}_{\mathrm{s}i}(t)$ is used to generate the control input $\boldsymbol{u}_i(t)$ to subsystem Σ_i according to

$$\boldsymbol{u}_i(t) = -\boldsymbol{K}_i \boldsymbol{x}_{\mathrm{s}i}(t). \tag{5.39}$$

Decentralized Event Generation. An event is triggered in the i-th subsystem whenever the relation

$$\|\boldsymbol{x}_{\Delta i}(t)\|_\infty = \|\boldsymbol{x}_i(t) - \boldsymbol{x}_{\mathrm{s}i}(t)\|_\infty = \bar{e}_i \tag{5.40}$$

holds, where $\bar{e}_i \in \mathbb{R}_+$ denotes the event threshold for subsystem Σ_i. At this time t_k the current plant state $\boldsymbol{x}_i(t_k)$ is transmitted by the i-th event generator to the i-th control input generator and the state $\boldsymbol{x}_{\mathrm{s}i}$ of the model (5.38) used in both generators of subsystem Σ_i is reset.

Due to an initial reset of the model states $\boldsymbol{x}_{\mathrm{s}i}(0)$ at time $t = 0$, $\boldsymbol{x}_{\Delta i}(0) = \boldsymbol{0}$ holds for all $i \in \mathcal{N}$ and, in addition, each state reset leads to the equality $\boldsymbol{x}_{\Delta i}(t_k^+) = \boldsymbol{0}$ which implies the boundedness of the difference state.

Lemma 5.6. *The difference state $\boldsymbol{x}_{\Delta i}(t)$ is bounded by*

$$|\boldsymbol{x}_{\Delta i}(t)| \leq \left(\bar{e}_i \dots \bar{e}_i\right)^{\mathrm{T}} = \boldsymbol{e}_i, \quad \forall\, t \geq 0, \tag{5.41}$$

with $\boldsymbol{e}_i \in \mathbb{R}^{n_i}$.

The following paragraph presents a method for the stability analysis of the interconnected event-based control loops. The comparison principle is used to derive a criterion to test the stability of the overall system.

Analysis of the Overall Event-Based State-Feedback System. The subsequently presented stability analysis makes use of comparison principle which yield upper bounds on the signals of the respective subsystems [245].

Definition 5.2. *Consider the isolated subsystem (5.37) that has the state trajectory*

$$\begin{aligned}\boldsymbol{x}_i(t) &= \mathrm{e}^{\bar{\boldsymbol{A}}_i t}\boldsymbol{x}_{i0} + \int_0^t \mathrm{e}^{\bar{\boldsymbol{A}}_i(t-\tau)}\boldsymbol{B}_i\boldsymbol{u}_i(\tau)\,\mathrm{d}\tau \\ &= \boldsymbol{F}_i(t)\boldsymbol{x}_{i0} + \boldsymbol{G}_{\mathrm{xu}i} * \boldsymbol{u}_i.\end{aligned}$$

The system

$$\boldsymbol{r}_i(t) = \bar{\boldsymbol{F}}_i(t)\,|\boldsymbol{x}_{i0}| + \bar{\boldsymbol{G}}_{\mathrm{xu}i} * |\boldsymbol{u}_i| \tag{5.42}$$

with $\boldsymbol{r}_i \in \mathbb{R}^{n_i}$ is called a comparison system *of subsystem (5.37) if it satisfies the inequality*

$$\boldsymbol{r}_i(t) \geq |\boldsymbol{x}_i(t)|\,, \quad \forall\, t \geq 0$$

for an arbitrary bounded input $\boldsymbol{u}_i(t)$.

In the following, a sufficient condition on the interconnection relation (5.27) between the subsystems is presented which ensures the overall event-based state-feedback system to be *GUUB*. The i-th subsystems (5.26), (5.27) with $\boldsymbol{d}_i(t) = \boldsymbol{0}$, $(i \in \mathcal{N})$ together with the event-based control (5.38), (5.39) is described by the state-space model

$$\begin{aligned}\dot{\boldsymbol{x}}_i(t) &= \bar{\boldsymbol{A}}_i\boldsymbol{x}_i(t) + \boldsymbol{B}_i\boldsymbol{K}_i\boldsymbol{x}_{\Delta i}(t) + \boldsymbol{E}_{\mathrm{s}i}\boldsymbol{s}_i(t), \quad \boldsymbol{x}_i(0) = \boldsymbol{x}_{i0} \\ \boldsymbol{z}_i(t) &= \boldsymbol{C}_{\mathrm{z}i}\boldsymbol{x}_i(t),\end{aligned} \tag{5.43}$$

which yields the state trajectory

$$\boldsymbol{x}_i(t) = \mathrm{e}^{\bar{\boldsymbol{A}}_i t}\boldsymbol{x}_{i0} + \boldsymbol{G}_{\mathrm{xx}i} * \boldsymbol{x}_{\Delta i} + \boldsymbol{G}_{\mathrm{xs}i} * \boldsymbol{s}_i$$

with

$$\boldsymbol{G}_{\mathrm{xx}i}(t) = \mathrm{e}^{\bar{\boldsymbol{A}}_i t}\boldsymbol{B}_i\boldsymbol{K}_i, \quad \boldsymbol{G}_{\mathrm{xs}i}(t) = \mathrm{e}^{\bar{\boldsymbol{A}}_i t}\boldsymbol{E}_{\mathrm{s}i}$$

and $\boldsymbol{x}_{\Delta i}(t) = \boldsymbol{x}_i(t) - \boldsymbol{x}_{\mathrm{s}i}(t)$. An upper bound on the state $\boldsymbol{x}_i(t)$ is obtained by means of the comparison system

$$\boldsymbol{r}_{\mathrm{x}i}(t) = \bar{\boldsymbol{F}}_i(t)\,|\boldsymbol{x}_{i0}| + \bar{\boldsymbol{G}}_{\mathrm{xx}i} * |\boldsymbol{x}_{\Delta i}| + \bar{\boldsymbol{G}}_{\mathrm{xs}i} * |\boldsymbol{s}_i| \tag{5.44}$$

with

$$\bar{\boldsymbol{F}}_i(t) = \left|\mathrm{e}^{\bar{\boldsymbol{A}}_i t}\right|, \quad \bar{\boldsymbol{G}}_{\mathrm{xx}i}(t) = |\boldsymbol{G}_{\mathrm{xx}i}(t)|\,, \quad \bar{\boldsymbol{G}}_{\mathrm{xs}i}(t) = |\boldsymbol{G}_{\mathrm{xs}i}(t)|\,.$$

Here, the absolute-value operator $|.|$ is applied elementewise to vectors and matrices. Since the difference state $\boldsymbol{x}_{\Delta i}(t)$ is bounded according to Eq. (5.41), the relation

$$\bar{\boldsymbol{G}}_{\mathrm{xx}i} * |\boldsymbol{x}_{\Delta i}| = \int_0^t \bar{\boldsymbol{G}}_{\mathrm{xx}i}(t-\tau)\, |\boldsymbol{x}_{\Delta i}(\tau)|\, \mathrm{d}\tau \le \boldsymbol{e}_{\max i}$$

holds with

$$\boldsymbol{e}_{\max i} = \int_0^\infty \bar{\boldsymbol{G}}_{\mathrm{xx}i}(\tau)\, \mathrm{d}\tau \cdot \boldsymbol{e}_i. \tag{5.45}$$

Given (5.45) the comparison system (5.44) can be reformulated as

$$\boldsymbol{r}_{\mathrm{x}i}(t) = \bar{\boldsymbol{F}}_i(t)\, |\boldsymbol{x}_{i0}| + \boldsymbol{e}_{\max i} + \bar{\boldsymbol{G}}_{\mathrm{xs}i} * |\boldsymbol{s}_i|\,.$$

In order to arrive at a comparison system for the overall event-based control system let

$$\boldsymbol{r}_{\mathrm{x}}(t) = \begin{pmatrix} \boldsymbol{r}_{\mathrm{x}1}(t) \\ \vdots \\ \boldsymbol{r}_{\mathrm{x}N}(t) \end{pmatrix}, \quad \boldsymbol{e}_{\max} = \begin{pmatrix} \boldsymbol{e}_{\max 1} \\ \vdots \\ \boldsymbol{e}_{\max N} \end{pmatrix}, \quad \boldsymbol{x}_0 = \begin{pmatrix} \boldsymbol{x}_{10} \\ \vdots \\ \boldsymbol{x}_{N0} \end{pmatrix}$$

and

$$\bar{\boldsymbol{F}}(t) = \mathrm{diag}\left(\bar{\boldsymbol{F}}_i(t)\right), \quad \bar{\boldsymbol{G}}_{\mathrm{xs}}(t) = \mathrm{diag}\left(\bar{\boldsymbol{G}}_{\mathrm{xs}i}(t)\right),$$

which yields

$$\boldsymbol{r}_{\mathrm{x}}(t) = \bar{\boldsymbol{F}}(t)\, |\boldsymbol{x}_0| + \boldsymbol{e}_{\max} + \bar{\boldsymbol{G}}_{\mathrm{xs}} * |\boldsymbol{s}|\,. \tag{5.46}$$

With the upper bound $|\boldsymbol{s}(t)| \le |\boldsymbol{L}|\, |\boldsymbol{z}(t)|$ on the interconnection relation (5.27) and $|\boldsymbol{z}(t)| \le |\boldsymbol{C}_{\mathrm{z}}|\, |\hat{\boldsymbol{x}}(t)|$ with $\boldsymbol{C}_{\mathrm{z}} = \mathrm{diag}\,(\boldsymbol{C}_{\mathrm{z}i})$, the relation

$$\boldsymbol{r}_{\mathrm{x}}(t) = \bar{\boldsymbol{F}}\, |\boldsymbol{x}_0| + \boldsymbol{e}_{\max} + \bar{\boldsymbol{G}}_{\mathrm{xs}}\, |\boldsymbol{L}|\, |\boldsymbol{C}_{\mathrm{z}}| * |\boldsymbol{x}| \ge |\boldsymbol{x}(t)|$$

follows from (5.46). The last inequality is an implicit bound on the overall state $\boldsymbol{x}(t)$.

An explicit statement in terms of the initial condition is obtained by means of the comparison principle [245]. Hence,

$$\boldsymbol{r}_{\mathrm{x}}(t) = \boldsymbol{G} * \left(\bar{\boldsymbol{F}}\, |\boldsymbol{x}_0| + \boldsymbol{e}_{\max}\right) \ge |\boldsymbol{x}(t)| \tag{5.47}$$

holds, where the impulse response matrix $\boldsymbol{G}(t)$ satisfies the relation

$$\boldsymbol{G}(t) = \delta(t)\boldsymbol{I} + \bar{\boldsymbol{G}}_{\mathrm{xs}}\, |\boldsymbol{L}|\, |\boldsymbol{C}_{\mathrm{z}}| * \boldsymbol{G}$$

with $\delta(t)$ representing the Dirac impulse. Note that in Eq. (5.47) the term $\bar{\boldsymbol{F}}\, |\boldsymbol{x}_0|$ asymptotically vanishes, whereas $\boldsymbol{e}_{\max}$ is non-vanishing. Therefore, the state of the event-based overall system is *GUUB* if the impulse response matrix $\boldsymbol{G}(t)$ satisfies the inequality

$$\int_0^\infty \boldsymbol{G}(t)\,\mathrm{d}t < \infty. \tag{5.48}$$

Theorem 5.3. *Consider the decentralized event-based control loops* (5.26), (5.27), (5.38) – (5.40). *If the inequality*

$$\lambda_{\mathrm{P}}\left(\int_0^\infty \bar{\boldsymbol{G}}_{\mathrm{xs}}(t)\,|\boldsymbol{L}|\,|\boldsymbol{C}_{\mathrm{z}}|\,\mathrm{d}t\right) < 1 \tag{5.49}$$

is satisfied, the relation (5.48) *holds and, hence, the overall system is* GUUB. λ_{P} *denotes the largest eigenvalue of the nonnegative matrix ("Perron root").*

The inequality (5.49) is a sufficient condition for the state of the overall event-based state-feedback loop to be *GUUB*. Note that this stability criterion is a small-gain theorem, which is satisfied only if the gains between the signals $\boldsymbol{s}_i(t)$ and $\boldsymbol{z}_i(t)$ for all $i \in \mathcal{N}$ are sufficiently small. To see this property, replace the matrices $\boldsymbol{E}_{\mathrm{s}i}$ in (5.43) by $\kappa \boldsymbol{E}_{\mathrm{s}i}$. Then the condition (5.49) must be reformulated as

$$\lambda_{\mathrm{P}}\left(\int_0^\infty \bar{\boldsymbol{G}}_{\mathrm{xs}}(t)\,|\boldsymbol{L}|\,|\boldsymbol{C}_{\mathrm{z}}|\,\mathrm{d}t\right) < \frac{1}{\kappa}$$

which is satisfied only for sufficiently small coupling strength κ.

Generalization of the Stability Condition. If instead of event-based sampling a continuous-time state-feedback control scheme is applied to the plant (5.26), (5.27) with $\boldsymbol{d}_i(t) = \boldsymbol{0}$, $(i \in \mathcal{N})$ the stability condition (5.49) ensures asymptotic stability of the overall control system. As shown in [353], this conclusion can be made more general as summarized in the following theorem.

Theorem 5.4. *Given a decentralized state-feedback gain* $\boldsymbol{K}$ *that yields a continuous closed-loop system which is asymptotically stable. Then the event-based implementation of this controller in form of decentralized event-based control stations* (5.26), (5.39), *that are reset each time the condition* (5.40) *holds, yields a control system that is* GUUB.

Theorem 5.4 states that the event-based implementation of a decentralized state-feedback controller as proposed in this section does not impose more restrictions on the interconnections between the subsystems than are claimed for asymptotic stability of the continuous overall system. Any condition that is sufficient to guarantee asymptotic stability of the continuous system is concomitantly sufficient to ensure ultimate boundedness of the event-based system.

Example 5.2 *Event-based control of interconnected thermofluid processes*

The following example illustrates the behavior of the two approaches to event-based control of interconnected systems which are applied to a thermofluid process explained in Section 5.8.

The overall system is composed of two subsystems. Subsystem Σ_1 is characterized by the behavior of the level l_1 and temperature ϑ_1 of the liquid in reactor TB while subsystem Σ_2 represents the dynamics of the level l_2 and temperature ϑ_2 of the content in reactor TS. Defining

$$\boldsymbol{x}_{\Sigma 1} = (x_1 \;\; x_2)^{\mathrm{T}} = (l_1 \;\; \vartheta_1)^{\mathrm{T}}, \qquad \boldsymbol{u}_{\Sigma 1} = (u_1 \;\; u_2)^{\mathrm{T}}$$
$$\boldsymbol{x}_{\Sigma 2} = (x_3 \;\; x_4)^{\mathrm{T}} = (l_2 \;\; \vartheta_2)^{\mathrm{T}}, \qquad \boldsymbol{u}_{\Sigma 2} = (u_3 \;\; u_4)^{\mathrm{T}}$$

the subsystems can be described by the linearized state-space models

$$\Sigma_1 : \begin{cases} \dot{\boldsymbol{x}}_{\Sigma 1}(t) &= \boldsymbol{A}_1 \boldsymbol{x}_{\Sigma 1}(t) + \boldsymbol{B}_1 \boldsymbol{u}_{\Sigma 1}(t) + \boldsymbol{E}_1 d_1(t) + \boldsymbol{E}_{\mathrm{s}1} s_1(t) \\ \boldsymbol{z}_1(t) &= \boldsymbol{C}_{\mathrm{z}1} \boldsymbol{x}_{\Sigma 1}(t) \end{cases}$$

$$\Sigma_2 : \begin{cases} \dot{\boldsymbol{x}}_{\Sigma 2}(t) &= \boldsymbol{A}_2 \boldsymbol{x}_{\Sigma 2}(t) + \boldsymbol{B}_2 \boldsymbol{u}_{\Sigma 2}(t) + \boldsymbol{E}_2 d_2(t) + \boldsymbol{E}_{\mathrm{s}2} s_2(t) \\ \boldsymbol{z}_2(t) &= \boldsymbol{C}_{\mathrm{z}2} \boldsymbol{x}_{\Sigma 2}(t) \end{cases}$$

with

$$\boldsymbol{A}_1 = 10^{-3} \begin{pmatrix} -2.08 & 0 \\ -4.96 & -3.35 \end{pmatrix}, \qquad \boldsymbol{A}_2 = 10^{-3} \begin{pmatrix} -2.43 & 0 \\ 23.6 & -4.66 \end{pmatrix},$$
$$\boldsymbol{B}_1 = 10^{-3} \begin{pmatrix} 2.30 & 0 \\ 0 & -38.9 \end{pmatrix}, \qquad \boldsymbol{B}_2 = 10^{-3} \begin{pmatrix} 2.59 & 0 \\ 0 & 35.0 \end{pmatrix},$$
$$\boldsymbol{E}_1 = 10^{-3} \begin{pmatrix} 0 \\ 168.9 \end{pmatrix}, \qquad \boldsymbol{E}_2 = 10^{-3} \begin{pmatrix} 1.16 \\ -20.7 \end{pmatrix},$$
$$\boldsymbol{E}_{\mathrm{s}1} = 10^{-3} \begin{pmatrix} 0 \\ 3.62 \end{pmatrix}, \qquad \boldsymbol{E}_{\mathrm{s}2} = 10^{-3} \begin{pmatrix} 0 \\ 4.66 \end{pmatrix},$$
$$\boldsymbol{C}_{\mathrm{z}1} = \begin{pmatrix} 0 & 1 \end{pmatrix}, \qquad \boldsymbol{C}_{\mathrm{z}2} = \begin{pmatrix} 0 & 1 \end{pmatrix}.$$

The subsystems are coupled according to the relation

$$\begin{pmatrix} s_1(t) \\ s_2(t) \end{pmatrix} = \begin{pmatrix} 0 & l_{12} \\ l_{21} & 0 \end{pmatrix} \begin{pmatrix} z_1(t) \\ z_2(t) \end{pmatrix}$$

with $l_{12}, l_{21} \in [0, 1]$. $l_{12} = l_{21} = 0$ means that the valves between the tanks are closed and, hence, the subsystems are not interconnected.

The control objective is to keep the level and the temperature in both reactors constant. In the case of continuous-time control the decentralized feedback matrix

$$\boldsymbol{K} = \begin{pmatrix} 4.26 & 0 & 0 & 0 \\ 0.69 & -1.06 & 0 & 0 \\ 0 & 0 & 4.76 & 0 \\ 0 & 0 & 0.60 & 0.26 \end{pmatrix}$$

yields a desired behavior with respect to the disturbance rejection and is, hence, also applied for the event-based control. The event threshold $\bar{e}_1$ for subsystem Σ_1 and $\bar{e}_2$ for subsystem Σ_2 is set to $\bar{e}_1 = \bar{e}_2 = 0.5$ which results in the triggering conditions

$$\left\| \begin{pmatrix} x_{\Delta 1}(t) \\ x_{\Delta 2}(t) \end{pmatrix} \right\| = \left\| \begin{pmatrix} x_1(t) \\ x_2(t) \end{pmatrix} - \begin{pmatrix} x_{s1}(t) \\ x_{s2}(t) \end{pmatrix} \right\| = 0.5 \qquad \text{for subsystem } \Sigma_1,$$

$$\left\| \begin{pmatrix} x_{\Delta 3}(t) \\ x_{\Delta 4}(t) \end{pmatrix} \right\| = \left\| \begin{pmatrix} x_3(t) \\ x_4(t) \end{pmatrix} - \begin{pmatrix} x_{s3}(t) \\ x_{s4}(t) \end{pmatrix} \right\| = 0.5 \qquad \text{for subsystem } \Sigma_2.$$

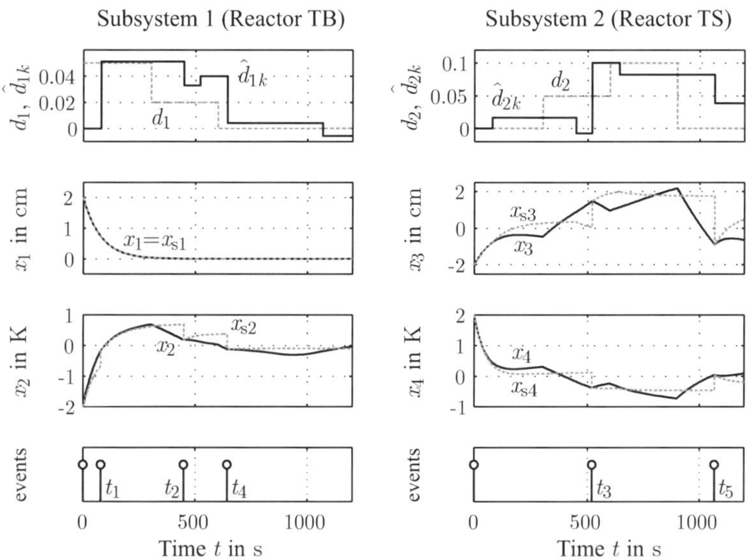

Fig. 5.14 Behavior of the distributed realization of the event-based state-feedback approach subject to a piecewise constant disturbance. Solid lines: plant state $\boldsymbol{x}(t)$; dashed lines: model state $\boldsymbol{x}_s(t)$

Distributed Realization of the Event-Based State-Feedback. The following analysis illustrates the behavior of the distributed realization of the event-based state-feedback approach subject to a piecewise constant disturbance and an initial condition $\boldsymbol{x}_1(0) = (0.02 \quad -2)^{\mathrm{T}}$ and $\boldsymbol{x}_2(0) = (-0.02 \quad 2)^{\mathrm{T}}$. Figure 5.14 shows the simulation results, where the plots on the left-hand side illustrate the disturbance and its estimate, the level $x_1(t)$ and the temperature $x_2(t)$ in subsystem Σ_1. The figure at the bottom indicates the event times. The figures on the right-hand side of Fig. 5.14 show the corresponding signals of the subsystem Σ_2.

The first event is triggered at $t = t_1$ by subsystem Σ_1 owing to the disturbance d_1 that affects this subsystem. At this time the model states x_{s1} and x_{s2} in all generators are reset and a new disturbance estimate is derived according to (5.36). While the estimate $\hat{d}_{11}$ almost coincides with the actual disturbance d_1, the estimate $\hat{d}_{21}$ deviates from the signal d_2, which is due to the incorrect assumption that $(x_{\Delta 3}(t_1) \;\; x_{\Delta 4}(t_1))^{\mathrm{T}} = \mathbf{0}$ (cf. Eq. (5.34)). After the reinitialization and disturbance estimation, both the model state $\boldsymbol{x}_{\mathrm{s}}$ and plant state $\boldsymbol{x}$ behave almost identical in subsystem Σ_1, until the disturbance changes again which leads to the next event at time $t = t_2$. The disturbance estimate in subsystem Σ_1 enlarges the deviation between the disturbance d_2 and the estimate $\hat{d}_{22}$ such that a new event is generated at $t = t_3$ by subsystem Σ_2. The estimate at $t = t_4$ yields a good result such that no further events are triggered until the disturbance d_2 changes which causes the event at $t = t_5$.

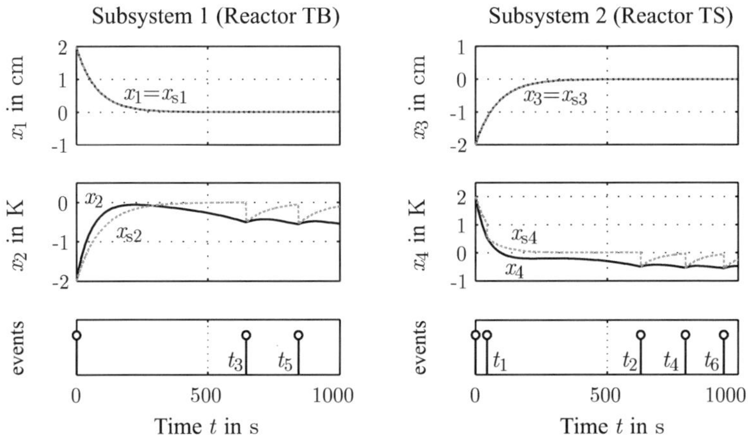

Fig. 5.15 Behavior of the decentralized event-based state-feedback controllers. Solid lines: plant state $\boldsymbol{x}(t)$; dashed lines: model state $\boldsymbol{x}_{\mathrm{s}}(t)$.

Decentralized Event-Based State-Feedback. In contrast to the previous example, the following analysis studies the behavior of the decentralized event-based state-feedback subject to an initial condition $\boldsymbol{x}_1(0) = (0.02 \;\; -2)^{\mathrm{T}}$ and $\boldsymbol{x}_2(0) = (-0.02 \;\; 2)^{\mathrm{T}}$ but no disturbance. The state-feedback controllers

$$\boldsymbol{K}_1 = \begin{pmatrix} 4.26 & 0 \\ 0.69 & -1.06 \end{pmatrix}, \quad \boldsymbol{K}_2 = \begin{pmatrix} 4.76 & 0 \\ 0.6 & 0.26 \end{pmatrix}$$

stabilize the uncoupled subsystems and this analysis now investigates the behavior of the overall system subject to a non-vanishing interconnection. Figure 5.15 illustrates the behavior of the overall system: The left-hand side shows the level x_1 and temperature x_2 in subsystem Σ_1. The event times are again indicated in the bottom figure. The right-hand side of Fig. 5.15 shows the corresponding signals of subsystem Σ_2. It can be seen that the interconnection only has an effect on the temperatures x_2 and x_4 in subsystem Σ_1 or subsystem Σ_2, respectively. As the states x_1 and x_3 always coincide with the corresponding model states x_{s1} and x_{s3}, events are only generated due to the behavior of the temperatures.

The first event is generated at $t = t_1$ caused by a deviation between $x_4(t_1)$ and $x_{s4}(t_1)$ which is due to the interconnection between both subsystems which, however, is not regarded in the models that are applied in each generator. After the states x_1 and x_3 approach the setpoint the subsystem Σ_2 triggers the next event at $t = t_2$. In the ensuing time both subsystems keep triggering events and the states x_2 and x_4 remain in a range between $-0.5\,\mathrm{K}$ and $-0.6\,\mathrm{K}$. □

5.4 Optimization-Based Control

An elegant way to construct a globally stabilizing controller for a generic control system is to apply Bellman's *optimality principle*. This approach will be used in this section for event-based control.

Bellman's optimality principle is a fixed point equation which, together with a suitable boundary condition, characterizes the minimal cost accumulated along any controlled trajectory (the so called *value function*) V, where the associated instantaneous cost may be given in the problem description or may have to be suitably defined. From this fixed point equation, the associated optimally stabilizing controller can be computed by repeatedly solving a finite-dimensional optimization problem. The value function will then act as a Lyapunov function for the closed-loop system.

If the dynamics is linear and the instantaneous cost is quadratic, an explicit formula for the function V and the controller can be derived by solving an associated *Ricatti equation*. In the nonlinear case, one typically has to resort to a numerical approximation of V. In this case, one needs to solve Bellman's equation on a finite-dimensional approximation space. Typically, piecewise (multi)linear functions are employed for the approximation [17, 145]. More recently, piecewise constant functions have been proposed for this purpose [147–150, 195]. This approach leads to a shortest path problem on a finite graph, which can be solved by corresponding algorithms from graph theory (like Dijkstra's algorithm). The approach is particularly well suited for problems with highly irregular value function, complicated state constraints and also naturally handles hybrid systems. Its extension to perturbed systems conceptually enables the treatment of systems with quantized states and also discrete-event systems in a natural way.

In this section, we describe this approach as well as its extension to state-quantized and event-based control (cf. Sections 5.4.5 and 5.4.6). We show how to treat systems with very coarsely quantized state space in Section 5.4.7. Motivated by *networked control systems* we describe how to deal with effects like a delayed transmission of the state information in Section 5.4.8 as well as how to use the optimization framework in order to minimize the number of times that information has to be transmitted from the controller to the plant.

5.4.1 Problem Formulation

Our goal is to globally and optimally stabilize a given subset $\mathcal{T}$ of the state space $\mathcal{X}$ of the perturbed discrete-time control system

$$\boldsymbol{x}(k+1) = \boldsymbol{f}(\boldsymbol{x}(k), \boldsymbol{u}(k), \boldsymbol{w}(k)), \quad k = 0, 1, \ldots, \tag{5.50}$$

$\boldsymbol{x}(k) \in \mathcal{X}$, subject to the continuous instantaneous cost

$$g(\boldsymbol{x}(k), \boldsymbol{u}(k)) \geq 0.$$

$\boldsymbol{u}(k)$ is the control input and $\boldsymbol{w}(k)$ the perturbation. To this end, we would like to construct an approximate *optimal feedback* $\boldsymbol{u}(\boldsymbol{x}(k))$ such that $\mathcal{T}$ is an asymptotically stable set for the resulting closed-loop system

$$\boldsymbol{x}(k+1) = \boldsymbol{f}(\boldsymbol{x}(k), \boldsymbol{u}(\boldsymbol{x}(k)), \boldsymbol{w}(k)), \quad k = 0, 1, \ldots \tag{5.51}$$

for *any* sequence $(\boldsymbol{w}(k))_{k \in \mathbb{N}}$ of perturbations and such that the accumulated cost $\sum_{k=0}^{\infty} g(\boldsymbol{x}(k), \boldsymbol{u}(\boldsymbol{x}(k)))$ along a feedback trajectory is minimal. The controls $\boldsymbol{u}(k)$ and perturbations $\boldsymbol{w}(k)$ have to be chosen from compact metric spaces $\mathcal{U}$ or $\mathcal{W}$, respectively (which may in particular be discrete).

5.4.2 Optimality Principle

The construction of the feedback controller will be based on the optimality principle. In order to introduce the basic ideas in this section we consider problems without perturbations,

$$\boldsymbol{x}(k+1) = \boldsymbol{f}(\boldsymbol{x}(k), \boldsymbol{u}(k)), \quad k = 0, 1, \ldots \tag{5.52}$$

and the simple target set $\mathcal{T} = \{0\}$. We assume the map $\boldsymbol{f} : \mathcal{X} \times \mathcal{U} \to \mathbb{R}^d$ to be continuous on some compact subset $\mathcal{X} \times \mathcal{U} \subset \mathbb{R}^d \times \mathbb{R}^m$ which contains the origin, $\boldsymbol{f}(\boldsymbol{0}, \boldsymbol{0}) = \boldsymbol{0}$ and $\inf_{u \in \mathcal{U}} g(\boldsymbol{x}, \boldsymbol{u}) > 0$ for all $\boldsymbol{x} \neq \boldsymbol{0}$. For a given initial state $\boldsymbol{x}(0) \in \mathcal{X}$ and a given control sequence $(\boldsymbol{u}(k))_{k \in \mathbb{N}} \in \mathcal{U}^{\mathbb{N}}$ there is a unique associated trajectory $(\boldsymbol{x}(k))_{k \in \mathbb{N}}$ of (5.52).

The total cost along a controlled trajectory is given by

$$J(\boldsymbol{x}(0), (\boldsymbol{u}(k))_{k \in \mathbb{N}}) = \sum_{k=0}^{\infty} g(\boldsymbol{x}(k), \boldsymbol{u}(k)) \in [0, \infty]. \tag{5.53}$$

The *(optimal) value function*

$$V(\boldsymbol{x}) = \inf_{(\boldsymbol{u}(k))_k \in \mathcal{U}^{\mathbb{N}}} J(\boldsymbol{x}, (\boldsymbol{u}(k))_k)$$

satisfies the *optimality principle*

$$V(\boldsymbol{x}) = \inf_{\boldsymbol{u}\in\mathcal{U}} \{g(\boldsymbol{x},\boldsymbol{u}) + V(\boldsymbol{f}(\boldsymbol{x},\boldsymbol{u}))\} \tag{5.54}$$

on the *stabilizable set* $\mathcal{S} = \{\boldsymbol{x} \in \mathcal{X} \mid V(\boldsymbol{x}) < \infty\}$. The operator

$$L[v](\boldsymbol{x}) := \inf_{\boldsymbol{u}\in\mathcal{U}} \{g(\boldsymbol{x},\boldsymbol{u}) + V(\boldsymbol{f}(\boldsymbol{x},\boldsymbol{u}))\}$$

acting on the space of real valued functions v on $\mathcal{S}$ with $v(0) = 0$ is called the *dynamic programming operator* and V is the unique fixed point of the equation $v = L[v]$ on this space [30].

Using the optimality principle, an optimal stabilizing feedback is given by

$$\boldsymbol{u}(\boldsymbol{x}) = \text{argmin}_{\boldsymbol{u}\in\mathcal{U}} \{g(\boldsymbol{x},\boldsymbol{u}) + V(\boldsymbol{f}(\boldsymbol{x},\boldsymbol{u}))\}, \quad \boldsymbol{x} \in \mathcal{S}, \tag{5.55}$$

whenever this minimum exists. The key property in order to prove asymptotic stability of the closed-loop system is the fact that by the (obvious) inequality

$$V(\boldsymbol{x}) \geq g(\boldsymbol{x},\boldsymbol{u}(\boldsymbol{x})) + V(\boldsymbol{f}(\boldsymbol{x},\boldsymbol{u}(\boldsymbol{x}))), \quad \boldsymbol{x} \in \mathcal{S} \tag{5.56}$$

the optimal value function is a Lyapunov function of the closed-loop system, provided that V is finite, positive definite and proper – properties which can be ensured by suitable asymptotic controllability properties and bounds on g, see Chapter 4 of [154].

5.4.3 Discretization of the State Space

We are going to approximate V by piecewise constant functions [195]. This approach is motivated on the one hand by the fact that the resulting discrete problem can be solved by fast shortest-path algorithms and on the other hand by the observation that – via a suitable generalization to perturbed systems (cf. Section 5.4.5) – the resulting feedback is piecewise constant, too, and thus directly applicable to quantized or event-based systems, respectively.

Let P be a partition of the stabilizable set $\mathcal{S}$, i.e. a collection of pairwise disjoint subsets of $\mathcal{S}$ whose union is equal to $\mathcal{S}$. For a state $\boldsymbol{x} \in \mathcal{S}$ we let $[\boldsymbol{x}] \in P$ denote the partition element which contains $\boldsymbol{x}$. In order to simplify the notation, in the sequel we identify any subset $\{\mathcal{P}_1, \ldots, \mathcal{P}_k\} \subseteq P$ with the corresponding subset $\cup_{i=1,\ldots k}\mathcal{P}_i \subseteq \mathcal{S}$.

Let $\mathbb{R}^P$ be the subspace of the space $\mathbb{R}^{\mathcal{S}}$ of all real-valued functions on $\mathcal{S}$ which are piecewise constant on the elements of the partition P. The map

$$\psi[V](\boldsymbol{x}) := \inf_{\boldsymbol{x}'\in[\boldsymbol{x}]} V(\boldsymbol{x}')$$

is a projection from $\mathbb{R}^{\mathcal{S}}$ onto $\mathbb{R}^{P}$. Using this projection, we can define the *discretized dynamic programming operator* $L_P : \mathbb{R}^P \to \mathbb{R}^P$ by

$$L_P := \psi \circ L.$$

Under the boundary condition $V_P(\boldsymbol{x}) = 0$ for all $\boldsymbol{x} \in [0]$ this operator has a unique fixed point V_P – the *approximate (optimal) value function.*

Since V_P is constant on each partition element $\mathcal{P} \in P$, we write $V_P(\mathcal{P})$ for the value of V_P on P. Using this notation, one can show [150] that the fixed point equation $V_P = L_P[V_P]$ is equivalent to the *discrete optimality principle*

$$V_P(\mathcal{P}) = \min_{\mathcal{P}' \in \mathcal{F}(\mathcal{P})} \{\mathcal{G}(\mathcal{P}, \mathcal{P}') + V_P(\mathcal{P}')\}, \tag{5.57}$$

where the map $\mathcal{F}$ is given by

$$\mathcal{F}(\mathcal{P}) = \{\mathcal{P}' \in P : \mathcal{P}' \cap \boldsymbol{f}(\mathcal{P}, \mathcal{U}) \neq \emptyset\} \tag{5.58}$$

and the cost function $\mathcal{G}$ by

$$\mathcal{G}(\mathcal{P}, \mathcal{P}') = \inf_{\boldsymbol{u} \in \mathcal{U}} \{g(\boldsymbol{x}, \boldsymbol{u}) \mid \boldsymbol{x} \in \mathcal{P}, \boldsymbol{f}(\boldsymbol{x}, \boldsymbol{u}) \in \mathcal{P}'\}. \tag{5.59}$$

The value $V_P(\mathcal{P})$ is the *length of the shortest path* from $\mathcal{P}$ to $[0]$ in the weighted directed graph (P, E), where the set of edges is defined by

$$E = \{(\mathcal{P}, \mathcal{P}') : \mathcal{P}' \in \mathcal{F}(\mathcal{P})\}$$

and the edge $(\mathcal{P}, \mathcal{P}')$ is weighted by $\mathcal{G}(\mathcal{P}, \mathcal{P}')$ (Fig. 5.16). As such, it can be computed by, e.g., Dijkstra's algorithm [195].

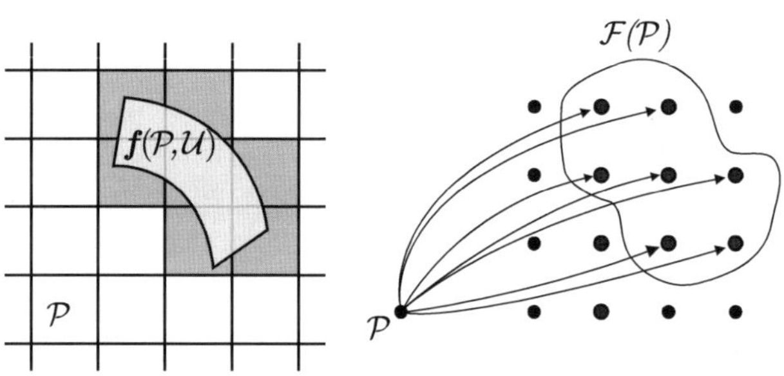

Fig. 5.16 Partition of the state space, image of an element (left) and corresponding edges in the induced graph (right)

In the implementation, the images $\boldsymbol{f}(\mathcal{P}, \mathcal{U})$ are approximated by evaluating $\boldsymbol{f}$ at a set of test points. While this introduces a discretization error, numerical tests have shown that usually this error is negligible. If desired or needed, rigorous discretization techniques could be employed which ensure that the numerically computed image covers the exact image.

Properties of V_P. From the definition of the projection ψ and by an induction argument on the elements of the partition one can conclude that

$$V_P(\boldsymbol{x}) \leq V(\boldsymbol{x}) \tag{5.60}$$

for any $\boldsymbol{x} \in \mathcal{S}$ and every partition P, i.e. the approximate value function V_P yields a *lower bound* on the true value function. As a result, the residual

$$e_{\mathcal{P}} = L[V_P] - V_P$$

is an efficient pointwise a-posteriori estimator for the error $e = V - V_P$, i.e. one has $e_P \leq e$ pointwise on $\mathcal{S}$ [147].

Example 5.3 *A simple first-order system*

Consider the system

$$x(k+1) = x(k) + (1-a)u(k)x(k), \quad k = 0, 1, \ldots \tag{5.61}$$

with the scalar state $x(k) \in \mathcal{X} = [0,1]$, scalar input $u(k) \in \mathcal{U} = [-1,1]$ and fixed parameter $a \in (0,1)$. Let

$$g(x,u) = (1-a)x$$

such that the optimal control policy is to steer the systems to the origin as fast as possible. Hence, for every x, the optimal sequence of controls is $(-1,-1,\ldots)$, which yields $V(x) = x$ as the optimal value function.

For the following computations we consider $a = 0.8$ and use a partition of 64 equally sized subintervals of $[0,1]$. The weights (5.59) are approximated by minimizing over 100 equally spaced test points in each subinterval and 10 equally spaced points in $\mathcal{U}$. Figure 5.17 shows V and the approximate optimal value function V_P. □

5.4.4 Dynamic Games

In this section we describe how our approach can be extended to dynamic games (for details cf. [150]). This extension ultimately enables the application of this approach to quantized systems in the following sections.

We additionally include a second control (aka perturbation) parameter and consider a map $\boldsymbol{f} : \mathcal{X} \times \mathcal{U} \times \mathcal{W} \to \mathcal{X}$, where $\mathcal{X}$ and $\mathcal{U}$ are as above and $\mathcal{W}$ is a compact metric space, together with a cost function $g : \mathcal{X} \times \mathcal{U} \times \mathcal{W} \to [0,\infty)$. For a given initial state $\boldsymbol{x} \in \mathcal{X}$, a given control sequence $(\boldsymbol{u}(k))_k \in \mathcal{U}^{\mathbb{N}}$ and a given perturbation sequence $(\boldsymbol{w}(k))_k \in \mathcal{W}^{\mathbb{N}}$, the associated trajectory of the game is given by the sequence $(\boldsymbol{x}(k))_k$ with

$$\boldsymbol{x}(k+1) = \boldsymbol{f}(\boldsymbol{x}(k), \boldsymbol{u}(k), \boldsymbol{w}(k)), \quad k = 0, 1, 2, \ldots.$$

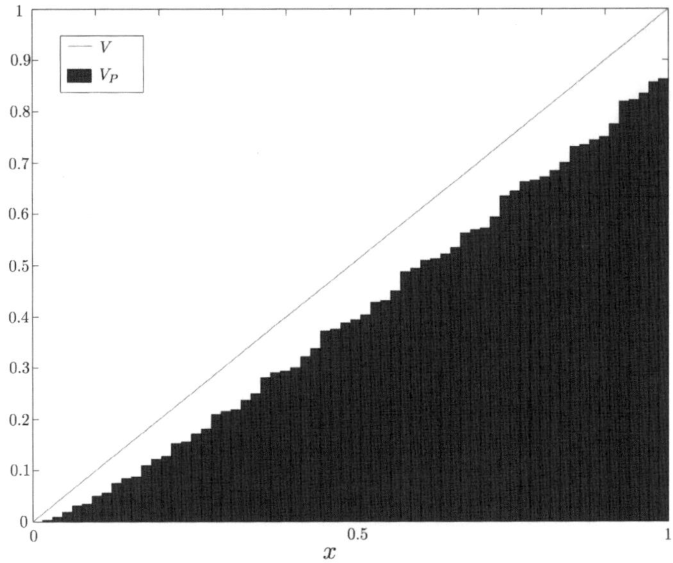

Fig. 5.17 True and approximate optimal value function for the example with 64 subintervals, $a = 0.8$

Specifying a target set $\mathcal{T} \subset \mathcal{X}$, the total cost accumulated along a trajectory is

$$J(\boldsymbol{x}(0), (\boldsymbol{u}(k))_k, (\boldsymbol{w}(k))_k) = \sum_{k=0}^{K} g(\boldsymbol{x}(k), \boldsymbol{u}(k), \boldsymbol{w}(k)),$$

with $K := \inf\{k \geq 0 \,|\, \boldsymbol{x}(k) \in \mathcal{T}\}$. We are interested the so-called *upper value function* of the game that is defined by

$$V(\boldsymbol{x}) = \sup_{\beta \in \mathcal{B}} \inf_{(\boldsymbol{u}(k))_k \in \mathcal{U}^{\mathbb{N}}} J(\boldsymbol{x}, \bar{\boldsymbol{u}}, \beta(\bar{\boldsymbol{u}})). \tag{5.62}$$

Here, $\mathcal{B}$ denotes the set of all nonanticipating strategies $\beta : \mathcal{U}^{\mathbb{N}} \to \mathcal{W}^{\mathbb{N}}$ (a strategy β is *nonanticipating* if from $\boldsymbol{u}(k) = \boldsymbol{u}'(k)$ for all $k \leq K$ it follows that $\beta[\bar{\boldsymbol{u}}](k) = \beta[\bar{\boldsymbol{u}}'](k)$ for all $k \leq K$ for any two control sequences $\bar{\boldsymbol{u}} = (\boldsymbol{u}(k))_k, \bar{\boldsymbol{u}}' = (\boldsymbol{u}'(k))_k \in \mathcal{U}^{\mathbb{N}}$). Again, we denote by $\mathcal{S} = \{\boldsymbol{x} \in \mathcal{X} \mid V(\boldsymbol{x}) < \infty\}$ the stabilizable set.

By standard dynamic programming arguments [30] one sees that this function is the unique solution to the optimality principle

$$V(\boldsymbol{x}) = \inf_{\boldsymbol{u} \in \mathcal{U}} \sup_{\boldsymbol{w} \in \mathcal{W}} \{g(\boldsymbol{x}, \boldsymbol{u}, \boldsymbol{w}) + V(\boldsymbol{f}(\boldsymbol{x}, \boldsymbol{u}, \boldsymbol{w}))\} \tag{5.63}$$

for $\boldsymbol{x} \in \mathcal{S} \backslash \mathcal{T}$ together with the boundary condition $V|_{\mathcal{T}} \equiv 0$. If the cost function g does not depend on $\boldsymbol{w}$, then this equation can be written as

$$V(\boldsymbol{x}) = \inf_{\boldsymbol{u} \in \mathcal{U}} \left\{ g(\boldsymbol{x}, \boldsymbol{u}) + \sup_{\boldsymbol{x}' \in \boldsymbol{f}(\boldsymbol{x}, \boldsymbol{u}, \mathcal{W})} V(\boldsymbol{x}') \right\}. \tag{5.64}$$

Note that in this equation it is sufficient to know the set-valued image $\boldsymbol{f}(\boldsymbol{x}, \boldsymbol{u}, \mathcal{W})$. The discretization described in the following section will be based on this observation.

Discretization. We employ the same approach as in Section 5.4.3 in order to discretize (5.63). Note that the setting in Section 5.4.3 can be seen as a special case of the more general situation using $\mathcal{W} = \{0\}$ and $\boldsymbol{f}(\boldsymbol{x}, \boldsymbol{u}, \boldsymbol{w}) = \boldsymbol{f}(\boldsymbol{x}, \boldsymbol{u})$. In our perturbed setup, one can show [150] that the corresponding discrete optimality principle is given by

$$V_P(\mathcal{P}) = \inf_{N \in \mathcal{F}(\mathcal{P})} \left\{ \mathcal{G}(\mathcal{P}, N) + \sup_{\mathcal{P}' \in N} V_P(\mathcal{P}') \right\} \tag{5.65}$$

for $\mathcal{P} \cap \mathcal{T} = \emptyset$, with boundary condition $V_P(\mathcal{P}) = 0$ for $\mathcal{P} \cap \mathcal{T} \neq \emptyset$, where

$$\begin{aligned} \mathcal{F}(\mathcal{P}) &= \{[\boldsymbol{f}(\boldsymbol{x}, \boldsymbol{u}, \mathcal{W})] \subset P \mid (\boldsymbol{x}, \boldsymbol{u}) \in \mathcal{P} \times \mathcal{U}\} \\ \mathcal{G}(\mathcal{P}, N) &= \inf \left\{ g(\boldsymbol{x}, \boldsymbol{u}) : \begin{array}{c} (\boldsymbol{x}, \boldsymbol{u}) \in \mathcal{P} \times \mathcal{U}, \\ {[\boldsymbol{f}(\boldsymbol{x}, \boldsymbol{u}, \mathcal{W})] = N} \end{array} \right\} \end{aligned} \tag{5.66}$$

(where for some set $\mathcal{Y} \subset \mathcal{X}$ we let $[\mathcal{Y}] = \{[\boldsymbol{y}] \in P \mid \boldsymbol{y} \in \mathcal{Y}\}$). Note the difference of $\mathcal{F}(\mathcal{P})$ compared to (5.58): while in (5.58) $\mathcal{F}(\mathcal{P})$ was a subset of P, in (5.66) $\mathcal{F}(\mathcal{P})$ is a set of subsets $N \subset P$. Thus, the map $\mathcal{F}$, together with the cost function $\mathcal{G}$ can be interpreted as a directed weighted hypergraph (P, E) with the set $E \subset P \times 2^P$ of hyperedges given by $E = \{(\mathcal{P}, N) : N \in \mathcal{F}(\mathcal{P})\}$ (Fig. 5.18).

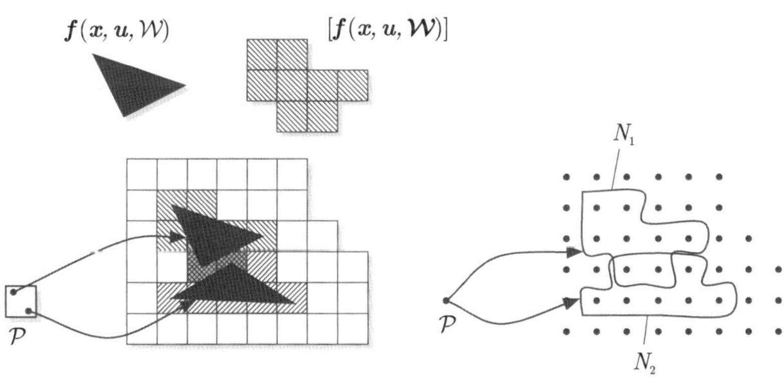

Fig. 5.18 Partition of the state space, images of an element (left) and corresponding edges in the induced hypergraph (right)

Once the graph is constructed, V_P can be computed by a min-max version of the well known Dijkstra Algorithm for computing shortest paths in graphs (for details cf. [150] and [375]).

5.4.5 Discretization as Perturbation

It was shown in [147] that if we compute a feedback law u_P similar to (5.55) but using the approximate value function V_P from (5.57) instead of V, then this feedback law will practically stabilize the system under suitable conditions. Numerical experiments, however, reveal that typically a rather fine partition is needed in order to achieve stability of the closed-loop system. Furthermore, even on this fine partition, the approximate value function does not decay monotonically along system trajectories. Formally, this is reflected by the two inequalities

$$V_P(\boldsymbol{x}) \leq V(\boldsymbol{x}) \tag{5.67}$$

and

$$\begin{aligned} V_P(\boldsymbol{x}) &\leq \min_{\boldsymbol{u}\in\mathcal{U}}\{g(\boldsymbol{x},\boldsymbol{u}) + V_P(\boldsymbol{f}(\boldsymbol{x},\boldsymbol{u}))\} \\ &= g(\boldsymbol{x},\boldsymbol{u}_P(\boldsymbol{x})) + V_P(\boldsymbol{f}(\boldsymbol{x},\boldsymbol{u}_P(\boldsymbol{x}))). \end{aligned} \tag{5.68}$$

A comparison with the Lyapunov function property (5.56) reveals that inequality (5.68) delivers exactly the opposite of what is needed in order to prove asymptotic stability.

Construction of the Dynamic Game. In order to cope with this phenomenon we are going to use the dynamic game formulation outlined above. The idea is to additionally incorporate the discretization error as a perturbation of the original control system. Hence, instead of dealing with the single-valued system $\boldsymbol{x} \mapsto \boldsymbol{f}(\boldsymbol{x},\boldsymbol{u})$, we consider the multi-valued system $\boldsymbol{x} \mapsto \boldsymbol{f}([\boldsymbol{x}],\boldsymbol{u}) \subset \mathcal{X}$, where, as before, $[\boldsymbol{x}]$ denotes the partition element containing $\boldsymbol{x}$. Likewise, we consider the "worst case" stage cost $\sup_{x\in[\boldsymbol{x}]} g(\boldsymbol{x},\boldsymbol{u})$. Formally, given a control system (5.52) and an associated cost function g, we consider the dynamic game

$$\boldsymbol{f}(\boldsymbol{x},\boldsymbol{u},\mathcal{W}) := \boldsymbol{f}([\boldsymbol{x}],\boldsymbol{u}) \tag{5.69}$$

for $\boldsymbol{x} \in \mathcal{X}, \boldsymbol{u} \in \mathcal{U}$ and the stage cost

$$G(\boldsymbol{x},\boldsymbol{u}) = \sup_{\boldsymbol{x}'\in[\boldsymbol{x}]} g(\boldsymbol{x}',\boldsymbol{u}). \tag{5.70}$$

Note that for our implementation the explicit knowledge or a parametrization of $\mathcal{W}$ are not needed. Hence, the implicit definition of $\mathcal{W}$ in (5.69) does not complicate the computation of V_P.

Properties of the Approximate Value Function. The following theorem shows that the associated approximate value function V_P satisfies the opposite inequalities compared to (5.67), (5.68), when the target set $\mathcal{T}$ is appropriately included in the formulation.

Theorem 5.5. [149] *Let V denote the optimal value function of the optimal control problem described in Section 5.4.2 and let V_P denote the approximate optimal value function of the game* (5.69), (5.70) *on a given partition P with target set $\mathcal{T} \subset P$. Then,*

$$V(\boldsymbol{x}) - \max_{\boldsymbol{x}' \in \mathcal{T}} V(\boldsymbol{x}') \leq V_P(\boldsymbol{x}), \tag{5.71}$$

i.e. V_P is an upper bound for $V - \max V|_{\mathcal{T}}$. Furthermore, V_P satisfies

$$V_P(\boldsymbol{x}) \geq \min_{\boldsymbol{u} \in U} \{g(\boldsymbol{x}, \boldsymbol{u}) + V_P(\boldsymbol{f}(\boldsymbol{x}, \boldsymbol{u}))\} \tag{5.72}$$

for all $\boldsymbol{x} \in \mathcal{X} \setminus \mathcal{T}$.

Properties of the Feedback Law. As a direct extension of (5.55), the approximately optimal feedback law based on the approximation V_P is defined as

$$\boldsymbol{u}_P(\boldsymbol{x}) = \operatorname*{argmin}_{\boldsymbol{u} \in \mathcal{U}} \left\{ G(\boldsymbol{x}, \boldsymbol{u}) + \sup_{\boldsymbol{x}' \in \boldsymbol{f}([\boldsymbol{x}], \boldsymbol{u})} V_P(\boldsymbol{x}') \right\}. \tag{5.73}$$

Note that we can directly extract $\boldsymbol{u}_P$ from the min-max Dijkstra Algorithm, used for computing V_P, for details cf. [149].

Due to the construction of the game (5.69), (5.70), the feedback $\boldsymbol{u}_P$ is constant on each partition element, i.e., for implementing $\boldsymbol{u}_P$ only the partition element $[\boldsymbol{x}]$ containing $\boldsymbol{x}$ needs to be known. Hence, the partition elements can be used as quantization regions and the control input only changes when the system state moves from one quantization region to another.

5.4.6 Event-Based Implementation

Since the feedback law u_P only changes when the state crosses the boundary of a quantization region, it can straightforwardly be implemented in an event-based fashion. In this implementation, an event is triggered each time the state moves from one quantization region to another and the information which needs to be transmitted at each event time is the new quantization region the state is contained in.

Hence, in line with the event-based schemes developed in this chapter, the partition P introduced defines the event generator. An event is triggered whenever $\boldsymbol{x}(k)$ and $\boldsymbol{x}(k+1)$ belong to different partition elements: $[\boldsymbol{x}(k)] \neq [\boldsymbol{x}(k+1)]$. The control input generator is a zero-order hold that applies the constant input $\boldsymbol{u}(k) = \boldsymbol{u}_P$ that belongs to the current partition element $[\boldsymbol{x}(k)]$.

However, so far this event-based structure is not adequately reflected in the construction of V_P and thus $\boldsymbol{u}_P$. Indeed, on coarse quantizations with large quantization regions $\mathcal{P} \in P$, it is quite likely that for each $\boldsymbol{u} \in \mathcal{U}$ there exists $\boldsymbol{x} \in \mathcal{P}$ such that $\boldsymbol{f}(\boldsymbol{x}, \boldsymbol{u}) \in \mathcal{P}$. In this case, it follows immediately from the definition of the game and V_P that $V_P([\boldsymbol{x}]) = \infty$, i.e. $\boldsymbol{x} \notin \mathcal{S}$. Consequently, V_P cannot be used as a Lyapunov function anymore and thus convergence of the closed loop to $\mathcal{T}$ is no longer guaranteed (and, in fact, typically indeed fails to hold).

A remedy to this problem is to modify the control system used for computing V_P and $\boldsymbol{u}_P$ such that it better reflects the event-based structure. To this end, for each partition element $\mathcal{P} \in P$, each $\boldsymbol{x} \in \mathcal{P}$ and each $\boldsymbol{u} \in \mathcal{U}$ we define the value $r(\boldsymbol{x}, \boldsymbol{u})$ as the smallest number $r \in \mathbb{N}$ such that the condition

$$\boldsymbol{f}^r(\boldsymbol{x}, \boldsymbol{u}) \notin \mathcal{P} \tag{5.74}$$

holds, where we set $r := \infty$ if (5.74) is never satisfied. Here $\boldsymbol{f}^r$ is defined recursively via $\boldsymbol{f}^0(\boldsymbol{x}, \boldsymbol{u}) := \boldsymbol{x}$ and $\boldsymbol{f}^{r+1}(\boldsymbol{x}, \boldsymbol{u}) := \boldsymbol{f}(\boldsymbol{f}^r(\boldsymbol{x}, \boldsymbol{u}), \boldsymbol{u})$. That is, the dynamics is iterated until the current quantization region is left (in our practial implementation, we specify a finite upper bound r_0 of the iterations and set $r(\boldsymbol{x}, \boldsymbol{u}) = \infty$ if (5.74) is violated for $r = 1, \dots, r_0$). We then define a new game using the iterated dynamics

$$\tilde{\boldsymbol{f}}(\boldsymbol{x}, \boldsymbol{u}) := \begin{cases} \boldsymbol{f}^{r(\boldsymbol{x}, \boldsymbol{u})}(\boldsymbol{x}, \boldsymbol{u}), & r(\boldsymbol{x}, \boldsymbol{u}) < \infty \\ \boldsymbol{x}, & r(\boldsymbol{x}, \boldsymbol{u}) = \infty, \end{cases}$$

the corresponding stage cost

$$\tilde{g}(\boldsymbol{x}, \boldsymbol{u}) = \sum_{r=0}^{r(\boldsymbol{x}, \boldsymbol{u})-1} g(\boldsymbol{f}^r(\boldsymbol{x}, \boldsymbol{u}), \boldsymbol{u})$$

and (5.69), (5.70) with $\tilde{\boldsymbol{f}}$ and $\tilde{g}$ in place of $\boldsymbol{f}$ and g.

It can be shown [153] that the optimal value function of the new problem (provided that it is finite) strictly decreases along solutions at event times and stays constant in between until it reaches the target $\mathcal{T}$. This allows to conclude that the closed-loop solution converges to the target.

Example 5.4 *Event-based control of a thermofluid process*

Figure 5.19 shows the benefit of this approach for a simulation of a thermofluid process described in Example 5.1. On the left we see the computed V_P for a sampled-data model of the process with sampling time $T = 6.0$, on the right we see V_P for the event-based model. In both cases, the minimal number of (uniform) quantization regions was chosen such that V_P is finite on the whole computational region. While for the sampled-data model 128^2 regions were necessary, for the event-based model only 16^2 regions were needed, which shows a considerable reduction of the off-line computation and the event-based implementation of the feedback. □

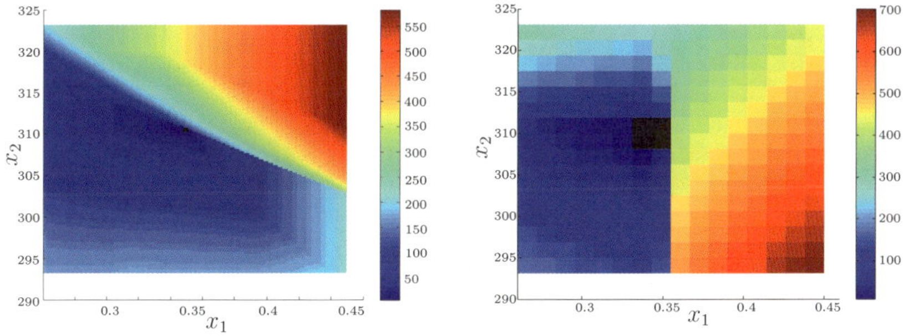

Fig. 5.19 Value function V_P for a sampled data model with 128^2 quantization regions (left) and the corresponding event-based model with 16^2 quantization regions (right)

5.4.7 Including Past Information

Even with the event-based setup just described, the use of large quantization regions induces large uncertainties into the model, because the sets $[\boldsymbol{x}]$ over which the maximization in (5.64) is performed when using (5.69) become large. This uncertainty can be considerably reduced if we let V_P and consequently also $\boldsymbol{u}_P$ depend not only on the current quantization region $[\boldsymbol{x}(k)]$ but also on past quantization regions $[\boldsymbol{x}(k-m)], \dots, [\boldsymbol{x}(k-1)]$. In order to describe how this past information can be incorporated into the computation of V_P, we consider the case $m = 1$ and note that the extension to $m > 1$ is straightforward.

We rewrite (5.69) by defining a map $\boldsymbol{f} : 2^{\mathcal{X}} \times \mathcal{U} \times \Gamma \to P$ via

$$\boldsymbol{f}(X, \boldsymbol{u}, \gamma) := [\boldsymbol{f}(\gamma(X), \boldsymbol{u})], \tag{5.75}$$

where $\Gamma := \{\gamma : 2^{\mathcal{X}} \to \mathcal{X} \mid \gamma(X) \in X\}$ now plays the role of the perturbation which for each $X \subset \mathcal{X}$ selects an arbitrary point $\boldsymbol{x} \in X$ and, as usual, $[\boldsymbol{f}(\gamma(X), \boldsymbol{u})]$ denotes the partition element containing the point $\boldsymbol{f}(\gamma(X), \boldsymbol{u})$. Of course, in this definition $\boldsymbol{f}$ can be either a sampled-data or an event-based model as introduced in Section 5.4.6. Note that the graph needed for the computation of V_P can directly be constructed from $\boldsymbol{f}$ defined in (5.75).

Next we define the set $P_2 := (P \cup \{\delta\}) \times P$ of all pairs of quantization regions, where δ stands for an undefined region (this is needed at initial time $k = 0$ because at this time no previous region $[\boldsymbol{x}(-1)]$ is known). We denote the elements from P_2 by $Z = (\mathcal{P}_1, \mathcal{P}_2)^{\mathrm{T}}$, where $\mathcal{P}_1$ represents the "past" region $[\boldsymbol{x}(k-1)]$ and $\mathcal{P}_2$ the "current" region $[\boldsymbol{x}(k)]$. We then define a perturbed control system on P_2 by setting

$$\boldsymbol{f}_2(Z, \boldsymbol{u}, \gamma) := \begin{pmatrix} \mathcal{P}_2 \\ \boldsymbol{f}(X(Z), \boldsymbol{u}, \gamma) \end{pmatrix}$$

and a corresponding stage cost via

$$G_2(Z, \boldsymbol{u}) := \sup_{\boldsymbol{x}' \in X(Z)} g(\boldsymbol{x}', \boldsymbol{u}).$$

Here $X(Z) \subset \mathcal{X}$ is defined by

$$X(Z) := \begin{cases} \mathcal{P}_2, & \text{if } \mathcal{P}_1 = \delta \\ \bigcup_{\boldsymbol{u} \in \mathcal{U}, \gamma \in \Gamma} \boldsymbol{f}(\gamma(\mathcal{P}_1), \boldsymbol{u}) \cap \mathcal{P}_2, & \text{else.} \end{cases}$$

The set $X(Z)$ used for computing the the second component of $\boldsymbol{f}_2$ contains only those points in $\mathcal{P}_2$ which can be reached from $\mathcal{P}_1$. For this reason, the image of the second component of $\boldsymbol{f}_2$ is always contained in $\boldsymbol{f}(\mathcal{P}_2, \boldsymbol{u}, \gamma)$ and typically considerably smaller. Consequently, when computing $V_{\mathcal{P}_2}$ from $\boldsymbol{f}_2$, the resulting values will be smaller than the original value function $V_{\mathcal{P}}$ and may in particular assume finite values on regions on which $V_{\mathcal{P}}$ exhibits infinite values (for a rigorous proof of this property see [152, 153]). Unfortunately, the attempt to quantify the difference between $V_{\mathcal{P}_2}$ and $V_{\mathcal{P}}$ has so far not been met with success. Nevertheless, simulations as well as experimental tests have shown that the reduction of the value and the increase of the regions where the optimal value function is finite is significant.

Example 5.4 (cont.) *Event-based control of a thermofluid process*

For an illustration of this fact, consider again the thermofluid process described in Example 5.1. Here we consider the event-based model for which a finite value function $V_{\mathcal{P}}$ could be obtained using a uniform quantization with 16^2 regions. Using one step of past information, a finite optimal value function $V_{\mathcal{P}_2}$ can now be computed with only 8^2 quantization regions (Fig. 5.20). Moreover, the resulting value function $V_{\mathcal{P}_2}$ is much more regular than $V_{\mathcal{P}}$ and assumes smaller values. □

5.4.8 Including Network Effects

If the loop between the plant and the controller is closed via a digital network (most prominently, a TCP/IP network), typical network effects like packet delays and dropouts have to be taken into account. We model these effects by introducing a *random* parameter δ, i.e. we consider a event-based model of the form

$$\boldsymbol{x}(k+1) = \tilde{\boldsymbol{f}}(\boldsymbol{x}(k), \boldsymbol{u}(k), \delta(k)), \quad k = 0, 1, 2, \ldots, \tag{5.76}$$

where at each time instance k the parameter $\delta(k) \in \mathbb{N}_\infty$ is chosen independently of a given distribution $\pi : \mathbb{N}_\infty \to [0, 1]$ (where $\mathbb{N}_\infty := \mathbb{N}_0 \cup \{\infty\} = \{0, 1, 2, \ldots\} \cup \{\infty\}$). Correspondingly, the game (5.75) takes the form

$$\mathcal{P}(k+1) = \boldsymbol{f}(\mathcal{P}(k), \boldsymbol{u}(k), \gamma(k), \delta(k)), \quad k = 0, 1, 2, \ldots,$$

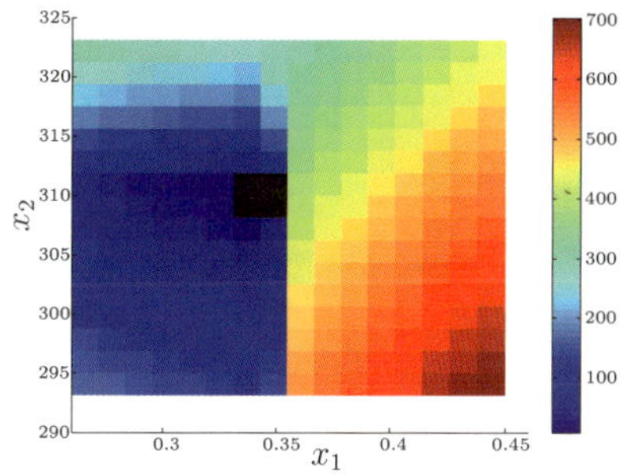

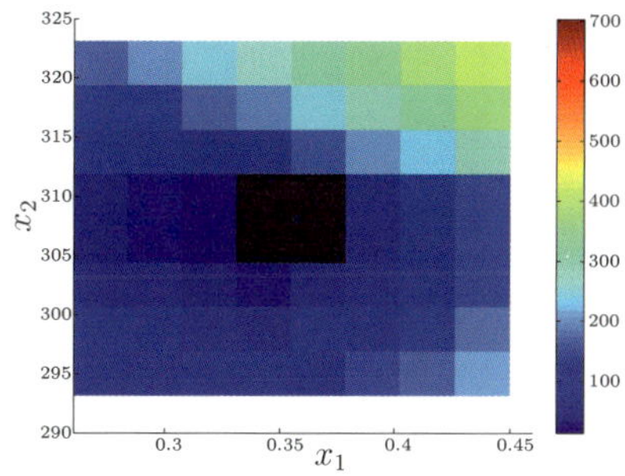

Fig. 5.20 Value function V_P for an event-based model with 16^2 quantization regions (left) and value function V_{P_2} for the same model with 8^2 quantization regions (right)

where $\boldsymbol{f}(\mathcal{P},\boldsymbol{u},\gamma,\delta) = [\tilde{\boldsymbol{f}}(\gamma(\mathcal{P}),\boldsymbol{u},\delta)]$ with associated cost function $G(\mathcal{P},\boldsymbol{u}) = \sup_{\boldsymbol{x}\in\mathcal{P}} \tilde{g}(\boldsymbol{x},\boldsymbol{u})$. Since the parameter δ is stochastic we define the accumulated cost as

$$J(\mathcal{P}(0),(\boldsymbol{u}(k))_k,(\gamma(k))_k) = \mathbb{E}_\delta\left\{\sum_{k=0}^{K-1} G(\mathcal{P}(k),\boldsymbol{u}(k))\right\} \in [0,\infty],$$

where $K = \inf\{k \geq 0 : \mathcal{P}(k) \in \mathcal{T}\}$. The (upper) optimal value function is defined as in Section 5.4.4. On the stabilizable set $S_P = \{\mathcal{P} \in P \mid 0 < V_P(\mathcal{P}) < \infty\}$, V_P fulfills the optimality principle

$$V_P(\mathcal{P}) = \inf_{\boldsymbol{u}\in\mathcal{U}}\left\{G(\mathcal{P},\boldsymbol{u}) + \sup_{\gamma\in\Gamma}\mathbb{E}_\delta\{V_P(\boldsymbol{f}(\mathcal{P},\boldsymbol{u},\gamma,\delta))\}\right\} \tag{5.77}$$

together with the boundary condition $V_P(\mathcal{P}) = 0$ for $\mathcal{P} \in \mathcal{T}$. Given V_P, we construct a feedback for (5.76) by setting

$$\boldsymbol{u}_P(\boldsymbol{x}) = \operatorname*{argmin}_{\boldsymbol{u}\in\mathcal{U}}\left\{G([\boldsymbol{x}],\boldsymbol{u}) + \sup_{\gamma\in\Gamma}\mathbb{E}_\delta\left\{V_P(\boldsymbol{f}([\boldsymbol{x}],\boldsymbol{u},\gamma,\delta))\right\}\right\}, \tag{5.78}$$

for $\boldsymbol{x} \in \mathcal{S} := \cup_{\mathcal{P}\in S_P}\mathcal{P}$.

We have the following stability theorem: For $\lambda > 0$ let $\mathcal{S}_\lambda = \{\boldsymbol{x} \in \mathcal{X} : V_P([\boldsymbol{x}]) \leq \lambda\}$ and consider the closed-loop system

$$\boldsymbol{x}(k+1) = \tilde{\boldsymbol{f}}(\boldsymbol{x}(k),\boldsymbol{u}_P(\boldsymbol{x}(k)),\delta(k)), \quad \ell = 0,1,2,\ldots. \tag{5.79}$$

Theorem 5.6. [191] *If $\boldsymbol{x}_0 \in \mathcal{S}_\lambda$ then with probability $1 - V_P([\boldsymbol{x}_0])/\lambda$, a (random) trajectory of (5.79) stays in $\mathcal{S}_\lambda$. Furthermore, for almost all trajectories $(\boldsymbol{x}(k))_k$ which stay in $\mathcal{S}_\lambda$, we have that $\boldsymbol{x}(k) \to \mathcal{T}$ as $k \to \infty$.*

In order to compute the optimal value function defined in the previous section we perform a standard value iteration:

$$V_P^{(j+1)}(\mathcal{P}) = \inf_{\boldsymbol{u}\in\mathcal{U}} \left\{ G(\mathcal{P},\boldsymbol{u}) + \sup_{\gamma\in\Gamma} \mathbb{E}_\delta \left\{ V_P^{(j)}(\boldsymbol{f}(\mathcal{P},\boldsymbol{u},\gamma,\delta)) \right\} \right\} \tag{5.80}$$

with $V_P^{(0)}(\mathcal{P}) = 0$ if $\mathcal{P} \in \mathcal{T}$ and $V_P^{(0)}(\mathcal{P}) = \infty$ else. As in similar situations (cf., e.g., [149]) the use of graph algorithms still proves helpful here [191].

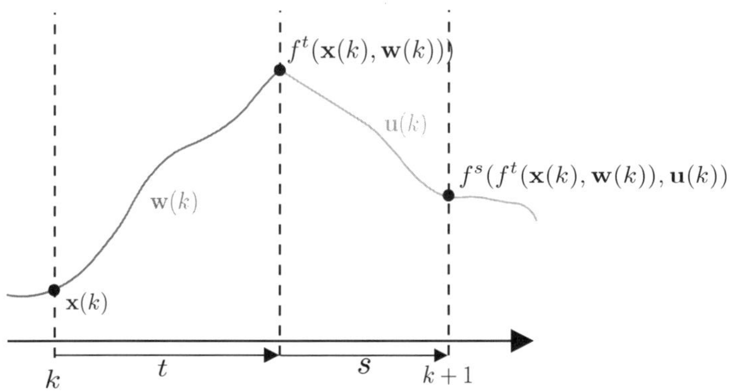

Fig. 5.21 Delay model: At time k, an event is generated and the system is in state $\boldsymbol{x}(k)$. The transmission of the state information from the plant to the controller is delayed by $t \in \mathbb{N}_\infty$ time units, during which the old control input $\boldsymbol{w}(k)$ is still operational. At time $t(k) + t$ (when the plant is already in state $\boldsymbol{f}^t(\boldsymbol{x}(k),\boldsymbol{w}(k))$) the state information $\boldsymbol{x}(k)$ reaches the controller, changing the input to its new value $\boldsymbol{u}(\boldsymbol{x}(k))$.

Network Model for Delays and Packet Loss. We now assume that whenever the state information is transmitted from the plant to the controller, this transmission is subject to a delay $\delta \in \mathbb{N}_\infty$ (where $\delta = \infty$ corresponds to the possibility that the information does not reach the controller at all, i.e. a "packet loss"). Thus, until the new state information reaches the controller, the plant will still operate for some time with the old control input computed from the previous state information (Fig. 5.21). Formally, we model this situation by a system of the form

$$\begin{pmatrix} \boldsymbol{x}(k+1) \\ \boldsymbol{w}(k+1) \end{pmatrix} = \begin{pmatrix} \boldsymbol{f}^s\left(\boldsymbol{f}^t(\boldsymbol{x}(k),\boldsymbol{w}(k)),\boldsymbol{u}(k)\right) \\ \boldsymbol{w}'(k) \end{pmatrix}, \quad k = 0,1,2,\ldots, \tag{5.81}$$

where $\boldsymbol{f}$ is an unperturbed, sampled-data model (5.52) and

$$t = \min\{\delta(k), r(\boldsymbol{x}(k),\boldsymbol{w}(k))\},$$
$$s = \begin{cases} r\left(\boldsymbol{f}^t(\boldsymbol{x}(k),\boldsymbol{w}(k)),\boldsymbol{u}(k)\right), & \text{if } \delta(k) < r(\boldsymbol{x}(k),\boldsymbol{w}(k)), \\ 0 & \text{else} \end{cases}$$

$$w'(k) = \begin{cases} u(k) & \text{if } \delta(k) < r(x(k), w(k)), \\ w(k) & \text{else.} \end{cases}$$

In this model, any delay $\delta(k) \geq r(x(k), w(k))$ is treated as $\delta = \infty$, i.e. as if the corresponding data would never reach the controller.

Example 5.5 *The delayed inverted pendulum*

We consider an inverted pendulum [191] using a feedback law based on the framework of this section. Figure 5.22 illustrates that by incorporating the delay into the construction of the controller a much larger region of the state space $\mathcal{X}$ remains stabilizable. In fact, the stabilizable set for the delayed system with delay-based controller almost does not deteriorate from the undelayed system with the standard controller. □

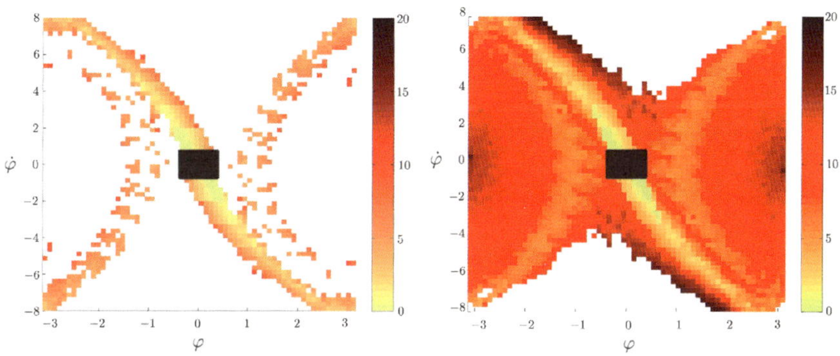

Fig. 5.22 The inverted pendulum controlled by the feedback construction from Section 5.4.5 (left) and from this section (right): In color (gray scale) are the regions of state space which are stabilizable to a neighborhood $\mathcal{T}$ of the origin (black centered rectangle) for a simulation with stochastic delays: The color of a cell indicates the average accumulated cost for initial states from that cell.

5.4.9 Construction of a Lazy Feedback

When the data transmission between the plant and the controller is realized via a digital network it is often desirable to minimize the amount of transmitted information in order to reduce the overall network load. More specifically, here we treat the question of how to minimize the number of times that a new control value has to be transmitted from the controller to the plant. Using an optimization-based feedback construction, this goal can directly be modeled by suitably defining the running cost function.

In order to detect a change in the control value generated by the controller we need to be able to compare to its value from the previous time step (resp. event). Based on a discrete-time model (5.52) with cost function g (which may either be a sampled-data or event-based model) we consider the model

$$\begin{pmatrix} \boldsymbol{x}(k+1) \\ \boldsymbol{u}(k+1) \end{pmatrix} = \begin{pmatrix} \boldsymbol{f}(\boldsymbol{x}(k), \boldsymbol{u}(k)) \\ \boldsymbol{u}(k) \end{pmatrix}, \quad k = 0, 1, 2, \dots. \tag{5.82}$$

We use $\mathcal{T} \times \mathcal{U}$ as the target set in the extended state space $\mathcal{Z} = \mathcal{X} \times \mathcal{U}$ so that reaching the target only depends on the state $\boldsymbol{x}$. We further define an associated running cost function $d : \mathcal{Z} \times \mathcal{U} \to [0, \infty)$ by

$$d((\boldsymbol{x}, \boldsymbol{w}), \boldsymbol{u}) = (1 - \lambda) g(\boldsymbol{x}, \boldsymbol{u}) + \lambda(1 - \delta(\boldsymbol{u} - \boldsymbol{w})) \tag{5.83}$$

with

$$\delta(\boldsymbol{u}) := \begin{cases} 1, \text{ if } \boldsymbol{u} = \boldsymbol{0}, \\ 0, \text{ else.} \end{cases} \tag{5.84}$$

Here, $\lambda \in [0, 1)$ must be strictly < 1 in order to guarantee that $d(\boldsymbol{z}, \boldsymbol{u}) = 0$ iff $\boldsymbol{z} \in \mathcal{T} \times \mathcal{U}$.

We can now apply the construction from the previous section to the system (5.82) with cost function (5.83). To this end, we would need to construct a partition of $\mathcal{U}$. Instead, in order to simplify the exposition, here we simply assume that $\mathcal{U}$ is discrete, i.e. contains only finitely many elements. We then use $P \times \mathcal{U}$ as the underlying partition for the quantization. We denote the resulting optimal value function by $V_\lambda : \mathcal{Z} \to [0, \infty]$, the stabilizable subset by $S_\lambda := \{\boldsymbol{z} \in \mathcal{Z} : V_\lambda([\boldsymbol{z}]) < \infty\}$ and the associated feedback by $u_\lambda : S_\lambda \to \mathcal{U}$.

Theorem 5.7. [190] *For a sufficiently large $\lambda < 1$ the associated closed-loop system is asymptotically stable to $\mathcal{T} \times \mathcal{U}$ on S_λ. Furthermore, the number of control value changes along any trajectory to a given initial condition of the closed-loop system is minimal among all feedbacks which stabilize this initial condition.*

Example 5.6 *Lazy control of the thermofluid process*

We reconsider the thermofluid process introduced in Example 5.1. The goal is to steer the system into a neighbourhood of a given operating point with only a minimum number of control changes. We use a weighted sum of time and a quadratic function in $\boldsymbol{x}$ and $\boldsymbol{u}$ with a small gain as cost function. For a specific initial state we compare two trajectories associated, respectively, to the ordinary feedback described in Section 5.4.6 and the lazy one as described in this section with $\lambda = 0.9$ (Fig. 5.23). Using the lazy feedback, the number of control value changes is reduced dramatically from 13 to only 2. However, the time required for the system to reach the target set remains almost the same. □

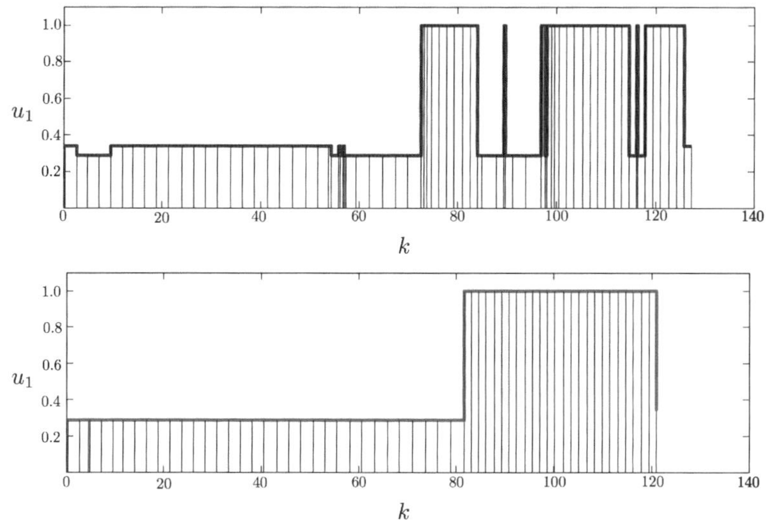

Fig. 5.23 Output of the controller (first component) over time for the thermofluid process with initial state (0.275m, 295K). Upper figure: using the standard cost function, lower figure: lazy feedback, with minimum number of control value changes ($\lambda = 0.9$)

5.4.10 Combination of the Global and the Local Approach

The problem of finding an event-based controller that renders the closed-loop system ultimately bounded can be divided into a global and a local approach, where the aim of the global approach is to drive the plant state $\boldsymbol{x}(t)$ from an initial condition $\boldsymbol{x}(0)$ into a target region $\Omega_{\rm t}$ (Fig. 5.24). Once the state $\boldsymbol{x}(\bar{t})$ enters $\Omega_{\rm t}$ the local approach should keep the state $\boldsymbol{x}(t)$ in this region for all $t \geq \bar{t}$ despite exogenous disturbances $\boldsymbol{d}(t)$.

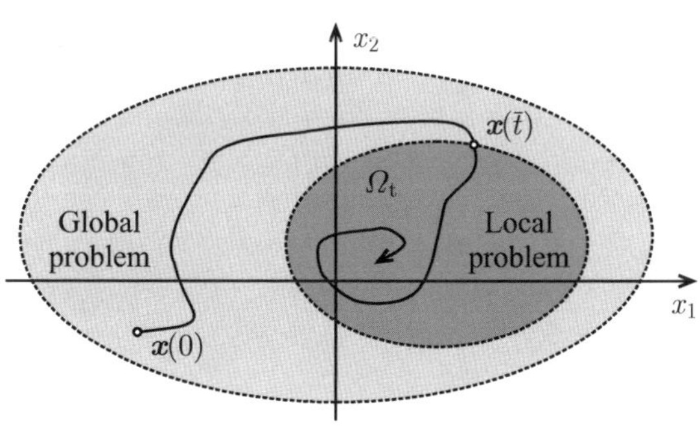

Fig. 5.24 Global and local control problem

For the transition of the state $\boldsymbol{x}(t)$ from an arbitrary initial condition $\boldsymbol{x}(0)$ into the target region Ω_{t} the nonlinearity of the plant dynamics must be explicitly taken into account for a controller design. On the other hand, a rough description of the state suffices in order to drive the state $\boldsymbol{x}(t)$ into the target region Ω_{t}. Therefore, the event-based controller can work with quantized state information $[\boldsymbol{x}(t)]$. The global approaches described in this section to optimally driving the plant state $\boldsymbol{x}(t)$ into the target set Ω_{t} solve this problem. For keeping the plant state $\boldsymbol{x}(t)$ in a target region Ω_{t} that contains the setpoint $\bar{\boldsymbol{x}}$, a linearized model of the plant can be used for the controller design. Sections 5.2 and 5.3 have presented two local approaches to event-based state-feedback which aim at disturbance rejection and, thus, maintaining the plant state $\boldsymbol{x}(t)$ in the target set Ω_{t}.

The local event-based control approaches that were presented in Sections 5.2 and 5.3 can be combined with the global approaches proposed in Sec. 5.4 in order to obtain an event-based controller that yields and ultimately bounded closed-loop system. To this end, the control input generator and the event generator have to include the procedure for generating the control input or for determining the event times, respectively, for both the nonlinear and the linear case. As long as the state $\boldsymbol{x}(t)$ is outside the target region Ω_{t} the nonlinear event-based controller is applied. The event generator detects the time instant $\bar{t}$ at which the state $\boldsymbol{x}(t)$ enters the set Ω_{t} and informs the control input generator. At this time both generators switch to the linear controller in order to keep the plant state $\boldsymbol{x}(t)$ in Ω_{t} for all $t \geq \bar{t}$. For details about this approach the reader is referred to [158].

5.5 A Small-Gain Perspective on Event-Based Control

5.5.1 Problem Statement

In this section we present an approach to event-triggered control in the context of large-scale systems. As in Section 5.3 consider a system composed of several subsystems, labeled by Σ_i, $(i = 1, \ldots, N)$. Each subsystem has a local controller $\boldsymbol{k}_i$ that uses knowledge of some of the states of other subsystems. The corresponding state information is communicated over a channel with limited capacity. The approach to reducing the communication load is to let each subsystem decide based on local information, whether other subsystems require an update and to restrict information interchanges to these instances. Of course, some information is needed concerning the system structure that can be utilized; here this information is given by the existence of Lyapunov functions.

Using results from the theory on input-to-state stability (ISS) of large-scale systems we show that a triggering scheme using solely local information is sufficient for stabilization. We give conditions under which an interconnection of N (N possibly large) nonlinear systems (sufficiently smooth) with event-triggered control of the form given below is stable. In the following, $\boldsymbol{x}_i$ is the state of subsystem Σ_i, $(i = 1, \dots, N)$, $\hat{\boldsymbol{x}}$ is the information available at the controller $\boldsymbol{k}_i$ and the controller error is $\boldsymbol{e} = \hat{\boldsymbol{x}} - \boldsymbol{x}$. To formulate a general setup we assume that the triggering condition is given by functions T_i, which are jointly continuous in $\boldsymbol{x}_i$, $\boldsymbol{e}_i$ and satisfy $T_i(\boldsymbol{x}_i, 0) < 0$ for all $\boldsymbol{x}_i \neq 0$. The subsystem description is then

$$\begin{aligned} \Sigma_i :\ \ & \dot{\boldsymbol{x}}_i(t) = \boldsymbol{f}_i(\boldsymbol{x}(t), \boldsymbol{u}_i(t)), \qquad && \boldsymbol{u}_i(t) = \boldsymbol{k}_i(\boldsymbol{x}(t) + \boldsymbol{e}(t)) \\ & \dot{\hat{\boldsymbol{x}}}(t) = \boldsymbol{0}, && \boldsymbol{e}(t) = \hat{\boldsymbol{x}}(t) - \boldsymbol{x}(t) \end{aligned} \tag{5.85}$$

with triggering condition (event condition)

$$T_i(\boldsymbol{x}_i(t), \boldsymbol{e}_i(t)) \geq 0\ . \tag{5.86}$$

Solutions to such an event-based feedback are defined as follows. We assume that the initial controller error is $\boldsymbol{e}_0 = \boldsymbol{0}$. Given an initial condition $\boldsymbol{x}_0$ we define

$$t_1 := \inf\{t > 0\ :\ \exists i\ \text{ such that }\ T_i(\boldsymbol{x}_i(t), \boldsymbol{e}_i(t)) \geq 0\}$$

to be the first event time. At time instant t_1 the subsystems Σ_i for which $T_i(\boldsymbol{x}_i, \boldsymbol{e}_i) \geq 0$ broadcast their respective state $\boldsymbol{x}_i$ to all controllers $\boldsymbol{k}_j$, $(j = 1, \dots, N)$. In particular, $\boldsymbol{e}_i(t_1^+) = \boldsymbol{0}$ for these indices i.

Then inductively we set for $k = 1, 2, \dots$

$$t_{k+1} := \inf\{t > t_k\ :\ \exists i\ \text{ such that }\ T_i(\boldsymbol{x}_i(t), \boldsymbol{e}_i(t)) \geq 0\}\ .$$

We say that the triggering scheme induces Zeno behavior, if for a given initial condition $\boldsymbol{x}_0$ the event times t_k converge to a finite t^*. Of course, Zeno behavior is undesirable from several aspects: first the intention to reduce the communication load is defeated, if communication is necessary infinitely often in a finite time interval, secondly the concept of a solution beyond the Zeno point would have to be elaborated. In practical situations it is of course necessary to impose a lower bound on the time between two communication events. It is, therefore, of interest to investigate conditions that exclude the possibility of Zeno behavior. More generally, it is also of interest to formulate guaranteed lower bounds for inter-event times. However, this topic is not treated in the present section.

The organization of this section is as follows: We begin by an introduction of event generation based on nonlinear small-gain conditions. To begin with we do not worry about Zeno behavior, but concentrate on the description of the setup in Section 5.5.2. It turns out that Zeno behavior can only occur, if trajectories of certain subsystems reach the origin in finite time. A natural idea to avoid this situation is not to require that the trajectories converge asymptotically to the desired fixed point. This approach to *practical stabilization* is outlined in Section 5.5.3. A new approach to avoid Zeno behavior while ensuring asymptotic stability is given in Section 5.5.4. The triggering scheme presented therein is called *parsimonious triggering* as it uses more information available to avoid unnecessary communication. We compare the two approaches in Section 5.5.5.

5.5.2 Event Generation and Input-to-State Stability

Frequently in the nonlinear case, the influence of one subsystem to another is modeled via nonlinear (gain-)functions, which we now recall. A continuous function $\gamma : [0, \infty) \to [0, \infty)$ is said to be of class $\mathcal{K}$, if it is strictly increasing and $\gamma(0) = 0$. For brevity, we write $\gamma \in \mathcal{K}$. Furthermore, $\mathcal{K}_\infty$ consists of $\gamma \in \mathcal{K}$ that are unbounded. For our approach we need a robust stability notion, which is provided by the concept of input-to-state stability (ISS). In this context, we assume that each subsystem satisfies the following assumption:

Assumption 5.1. *For $i = 1, 2, \ldots, N$, there exists a differentiable function $V_i : \mathbb{R}^{n_i} \to \mathbb{R}_+$, and class-$\mathcal{K}_\infty$ functions α_{i1}, α_{i2} such that*

$$\alpha_{i1}(||\boldsymbol{x}_i(t)||) \leq V_i(\boldsymbol{x}_i(t)) \leq \alpha_{i2}(||\boldsymbol{x}_i(t)||) \,, \quad \forall \boldsymbol{x}_i(t) \in \mathbb{R}^{n_i} \,.$$

Moreover, there exist functions $\gamma_{ij}, \eta_{ij} \in \mathcal{K}_\infty \cup \{0\}$, α_i positive definite such that

$$\begin{aligned} V_i(\boldsymbol{x}_i(t)) \geq \max\{&\gamma_{i1}(V_1(\boldsymbol{x}_1(t))), \ldots, \gamma_{iN}(V_N(\boldsymbol{x}_N(t))), \\ &\eta_{i1}(||\boldsymbol{e}_1(t)||), \ldots, \eta_{iN}(||\boldsymbol{e}_N(t)||)\} \\ &\Rightarrow \nabla V_i(\boldsymbol{x}_i(t)) f_i(\boldsymbol{x}(t), k_i(\boldsymbol{x}(t) + \boldsymbol{e}(t))) \leq -\alpha_i(||\boldsymbol{x}_i(t)||) \,, \end{aligned} \tag{5.87}$$

with $\gamma_{ii} = 0$ for all i.

The γ_{ij}, $(i, j = 1, ..., N)$ describe the influence of the states of the other subsystems on system Σ_i and the η_{ij} model the influence of the imperfect knowledge of the states on the controller. It is easy to see that if for subsystem Σ_i all the other states as well as all the errors are zero, the subsystem Σ_i is asymptotically stable in $\boldsymbol{x}_i^*(t) = \boldsymbol{0}$.

To ease the presentation, we choose the maximum to compare the influence of the different gains in Assumption 5.1. This approach can also be used for the sum or any p-norm or even more general concepts.

The following example illustrates Assumption 5.1.

Example 5.7 *Linear interconnected system*

Consider the interconnection of N linear subsystems

$$\dot{\boldsymbol{x}}_i(t) = \sum_{j=1}^{N} \boldsymbol{A}_{ij} \boldsymbol{x}_j(t) + \boldsymbol{B}_i \boldsymbol{u}_i(t)$$

$$\boldsymbol{u}_i(t) = \sum_{j=1}^{N} \boldsymbol{K}_{ij}(\boldsymbol{x}_j(t) + \boldsymbol{e}_j(t))$$

with matrices of appropriate dimensions. For each index i, we assume that the pairs $(\boldsymbol{A}_{ii}, \boldsymbol{B}_i)$ are stabilizable and we let the matrix $\boldsymbol{K}_{ii}$ be such that $\bar{\boldsymbol{A}}_{ii} := \boldsymbol{A}_{ii} + \boldsymbol{B}_i \boldsymbol{K}_{ii}$ is Hurwitz. Then for each $\boldsymbol{Q}_i = \boldsymbol{Q}_i^{\mathrm{T}} > 0$ there exists a matrix $\boldsymbol{P}_i = \boldsymbol{P}_i^{\mathrm{T}} > 0$ such that

$$\bar{\boldsymbol{A}}_{ii}^{\mathrm{T}} \boldsymbol{P}_i + \boldsymbol{P}_i \bar{\boldsymbol{A}}_{ii} = -\boldsymbol{Q}_i .$$

Define $V_i(\boldsymbol{x}_i(t)) := \boldsymbol{x}_i^{\mathrm{T}}(t) \boldsymbol{P}_i \boldsymbol{x}_i(t)$. We consider now the expression $\nabla V_i(\boldsymbol{x}_i(t)) \dot{\boldsymbol{x}}_i(t)$ where

$$\dot{\boldsymbol{x}}_i(t) = \sum_{j=1}^{N} (\boldsymbol{A}_{ij} + \boldsymbol{B}_i \boldsymbol{K}_{ij}) \boldsymbol{x}_j(t) + \sum_{j=1}^{N} \boldsymbol{B}_i \boldsymbol{K}_{ij} \boldsymbol{e}_j(t) =: \sum_{j=1}^{N} \bar{\boldsymbol{A}}_{ij} \boldsymbol{x}_j(t) + \sum_{j=1}^{N} \bar{\boldsymbol{B}}_{ij} \boldsymbol{e}_j(t) ,$$

with $\bar{\boldsymbol{B}}_{ij} := \boldsymbol{B}_i \boldsymbol{K}_{ij}$.

Standard calculations lead to the following estimate

$$\begin{aligned} \nabla V_i(\boldsymbol{x}_i(t)) \dot{\boldsymbol{x}}_i(t) \leq &- c_i ||\boldsymbol{x}_i(t)||^2 + \\ & 2||\boldsymbol{x}_i(t)||\,||\boldsymbol{P}_i|| \left(\sum_{j=1, j\neq i}^{N} ||\bar{\boldsymbol{A}}_{ij}||\,||\boldsymbol{x}_j(t)|| + \sum_{j=1}^{N} ||\bar{\boldsymbol{B}}_{ij}||\,||\boldsymbol{e}_j(t)|| \right) , \end{aligned}$$

where $c_i = \lambda_{\min}(\boldsymbol{Q}_i)$ and $\lambda_{\min}(\boldsymbol{Q}_i)$ the smallest eigenvalue of $\boldsymbol{Q}_i$. For any $0 < \tilde{c}_i < c_i$, if

$$||\boldsymbol{x}_i(t)|| \geq \frac{2||\boldsymbol{P}_i||}{\tilde{c}_i} \left(\sum_{j=1, j\neq i}^{N} ||\bar{\boldsymbol{A}}_{ij}||\,||\boldsymbol{x}_j(t)|| + \sum_{j=1}^{N} ||\bar{\boldsymbol{B}}_{ij}||\,||\boldsymbol{e}_j(t)|| \right) \tag{5.88}$$

then

$$\nabla V_i(\boldsymbol{x}_i(t)) \dot{\boldsymbol{x}}_i(t) \leq -(c_i - \tilde{c}_i) ||\boldsymbol{x}_i(t)||^2 . \tag{5.89}$$

Note that (5.88) is implied by

$$\begin{aligned} V_i(\boldsymbol{x}_i(t)) \geq \frac{4}{\tilde{c}_i^2} \|\boldsymbol{P}_i\|^3 (2N-1)^2 \max_{j\neq i} \Big\{ & \|\bar{\boldsymbol{A}}_{ij}\|^2 \frac{V_j(\boldsymbol{x}_j(t))}{\lambda_{\min}(\boldsymbol{P}_i)}, \|\bar{\boldsymbol{B}}_{ij}\|^2 \|\boldsymbol{e}_j(t)\|^2, \\ & \|\bar{B}_{ii}\|^2 \|\boldsymbol{e}_i(t)\|^2 \Big\} . \end{aligned}$$

We conclude that Assumption 5.1 holds with $\gamma_{ii} = 0$ and

$$\begin{aligned} \gamma_{ij}(r) &= \frac{4\|\boldsymbol{P}_i\|^3 (2N-1)^2 \|\bar{\boldsymbol{A}}_{ij}\|^2}{\tilde{c}_i^2 \lambda_{\min}(\boldsymbol{P}_i)} r , \\ \eta_{ij}(r) &= \frac{4\|\boldsymbol{P}_i\|^3 (2N-1)^2 \|\bar{\boldsymbol{B}}_{ij}\|^2}{\tilde{c}_i^2} r^2 , \\ \alpha_{i1}(r) &= \lambda_{\min}(\boldsymbol{P}_i) r^2 , \qquad \text{and} \qquad \alpha_{i2}(r) = \|\boldsymbol{P}_i\| r^2 . \end{aligned}$$

□

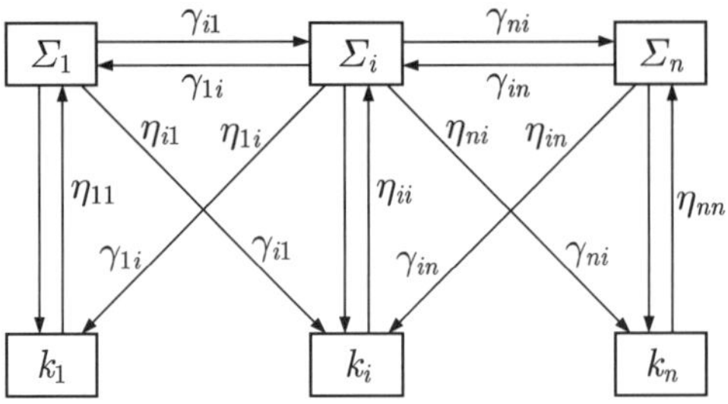

Fig. 5.25 Schematic representation of the interconnected system

If each subsystem satisfies Assumption 5.1, suitable event-triggering conditions together with a generalized small-gain condition are sufficient to infer stability of the interconnected system. In essence, a notion of weak couplings among the subsystems is given by the generalized small-gain condition. The influence of the other states are modeled as a disturbance. The small-gain condition ensures that the influence of the other subsystems does not destroy stability; whereas the triggering condition makes sure that the η part of the left-hand side of the implication in (5.87) does not influence the stability property of the overall system.

A schematic representation of such a large-scale system is given in Fig. 5.25. The plants are given by Σ_i and the controllers by $\boldsymbol{k}_i$. The interconnection can be direct (e.g., physically) or over the controllers. As the imperfect knowledge of the state only affects the interconnection over the controllers, the η functions appear only there.

To clarify the small-gain condition we define an operator $\Gamma : \mathbb{R}^n_+ \to \mathbb{R}^n_+$ with the gains (γ_{ij}) from Assumption 5.1 as

$$\Gamma(\boldsymbol{s}) := \begin{pmatrix} \max_{j=1,\ldots,N}\{\gamma_{1j}(s_j)\} \\ \vdots \\ \max_{j=1,\ldots,N}\{\gamma_{nj}(s_j)\} \end{pmatrix}. \tag{5.90}$$

We say that Γ satisfies the small-gain condition if $\Gamma(\boldsymbol{s}) \not\geq s$ for all $\boldsymbol{s} \neq \boldsymbol{0}, \boldsymbol{s} \in \mathbb{R}^n_+$. This can be equivalently rephrased to: for every nonnegative vector $\boldsymbol{s} \neq \boldsymbol{0}$ there exists an index i such that $(\Gamma(\boldsymbol{s}))_i < s_i$. Note that in two dimension the small-gain condition is equivalent to $\gamma_{12} \circ \gamma_{21} < \mathrm{id}$.

To construct a Lyapunov function for the interconnected system the following theorem is helpful.

Theorem 5.8. *For Γ defined in* (5.90) *the following statement are equivalent:*
(i) The small-gain condition $\Gamma(\boldsymbol{s}) \not\geq \boldsymbol{s}$ for all $\boldsymbol{s} \neq \boldsymbol{0}$ is satisfied.
(ii) There exists a continuous function $\sigma : \mathbb{R}_+ \to \mathbb{R}^n_+$ such that

$$\Gamma(\sigma(r)) < \sigma(r), \quad \textit{for all } r > 0\,.$$

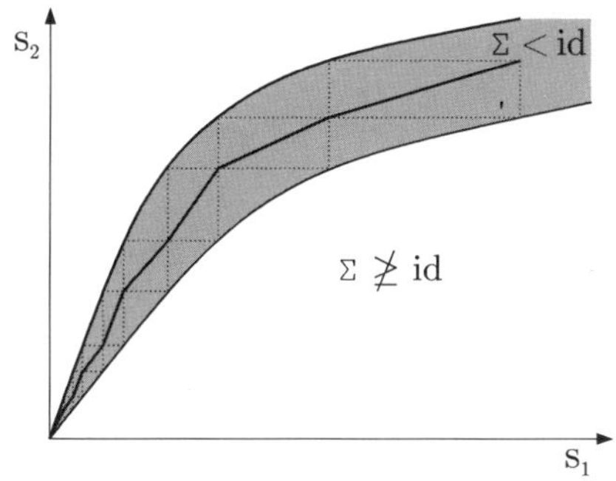

Fig. 5.26 Small-gain condition in two dimensions

In Fig. 5.26 a graphical representation of Theorem 5.8 for dimension two is given. If $\Gamma \ngeq \mathrm{id}$ holds, then there exists the shaded region, where $\Gamma < \mathrm{id}$. This region starts at $\mathbf{0}$ and is unbounded in every dimension. Now we can construct a path σ in the shaded region. Note that there are efficient numerical methods for constructing such a path.

Theorem 5.9. *Consider system* (5.85) *with* $\boldsymbol{e} \equiv \mathbf{0}$. *If the small-gain condition holds, a Lyapunov function for the interconnected system is given by*

$$V(\boldsymbol{x}(t)) := \max_{j=1,\ldots,N} \{\sigma_j^{-1}(V_j(\boldsymbol{x}_j(t)))\}\,. \tag{5.91}$$

By Theorem 5.9, if all the states have perfect knowledge of all the other states, a Lyapunov function for the interconnected system is given by the maximum of the scaled individual Lyapunov functions, where the scaling stems from the path σ.

Up to now we considered the gains from (5.87), which describe the influence of the different states. By Theorem 5.9 we know that if $\boldsymbol{e} \equiv \mathbf{0}$ the interconnected system is asymptotically stable. To handle the effect of the imperfect knowledge of the states ($\boldsymbol{e} \not\equiv \mathbf{0}$) we augment Γ to $\bar{\Gamma}(\boldsymbol{s}, \boldsymbol{w}) \;:\; \mathbb{R}_+^n \times \mathbb{R}_+^{n\times n} \rightarrow \mathbb{R}_+^n$ as

$$\bar{\Gamma}(s, A) := \begin{pmatrix} \max_{j=1,\ldots,N}\{\gamma_{1j}(s_j), w_{1j}\} \\ \vdots \\ \max_{j=1,\ldots,N}\{\gamma_{nj}(s_j), w_{nj}\} \end{pmatrix}.$$

As a consequence of the small-gain condition we can choose a $\varphi \;:\; \mathbb{R}_+ \rightarrow \mathbb{R}_+^{n\times n}$ such that

$$\bar{\Gamma}(\sigma(r), \varphi(r)) < \sigma(r)\,. \tag{5.92}$$

For instance, choose $\varphi_{ij} \leq \gamma_{ij} \circ \sigma_j$.

An event condition for each subsystem, which ensures that the error $\boldsymbol{e}$ is small enough to conserve the stability property of the system is given by

$$T_i^1(\boldsymbol{x}_i(t), \boldsymbol{e}_i(t)) = \hat{\eta}_i(\|\boldsymbol{e}_i(t)\|) - \sigma_i^{-1}(V_i(\boldsymbol{x}(t)_i)) \geq 0\,, \tag{5.93}$$

where $\hat{\eta}_i := \max_j \varphi_{ji}^{-1} \circ \eta_{ji}$ and σ_i^{-1} are scaling functions which come from (5.92) and (5.87). This means that each subsystem compares its error to its state (modulo proper scaling). If the error becomes too large, the subsystem transmits its state to all other subsystems resetting the error to zero. With this approach we obtain a stable event-triggering scheme as the next result shows.

Theorem 5.10. *Consider the interconnected system* (5.85) *with triggering conditions given by* (5.93)*. Assume that no Zeno behavior is induced. If the small-gain condition holds, then the origin is a globally uniformly asymptotically stable equilibrium.*

To discuss the extra condition on Zeno behavior, note that a system with triggered control as in (5.85) is a hybrid system, where Zeno behavior may occur. As event triggering tries to reduce the amount of information transmissions, we need a way to rule out the occurrence of Zeno behavior. This is the subject of Sections 5.5.3 and 5.5.4.

Example 5.8 *A second-order nonlinear system*

Although the presented approach can handle large-scale systems, for reasons of presentation consider the following two-dimensional example:

$$\dot{x}_1(t) = x_1(t)x_2(t) + x_1^2(t)u_1(t)$$
$$\dot{x}_2(t) = x_1^2(t) + u_2(t)$$

It is assumed that each controller can only access the state of the system it controls. The control laws are chosen accordingly as

$$u_1(t) = -(x_1(t) + e_1(t))\ , \ u_2(t) = -33(x_2(t) + e_2(t))\ .$$

If we choose a quadratic Lyapunov function (i.e., $V_i(\boldsymbol{x}_i) = \frac{1}{2}\boldsymbol{x}_i^2$, $(i = 1, 2)$ for each subsystem, it is easy to show that Assumption 5.1 is satisfied. For this particular example the triggering conditions are given by

$$T_1^1(x_1(t), e_1(t)) = \frac{2}{\bar{\sigma}\sqrt{2}} e_1^2(t) - \frac{1}{2} x_1^2(t)$$

and

$$T_2^1(x_2(t), e_2(t)) = \frac{33}{2}|e_2(t)| - \frac{1}{\bar{\sigma}\sqrt{2}}|x_2(t)|$$

with $\bar{\sigma} \in \mathbb{R}_+$. Appropriately chosen, $\bar{\sigma}$ ensures that the small-gain condition is fulfilled.

Simulation results for the initial condition $x_1(0) = -4$, $x_2(0) = 3$, $\hat{x}_1(0) = -4$ and $\hat{x}_2(0) = 3$ can be found for $t \in [0, 2]$ in Fig. 5.27 and 5.28. The trajectory of the first system is given by the lower dashed line, starting at -4 and the second system is the dashed line starting at 3. The input is calculated using the according horizontal lines.

Figure 5.27 shows the event triggering scheme from Theorem 5.10. Between $t = 0$ and $t = 2$, 39 events are triggered and the inputs change at the triggering events. For the upper curve the control inputs cannot be distinguished in the picture as in the early phase they are too frequent as suggested by Fig. 5.29 and in the later phase the trajectory is too flat.

In Fig. 5.28 a periodic sampling scheme was used with a sampling period equal to the shortest time between events from Fig. 5.27 resulting in 286 samples. No major difference in the behavior can be seen despite of the fact that more than 7 times the amount of information was transmitted. Note that because of the large number of sampling points we omitted the values that are used to close the loop. Using a periodic sampling scheme with 39 samples between $t = 0$ and $t = 2$ results in instability of the system. Numerical simulations suggests that the smallest period, which stabilizes the system results in 66 events between $t = 0$ and $t = 2$.

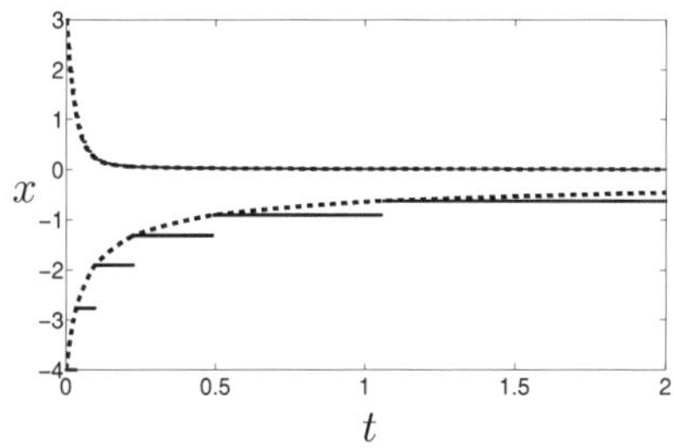

Fig. 5.27 Trajectories of the system with event triggered control

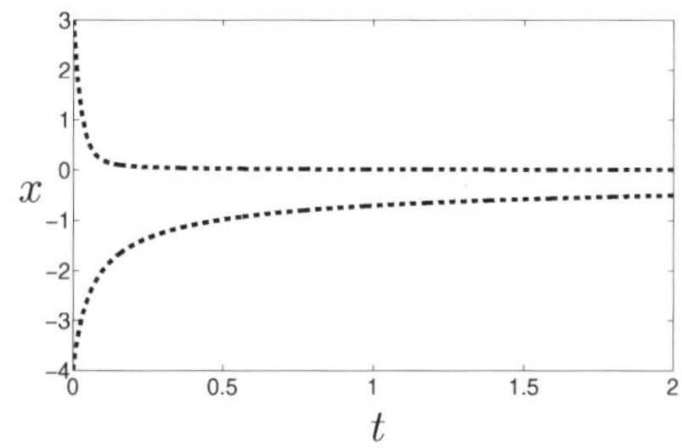

Fig. 5.28 Trajectories of the system with periodic sampling

In Figs. 5.29 and 5.30 the Lyapunov function and the evolution of the error can be seen. Whenever the solid curve (the error) hits the dashed line (the Lyapunov function) an event is triggered and the error is reset to 0. □

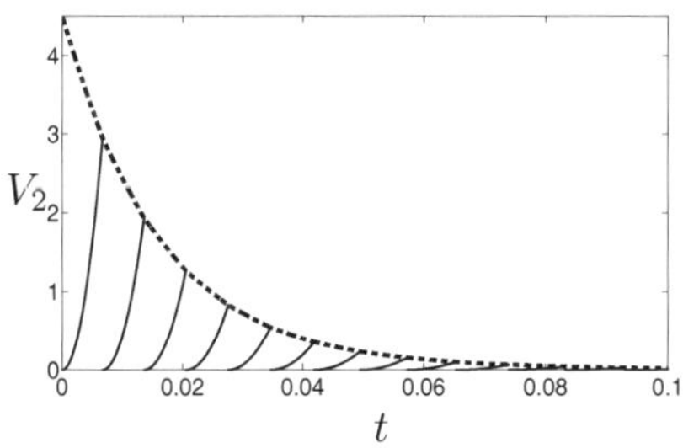

Fig. 5.29 Lyapunov function of the second subsystem together with the error

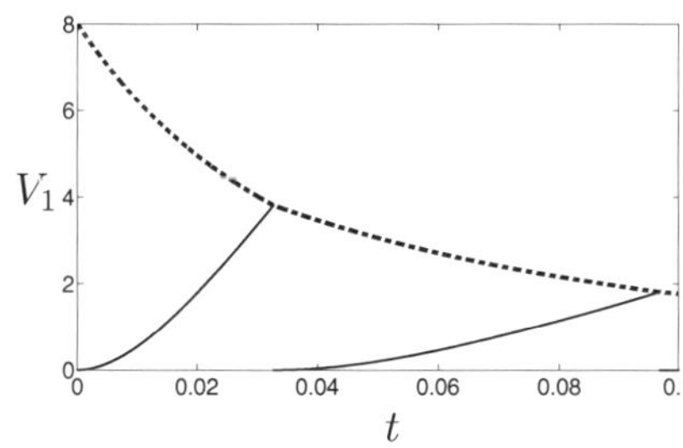

Fig. 5.30 Lyapunov function of the first subsystem together with the error

5.5.3 Practical Stabilization

If we change the initial condition from our example to $x_1(0) = 4$, $x_2(0) = -3$ the event triggering scheme (5.93) exhibits Zeno behavior and hence Theorem 5.10 is not applicable.

One possible way to deal with the Zeno phenomenon is to alter the assumption of ISS-Lyapunov functions to a notion of practical stability. This step introduces a new design parameter $c_i > 0$ which allows us a trade-off between the size of the region we are converging to and the minimal time between events. Of course, if we can bound the minimal time between events away from zero, Zeno cannot occur.

In this case, the triggering condition is

$$T_i^2(x(t), \boldsymbol{e}_i(t)) = \hat{\eta}_i(\|\boldsymbol{e}_i(t)\|) - \max\{\sigma_i^{-1}(V_i(\boldsymbol{x}_i(t))), c_i\} \geq 0\,. \qquad (5.94)$$

The only difference to (5.93) is that here after the error is reset to zero it must evolve at least until the norm of the error is larger than c_i before a new event is triggered.

Theorem 5.11. *Consider the interconnected system* (5.85) *with triggering conditions* (5.94). *If the small-gain condition holds, then the origin is a globally uniformly practically asymptotically stable equilibrium.*

In the triggering condition (5.93) or (5.94), each system compares its local error to its local Lyapunov function. Hence each system can decide based purely on local information when to trigger an event.

Example 5.8 (cont.) *A second-order nonlinear system*

For this example we choose $c_1 = 35$ and $c_2 = 1.86$. Because of the large offset of the first subsystem, after the initial broadcast no event is triggered. Therefore we discuss only the, more interesting, second subsystem. In Fig. 5.31 the trajectory of the second subsystem using (5.94) is depicted resulting in 10 events between $t = 0$ and $t = 80$. The system enters a stable limit cycle, the size of which depends on c_2.

In Fig. 5.32 the Lyapunov function of the second subsystem together with the error e_2 is given. Whenever the solid line (the error) hits the dashed line (the Lyapunov function) or the threshold c_2 (the dotted line), whichever is greater, an event is triggered and the error is reset to 0.

For the short time evolution of the system for $t \in [0, 0.5]$ (cf. Fig. 5.33 and for the corresponding Lyapunov function in Fig. 5.34). After $t \approx 0.6$ only two events are triggered, namely at $t \approx 58$ and $t \approx 80$.

An interesting question is how the threshold c_2 affects the number of triggered events. In Fig. 5.37 the threshold is plotted against the resulting number of events triggered by subsystem Σ_2. It can be seen that there is no heuristic that a larger c_2 leads to fewer events. The reason becomes apparent by inspecting Figs. 5.36 and 5.35, where we used $c_2 = 0.16$ resulting in 339 events. Compared to Fig. 5.33 or Fig. 5.34, respectively, the trajectory oscillates faster within the limit cycle while triggering more events. □

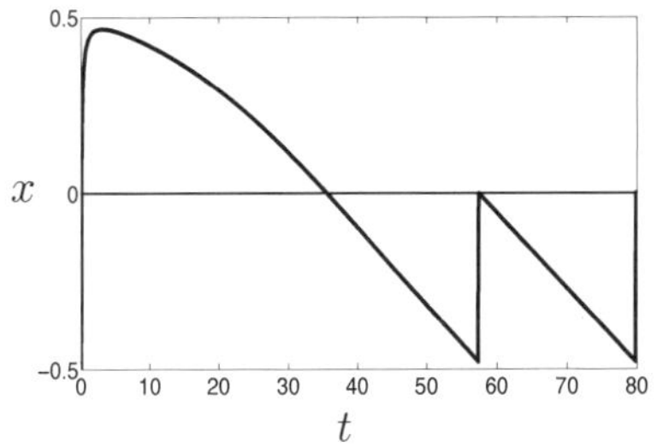

Fig. 5.31 Trajectory of the second subsystem with event triggering (5.94), $c_2 = 1.86$

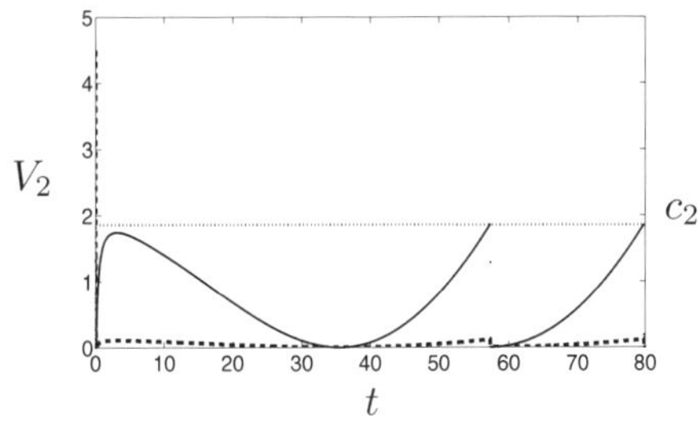

Fig. 5.32 Lyapunov function of the second subsystem together with the error and the threshold c_2

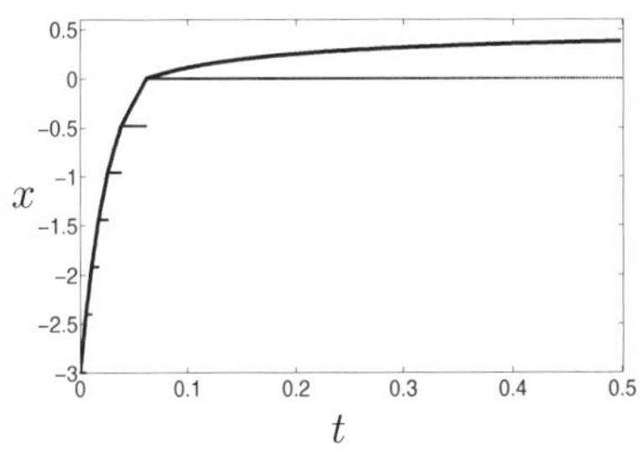

Fig. 5.33 Zoom into $t \in [0, 0.5]$ of Fig. 5.31

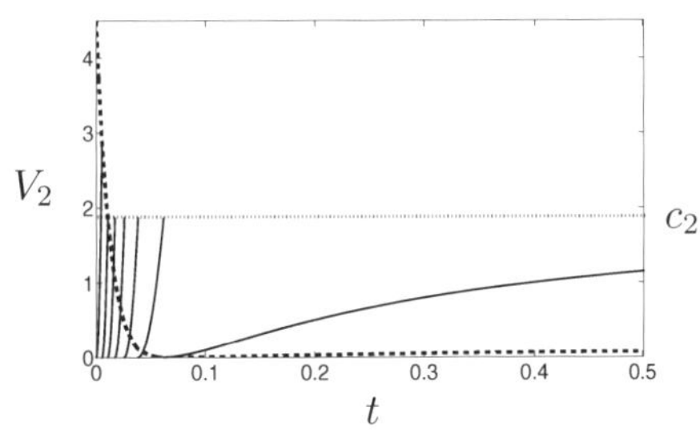

Fig. 5.34 Zoom into the Lyapunov function

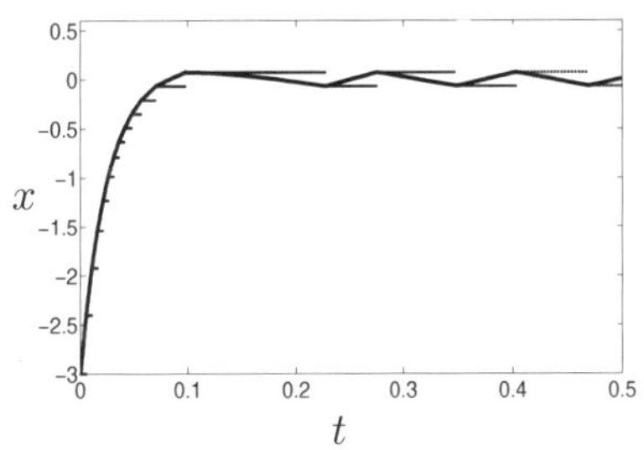

Fig. 5.35 Trajectories with event triggering (5.94), $c_2 = 0.16$

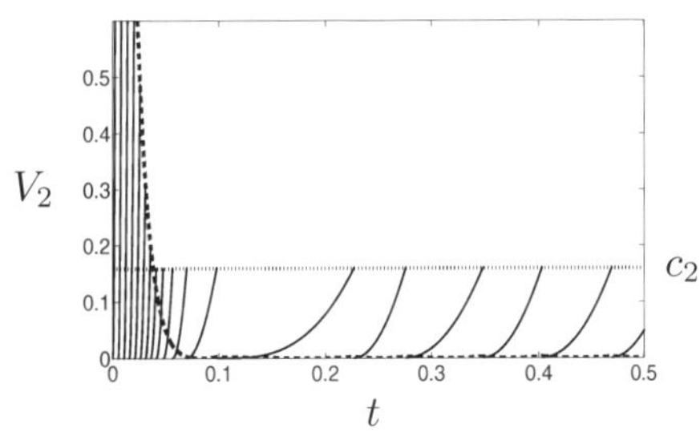

Fig. 5.36 Lyapunov function for $c_2 = 0.16$

5.5.4 Parsimonious Triggering and Asymptotic Stabilization

Although in many applications a notion of practical stability is enough, sometimes it is desired to have asymptotic stability. To achieve this we introduce a further mechanism for reducing the number of triggering times.

The approach is based on the observation that besides comparing each state to the corresponding error, it would be sufficient to compare each error to the largest Lyapunov function among the subsystems.

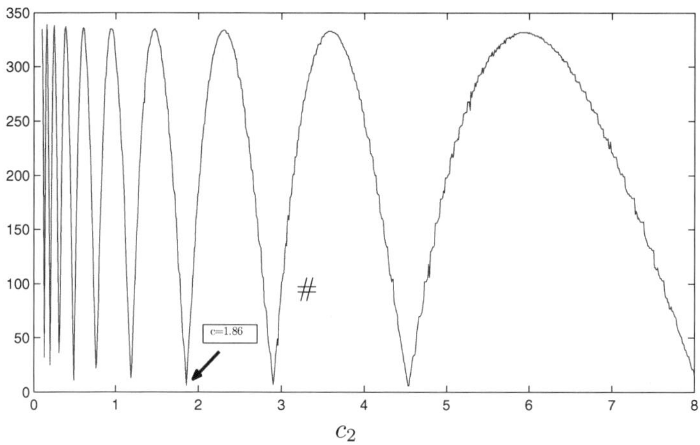

Fig. 5.37 Plot of the threshold c_2 against the number of triggered events for $t \leq 20$

Intuitively, if the errors of all subsystems are smaller than the largest Lyapunov function, then these errors cannot affect the stability of the system. In particular, implication (5.87) suggests that one can either compare each error to each state as in (5.93) and (5.94) or compare all the errors to the largest Lyapunov function to ensure that the maximum in (5.87) is attained by one of the states (the γ part) instead of the error.

A triggering condition following these ideas is of the form

$$T_i(\boldsymbol{x}_i(t), \boldsymbol{e}_i(t)) = \hat{\eta}_i(\|\boldsymbol{e}_i(t)\|) - \max_j \{\sigma_j^{-1}(V_j(\boldsymbol{x}_j(t)))\} \geq 0\,. \tag{5.95}$$

The drawback of this approach is that it requires the knowledge of all states, contradicting our wish for a decentralized setup.

However, under suitable regularity assumptions on the involved gains, respectively scaling functions, we can give an approximation on the size of the other Lyapunov functions, which can be used for a triggering condition; based only on local information.

The intuition behind this approximation is that a subsystem can decide based on its local dynamics (i.e., $\|\dot{\boldsymbol{x}}_i(t)\|$) whether there must be other states that are larger than the state of the i-th system. It is then reasonable to include this information in an augmented triggering condition. This can be used to obtain asymptotic stability but still rule out the occurrence of Zeno behavior.

We know that if Zeno occurs, the corresponding state must approach zero at the Zeno time. On the other hand, at least one other subsystem is bounded away from zero at the Zeno point, because the "strength" to force a Lipschitz continuous system to zero in finite time cannot result from its own state. Hence we know that at some point the largest Lyapunov function in (5.95) is not the Lyapunov function of the system that induces Zeno behavior. Moreover, this Lyapunov function is bounded away from zero. Therefore, no Zeno behavior can occur, because the time until the error evolves until it reaches the level of this larger Lyapunov function is bounded away from zero.

Theorem 5.12. *Consider a large scale system with triggered control of the form* (5.85). *Let the small-gain condition hold. There exists a triggering condition*

$$T_i(\boldsymbol{x}_i(t),\boldsymbol{e}_i(t),\boldsymbol{d}_i(t))=\min\{T_i^1(\boldsymbol{x}_i(t),\boldsymbol{e}_i(t)),T_i^3(\boldsymbol{x}_i(t),\boldsymbol{e}_i(t),\boldsymbol{d}_i(t))\}\geq 0, \tag{5.96}$$

where T_i^1 *is defined in* (5.93) *and* T_i^3 *is a condition involving a suitable approximation* $\boldsymbol{d}_i(t)$ *of* $\dot{\boldsymbol{x}}_i(t)$, *so that the origin is a globally uniformly asymptotically stable equilibrium. In particular, no Zeno behavior occurs.*

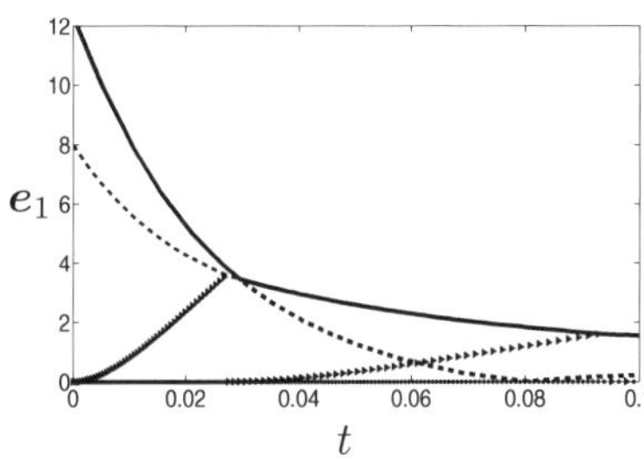

Fig. 5.38 Lyapunov function of both subsystems together with e_2 and the lower bound on V_1

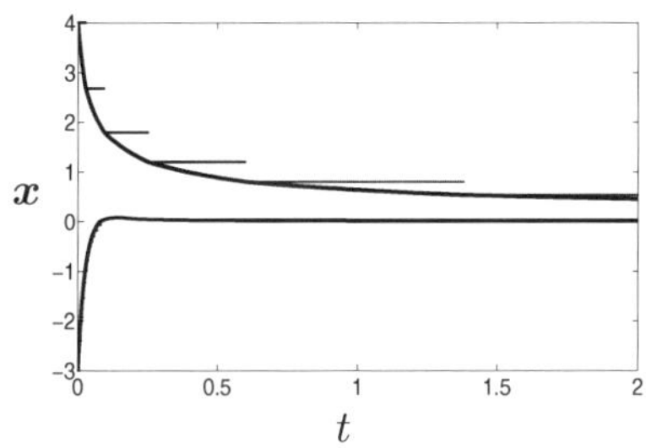

Fig. 5.39 Trajectories with event triggering from (5.96)

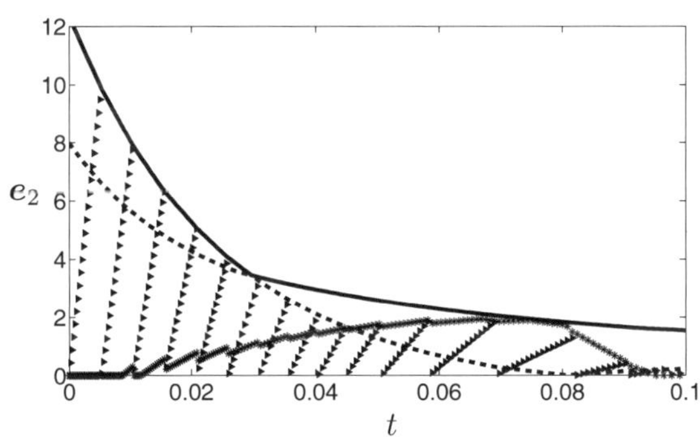

Fig. 5.40 Lyapunov function of both subsystems together with e_2 and the lower bound on V_1

Theorem 5.9 shows that if the small-gain condition holds, a Lyapunov function for the interconnected system is given by the maximum of the individual Lyapunov functions (properly scaled). We now employ parsimonious triggering for the example discussed before.

Example 5.8 (cont.) *A second-order nonlinear system*

In Fig. 5.38 the Lyapunov function of the first and second subsystem are given as dashed lines. Here the Lyapunov functions are already scaled as in (5.91), so that the Lyapunov function of the interconnected system is the maximum of the two, shown as a solid line. At approximately $t \approx 0.03$ the maximum of the Lyapunov functions changes. Before that time the Lyapunov function of the second subsystem is equal to the overall Lyapunov function. After the maximum changed, the Lyapunov function is equal to the Lyapunov function of the first subsystem.

In Fig. 5.38 the error of the first subsystem is given as small triangles. Again, whenever the error hits the Lyapunov function, an event is triggered.

Please note that the Lyapunov functions in Figs. 5.38 and 5.40 are the same. The only difference is that in Fig. 5.38 the error of the first subsystem is given, whereas the error of the second subsystem is depicted in Fig. 5.40. We know that if Zeno occurs, the corresponding state must approach zero at the Zeno time. In Fig. 5.40 it can be seen that the dashed line (the Lyapunov function of the second subsystem) hits 0 at $t \approx 0.08$. On the other hand, at least one other subsystem is bounded away from zero at the Zeno point, because the "strength" to force a Lipschitz continuous system to zero in finite time cannot result from its own state. Hence we know that at some point the largest Lyapunov function in (5.95) is not the Lyapunov function of the system that induces Zeno behavior. Moreover, this Lyapunov function is bounded away from zero.

In theory, it would be sufficient to trigger an event whenever the error (the triangles) hits the solid line (Lyapunov function of the overall system), ruling out the occurrence of Zeno behavior. But the knowledge of the Lyapunov function of the overall system would make the knowledge of all states to all systems necessary, contradicting our decentralized approach.

With the help of an approximation of the dynamics of the subsystems, we can give lower bounds of the Lyapunov functions of the other states (given as stars). Using this bound also rules out the occurrence of Zeno behavior. Basically, the theory predicts that the bound becomes tighter around the Zeno point ($t^* \approx 0.8$). Hence with the above considerations, we know that it is sufficient to trigger an event, if the error (triangles) hits the approximation (stars) to infer asymptotic stability of the system and ruling out Zeno behavior. □

5.5.5 Comparison of the Different Approaches

Both approaches that have been presented in the preceding sections have advantages and drawbacks, which will be discussed in this section. Both Theorem 5.11 and 5.12 can be regarded as generalizations of Theorem 5.10. The major drawback of Theorem 5.10 is that Zeno behavior may occur. The advantage of Theorem 5.11 lies in its simplicity. There is no difference in implementation and numerical complexity compared to Theorem 5.10. Note that the class of systems that are practical ISS is larger than the class of systems that are only ISS. Hence Theorem 5.11 is applicable to a larger number of systems than Theorems 5.10 and 5.12. On the other hand, the result is limited to practical stability of the interconnected system.

The advantage of Theorem 5.12 is that it rules out the Zeno phenomenon, while retaining asymptotic stability. The price to pay lies in a higher computational complexity and some regularity assumptions on the involved triggering functions. As event-triggering tries to reduce the amount of information transmissions, the question which of the approaches results in the fewest events is of interest.

In our example Theorem 5.10 behaves exactly like Theorem 5.12, if no Zeno behavior occurs. For the Zeno case, Theorem 5.11 can lead to fewer events, although it is not a priori clear how to choose the offset to achieve this. The wrong choice of the offset leads to an increase in the number of events beyond the number of events resulting from Theorem 5.12. Further investigations that analyze the impact of the choice of the offset are required here.

Bibliographical Notes

The references given below as well as references therein can be regarded as a starting point for further studies. Different approaches to event-triggering can be found in [107, 255, 354, 380]. In particular, the effect of Zeno behavior has been discussed in [4].

The approach studied in this chapter is based on nonlinear small-gain theorems formulated in the context of input-to-state stability. For an introduction to the concept of ISS and related notions, we refer to [343]. The investigation of small-gain theorems for large-scale systems started with [77]. For the event-triggering approach discussed here, the Lyapunov version of these results is fundamental [78]. Further insight into this type of small-gain results is presented in [79, 80, 186]. For large-scale systems the conditions need to be checked numerically. In particular, the construction of the path σ needed in (5.92) can be treated by simplicial fixed point algorithms that need no regularity assumptions on the gain functions, see [131, 132, 323].

The event-triggering approach using small-gain theorems is discussed in [87–89], where full proofs can be found.

5.6 Event-Based Control of Interconnected Nonlinear Systems

5.6.1 Problem Statement

In this section we consider N physically interconnected nonlinear control systems given by

$$\Sigma : \begin{cases} \dot{\boldsymbol{x}}_1(t) = \boldsymbol{f}_1\big(\boldsymbol{x}_1(t), \ldots, \boldsymbol{x}_N(t), \boldsymbol{u}_1(t), \boldsymbol{d}(t)\big) \\ \vdots \\ \dot{\boldsymbol{x}}_N(t) = \boldsymbol{f}_N\big(\boldsymbol{x}_1(t), \ldots, \boldsymbol{x}_N(t), \boldsymbol{u}_N(t), \boldsymbol{d}(t)\big) \end{cases} \tag{5.97}$$

where $\boldsymbol{x}_i(t) \in \mathcal{X}_i \subset \mathrm{I\!R}^{n_i}, \boldsymbol{u}_i(t) \in \mathcal{U}_i \subset \mathrm{I\!R}^{m_i}, \boldsymbol{d}(t) \in \mathrm{I\!R}^{l}$.

We denote the i-th subsystem by Σ_i and the overall system by Σ. The state and control input of the overall system are denoted by $\boldsymbol{x}(t) = (\boldsymbol{x}_1^{\mathrm{T}}(t), \ldots, \boldsymbol{x}_N^{\mathrm{T}}(t))^{\mathrm{T}}$ and $\boldsymbol{u} = (\boldsymbol{u}_1^{\mathrm{T}}(t), \ldots, \boldsymbol{u}_N^{\mathrm{T}}(t))^{\mathrm{T}}$. The disturbance is described by $\boldsymbol{d}(t)$. The full state space and set of controls are denoted by $\mathcal{X} = \bigotimes_{i=1}^{N} \mathcal{X}_i$ and $\mathcal{U} = \bigotimes_{i=1}^{N} \mathcal{U}_i$.

For a given set $\mathcal{T} \subset \mathcal{X}$, the so-called *target set*, we pursue the following goal. We would like to design feedback functions $\boldsymbol{F}_i : \mathcal{X} \rightarrow \mathcal{U}_i$ for the distributed controller which steer the system (globally, if possible) into the target set $\mathcal{T}$. Once the system enters $\mathcal{T}$ the controller could be switched, for example to the distributed local controller proposed in Section 5.3.

The following sections present two different approaches for constructing the feedback functions $\boldsymbol{F}_i$. The particular realizations of the developed methods take advantage of the robust, optimization-based design approach from Section 5.4. The feedback laws are then given as lookup tables.

- Section 5.6.2 illustrates a possible approach to construct a decentralized stabilizing controller by exploiting concepts of robust stability, input-to-state stability and a small gain condition.

- Section 5.6.3 is concerned with the construction of a centralized stabilizing feedback controller. By considering an adequate optimal control problem we analyze regularity properties of the corresponding optimal cost function, and then propose an iterative feedback computation scheme.

5.6.2 ISS-Based Decentralized Feedback Design

Our goal is to construct a controller $\boldsymbol{u}(t) = \boldsymbol{F}(\boldsymbol{x}(t))$ which renders Σ input-to-state stable (ISS). To this end we need the concept of robust stability.

Definition 5.3. *System Σ is called robustly stable if there exists a $\mathcal{K}_\infty$ function ρ (called stability margin), such that*

$$\dot{\boldsymbol{x}}_\rho(t) = \boldsymbol{f}(\boldsymbol{x}_\rho(t), \boldsymbol{w}(t)\rho(|\boldsymbol{x}_\rho(t)|)) \tag{5.98}$$

is globally asymptotically stable uniformly in $\boldsymbol{w}(t)$, where $\boldsymbol{w}(t) \in [-1,1]^l$. This means that there exists a $\beta \in \mathcal{KL}$, such that

$$|\boldsymbol{x}_\rho(t, \boldsymbol{x}_0, \boldsymbol{w}(t))| \leq \beta(|\boldsymbol{x}_0|, t) \tag{5.99}$$

for all $t \geq 0$ and $\boldsymbol{w}(t) \in [-1,1]^l$.

A possible approach for the construction of an ISS controller is provided by the following theorem (cf. [347] or [194], [193], [214] in discrete time).

Theorem 5.13. *A nonlinear control system is ISS if and only if it is robustly stable.*

Based on this theorem one can compute a controller rendering the origin ISS for the overall system Σ as follows:

1. Replace the disturbance $\boldsymbol{d}(t)$ by the scaled perturbation $\rho(|\boldsymbol{x}(t)|)\boldsymbol{w}(t)$ with $\boldsymbol{w}(t) \in [-1,1]^l$ and $\rho \in \mathcal{K}_\infty$ as described in Definition 5.3.
2. Apply an algorithm that calculates a controller which renders the system with scaled perturbation robustly asymptotically stable.
3. Using this controller, calculated by means of the scaled system, on the original non-scaled system Σ then yields the ISS property for system Σ via Theorem 5.13.

In Step 2, we are going to apply the algorithm described in Section 5.4. Besides the clarification of theoretical issues like stating and proving Theorem 5.13 in an event-based setting (which is subject of ongoing research), the main bottleneck of this approach is the fact that, with growing dimension of Σ, the problem will become computationally infeasible since the necessary memory and the running time of the algorithm increases exponentially due to the well known "curse of dimensionality". In order to circumvent this problem, we propose to implement this algorithm in a distributed fashion exploiting the structure of the interconnected system, i.e., we compute decentralized controllers $\boldsymbol{u}_i(t) = \boldsymbol{F}_i(\boldsymbol{x}_i(t))$ for the subsystems.

In the interest of ensuring that the overall system controlled by the distributed controllers is asymptotically stable, we introduce the concept of a nonlinear small-gain condition as described in [76], [78], which is similar to the one in Section 5.5. To this end, we define the diagonal operator $D : \mathbb{R}_+^N \to \mathbb{R}_+^N$ by

$$D\left((s_1, ..., s_N)^{\mathrm{T}}\right) := \begin{pmatrix} s_1 + \alpha_1(s_1) \\ \vdots \\ s_1 + \alpha_N(s_N) \end{pmatrix}, \tag{5.100}$$

where $\alpha_i \in \mathcal{K}_\infty$. Then, we say that a nonlinear gain matrix Γ satisfies the small-gain condition if an operator D defined as in (5.100) exists, such that

$$\Gamma \circ D(s) := \Gamma(D(s)) \not\geq s \tag{5.101}$$

holds for all $s \in \mathbb{R}_+^n \backslash \{0\}$. Here, "$\not\geq$" is to be understood componentwise, i.e., (5.101) is satisfied if there exists at least one index $i \in \{1, ..., N\}$ such that $\Gamma(D(s))_i < s_i$ holds.

With these definitions we are able to formulate the following theorem, cf. [76].

Theorem 5.14. *Let each subsystem Σ_i be ISS and the gain matrix Γ be given. If condition* (5.101) *is satisfied, then system* (5.97) *is ISS.*

The idea is now to compute a controller for every subsystem Σ_i as described above, which guarantees ISS of the subsystems by Theorem 5.13 at a desired equilibrium. As in Section 5.5 we model the influence of other states as disturbance. Then, ISS of the overall system Σ can be concluded by applying Theorem 5.14, using the controls calculated for the separate subsystems Σ_i.

Just as Theorem 5.13, when applying the event-based controller from Section 5.3, Theorem 5.14 needs to be stated and proved in the event-based setting. Again, this is subject of ongoing research.

Example 5.9 *Stability test for the thermofluid process*

In order to illustrate that the distributed approach just sketched can be successfully applied, we show numerical results for the thermofluid process described in Section 5.8. We set $u_2(t) = 0.21$, $u_3(t) = 0.19$, $u_6(t) = 0.29$, $u_7(t) = 0.22$, using the setpoint for the linearized model in Section 5.8, and the possible disturbances to zero, i.e. $d_1(t) = 0$, $d_2(t) = 0$.

We start by splitting up the four-dimensional system into two subsystems, Σ_1 consisting of the first two components $x_1(t)$ and $x_2(t)$ and Σ_2 consisting of $x_3(t)$ and $x_4(t)$. Following the small-gain paradigm to interpret the coupling states as perturbations and the scaling approach in order to obtain ISS from robust stability, we substitute each state $x_k(t)$ of subsystem Σ_j appearing in the dynamics of Σ_i, $j \neq i$. During our numerical experiments it turned out to be useful to choose separate scaling functions for each component, therefore we replace x_k by $x_k^\star + \rho_i^k(\boldsymbol{x}(t))w_k(t)$, where $w_k(t) \in [-1,1]$, ρ_i^k denotes the corresponding component of ρ_i and $x_k^\star$ stands for the corresponding coordinates of the desired equilibrium. This equilibrium is provided by the setpoint for the linearized model in Section 5.8, $\boldsymbol{x}^\star(t) = (0.33, 294.7, 0.34, 300.2)$. The scaling functions ρ_i are chosen to be

$$\rho_1(\boldsymbol{x}(t)) = \begin{pmatrix} \sqrt{1.24082(x_1 - 0.33)^2} \\ \sqrt{0.618194(x_2 - 294.7)^2} \end{pmatrix}$$

$$\rho_2(\boldsymbol{x}(t)) = \begin{pmatrix} \sqrt{0.727344(x_1 - 0.34)^2} \\ \sqrt{1.4599(x_2 - 300.2)^2} \end{pmatrix}.$$

We computed the controllers by applying the algorithm proposed in Section 5.4 using all the extensions described there, i.e., the event-based setting and the use of past information.

To this end, we define the cost functionals of the subsystems as follows

$$c(x(t), u(t)) = \frac{1}{\Delta z_1{}^2}(x_1(t) - x_1^\star)^2 + \frac{1}{\Delta z_2{}^2}(x_2(t) - x_2^\star)^2,$$

where Δz_1, Δz_2 describe the size of one cell in the grid.

For the numerical results we use a sampling time of $h = 2s$ in order to obtain the discrete-time model from Section 5.4, the bound $r_0 = 400$ steps for the computation of $r(\boldsymbol{x}(t), \boldsymbol{u}(t))$ and an 8×8 grid covering the desired operating region. We apply a discretization of 5 control values and 3 disturbances. The resulting optimal value functions for each of the two subsystems are depicted in Fig. 5.41 and Fig. 5.42.

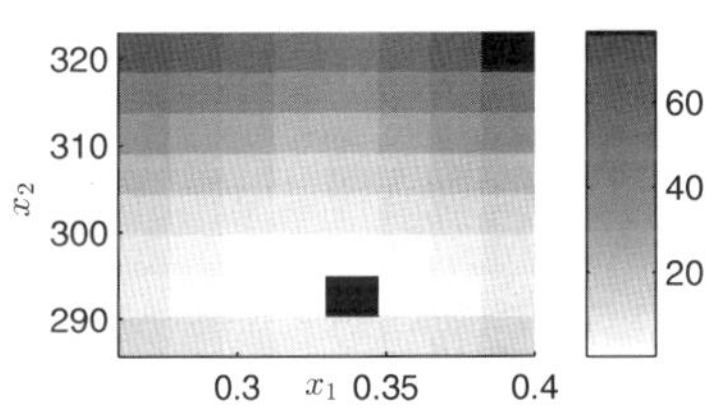

Fig. 5.41 Optimal value function for Σ_1

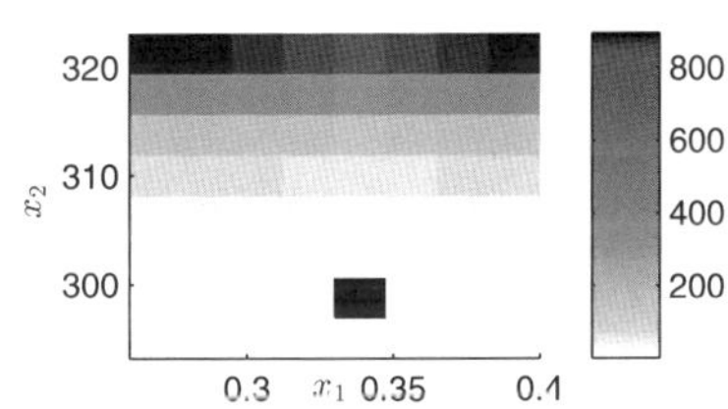

Fig. 5.42 Optimal value function for Σ_2

The black cell in the middle of each plot represents the target region which contains the equilibrium $\boldsymbol{x}^\star$. Since all the values on the grid are finite, the controllers are defined on the whole operating region. As expected, the values increase further away from the target area. For the subsequent trajectory simulations, once the target cell is reached (on which our controller is undefined) we set the control to the value for which $x^\star$ is an equilibrium, cf. the setpoint for the linearized model in Section 5.8. In a future implementation, this will be replaced by a distributed

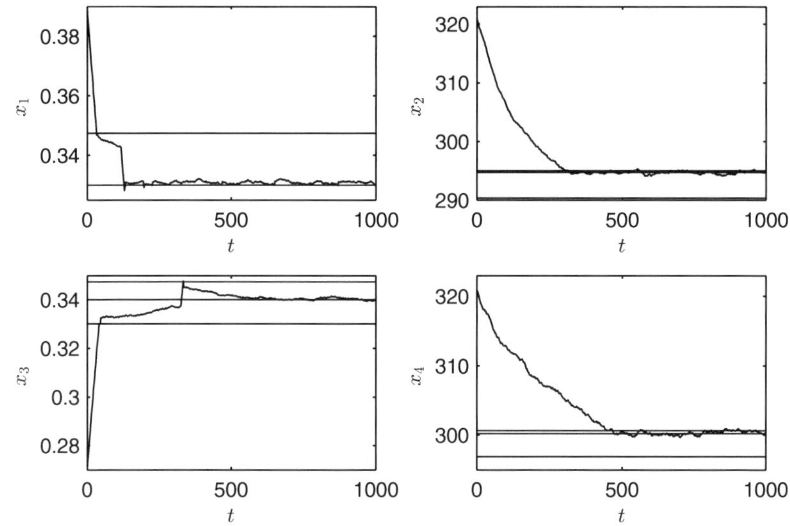

Fig. 5.43 Trajectories for the individual subsystems Σ_1 (top) and Σ_2 (bottom)

local controller obtained from linearization techniques as described in Section 5.2 and Section 5.3.

In the first set of simulations shown in Fig. 5.43, the subsystems have been simulated independently by replacing the coupling states with a randomly generated perturbation in the intervall $[-0.07, 0.07] \times [-28.45, 28.45]$ or $[-0.08, 0.08] \times [-22.95, 22.95]$ and initial value $\boldsymbol{x}_0 = (0.39;\ 321)^{\mathrm{T}}$ or $\boldsymbol{x}_0 = (0.27;\ 321)^{\mathrm{T}}$, respectively. The straight lines in the diagrams represent the borders of the target region or the equilibrium values, respectively. Note that the first component of the equilibrium $x_1^* = 0.33$ coincides with the boundary of the target set.

One observes that the first components, representing the water level, are steered to the target region very fast, while it takes somewhat longer to control the temperature. The nonsmoothness of the trajectories is due to the influence of the randomly generated disturbances.

Using the same controllers on the coupled system Σ, one obtains the trajectory shown in Fig. 5.44, starting again in $\boldsymbol{x}_0 = (0.39;\ 321;\ 0.27;\ 321)^{\mathrm{T}}$.

As in the previous figure the straight lines represent the targets. All four components are controlled fairly smoothly to the desired equilibrium, which demonstrates the effectiveness of our approach. The few still occurring bends are the result of insetting events. Sharp bends mainly appear in the trajectory of components which already reached the target set while other components are still further away. □

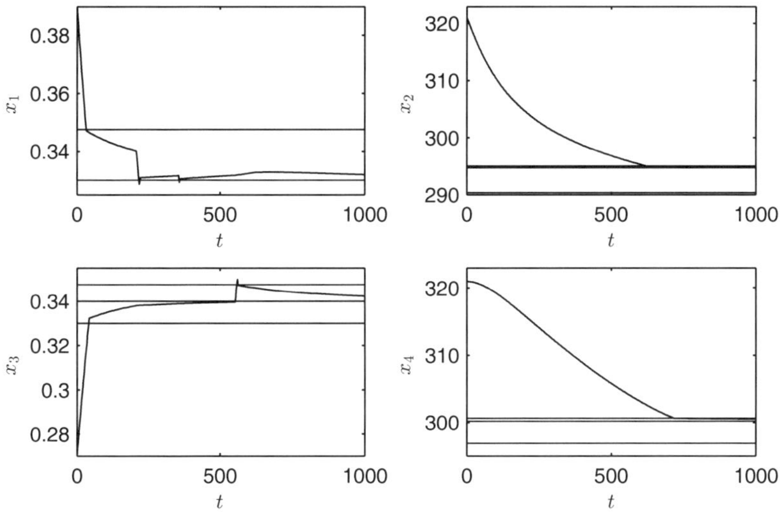

Fig. 5.44 Trajectory for the coupled system Σ

5.6.3 Optimization-Based Centralized Feedback Design

We are searching for a feedback law $\boldsymbol{F} = (\boldsymbol{F}_1, \ldots, \boldsymbol{F}_N)$ with $\boldsymbol{F}_i : \mathcal{X} \to \mathcal{U}_i$ that steers the nonlinear interconnected system Σ (if possible) globally into the target set $\mathcal{T}$. Note that all $\boldsymbol{F}_i$ map the whole $\mathcal{X}$ onto the respective $\mathcal{U}_i$, i.e. the controller is centralized, and it is assumed that it disposes of the whole state information from all subsystems.

In general, an arbitrary feedback that stabilizes the system can perform poorly, thus we will introduce appropriate cost functions with respect to which the controller should act optimally. With no further structural assumptions on the underlying nonlinear system Σ, the feedback law could be any (measurable) function $\boldsymbol{F} : \mathcal{X} \to \mathcal{U}$. In general, in order to approximate this function to an accuracy ε one will need $\mathcal{O}(\varepsilon^{-n})$ variables, where $n = \sum_i n_i$ is the dimension of the overall system. Even if this number is moderate, the computation and storage of these variables can pose considerable problems, if it gets large, those tasks turn impossible. This phenomenon is known as the *curse of dimension* [23].

Optimal Control and Its Regularity for Weakly Coupled Systems. In the following we consider a discrete time system, viewed as the time-sampled analog of (5.97), without disturbances:

$$\Sigma^{\#} : \begin{cases} \boldsymbol{x}_1(k+1) = \boldsymbol{f}_1\big(\boldsymbol{x}_1(k), \ldots, \boldsymbol{x}_N(k), \boldsymbol{u}_1(k)\big) \\ \qquad\vdots \\ \boldsymbol{x}_N(k+1) = \boldsymbol{f}_N\big(\boldsymbol{x}_1(k), \ldots, \boldsymbol{x}_N(k), \boldsymbol{u}_N(k)\big) \end{cases} \tag{5.102}$$

with subsystems $\Sigma_i^\#$. By defining a cost function $g : \mathcal{X} \times \mathcal{U} \to \mathbb{R}_{\geq 0}$, with $g(\boldsymbol{x}, \boldsymbol{u}) = 0$ whenever $\boldsymbol{x} \in \mathcal{T}$, the feedback should be constructed such that the accumulated costs $\sum_{k \geq 0} g(\boldsymbol{x}(k), \boldsymbol{u}(k))$, $\boldsymbol{u}(k) = \boldsymbol{F}(\boldsymbol{x}(k))$, are minimal. As discussed in Section 5.4.2, the minimal value of the accumulated costs is given by the *(optimal) value function* , denoted by V, which in turn is characterized by the Bellman equation (5.54). Once V is known, the feedback $\boldsymbol{F}$ can be computed from (5.55).

However, it is not just a large dimension of $\mathcal{X}$ that is prohibitive for representing the optimal feedback law. If one would choose to compute $\boldsymbol{F}$ through the Bellman equation, a minimization w.r.t. $\boldsymbol{u} \in \mathcal{U}$ would have to be carried out, which also would suffer from the curse of dimension. It is obvious that without further structural assumptions on the problem, i.e. without further regularity of the objects involved, we can hardly proceed.

One way to obtain regularity is to "decouple" the system, i.e. to consider the subsystems $\Sigma_i^\#$ without interconnections. A system with no interconnections satisfies $\boldsymbol{f}_i(\boldsymbol{x}, \boldsymbol{u}_i) = \boldsymbol{f}_i(\boldsymbol{x}_i, \boldsymbol{u}_i)$ and $g(\boldsymbol{x}, \boldsymbol{u}) = \sum_i g_i(\boldsymbol{x}_i, \boldsymbol{u}_i)$ for appropriate subsystem cost functions g_i (we call such cost functions separable). Under these assumptions we have $V(\boldsymbol{x}) = \sum_i V_i(\boldsymbol{x}_i)$, with V_i being the value functions of the respective (independent) subsystems. Observe that the representation of this V up to an error ε can be achieved with $\mathcal{O}\big(\sum_i \varepsilon^{-n_i}\big)$ variables, in contrast to $\mathcal{O}\big(\varepsilon^{-\sum_i n_i}\big)$ as before. Its computation by N independent Bellman equations (carried out for each subsystem) is also much cheaper computationally (for details on a possible method and its numerical costs see Section 5.4.2). Hence we make the following

Assumption 5.2 (Tractability of subsystems). *For all i, let $\widehat{\mathcal{U}}_i \subset \mathcal{U}_i$ be finite and let $\widehat{\Sigma}_i^\#$ denote the ith subsystem of the decoupled system where the admissible control parameters are chosen from the set $\widehat{\mathcal{U}}_i$. Further, let $\varepsilon > 0$ be such that if V_i is approximated to an accuracy $\mathcal{O}(\varepsilon)$, then the corresponding feedback law from* (5.55) *for $\widehat{\Sigma}_i^\#$ (considered as an individual system) renders the closed loop system asymptotically stable w.r.t. $\mathcal{T}$. Then, we assume that an algorithm which performs $\mathcal{O}\big(|\widehat{\mathcal{U}}_i|\, \varepsilon^{-n_i}\big)$ many evaluations of $\boldsymbol{f}$ is computationally tractable.*

If decoupled systems show a regularity which one can exploit, it is natural to look for a similar regularity in *weakly interconnected* systems. We call the system $\Sigma^\#$ (or the continuous-time version from (5.97)) weakly interconnected (weakly coupled), if the ratio

$$\frac{\big\|\partial_{\boldsymbol{x}_j} \boldsymbol{f}_i(\boldsymbol{x}, \boldsymbol{u}_i)\big\|}{\big\|\partial_{\boldsymbol{x}_i} \boldsymbol{f}_i(\boldsymbol{x}, \boldsymbol{u}_i)\big\|} \tag{5.103}$$

is small for all $i \neq j$, $\boldsymbol{x}$ and $\boldsymbol{u}_i$. This means that the dynamics of the ith subsystem is assumed to be much more sensitive to changes in $\boldsymbol{x}_i$ than to changes in other $\boldsymbol{x}_j$. The question is now whether the value function of $\Sigma^\#$

resembles the structure $V(\boldsymbol{x}) = \sum_i V_i(\boldsymbol{x}_i)$, since such functions can be efficiently represented. For this, we compare it to the value function of the decoupled system.[1]

Let the cost function be separable, let V^0 denote the value function of the decoupled system, and $\boldsymbol{F}^0$ its optimal feedback law (assumed to be globally stabilizing). The following statement holds for linear-quadratic regulator problems, and we have numerical evidence for the general non-linear case. Note that if the interconnected system is stabilized by the optimal feedback $\boldsymbol{F}^0$ of the decoupled system, the control problem is solved (unless the performance of this controller is not acceptable). This case is analyzed in Section 5.3.3 for linear systems. Roughly speaking,

> if the closed system $\boldsymbol{x} \mapsto \boldsymbol{f}(\boldsymbol{x}, \boldsymbol{F}^0(\boldsymbol{x}))$ is not globally stable, then we can not expect $V - V^0$ to be small compared with V^0, even if the dynamics is weakly coupled. The same holds for $\boldsymbol{F}$, the optimal feedback law of the interconnected system.

For the precise statement and quantitative estimates see [208].

Remark 5.1. Note that the above statement does not say that V can not be properly approximated by *some* $\sum_i W_i$, where $W_i : \mathcal{X}_i \to \mathbb{R}$. However, this problem seems to be hard, and is subject of future work.

These findings suggest that the approximation of the optimal controller is out of reach, and other strategies have to be applied. One is to relax the optimality property, and in that way retain the regularity needed for efficient numerical approximation.

Split Optimality Principle. As we have seen above, for all interesting cases where the control problem can not be solved by just using the feedback law of the decoupled problem, no regularity of the value function and the optimal feedback can be expected. Thus, our strategy is to relax the Bellman optimality, and search for *some* Lyapunov function (not necessarily the value function itself), and *some* feedback which satisfies prescribed regularity constraints, and is possibly optimal under these.

For this, let us pick one arbitrary $i \in \{1, \ldots, N\}$ and assume that the optimal feedback laws $\boldsymbol{F}_j$ for $j \neq i$ are known. Then, by explicitly including these into the Bellman equation, we can write

$$\begin{aligned} V(\boldsymbol{x}) &= \inf_{\boldsymbol{u} \in \mathcal{U}} \{g(\boldsymbol{x}, \boldsymbol{u}) + V(\boldsymbol{f}(\boldsymbol{x}, \boldsymbol{u}))\} \\ &= \inf_{\boldsymbol{u}_i \in \mathcal{U}_i} \{g_i[\hat{\boldsymbol{F}}_i](\boldsymbol{x}, \boldsymbol{u}_i) + V(\boldsymbol{f}_i[\hat{\boldsymbol{F}}_i](\boldsymbol{x}, \boldsymbol{u}_i))\} \end{aligned}$$

[1] Here, technically we assume that the system depends on a scalar parameter which homotopes between a decoupled and a coupled system. This is the case for the thermofluid process described in Sectoin 5.8 that we investigate below.

where $\hat{\boldsymbol{F}}_i = (\ldots, \boldsymbol{F}_{i-1}, \boldsymbol{F}_{i+1}, \ldots)$, $g_i[\hat{\boldsymbol{F}}_i](\boldsymbol{x}, \boldsymbol{u}_i) = g(\boldsymbol{x}, (\boldsymbol{u}_i, \hat{\boldsymbol{F}}_i(\boldsymbol{x})))$ and $\boldsymbol{f}_i[\hat{\boldsymbol{F}}_i](\boldsymbol{x}, \boldsymbol{u}_i) = \boldsymbol{f}(\boldsymbol{x}, (\boldsymbol{u}_i, \hat{\boldsymbol{F}}_i(\boldsymbol{x})))$. Hence, the value function and feedback satisfying (5.54) and (5.55) also satisfy

$$\left.\begin{aligned} V_i(\boldsymbol{x}) &= \quad \inf_{\boldsymbol{u}_i \in \mathcal{U}_i} \{g[\hat{\boldsymbol{F}}_i](\boldsymbol{x}, \boldsymbol{u}_i) + V_i(\boldsymbol{f}_i[\hat{\boldsymbol{F}}_i](\boldsymbol{x}, \boldsymbol{u}_i))\} \\ \boldsymbol{F}_i(\boldsymbol{x}) &= \arg\inf_{\boldsymbol{u}_i \in \mathcal{U}_i} \{g[\hat{\boldsymbol{F}}_i](\boldsymbol{x}, \boldsymbol{u}_i) + V_i(\boldsymbol{f}_i[\hat{\boldsymbol{F}}_i](\boldsymbol{x}, \boldsymbol{u}_i))\} \end{aligned}\right\} \tag{5.104}$$

for $i = 1, \ldots, N$, where the V_i are all copies of the value function V. Although the converse does not hold, Theorem 5.15 will give us the justification for working with this split form. We have thus split one problem into N nonlinearly coupled ones, with the advantage that now only minimizations w.r.t. $\boldsymbol{u}_i \in \mathcal{U}_i$ occur, which we assumed to be numerically tractable. Another advantage is that now the subproblems can be solved on different grids (partitions of the state space), thus *enforcing* the regularity we would like to have. We intend to solve (5.104) by cyclically solving the subproblems with the method developed in Section 5.4 on different grids: The ith subproblem is solved on a grid which is fine in the directions of the ith subsystem, and coarse in the other directions; for details see Fig. 5.45 and Algorithm 5.1. Thus, the curse of dimension is greatly reduced in the representation of the value function and of the feedback, as well as in the minimization w.r.t. the controls.

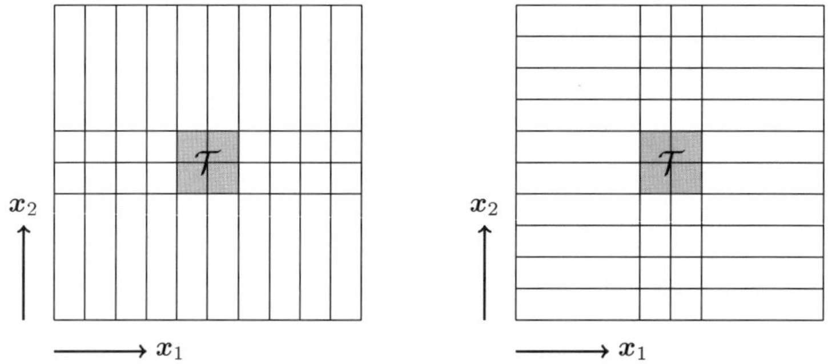

Fig. 5.45 Left: a possible partition P_1, which is fine in the $\boldsymbol{x}_1$-direction and coarse in the others. Right: a possible partition P_2, which is fine in the $\boldsymbol{x}_2$-direction, and coarse in the others. The gray rectangle in the center ($\mathcal{T}$) denotes the target set. The feedback law $\boldsymbol{F}_1$ is constant on elements of the P_1-partition, $\boldsymbol{F}_2$ is constant on elements of the P_2-partition.

Due to the sparse structure of the chosen partitions the controller is "ignorant": the controller of each subsystem cares more about the evolution of the own subsystem, than about the others.

Algorithm 5.1. Split optimal policy iteration

Given: The system $\Sigma^{\#}$, and N partitions of $\mathcal{X}$, the whole state space, $P_1, \ldots, P_N$.
Init: Set $\boldsymbol{F}_i^{(0)} = \boldsymbol{F}_i^0$, $i = 1, \ldots, N$, the feedback laws of the decoupled system.

1: **for** $k = 0, 1, \ldots$ **do**
2: **for** $i = 1, \ldots, N$ **do**
3: Compute $V_i^{(k+1)}$ and $\boldsymbol{F}_i^{(k+1)}$ by solving
$V_i^{(k+1)}(\boldsymbol{x}) = \inf_{\boldsymbol{u}_i \in \mathcal{U}_i} \{g[\hat{\boldsymbol{F}}_i^{(k)}](\boldsymbol{x}, \boldsymbol{u}_i) + V_i^{(k+1)}(\boldsymbol{f}_i[\hat{\boldsymbol{F}}_i^{(k)}](\boldsymbol{x}, \boldsymbol{u}_i))\}$, and
$\boldsymbol{F}_i^{(k+1)}(\boldsymbol{x}) = \arg\inf_{\boldsymbol{u}_i \in \mathcal{U}_i} \{g[\hat{\boldsymbol{F}}_i^{(k)}](\boldsymbol{x}, \boldsymbol{u}_i) + V_i^{(k+1)}(\boldsymbol{f}_i[\hat{\boldsymbol{F}}_i^{(k)}](\boldsymbol{x}, \boldsymbol{u}_i))\}$
on the partition P_i via (5.65).
4: **if** $\boldsymbol{F}_i^{(k+1)} = \boldsymbol{F}_i^{(k)}$ **then**
5: **return** the most actual $\boldsymbol{F}_j^{(\ell)}$ and $V_j^{(\ell)}$ for $j = 1, \ldots, N$
6: **end if**
7: **end for**
8: **end for**

Theorem 5.15. [209] *If the iteration in Algorithm 5.1 converges to a (approximate) solution of* (5.104) *with* $V_i(\boldsymbol{x}) < \infty$ *for* $i = 1, \ldots, N$ *and* $\boldsymbol{x} \in \mathcal{X}$*, then the associated feedback law* $(\boldsymbol{F}_1, \ldots, \boldsymbol{F}_N)$ *renders the closed system* $\Sigma^{\#}$ *asymptotically stable w.r.t.* $\mathcal{T}$.

The problem of solving (5.104) on different partitions for each subsystem is similar to a cooperative game: all players try to act together to achieve a common goal. However, the players only have accurate information about the state of their own subsystem, and less accurate about the others – this situation occurs when communication underlies restrictions like cost (Section 5.7), or delays and losses (Section 3.1). Since it is sufficient to send less accurate information, the communication between the subsystems (and with it the network load) decreases. Step 3 of Algorithm 5.1 resembles the choices which have to be made in multi-agent systems (Chapter 6): the own action has to be optimal w.r.t. the assumed/current action of the other agents.

Example 5.10 *Global stabilization of the thermofluid process*

We demonstrate the above approach on a modified version of the thermofluid process described in Section 5.8. The first subsystem Σ_1 describes the evolution of $(\boldsymbol{x}_1, \boldsymbol{x}_2)^{\mathrm{T}} = (l_{TB}, \vartheta_{TB})^{\mathrm{T}}$, i.e. the fill height and temperature in tank TB. The second subsystem Σ_2 describes the evolution of $\boldsymbol{x}_3 = \vartheta_{TS}$, the temperature in tank TS. The inflow from tank T_3 is adjusted by $\boldsymbol{u}_5$ such that the water level stays constant $\bar{\boldsymbol{x}}_4$. All other control parameters are set as at the setpoint, except the coupling constants; we set $\boldsymbol{u}_3 = 0.25$ and $\boldsymbol{u}_7 = 0.095$. The perturbations are set constant to $\boldsymbol{d}_1 = 0.015$ and $\boldsymbol{d}_2 = 0.1$. These choices are made such that the decoupled system (i.e. setting $\boldsymbol{u}_3 = \boldsymbol{u}_7 = 0$) is globally stabilizable, however the optimal feedback (w.r.t. the cost function given below) of the decoupled system does not stabilize the full coupled system Σ.

Algorithm 5.1 is applied to compute a feedback law steering the system $\Sigma^{\#}$ with sampling time 1500s globally into the target region

$$\mathcal{T} = [0.330, 0.348] \times [290.3, 295.0] \times [296.9, 300.6].$$

The cost function is penalizing deviations in the state from the center of the target region;

$$g(\boldsymbol{x}, \boldsymbol{u}) = 1000\,|\boldsymbol{x}_1 - 0.339| + |\boldsymbol{x}_2 - 292.7| + |\boldsymbol{x}_3 - 298.8|.$$

The partition P_1 is a uniform $32 \times 32 \times 16$ rectangular partition of the state space, while P_2 is chosen to be a uniform $16 \times 16 \times 32$ partition. The control sets are discretized by 20 equispaced control values. The split optimal policy iteration converges after 6 cycles (i.e. each subsystem is updated 6 times), and 97 % of the boxes (for each partition) is controllable for the model incorporating worst case perturbations. The algorithm fails to find a stabilizing feedback for a small region of the phase space with low temperatures in both tanks. Figure 5.46 shows the simulation of one trajectory starting at $\boldsymbol{x} = (0.29, 321, 295)^{\mathrm{T}}$ with the computed feedback law. □

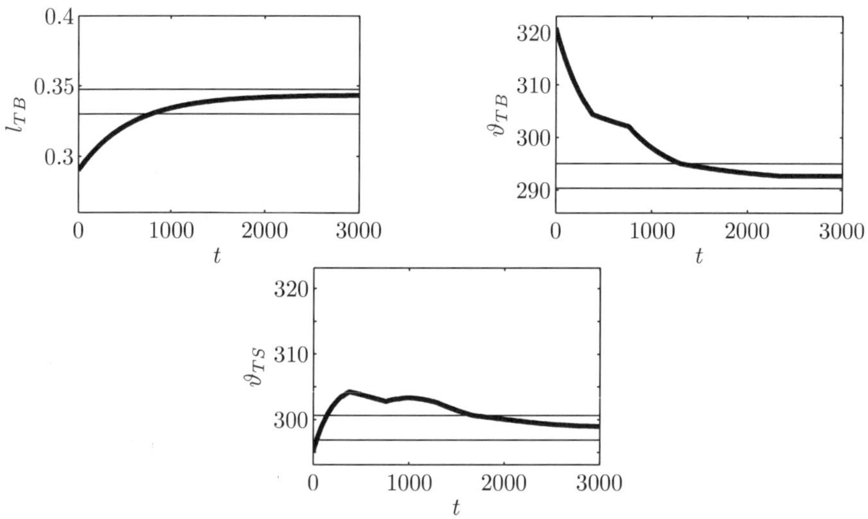

Fig. 5.46 Simulation of one trajectory with the computed feedback for 3000 seconds, with 5 seconds sampling time. The thin horizontal lines depict the borders of the target region.

The results in Example 5.10 are sensitive w.r.t. the chosen sampling time of the continuous-time system. An adequate event-based discretization which resolves the problem of choosing the proper integration time, similarly as in Section 5.4.6, is currently under development.

Bibliographical Notes

The efficient approximation of high dimensional scalar functions is a growing issue in many applications. Based on the different kind of regularity one can find, sparse grid approaches [42, 403], and even more recently compressed sensing techniques [48, 65, 109, 124] have been successful.

The sensitivity analysis of optimal control is widely represented in the literature. For perturbation analysis of LQR problems we refer the reader to [167, 180, 203, 210, 366].

If Algorithm 5.1 converges, it settles in a Nash equilibrium, known from game theory [280]. For more game theoretical considerations in optimal control see e.g. [235].

5.7 A Stochastic Approach to Event-Based Control

5.7.1 Event-Based Control as a Two-Person Team Problem

This section considers the design of event-triggered controllers in the context of stochastic linear systems. The problem setup is posed as a two-person team problem, where the two agents are given by the controller and the event-trigger. Both agents aim at minimizing conjointly a common cost function, which comprises a quadratic control cost and a penalty that is paid whenever an event is generated. Triggering an event evokes the transmission of the current state information to the controller that adjusts the control inputs. The event generator, which is situated at the sensor, must therefore decide carefully, whenever it is worth to pay the penalty in order to update the controller. As the information available differs among the agents, the possibility of signaling between the agents is an integral part of the stochastic optimal control problem.

In the present system signaling can occur in two ways. By choosing specific control inputs, the controller may invoke the event generator to send another state update. On the other hand, by not sending information to the controller, the event generator implicitly signals also information to the controller depending on the choice of the event-triggering law.

Signalling can improve the system performance, but it is generally undesired in the design process, as it does not admit efficient numerical algorithms to be developed that solve the optimization problem. The resulting optimal policies are usually non-linear and depend on the complete observation history. In the time-triggered case, when transmission times are chosen beforehand, signaling is not possible and, therefore, the optimal design can be carried out in a straightforward manner. Interestingly, it turns out for the event-triggered transmission scheme that under some mild assumptions, signaling is not beneficial. These mild assumptions presume that the policies are deterministic and that the distributions of the noise process and the initial condition are symmetric.

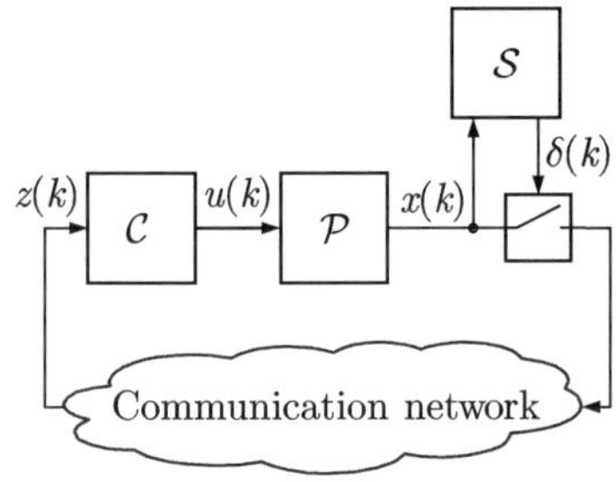

Fig. 5.47 Resource-constrained control system the with process $\mathcal{P}$, the sensor station $\mathcal{S}$ executing the event-triggering law, and the control station $\mathcal{C}$

The solution of the initial joint optimization problem reduces to an emulation-based approach: In the first step, the optimal controller is designed independently of the event generator and the communication penalty yielding a linear control law and a state estimator. In the second step, the event-triggering law is computed via stochastic dynamic programming, where events are triggered by the one-step ahead estimation error. Apart from the similar structure compared with the event-triggered controller in Section 5.2, it is interesting to note that both designs use an emulation-based approach despite of the differing theoretical backgrounds.

This section is separated into three parts. In Section 5.7.2 the problem setup is introduced. This problem is solved for time-triggered transmission schemes in Section 5.7.3. The main results for the event-triggered case are given in Section 5.7.4.

5.7.2 Problem Formulation

The resource-constrained networked control system under consideration is illustrated in Fig. 5.47. It consists of a process $\mathcal{P}$ to be regulated by a controller $\mathcal{C}$, which is implemented at the actuator. The control station $\mathcal{C}$ receives measurements from a sensor station $\mathcal{S}$ that have to be transmitted over a resource-constrained communication network. Within the sensor station an event generator is implemented that decides upon current observations whether or not to send information to the controller. Subsequently, the individual functional blocks in Fig. 5.47 are defined.

The process $\mathcal{P}$ is described by the following discrete-time stochastic difference equation

$$\boldsymbol{x}(k+1) = \boldsymbol{A}\boldsymbol{x}(k) + \boldsymbol{B}\boldsymbol{u}(k) + \boldsymbol{w}(k), \quad \boldsymbol{x}(0) = \boldsymbol{x}_0, \tag{5.105}$$

where $\boldsymbol{A} \in \mathbb{R}^{n\times n}$, $\boldsymbol{B} \in \mathbb{R}^{n\times d}$. The variables $\boldsymbol{x}(k)$ and $\boldsymbol{u}(k)$ denote the state and the control input and are taking values in $\mathbb{R}^n$ and $\mathbb{R}^d$, respectively. The initial state $\boldsymbol{x}_0$ is a random variable with finite mean and covariance $\boldsymbol{C}_{\boldsymbol{x}_0}$. The system noise process $\boldsymbol{w}(k)$ is i.i.d. (independent identically distributed), takes values in $\mathbb{R}^n$, and is normal distributed with zero-mean and covariance matrix $\boldsymbol{C}_{\boldsymbol{w}}$. The random variables $\boldsymbol{x}_0$ and $\boldsymbol{w}(k)$ are statistically independent for each k. Let $(\Omega, \mathcal{F}, \text{Prob})$ denote the probability space generated by the initial state $\boldsymbol{x}_0$ and noise sequence $\boldsymbol{W}^{N-1}$, where $\boldsymbol{W}^k = \{\boldsymbol{w}(0), \ldots, \boldsymbol{w}(k)\}$ denotes the truncated sequence up to time k. The variables $\boldsymbol{x}_0$ and $\boldsymbol{w}(k)$ are called the primitive random variables of the system. It is assumed that the statistics of the process $\mathcal{P}$ are known a-priori to both, the event generator and the controller.

Concerning our system model, it is needed to define the amount of information available at the control station at each time step k. The output signal $\delta(k)$ of the event generator takes values in $\{0, 1\}$ deciding whether information is transmitted at time k, i.e.,

$$\delta(k) = \begin{cases} 1, & \text{measurement } \boldsymbol{x}(k) \text{ is sent,} \\ 0, & \text{no measurement is transmitted.} \end{cases}$$

Therefore, the signal $\boldsymbol{z}(k)$ is defined as

$$\boldsymbol{z}(k) = \begin{cases} \boldsymbol{x}(k), & \delta(k) = 1 \\ \emptyset, & \delta(k) = 0. \end{cases} \tag{5.106}$$

As various steps of decisions are made within one time period k, a causal ordering is specified by the following sequence in which the events within the system occur:

$$\cdots \to \boldsymbol{x}(k) \to \delta(k) \to \boldsymbol{z}(k) \to \boldsymbol{u}(k) \to \boldsymbol{x}(k+1) \to \cdots$$

Note that the choice of $\delta(k)$ influences the amount of information at the controller station at time k.

We allow the control input $\boldsymbol{u}(k)$ and the event-trigger output $\delta(k)$ to depend on their complete past history. This implies in particular that they may have memory and are by themselves dynamic systems. Let the event-triggering law $\pi = \{\pi_0, \pi_1, \ldots, \pi_{N-1}\}$ and the control law $\gamma = \{\gamma_0, \gamma_1, \ldots \gamma_{N-1}\}$ denote admissible policies for the finite horizon N with

$$\delta(k) = \pi_k(\boldsymbol{X}^k), \quad \boldsymbol{u}(k) = \gamma_k(\boldsymbol{Z}^k). \tag{5.107}$$

We assume that the mappings π_k and γ_k are measurable mappings of their available information $\boldsymbol{X}^k$ or $\boldsymbol{Z}^k$, respectively. Let us denote $\mathcal{U}_{\text{ET}}$ to be the set of all admissible policy pairs (π, γ). As we will also consider time-triggered policies, we define the set $\mathcal{U}_{\text{TT}}$ of admissible policy pairs, where the controller is given by (5.107) and π is either 0 or 1 at each time k, i.e.,

$$\mathcal{U}_{\text{TT}} = \{(\pi, \gamma) \in \mathcal{U}_{\text{ET}} | \pi_k \equiv c_k, c_k \in \{0, 1\}, k \in \{0, \ldots, N-1\}\}. \tag{5.108}$$

The communication channel takes the role of restricting or penalizing transmissions in the feedback loop. This will be reflected in the optimization problem. Let $J_{\mathcal{C}}$ be the control objective defined as

$$J_{\mathcal{C}} = \boldsymbol{x}^{\mathrm{T}}(N)\boldsymbol{Q}_N\boldsymbol{x}(N) + \sum_{k=0}^{N-1} \boldsymbol{x}(k)^{\mathrm{T}}\boldsymbol{Q}\boldsymbol{x}(k) + \boldsymbol{u}(k)^{\mathrm{T}}\boldsymbol{R}\boldsymbol{u}(k), \tag{5.109}$$

where $\boldsymbol{Q}_N, \boldsymbol{Q} \in \mathbb{R}^{n\times n}$ are positive definite matrices and $\boldsymbol{R} \in \mathbb{R}^{d\times d}$ is positive semi-definite. Let r be the communication cost given by the number of transmissions, i.e.,

$$r = \sum_{k=0}^{N-1} \delta(k). \tag{5.110}$$

Problem 5.1. For a given $\lambda \geq 0$, find the optimal policies π^* and γ^*:

$$\inf_{(\pi,\gamma)\in\mathcal{U}_{\mathrm{ET}}} \mathrm{E}\left[J_{\mathcal{C}} + \lambda r\right].$$

The nonnegative value λ can be interpreted as the weight of penalizing transmissions over the communication channel. For notational convenience, we define the cost function $J(\pi, \gamma)$ for $(\pi, \gamma) \in \mathcal{U}_{\mathrm{ET}}$ to be

$$J(\pi, \gamma) = \mathrm{E}\left[J_{\mathcal{C}} + \lambda r\right]. \tag{5.111}$$

From the definition of $\mathcal{U}_{\mathrm{TT}}$ in (5.108), it can be observed that in our framework time-triggered controllers constitute a subset of event-triggered controllers, i.e., $\mathcal{U}_{\mathrm{TT}} \subset \mathcal{U}_{\mathrm{ET}}$. Therefore, we can state the following inequality

$$\inf_{(\pi,\gamma)\in\mathcal{U}_{\mathrm{ET}}} J(\pi, \gamma) \leq \inf_{(\pi,\gamma)\in\mathcal{U}_{\mathrm{TT}}} J(\pi, \gamma),$$

which says that the optimal event-based controller always outperforms the optimal time-triggered controller.

5.7.3 Optimal Time-Triggered Transmission

In the following we are concerned with the optimal design of the time-triggered controller. This means we want to minimize J within the admissible time-triggered control policies given by $\mathcal{U}_{\mathrm{TT}}$. First, we fix an arbitrary triggering sequence π^{TT} and investigate the corresponding optimal control law γ^* stated by the following problem:

Problem 5.2.

$$\inf_{\gamma} J(\pi^{\mathrm{TT}}, \gamma), \quad \text{s.t. } (\pi^{\mathrm{TT}}, \gamma) \in \mathcal{U}_{\mathrm{TT}}.$$

It can be observed that the communication cost r is constant and can, therefore, be omitted from the optimization for a fixed triggering sequence π^{TT}. What remains is the expected quadratic cost term $\mathrm{E}[J_{\mathcal{C}}]$. Second, the equation can be written as a linear time-varying measurement equation

$$\boldsymbol{z}(k) = \boldsymbol{C}(k)\boldsymbol{x}(k), \quad \boldsymbol{C}(k) = \begin{cases} \boldsymbol{I}_n, & \delta(k) = 1 \\ \boldsymbol{0}_n, & \delta(k) = 0. \end{cases} \tag{5.112}$$

Therefore, we can conclude that the problem of finding the optimal control law in Problem 5.2 reduces to a standard LQG problem without measurement noise, as the process evolves according to a linear difference equation given by (5.105) and the measurement equation given by (5.112) is also linear in the state $\boldsymbol{x}(k)$ and the costs are quadratic in the state and the control input. Hence, the separation principle of stochastic control is applicable and the solution is given by a certainty-equivalence controller consisting of a linear gain and a Kalman estimator. A certainty-equivalence controller is given by solving a related deterministic control problem, where all primitive random variables are set to their means, and by replacing the state variable by its least-squares estimate within the deterministic solution.

This result is summarized in the following theorem.

Theorem 5.16. *Let an arbitrary time-triggered transmission sequence π^{TT} be fixed. Then, the solution of Problem 5.2 is given by the certainty-equivalence controller*

$$\boldsymbol{u}(k) = \gamma_k^*(\boldsymbol{Z}^k) = -\boldsymbol{L}(k)\mathrm{E}[\boldsymbol{x}(k)|\boldsymbol{Z}^k], \quad k \in \{0, \ldots, N-1\} \tag{5.113}$$

with

$$\begin{aligned} \boldsymbol{L}(k) &= \left(\boldsymbol{R} + \boldsymbol{B}^{\mathrm{T}}\boldsymbol{P}(k+1)\boldsymbol{B}\right)^{-1} \boldsymbol{B}^{\mathrm{T}}\boldsymbol{P}(k+1)\boldsymbol{A}, \\ \boldsymbol{P}(k) &= \boldsymbol{A}^{\mathrm{T}}\boldsymbol{P}(k+1)\boldsymbol{A} + \boldsymbol{Q} \\ &\quad - \boldsymbol{A}^{\mathrm{T}}\boldsymbol{P}(k+1)\boldsymbol{B}\left(\boldsymbol{R} + \boldsymbol{B}^{\mathrm{T}}\boldsymbol{P}(k+1)\boldsymbol{B}\right)^{-1} \boldsymbol{B}^{\mathrm{T}}\boldsymbol{P}(k+1)\boldsymbol{A}, \end{aligned} \tag{5.114}$$

where $\boldsymbol{P}(N) = \boldsymbol{Q}_N$ and $\boldsymbol{P}(k) \in \mathbb{R}^{n \times n}$ is non-negative definite for $k \in \{0, \ldots, N\}$. The estimator $\mathrm{E}[\boldsymbol{x}(k)|\boldsymbol{Z}^k]$ is given by the following recursive form

$$\mathrm{E}[\boldsymbol{x}(k)|\boldsymbol{Z}^k] = \begin{cases} \boldsymbol{x}(k), & \delta(k) = 1 \\ (\boldsymbol{A} - \boldsymbol{B}\boldsymbol{L}(k))\mathrm{E}[\boldsymbol{x}(k-1)|\boldsymbol{Z}^{k-1}], & \delta(k) = 0. \end{cases} \tag{5.115}$$

Having obtained the optimal controller for a given transmission sequence, we focus now on the calculation of the optimal time-triggered transmission scheme $\pi^{\mathrm{TT},*}$. For that reason, let $\boldsymbol{e}(k)$ be the estimation error at time k defined as

$$\boldsymbol{e}(k) = \boldsymbol{x}(k) - \mathrm{E}[\boldsymbol{x}(k)|\boldsymbol{Z}^k].$$

By using Lemma 6.1 in Chapter 8 of [12] and a couple of straightforward reformulations, the cost function can be rewritten in the following form.

$$\begin{aligned} J =& \lambda \mathrm{E}\left[\sum_{k=0}^{N} \delta(k)\right] + \mathrm{E}[\boldsymbol{x}^{\mathrm{T}}(0)\boldsymbol{P}(0)\boldsymbol{x}(0)] + \mathrm{E}\left[\sum_{k=0}^{N-1} \boldsymbol{w}(k)^{\mathrm{T}}\boldsymbol{P}_{k+1}\boldsymbol{w}(k)\right] \\ &+ \mathrm{E}\left[\sum_{k=0}^{N-1} \boldsymbol{e}(k)^{\mathrm{T}}\boldsymbol{L}(k)^{\mathrm{T}}\boldsymbol{\Gamma}(k)\boldsymbol{L}(k)\boldsymbol{e}(k)\right] \\ &+ \mathrm{E}\left[\sum_{k=0}^{N-1} (\boldsymbol{u}(k) + \boldsymbol{L}(k)\mathrm{E}[\boldsymbol{x}(k)|\boldsymbol{Z}^k])^{\mathrm{T}}\boldsymbol{\Gamma}(k)(\boldsymbol{u}(k) + \boldsymbol{L}(k)\mathrm{E}[\boldsymbol{x}(k)|\boldsymbol{Z}^k)\right], \end{aligned} \tag{5.116}$$

where $\boldsymbol{\Gamma}(k)$ is defined as

$$\boldsymbol{\Gamma}(k) = \boldsymbol{B}^{\mathrm{T}}\boldsymbol{P}(k+1)\boldsymbol{B} + \boldsymbol{R}, \quad k \in \{0, \ldots, N-1\}.$$

Obviously, the second and the third term on the right-hand side of (5.116) are constant. For a fixed time-triggered transmission sequence, we observe that the first term is constant. In addition, it can be shown that $\boldsymbol{e}(k)$ is a random variable that is independent of the policy γ. This gives the optimal control law γ^* in (5.113) for a fixed transmission sequence as already stated in Theorem 5.16. On the other hand, only the first and the forth term are varying with different transmission sequences when assuming that the control law is given by (5.113). In order to calculate the optimal time-triggered transmission sequence, we define the one-step ahead estimation error $\boldsymbol{e}_1(k)$ by

$$\boldsymbol{e}_1(k) = \boldsymbol{x}(k) - \mathrm{E}[\boldsymbol{x}(k)|\boldsymbol{Z}^{k-1}]. \tag{5.117}$$

From this definition, we have the following connection to the estimation error:

$$\boldsymbol{e}(k) = \begin{cases} \boldsymbol{0}, & \delta(k) = 1 \\ \boldsymbol{e}_1(k), & \delta(k) = 0. \end{cases}$$

The evolution of $\boldsymbol{e}_1(k)$ can be derived by

$$\begin{aligned} \boldsymbol{e}_1(k+1) &= \boldsymbol{x}(k+1) - \mathrm{E}[\boldsymbol{x}(k+1)|\boldsymbol{Z}^k] \\ &= \boldsymbol{A}\boldsymbol{x}(k) + \boldsymbol{B}\boldsymbol{u}(k) + \boldsymbol{w}(k) - \mathrm{E}[\boldsymbol{A}\boldsymbol{x}(k) + \boldsymbol{B}\boldsymbol{u}(k) + \boldsymbol{w}(k)|\boldsymbol{Z}^k] \\ &= \boldsymbol{A}(\boldsymbol{x}(k) - \mathrm{E}[\boldsymbol{x}(k)|\boldsymbol{Z}^k]) + \boldsymbol{w}(k) \\ &= (1-\delta_k)\boldsymbol{A}\boldsymbol{e}_1(k) + \boldsymbol{w}(k). \end{aligned}$$

The remaining optimization problem has then the following form

$$\begin{aligned}\pi^{\mathrm{TT},*} &= \arg \inf_{\delta_0,\ldots,\delta_{N-1}} \mathrm{E}\left[\sum_{k=0}^{N-1}(1-\delta(k))\boldsymbol{e}_1(k)^{\mathrm{T}}\boldsymbol{L}(k)^{\mathrm{T}}\boldsymbol{\Gamma}(k)\boldsymbol{L}(k)\boldsymbol{e}_1(k)+\lambda\delta(k)\right] \\ &\text{s.t.} \quad \boldsymbol{e}_1(k+1)=(1-\delta(k))\boldsymbol{A}\boldsymbol{e}_1(k)+\boldsymbol{w}(k) \end{aligned} \tag{5.118}$$

Since the triggering variable δ_k is chosen before execution, i.e., it is independent of $\boldsymbol{e}_1(k)$, it is possible to rewrite above optimization problem in order to apply dynamic programming. For that reason, we define the error covariance

$$\boldsymbol{\Phi}(k)=\mathrm{E}[\boldsymbol{e}_1(k)\boldsymbol{e}_1^{\mathrm{T}}(k)].$$

The evolution of $\boldsymbol{\Phi}(k)$ is given by

$$\boldsymbol{\Phi}(k+1)=(1-\delta(k))\boldsymbol{A}\boldsymbol{\Phi}(k)\boldsymbol{A}^{\mathrm{T}}+\boldsymbol{C}_{\boldsymbol{w}}, \quad \boldsymbol{\Phi}(0)=\boldsymbol{C}_{\boldsymbol{x}_0}.$$

Then, the optimization problem in (5.118) can be written as

$$\begin{aligned}\pi^{\mathrm{TT},*} &= \arg \inf_{\delta_0,\ldots,\delta_{N-1}} \sum_{k=0}^{N-1}(1-\delta(k))\mathrm{tr}[\boldsymbol{\Phi}(k)\boldsymbol{L}(k)^{\mathrm{T}}\boldsymbol{\Gamma}(k)\boldsymbol{L}(k)]+\lambda\delta(k) \\ &\text{s.t.} \quad \boldsymbol{\Phi}(k+1)=(1-\delta(k))\boldsymbol{A}\boldsymbol{\Phi}(k)\boldsymbol{A}^{\mathrm{T}}+\boldsymbol{C}_{\boldsymbol{w}}, \quad \boldsymbol{\Phi}(0)=\boldsymbol{C}_{\boldsymbol{x}_0} \end{aligned} \tag{5.119}$$

We observe that the initially stochastic optimization problem reduces to a deterministic optimal control problem with state variable $\boldsymbol{\Phi}(k)$. This implies that the calculation of the optimal time-triggered transmission sequence $\pi^{\mathrm{TT},*}$ can be performed by deterministic dynamic programming. In summary, the optimal time-triggered controller within the set $\mathcal{U}^{\mathrm{TT}}$ can be calculated in two steps:

1. Obtain the optimal control gain $L(k)$ from the discrete-time Riccati equation in (5.114)
2. Solve optimization problem (5.119) that yields the optimal transmission timings.

Inspired by this design approach, the more challenging problem of event-triggered transmission strategies is studied in the next section.

5.7.4 Optimal Event-Triggered Transmission

What makes the derivation in the previous section appealing relies on the fact that the cost function J is completely separable with respect to the control law and the transmission times. This becomes evident when regarding (5.116) that results from reordering the cost terms of J. When allowing the transmissions to be triggered by events rather than by a-priori fixed timings, the separation does not hold in the way as for the time-triggered mechanism. This is due to the fact that the estimation error $\boldsymbol{e}_k$ is generally not independent of the control law anymore when assuming a fixed event-triggering law π. In other words, the controller is able to signal through the plant to the event-trigger that it may want to receive another state update.

Such signaling is called the dual effect of control and refers to the dual role of control: (i) influencing the state evolution and (ii) decreasing the estimation error. When the second phenomenon is not present, which is also referred to as the absence of the dual effect, then the optimal control law is given by (5.113). On the other hand, in our case the dual effect is present in general, which implies that the optimal control law will be a nonlinear function of the complete history $\boldsymbol{Z}^k$, which highly depends on the choice of the fixed event generator.

Another approach that might be taken is the direct optimization of both the control law and the event-triggering law at the same time. But as the information available at the controller and at the event generator differ, the optimization problem has a non-classical information pattern, whose solution is very hard to find and no systematic algorithms are available, even for simple cases. In fact, the joint optimization problem under consideration falls into the category of sequential stochastic control problems, for which a dynamic programming formulation is possible. But the value function must be parameterized by the distribution of the state, which implies an infinite dimensional state space, and the minimum is taken over all control laws rather than over the inputs. Obviously, this formulation does not allow efficient numerical algorithms and also restrains us from getting new insights.

The aforementioned arguments suggest that only little can be said about the optimal event-triggered controller that solves Problem 5.1. It is, therefore, even more surprising that the actual optimal event-triggered controller is a certainty equivalence controller given by (5.113), i.e., it takes the same form as in the time-triggered case.

The crucial property that is exploited to show the above statement is the nestedness property of the information pattern. The information pattern is nested because the information available at the controller represented by the sigma algebra of $\boldsymbol{Z}^k$ is a subset of the information available at the event-trigger given by the sigma algebra $\boldsymbol{X}^k$ for any k, i.e.,

$$\sigma(\boldsymbol{Z}^k) \subset \sigma(\boldsymbol{X}^k) \subset \mathcal{F}, \quad k \in \{0, \ldots, N-1\}.$$

The key idea to show that certainty-equivalence controller are optimal bases on the following common concept in optimal control.

Definition 5.4 (Dominating Policies). *A set of policies $\mathcal{U}'_{\mathrm{ET}} \subset \mathcal{U}_{\mathrm{ET}}$ is called a dominating class of policies for Problem 5.1 if for any feasible $(\pi, \gamma) \in \mathcal{U}_{\mathrm{ET}}$, there exists a feasible $(\pi', \gamma') \in \mathcal{U}'_{\mathrm{ET}}$, such that*

$$J(\pi', \gamma') \leq J(\pi, \gamma),$$

where J is the cost function defined by (5.111) for the corresponding problem.

Once a dominating class of policies is found, the above definition implies that we can restrict the solutions of the optimization problem to such policies. In the following, we show that the set of policy pairs where the controller is a certainty-equivalence controller denoted by γ^* is a dominating class of policies. Therefore, the remaining goal is to prove that for any pair (π, γ), we can find a pair (π', γ^*) whose costs are at most that of (π, γ).

In order to achieve this, we introduce a suitable reparametrization of the triggering law. Given a policy (π, γ), we define another policy (ρ, γ) where $\rho = \{\rho_0, \ldots, \rho_{N-1}\}$ is the triggering law and ρ_k is a function of $\{\boldsymbol{x}_0, \boldsymbol{W}^{k-1}\}$, such that

$$\rho_k(\boldsymbol{x}_0, \boldsymbol{W}^{k-1}) = \pi_k(\boldsymbol{X}^k), \quad k \in \{0, \ldots, N-1\},\ \omega \in \Omega, \tag{5.120}$$

when both systems use the control law γ. As the control inputs $\boldsymbol{U}^{k-1}$ are known at the event time k by the law γ due to $\sigma(\boldsymbol{Z}^k) \subset \sigma(\boldsymbol{X}^k)k$, the variables $\{\boldsymbol{x}_0, \boldsymbol{W}^{k-1}\}$ can be fully recovered by the state sequence $\boldsymbol{X}^k$ and vice versa. Therefore, the triggering law ρ satisfying (5.120) always exists. On the other hand, this also implies that given (ρ, γ), there is always a (π, γ) satisfying (5.120).

The next auxiliary result gives a statement on the optimal control law for fixed ρ.

Lemma 5.7. *Let the triggering law ρ be a function of primitive variables given by*

$$\delta_k = \rho_k(\boldsymbol{x}_0, \boldsymbol{W}^{k-1}), \quad k \in \{0, \ldots, N-1\}. \tag{5.121}$$

If the triggering law ρ is fixed, then the optimal control law γ^* minimizing $J(\rho, \gamma)$ is a certainty-equivalence controller defined in (5.113).

Proof. The proof can be outlined as follows. First, it is shown that the estimation error $\boldsymbol{e}_k$ is a random variable independent of the control law chosen for a fixed event-triggering law ρ. Inspecting Eq. (5.116), which is also valid for event-triggering law ρ, it can be seen that unlike the last expression all others are constants. Hence, it can be concluded that the certainty equivalence controller defined by (5.113) is optimal. The details of the proof can be found in [260]. □

Lemma 5.7 enables us to show that the certainty-equivalence controller is optimal for Problem 5.1, which is stated in the following theorem.

Theorem 5.17. *Let the system be given by (5.105) and (5.106). The class of policies* $\mathcal{U}_{\mathrm{CE}} \subset \mathcal{U}_{\mathrm{ET}}$ *defined by*

$$\mathcal{U}_{\mathrm{CE}} = \{(\pi, \gamma^*) \in \mathcal{U}_{\mathrm{ET}} \,|\, \gamma^* = -\boldsymbol{L}(k)\mathrm{E}[\boldsymbol{x}(k)|\boldsymbol{Z}^k],\ \boldsymbol{L}(k) \text{ given by (5.114)}\}$$

is a dominating class of policies for Problem 5.1.

Proof. According to Def. 5.4, it suffices to show that for any feasible pair $(\pi, \gamma) \in \mathcal{U}_{\mathrm{ET}}$, there is a feasible policy $(\pi', \gamma^*) \in \mathcal{U}_{\mathrm{CE}}$ whose costs are at most that of (π, γ).

Given an admissible policy (π, γ), there exists an admissible policy (ρ, γ) with ρ_k being a function of primitive variables that satisfies (5.120). Condition (5.120) implies that for (π, γ) and (ρ, γ), we have identical random variables $\boldsymbol{u}_k$ and δ_k for $k \in \{0, \ldots, N-1\}$ and, therefore, identical costs. In the same way for the pair (ρ, γ^*), we find a triggering law π' being a function of $\boldsymbol{X}^k$, such that both (ρ, γ^*) and (π', γ^*) output identical random variables $\boldsymbol{u}_k$ and δ_k for $k \in \{0, \ldots, N-1\}$. Due to Lemma 5.7, we obtain

$$J(\pi, \gamma) = J(\rho, \gamma) \geq \min_{\gamma} J(\rho, \gamma) = J(\rho, \gamma^*) = J(\pi', \gamma^*).$$

This concludes the proof. □

Theorem 5.17 implies that we can characterize optimal control policies to be certainty-equivalent control laws given by (5.113). The remaining problem is to design the optimal event generator π^*. Opposed to the time-triggered case, special care needs to be taken, as the estimator $\mathrm{E}[\boldsymbol{x}(k)|\boldsymbol{Z}^k]$ depends on the choice of the event-triggering policy π. This is because not sending update information to the controller may still be useful information for the state estimate $\mathrm{E}[\boldsymbol{x}(k)|\boldsymbol{Z}^k]$ and constitutes another type of signaling between the event generator and the controller. Throughout the subsequent paragraph, suppose that a particular event-triggering law π has been selected. We define $\tau(k)$ to be the last time an update was sent, i.e.

$$\tau(k) = \max\{\ell | \delta_\ell, \ell < k\}$$

with $\tau(k) = -1$, if there were no transmissions before k. Obviously, for $\delta(k) = 1$, we get $\mathrm{E}[\boldsymbol{x}(k)|\boldsymbol{Z}^k] = \boldsymbol{x}_k$. However, for $\delta(k) = 0$, we have

$$\mathrm{E}[\boldsymbol{x}(k)|\boldsymbol{Z}^k] = \boldsymbol{A}^{k-\tau(k)}\boldsymbol{x}(\tau(k)) + \sum_{\ell=\tau(k)}^{k-1} \boldsymbol{A}^{k-\ell-1}\boldsymbol{B}\boldsymbol{u}_\ell + \mathrm{E}\left[\sum_{\ell=\tau(k)}^{k-1} \boldsymbol{A}^{k-\ell-1}\boldsymbol{w}(\ell)|\boldsymbol{Z}^k\right] \tag{5.122}$$

In the time-triggered case, the noise process $\boldsymbol{w}(\ell)$ for $\ell \in \{\tau(k), \dots, k-1\}$ is statistically independent of $\boldsymbol{Z}^k$, as no information is gathered after $\tau(k)$ at the controller. Therefore, the last term vanishes, because $\boldsymbol{w}(k)$ is zero-mean. On the other hand, by not sending information, when having an event-triggering law π_k that depends on $\boldsymbol{X}^k$, it also depends on $\boldsymbol{w}_{k-1}$. Therefore, the last term may not be zero in the event-triggering case due to the statistical dependence between $\boldsymbol{Z}^k$ and $\boldsymbol{w}(\ell)$ for $\ell \in \{\tau(k), \dots, k-1\}$. The additional bias represented by the last term in (5.122) is constant for fixed $\tau(k)$ at any time $k \in \{0, \dots, N-1\}$. Therefore, the estimator eventually takes the following form

$$\mathrm{E}[\boldsymbol{x}(k)|\boldsymbol{Z}^k] = \begin{cases} \boldsymbol{x}(k), & \delta(k) = 1 \\ (\boldsymbol{A} - \boldsymbol{B}\boldsymbol{L}(k))\mathrm{E}\left[\boldsymbol{x}(k-1)|\boldsymbol{Z}^{k-1}\right] + \boldsymbol{\alpha}(\tau(k), k), & \delta(k) = 0, \end{cases} \tag{5.123}$$

which is similar to the least-squares estimator for the time-triggered case given by (5.115), but differs by the additional bias term $\boldsymbol{\alpha}(\tau(k), k)$.

This formulation of the optimal estimator allows us to recover a similar optimization as for the optimal time-triggered transmission sequence given by (5.118).

$$\begin{aligned} \pi^* = \arg\inf_{\pi} \mathrm{E} & \left[\sum_{k=0}^{N-1} (1 - \delta(k))(\boldsymbol{e}_1(k) - \boldsymbol{\alpha}(\tau(k), k))^{\mathrm{T}} \boldsymbol{L}(k)^{\mathrm{T}} \boldsymbol{\Gamma}(k) \boldsymbol{L}(k) \right] \times \\ & \times (\boldsymbol{e}_1(k) - \boldsymbol{\alpha}(\tau(k), k)) + \lambda\delta(k)] \\ & \text{s.t.} \quad \boldsymbol{e}_1(k+1) = (1 - \delta(k))\boldsymbol{A}\boldsymbol{e}_1(k) + \boldsymbol{w}(k). \end{aligned} \tag{5.124}$$

Assuming that the distributions of the primitive random variables are symmetric, as in our case, it can be shown for first-order linear systems that the optimal event generator is a symmetric threshold function and the state estimator is given by the optimal least-squares estimator for the time-triggered case, i.e., $\boldsymbol{\alpha} \equiv \boldsymbol{0}$. Symmetry means that the event-triggering mapping π_k is an even function in $\boldsymbol{e}_1(k)$. The question whether symmetric policies are also optimal for higher-order systems remains an open problem. In the following, we assume that also for higher dimensional systems symmetric even-triggering policies are optimal. Then, the optimization problem takes the standard form of a optimal stochastic control problem with state $\boldsymbol{e}_1(k)$, which can be solved by means of dynamic programming. As the terminal and the running costs are radially increasing functions and the noise distributions are symmetric, it can be concluded for the scalar case that the event-triggering law will be a time-varying threshold function of the one-step ahead estimation error $\boldsymbol{e}_1(k)$.

In summary, it has been shown that the initial Problem 5.1, which was supposed to be hard to solve, turns out to be numerically tractable, as the optimal solution can be constrained to a certain structure without loosing optimality. Fig. 5.48 illustrates the structure of the optimal event-triggered

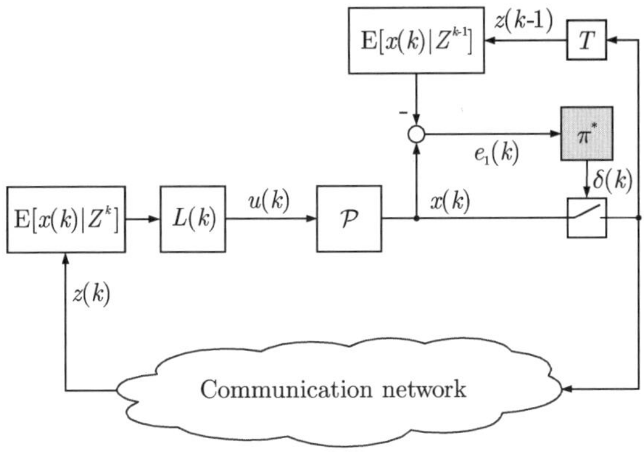

Fig. 5.48 Complete structure of the optimal event based control system with event-trigger π^* and certainty-equivalence controller represented by $\boldsymbol{L}(k)$. System block T denotes a 1-step delay element

controller. The steps involved in the computation, which can be performed offline, can be summarized as follows.

1. Obtain the optimal control gain $\boldsymbol{L}(k)$ from the discrete-time Riccati equation in (5.114).
2. Solve optimization problem (5.119) that yields the optimal event-trigger as a function of $\boldsymbol{e}_1(k)$ by stochastic dynamic programming.

In terms of numerical complexity, the first step can be solved for higher dimensional processes, as it has polynomial complexity with respect to the state dimension n. Therefore, this step does not put severe requirements on the computations. On the other hand, it is well known that the computational complexity of the dynamic programming algorithm grows exponentially with the state dimension when the cost-to-go function has no closed form solution, which is also the case for (5.124) with $\boldsymbol{\alpha} \equiv \boldsymbol{0}$. However, near-optimal solutions, which can be solved in polynomial-time, can be obtained by using approximate dynamic programming.

Bibliographical Notes

A proof for the time-triggered case in Section 5.7.3 that shows that the estimation error is a random variable that is independent of the control policy can be found in section 5.2 of [30].

The fact that stochastic optimal control problems with non-classical information pattern are hard to solve is shown in [387] for a very elementary problem setting and is also supported by the complexity theoretic viewpoint taken in [283]. The dynamic programming formulation for problems with a certain non-classical information pattern has been discussed in [388].

The optimality of the certainty equivalence controller for event-triggered systems has been discussed in [260], where also problem settings with transmission constraints rather than communication penalties have been analyzed.

The problem of event-triggered estimation for scalar systems has been extensively studied in [234, 259]. They both show that the optimal event-trigger is a symmetric threshold function, when the distributions of the primitive random variables are symmetric. While this fact is the consequence of results in majorization theory and the Riesz rearrangement inequality in [234], the work in [259] bases on global convergence properties of the proposed iterative algorithm and uses Lyapunov methods. In [259], the bias parameter $\boldsymbol{\alpha}$ is considered as a free optimization parameter and an iterative method is developed that alternates between optimizing the event-triggering policy while fixing the estimator and vice versa.

The suboptimal design of the event generator is studied in [64] by means of approximate dynamic programming.

The obtained results can be used to extend the design of event-triggered controllers for multiple feedback loops over a common communication medium analyzed in Section 7.3 beyond the consideration of integrator subsystems. This is demonstrated in [261] through a bi-level design approach, where the communication penalty functions as a Lagrange multiplier constraining the average number of transmissions.

5.8 Appendix: The Thermofluid Process used for Experiments

This appendix summarizes the models of a thermofluid process, which has been used to evaluate experimentally the methods for event-based control proposed in Sections 5.3 and 5.6. The experiments have been carried out at the laboratory set-up VERA at the Institute of Automation and Computer Control at Ruhr-Universität Bochum, Germany (Fig. 5.49).

The process consists of the two reactors TB and TS in which the levels $l_{\rm TB}(t)$ and $l_{\rm TS}(t)$ as well as the temperatures $\vartheta_{\rm TB}(T)$ and $\vartheta_{\rm TS}(t)$ of the liquid can be measured and controlled (Fig. 5.50). The reactor TB is connected to the spherical tank $\rm T_1$ and the flow from $\rm T_1$ to TB can be controlled by means of the opening angle u_1. Via PB a part of the outflow is pumped back into $\rm T_1$ while the remaining outflow is conducted to the reactor TS. The temperature $\vartheta_{\rm TB}(t)$ of the water in TB can be influenced by the cooling unit using the input u_4 or by the heating rods that are driven by the signal d_1. The inflow from the spherical tank $\rm T_3$ to the reactor TS can be adjusted by means of the opening angle u_5. TS is also fed by another supply HW that can be controlled by the opening angle d_2. Like in TB the outflow is split and one part is pumped via PS back to $\rm T_3$ and the other part is pumped to the reactor TB. The temperature $\vartheta_{\rm TS}(t)$ of the liquid in TS can be increased by the heating rods that are controlled by the signal u_8. The two reactors are coupled by the flow from TB to TS and vice versa. The coupling strength can be adjusted by means of the opening angles l_{12} and l_{21}.

Fig. 5.49 VERA: Experimental set-up

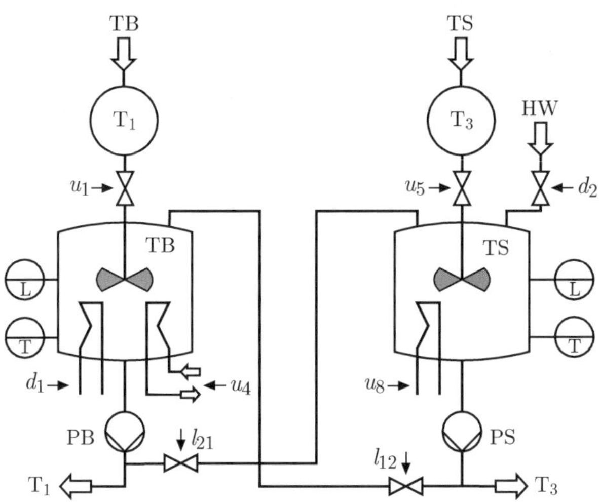

Fig. 5.50 Interconnected tank system

Nonlinear Model. With the state

$$\left(x_1(t)\ x_2(t)\ x_3(t)\ x_4(t)\right)^{\mathrm{T}} = \left(l_{\mathrm{TB}}(t)\ \vartheta_{\mathrm{TB}}(t)\ l_{\mathrm{TS}}(t)\ \vartheta_{\mathrm{TS}}(t)\right)^{\mathrm{T}}$$

the behavior of the interconnected tank system is described by the nonlinear state-space model

$$\begin{aligned}
\dot{x}_1(t) &= \frac{1}{A_{\mathrm{TB}}(x_1(t))}\Big(q_{\mathrm{T1TB}}(u_1(t)) + q_{\mathrm{TSTB}}(x_3(t), u_7(t)) - q_{\mathrm{TBT1}}(x_1(t), u_2(t)) \\
&\qquad\qquad - q_{\mathrm{TBTS}}(x_1(t), u_3(t))\Big) \\
\dot{x}_2(t) &= \frac{1}{V_{\mathrm{TB}}(x_1(t))}\Big(q_{\mathrm{TSTB}}(x_3(t), u_7(t))(x_4(t) - x_2(t)) + H_{\mathrm{TB}} d_1(t) \\
&\qquad\qquad + q_{\mathrm{C}}(u_4(t))(\vartheta_{\mathrm{C}} - x_2(t))\Big) \\
\dot{x}_3(t) &= \frac{1}{A_{\mathrm{TS}}(x_3(t))}\Big(q_{\mathrm{HW}}(d_2(t)) + q_{\mathrm{T3TS}}(u_5(t)) + q_{\mathrm{TBTS}}(x_1(t), u_3(t)) \\
&\qquad\qquad - q_{\mathrm{TSTB}}(x_3(t), u_7(t)) - q_{\mathrm{TST3}}(x_3(t), u_6(t))\Big) \\
\dot{x}_4(t) &= \frac{1}{V_{\mathrm{TS}}(x_3(t))}\Big(q_{\mathrm{HW}}(d_2(t))(\vartheta_{\mathrm{HW}} - x_4(t)) + H_{\mathrm{TS}} u_8(t) \\
&\qquad\qquad + q_{\mathrm{TBTS}}(x_1(t), u_3(t))(x_2(t) - x_4(t))\Big) \\
\boldsymbol{y}(t) &= \left(x_1(t)\ x_2(t)\ x_3(t)\ x_4(t)\right)^{\mathrm{T}}.
\end{aligned}$$

The input signals are assembled in the vector

$$\boldsymbol{u}(t) = \left(u_1(t)\ u_2(t)\ u_3(t)\ u_4(t)\ u_5(t)\ u_6(t)\ u_7(t)\ u_8(t)\right)^{\mathrm{T}}$$

where

$$u_i(t) \in [0, 1], \quad \forall\, i = 1, \ldots, 8.$$

Note that only the signals u_1, u_4, u_5 and u_8 are control inputs, whereas the remaining inputs are used in order to adjust a setpoint. The signals

$$d_1(t), d_2(t) \in [0, 1]$$

are used to realize desired characteristics. The flows used in the nonlinear state-space model are given as follows:

$$q_{\mathrm{TBT1}}(x_1(t), u_2(t)) = K_{\mathrm{TBT1}}(u_2(t))\sqrt{2 g x_1(t)}$$

$$K_{\mathrm{TBT1}}(u_2(t)) = 10^{-6} \begin{cases} 96.10 \cdot u_2(t) & \text{if } u_2(t) \le 0.1 \\ 200.18 \cdot u_2(t) - 10.41 & \text{if } 0.1 < u_2(t) \le 1 \end{cases}$$

$$q_{\mathrm{TBTS}}(x_1(t), u_3(t)) = K_{\mathrm{TBTS}}(u_3(t))\sqrt{2 g x_1(t)}$$

$$K_{\mathrm{TBTS}}(u_3(t)) = 10^{-6} \begin{cases} 102.04 \cdot u_3(t) & \text{if } u_3(t) \le 0.1 \\ 212.96 \cdot u_3(t) - 11.09 & \text{if } 0.1 < u_3(t) \le 1 \end{cases}$$

$$q_{\mathrm{TST3}}(x_3(t), u_6(t)) = K_{\mathrm{TST3}}(u_6(t))\sqrt{2gx_3(t)}$$

$$K_{\mathrm{TST3}}(u_6(t)) = 10^{-6}\begin{cases} 78.99 \cdot u_6(t) & \text{if } u_6(t) \leq 0.1 \\ 141.68 \cdot u_6(t) - 6.27 & \text{if } 0.1 < u_6(t) \leq 1 \end{cases}$$

$$q_{\mathrm{TSTB}}(x_3(t), u_7(t)) = K_{\mathrm{TSTB}}(u_7(t))\sqrt{2gx_3(t)}$$

$$K_{\mathrm{TSTB}}(u_7(t)) = 10^{-6}\begin{cases} 90.26 \cdot u_7(t) & \text{if } u_7(t) \leq 0.1 \\ 168.22 \cdot u_7(t) - 7.80 & \text{if } 0.1 < u_7(t) \leq 1 \end{cases}$$

$$q_{\mathrm{T1TB}}(u_1(t)) = 161.15 \cdot 10^{-6} \cdot u_1(t)$$

$$q_{\mathrm{T3TS}}(u_5(t)) = 181.08 \cdot 10^{-6} \cdot u_5(t)$$

$$q_{\mathrm{HW}}(d_2(t)) = 81.06 \cdot 10^{-6} \cdot d_2(t)$$

$$q_{\mathrm{C}}(u_4(t)) = 96.84 \cdot 10^{-6} \cdot u_4(t).$$

The cross sectional areas of the tanks TB and TS as well as the volume of their content depend upon the respective filling level:

$$A_{\mathrm{TB}}(x_1(t)) = \begin{cases} 0.07 & \text{if } x_1(t) > 0.26 \\ 0.0626 & \text{else} \end{cases}$$

$$V_{\mathrm{TB}}(x_1(t)) = \begin{cases} 0.07 \cdot x_1(t) - 1.9 \cdot 10^{-3} & \text{if } x_1(t) > 0.26 \\ 0.0626 \cdot x_1(t) & \text{else} \end{cases}$$

$$A_{\mathrm{TS}}(x_3(t)) = \begin{cases} 0.07 & \text{if } x_3(t) > 0.26 \\ 0.069 & \text{else} \end{cases}$$

$$V_{\mathrm{TS}}(x_3(t)) = \begin{cases} 0.07 \cdot x_3(t) - 0.26 \cdot 10^{-3} & \text{if } x_3(t) > 0.26 \\ 0.069 \cdot x_3(t) & \text{else.} \end{cases}$$

All remaining parameters of the model are summarized in Table 5.1. Due to physical restrictions the state-space is constrained as follows:

$$0.26\,\mathrm{m} \leq x_1 \leq 0.4\,\mathrm{m}, \quad 285.65\,\mathrm{K} \leq x_2 \leq 323.15\,\mathrm{K},$$
$$0.26\,\mathrm{m} \leq x_3 \leq 0.4\,\mathrm{m}, \quad 293.15\,\mathrm{K} \leq x_4 \leq 323.15\,\mathrm{K}.$$

Linearized Model. In a surrounding of the setpoint

$$\bar{x}_1 = 0.33\,\mathrm{m}, \quad \bar{x}_2 = 294.7\,\mathrm{K}, \quad \bar{x}_3 = 0.34\,\mathrm{m}, \quad \bar{x}_4 = 300.2\,\mathrm{K}$$

Table 5.1 Parameters of the nonlinear model of the thermofluid process

Parameter	Value	Unit	Meaning
ϑ_{C}	286.15	K	Temperature of the cooling medium
ϑ_{HW}	294.15	K	Temperature of the fresh water supply
H_{TB}	$35.8 \cdot 10^{-4}$	$\frac{\mathrm{Km}^3}{\mathrm{s}}$	Heating coefficient of the heating rods in TB
H_{TS}	$8.25 \cdot 10^{-4}$	$\frac{\mathrm{Km}^3}{\mathrm{s}}$	Heating coefficient of the heating rods in TS

and

$$\begin{aligned}
&\bar{u}_1 = 0.5, \quad \bar{u}_2 = 0.21, \quad \bar{u}_3 = 0.19, \quad \bar{u}_4 = 0.5, \\
&\bar{u}_5 = 0.5, \quad \bar{u}_6 = 0.29, \quad \bar{u}_7 = 0.22, \quad \bar{u}_8 = 0.5
\end{aligned}$$

the behavior of the interconnected tank system is described by the linearized model

$$\begin{aligned}
\dot{\boldsymbol{x}}(t) &= \boldsymbol{A}\boldsymbol{x}(t) + \boldsymbol{B}\boldsymbol{u}(t) + \boldsymbol{E}\boldsymbol{d}(t) \\
\boldsymbol{y}(t) &= \boldsymbol{x}(t)
\end{aligned}$$

with

$$\boldsymbol{A} = 10^{-3} \begin{pmatrix} -5.74 & 0 & 2.42 & 0 \\ -34.5 & -8.58 & 43.9 & 5.44 \\ 2.85 & 0 & -5.01 & 0 \\ -46.5 & 5.58 & 39.2 & -5.58 \end{pmatrix}$$

$$\boldsymbol{B} = 10^{-3} \begin{pmatrix} 2.30 & 0 & 0 & 0 \\ 0 & -38.9 & 0 & 0 \\ 0 & 0 & 2.59 & 0 \\ 0 & 0 & 0 & 35.0 \end{pmatrix}$$

$$\boldsymbol{E} = 10^{-3} \begin{pmatrix} 0 & 0 \\ 168.9 & 0 \\ 0 & 1.16 \\ 0 & -20.7 \end{pmatrix}.$$

6

Multi-agent Systems

J. Lunze, F. Allgöwer, M. Bürger, O. Demir, U. Helmke, A. von Heusinger, and R. Schuh

This chapter presents methods for the analysis of networked dynamic systems and the design of controllers for systems consisting of identical or similar subsystems. For multi-agent systems, the controllers have to introduce couplings into the overall system in order to be able to satisfy a common goal of the isolated subsystems. For physically coupled systems, distributed controllers should communicate to improve the overall performance.

Chapter Contents

6.1 Control Problems for Multi-agent Systems

Communication networks can be used as an important means to coordinate the activities within interconnected systems in order to make the subsystems reach a common goal or to improve the overall system performance. The structure of the systems considered in this chapter is depicted in Figs. 6.1 and 6.2. The subsystems ("agents") Σ_i have decentralized controllers C_i that represent local feedbacks with the aim to satisfy local goals prescribing the subsystem performance. The communication network is used to couple the decentralized controllers in order to improve the overall system behavior. The local controllers together with the communication network constitute a networked controller.

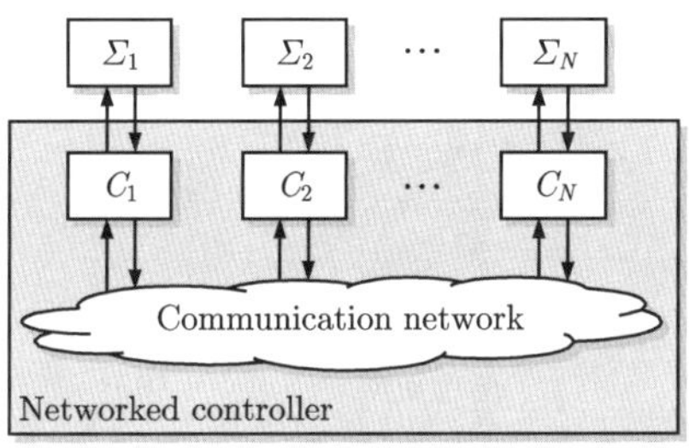

Fig. 6.1 Networked control of a multi-agent system

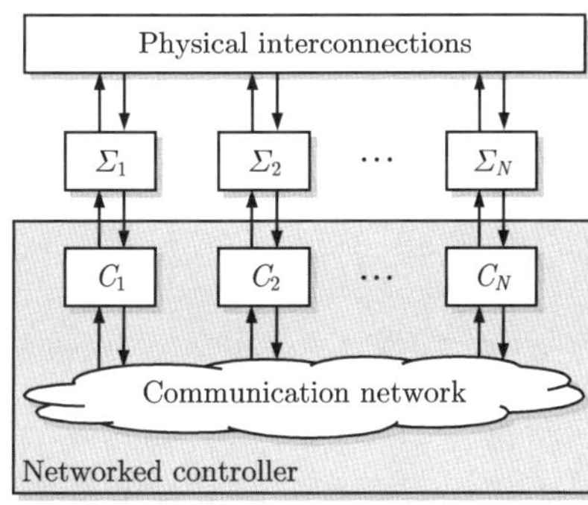

Fig. 6.2 Networked control of interconnected systems

An important aspect of the current research on multi-agent systems concerns the fact that the subsystems should be coordinated without an explicit "coordinator" component. No pre-decision should be made on the distribution of the functions of the individual agents within the group of all agents. Furthermore, each agent has only a limited access to sensor information, limited computing facilities and limited communication capabilities. These circumstances require both a distributed control law and distributed methods for designing the individual controllers of the agents (cf. the explanation of decentralized design in Section 3.1).

The main ideas have been developed in the recent literature for multi-agent systems that do not have physical interconnections but represent individual dynamic systems like robots or vehicles that have to accomplish a common task (Fig. 6.1). *Multi-agent systems* usually do not have physical couplings, but the agents have to be coupled by the networked controller because they have to satisfy a common goal. The main question asks how to structure the communication among the agents so as to ensure that the agents eventually assume a common state (consensus problem), follow a synchronous trajectory (synchronization problem) or achieve a common task (coordination problem). In literature, these problems are usually considered for agents with identical dynamics.

Physically coupled systems likewise can be controlled by a networked controller consisting of local control stations C_i at the subsystems and a communication network (Fig. 6.2). Here the networked controller introduces additional couplings into the overall system. Examples for physically interconnected systems with similar (ideally identical) subsystems are energy networks, multizone furnaces or adaptive optics.

The great challenge underlying the mentioned control problems for both classes of systems is to understand how the properties of the dynamic systems and the topological properties of the network influence the behavior of the overall networked system.

Consensus and Synchronization Problems. *Synchronization* is a very general notion to describe the phenomenon that the outputs of autonomous subsystems are adjusted by the interactions among the subsystems to generate a common behavior mode. In physics, synchronization is mainly dealt with as an analysis problem of nonlinear systems theory with an emphasis on the frequency entrainment of nonlinear or chaotic oscillators [281]. In the control literature, the focus has been laid on the design of distributed controllers. The synchronization problem has been dealt with mainly for a set of agents with identical dynamics, which start in different initial states $\boldsymbol{x}_{i0}$ and should join a common state trajectory $\boldsymbol{x}_{\rm s}(t)$, which is a solution of the agent model [328, 363]. The leader-follower structure has been investigated in [162, 242].

The control-theoretic view on synchronization problems, where the interactions among the agents are not pre-defined but should be found as the result of a design process, has brought about two interesting results:

- **Synchronizability:** The agents have to possess certain properties in order to be synchronizable by an appropriately chosen network. The property of synchronizability has been defined in [242, 398] for linear agents in such a way that a multi-agent system can be synchronized if and only if the agents are synchronizable. If this requirement is not satisfied, dynamic local controllers may make the extended agents synchronizable. The internal-reference principle proved in [243, 385] (cf. Section 6.2) shows which properties the local controllers have to introduce into the extended agents to satisfy the synchronizability requirement.
- **Connectedness of the communication topology:** The second prerequisite for synchronization is the existence of a communication network in which all agents are coupled with all other ones. In graph-theoretic terms, synchronization is related to connectivity properties of the graph [263, 278].

Distributed Control of Interconnected Systems. General control problems including the problems of stabilization of interconnected systems and disturbance attenuation have been considered for networked systems with the aim to find design methods for *distributed controllers*, where the controllers – like in a multi-agent system – introduce new couplings into the overall system with the aim to improve the overall system performance. Unlike in the methods that have been developed in the 1980's [188, 245, 339] for large-scale systems with decentralized control structures, information exchange among the control stations are allowed and the question has to be answered which information links should be implemented by the communication network in order to considerably improve the overall performance.

Chapter Overview. This chapter presents several new results on the control of multi-agent and interconnected systems:

- **Synchronization of agents with individual dynamics:** Section 6.2 extends the methods for the design of synchronizing controllers for identical agents towards agents with individual dynamics, linear parameter-variable agents and nonlinear systems. The basic idea results from an investigation of the synchronizability of linear agents for controllers that impose diffusive couplings among the agents. Accordingly, the agents have to satisfy an internal-reference principle or have to be made to do so by introducing appropriate local controllers.

- **Synchronizable subspaces** (Section 6.3): For agents with identical dynamics, the stabilization and synchronization problems can be dealt with in terms of the transfer functions of the subsystems and of the network. Conditions on the network structure in terms of the eigenvalues of the Laplacian matrix are derived for the stability or the synchronization of the overall networked system.

- **Cluster synchronization** (Section 6.4): Whereas for the networks with linear node dynamics and linear couplings investigated in Sections 6.2 and 6.3 synchronization refers to the overall systems as a whole, nonlinear saturating couplings may bring about clusters of nodes that synchronize locally while nodes in different clusters follow different trajectories. Section 6.4 describes conditions under which cluster synchronization occurs and methods for finding synchronized clusters and communities.

- **Autonomy and cooperation in networked systems:** Section 6.5 concerns multi-agent systems and physically coupled systems with identical dynamics. Methods are elaborated for describing which and when some information exchange should be temporarily invoked among the decentralized controllers. It is shown that the design problem for the overall controller can be reduced to design problems of the dimension of the subsystems.

6.2 Synchronization of Agents with Individual Dynamics

6.2.1 Synchronization Problem

Synchronization is an important phenomenon occurring in multi-agent systems. The overall system consists of a (large) number of subsystems that should have a coherent behavior, in which the subsystems follow the same dynamic pattern. In this section, synchronization is investigated as the control task of moving the subsystem outputs $\boldsymbol{y}_i(t)$, $(i = 1, 2, ..., N)$ onto a common trajectory $\boldsymbol{y}_{\rm s}(t)$:

$$\lim_{t \to \infty} \|\boldsymbol{y}_i(t) - \boldsymbol{y}_{\rm s}(t)\| = 0, \quad i = 1, 2, ..., N.$$

The research aim is to find methods for choosing the interactions among the autonomous agents, which are implemented by a digital communication network, for which the overall system becomes synchronized. Furthermore, the notion of synchronizability should be introduced as a property of the agents and methods to find local controllers that render the agents synchronizable are to be developed.

An important issue of synchronization is given by the fact that the agents should be able to follow the synchronous trajectory $\boldsymbol{y}_{\rm s}(t)$ due to their internal dynamics [294]. On the synchronous trajectory the interactions among the agents are not active and the agents generate their synchronous outputs autonomously. This characteristic distinguishes synchronization from many control problems, where a system is forced by an appropriate choice of the control input to follow a nominal trajectory. Nevertheless, this section will show that there is a direct connection between the internal-model principle of multivariable feedback control and a similar principle that enables synchronization. A necessary condition for synchronization requires the extended agents to include the dynamics of the synchronous output.

Synchronization Structures. The basic structures of synchronization are shown in Figs. 6.3 and 6.4. The agents Σ_i have local controllers C_i, which are interconnected over a digital communication network. The notion of the autonomous agent results from application fields, where the subsystems Σ_i are moving objects that collectively accomplish a common task. Application examples are multi-vehicle and multi-robot systems, reconnaissance systems, adaptive optics or sensor networks. As the agents do not physically interact, a coordinated movement can only be ensured if the local controllers communicate. As an important issue of many applications, the interactions among the local controllers should occur without a centralized coordinator. Thus, the overall system has a self-organized dynamics.

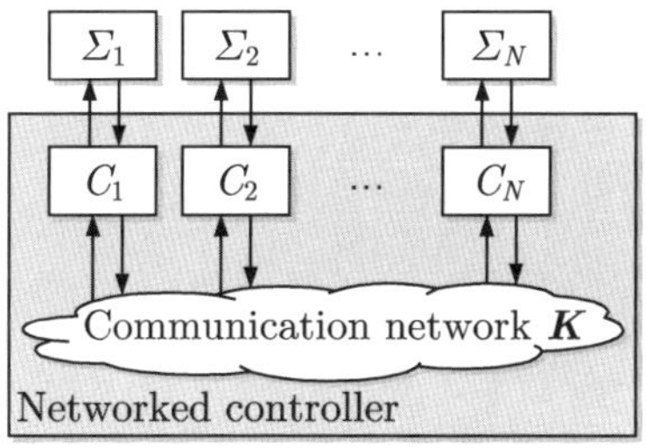

Fig. 6.3 Leaderless synchronization of autonomous agents

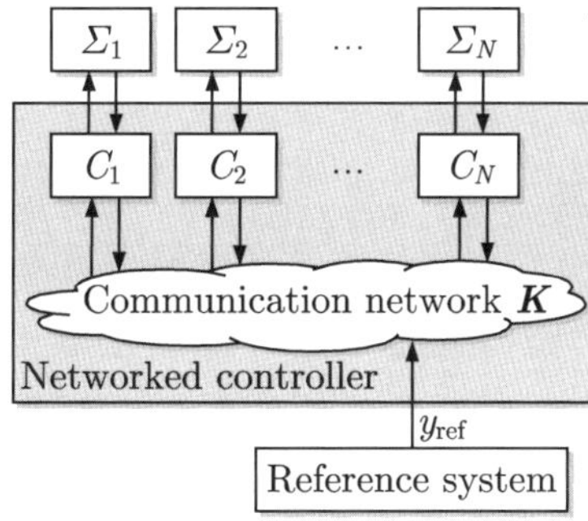

Fig. 6.4 Leader-follower synchronization of autonomous agents

The class of systems characterized so far is investigated here from the viewpoint of networked control, where the main questions ask which properties should the autonomous agents including their local controllers have and which communication is necessary for ensuring a synchronized movement of the agents. The communication structure is not predefined, but the analysis method developed in this section should show which communication links are necessary for synchronization. Thus, the synchronization problem has the interesting aspect that the choice of the communication structure is an important design step.

Two synchronization schemes will be considered:

- **Leaderless systems.** All agents participate in defining the synchronous trajectory $\boldsymbol{y}_{\mathrm{s}}(t)$, which occurs as a result of self-organization and which depends upon the communication structure (Fig. 6.3). Hence, one speaks about *leaderless synchronization* or *mutual synchronization* .

- **Leader-follower systems.** The agents should follow the synchronous trajectory $\boldsymbol{y}_{\mathrm{s}}(t)$, which is prescribed by a reference system (leader) (Fig. 6.4). Then the controlled agents (Σ_i, C_i) are called the followers. In literature, this structure is also referred to as *master-slave synchronization.*

Section Overview. In contrast to the majority of publications on consensus and synchronization of identical agents, this section deals with multi-agent systems in which the agents Σ_i have individual dynamics. Such systems cannot be synchronized by some static networked controller $\boldsymbol{K}$, but the synchronization problem has to be posed in the more general setting shown in Fig. 6.5. The agents Σ_i are extended by dynamic local units C_i, which are interconnected over a digital communication network $\boldsymbol{K}$. The local units together with the network represent the networked controller.

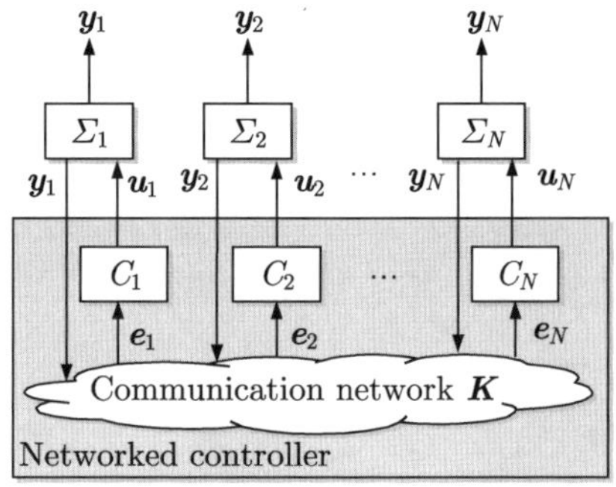

Fig. 6.5 Synchronization of autonomous agents

The section tackles three important problems:

- **Internal-reference principle for synchronization.** Considering agents with arbitrary linear dynamics, this section answers the question under what conditions these agents can be synchronized with a reference trajectory $\boldsymbol{y}_{\rm s}(t)$. The result is an internal-reference principle for synchronization, which claims that each agent Σ_i together with its local controller C_i has to include the model of the reference trajectory $\boldsymbol{y}_{\rm s}(t)$. In the leader-follower structure the reference system is defined by the leading agent, whereas in leaderless systems, the reference model is represented by the intersection of the agent dynamics (according to Definition 6.2). If an agent does not include the reference model, its local controller C_i can be structured so as to introduce this model into the controlled agent.

- **Synchronization test.** Necessary and sufficient conditions are derived under which the overall system is synchronized. In the general set-up considered here, this condition concerns the model of the overall system, but for cycle-free communication topologies this condition breaks down to separate conditions on the controlled agents.

- **Design of the communication structure.** The section derives requirements on the communication structure, which have to be satisfied in order to synchronize agents. In leaderless synchronization, the communication graph needs to have a spanning tree. Then the synchronous trajectory $\boldsymbol{y}_{\rm s}(t)$ also depends upon the communication topology. In leader-follower synchronization the reference system has to be the root node of a spanning tree. These structural constraints are known from the literature about consensus and the synchronization of agents with identical dynamics, but they are shown here to be necessary and sufficient for the synchronization of agents with individual linear dynamics.

6.2.2 Models

Agent Model and Networked Controller. The agents Σ_i have linear dynamics

$$\Sigma_i : \begin{cases} \dot{\boldsymbol{x}}_i(t) = \boldsymbol{A}_i \boldsymbol{x}_i(t) + \boldsymbol{B}_i \boldsymbol{u}_i(t), \quad \boldsymbol{x}_i(0) = \boldsymbol{x}_{i0}, \quad i = 1, 2, ..., N \\ \boldsymbol{y}_i(t) = \boldsymbol{C}_i \boldsymbol{x}_i(t) \end{cases} \tag{6.1}$$

with $\boldsymbol{u}_i(t) \in \mathbb{R}^{m_i}$ denoting the input, $\boldsymbol{x}_i(t) \in \mathbb{R}^{n_i}$ the state, and $\boldsymbol{y}_i(t) \in \mathbb{R}^r$ the output. Note that all agents have outputs of the same dimension. The matrices $\boldsymbol{A}_i$ are assumed to be diagonalizable.

> **Assumption 6.1.** *The communication among the agents is restricted to a transfer of the agent outputs $\boldsymbol{y}_i(t)$, $(i = 1, 2, ..., N)$ to other agents.*

The generalized synchronization error of the i-th agent is defined by

$$\boldsymbol{e}_i(t) = \sum_{j=1, j\neq i}^{N} k_{ij}(\boldsymbol{y}_j(t) - \boldsymbol{y}_i(t)) = \sum_{j=1}^{N} k_{ij} \boldsymbol{y}_j(t), \quad i = 1, 2, ..., N. \tag{6.2}$$

Non-vanishing parameters k_{ij} indicate communication links from Agent Σ_j towards Agent Σ_i and

$$k_{ii} = - \sum_{j=1, j\neq i}^{N} k_{ij} \tag{6.3}$$

holds. The elements k_{ij} introduced above form the network matrix $\boldsymbol{K} = (k_{ij})$. If the agents are synchronized, the synchronization errors $\boldsymbol{e}_i(t)$, $(i = 1, 2, ..., N)$ vanish and the agents have to generate the synchronous trajectory $\boldsymbol{y}_{\rm s}(t)$ without interactions (for $\boldsymbol{u}_i(t) = \boldsymbol{0}$).

The local units C_i are dynamic systems with $n_{{\rm r}i}$-dimensional internal state $\boldsymbol{x}_{{\rm r}i}(t)$:

$$C_i : \begin{cases} \dot{\boldsymbol{x}}_{{\rm r}i}(t) = \boldsymbol{A}_{{\rm r}i} \boldsymbol{x}_{{\rm r}i}(t) + \boldsymbol{B}_{{\rm r}i} \boldsymbol{e}_i(t), \; \boldsymbol{x}_{{\rm r}i}(0) = \boldsymbol{x}_{{\rm r}i0}, \; i = 1, 2, ..., N \\ \boldsymbol{u}_i(t) = \boldsymbol{K}_{{\rm r}i} \boldsymbol{x}_{{\rm r}i}(t) + \boldsymbol{K}_{{\rm e}i} \boldsymbol{e}_i(t). \end{cases} \tag{6.4}$$

The synchronous trajectory $\boldsymbol{y}_{\rm s}(t)$ is described by the *virtual reference system*

$$\Sigma_{\rm s} : \begin{cases} \dot{\boldsymbol{x}}_{\rm s}(t) = \boldsymbol{A}_{\rm s} \boldsymbol{x}_{\rm s}(t), \quad \boldsymbol{x}_{\rm s}(0) = \boldsymbol{x}_{{\rm s}0} \\ \boldsymbol{y}_{\rm s}(t) = \boldsymbol{C}_{\rm s} \boldsymbol{x}_{\rm s}(t) \end{cases} \tag{6.5}$$

with the $n_{\rm s}$-dimensional state $\boldsymbol{x}_{\rm s}(t)$, which defines the class of synchronous trajectories

$$\mathcal{Y}_{\rm s} = \left\{ \boldsymbol{C}_{\rm s} {\rm e}^{\boldsymbol{A}_{\rm s} t} \boldsymbol{x}_{{\rm s}0} \,|\, \boldsymbol{x}_{{\rm s}0} \in \mathbb{R}^{n_{\rm s}} \right\}. \tag{6.6}$$

To avoid trivial situations, $\boldsymbol{A}_{\rm s}$ is assumed to have eigenvalues with nonnegative real part.

The agent model (6.1) together with the local unit (6.4) yield

$$\Sigma_{0i}: \begin{cases} \begin{pmatrix} \dot{\boldsymbol{x}}_i(t) \\ \dot{\boldsymbol{x}}_{\mathrm{r}i}(t) \end{pmatrix} = \underbrace{\begin{pmatrix} \boldsymbol{A}_i & \boldsymbol{B}_i \boldsymbol{K}_{\mathrm{r}i} \\ \boldsymbol{O} & \boldsymbol{A}_{\mathrm{r}i} \end{pmatrix}}_{\boldsymbol{A}_{0i}} \underbrace{\begin{pmatrix} \boldsymbol{x}_i(t) \\ \boldsymbol{x}_{\mathrm{r}i}(t) \end{pmatrix}}_{\bar{\boldsymbol{x}}_i(t)} + \underbrace{\begin{pmatrix} \boldsymbol{B}_i \boldsymbol{K}_{\mathrm{e}i} \\ \boldsymbol{B}_{\mathrm{r}i} \end{pmatrix}}_{\boldsymbol{B}_{0i}} \boldsymbol{e}_i(t), \ \bar{\boldsymbol{x}}_i(0) = \bar{\boldsymbol{x}}_{i0} \\ \boldsymbol{y}_i(t) = \underbrace{(\boldsymbol{C}_i \ \ \boldsymbol{O})}_{\boldsymbol{C}_{0i}} \begin{pmatrix} \boldsymbol{x}_i(t) \\ \boldsymbol{x}_{\mathrm{r}i}(t) \end{pmatrix}, \end{cases} \tag{6.7}$$

which is called the *extended agent* and is assumed to be completely controllable and completely observable.

Model of the Overall System. The overall system is described by (6.2) and (6.7)

$$\Sigma_{\mathrm{LL}}: \begin{cases} \frac{\mathrm{d}}{\mathrm{d}t}\bar{\boldsymbol{x}}(t) = \bar{\boldsymbol{A}}\bar{\boldsymbol{x}}(t), \quad \bar{\boldsymbol{x}}(0) = \bar{\boldsymbol{x}}_0 \\ \boldsymbol{y}(t) = \bar{\boldsymbol{C}}\bar{\boldsymbol{x}}(t) \end{cases} \tag{6.8}$$

with

$$\bar{\boldsymbol{x}} = (\bar{\boldsymbol{x}}_1^{\mathrm{T}}, \ \bar{\boldsymbol{x}}_2^{\mathrm{T}}, \ ..., \ \bar{\boldsymbol{x}}_N^{\mathrm{T}})^{\mathrm{T}}$$

$$\bar{\boldsymbol{A}} = \begin{pmatrix} \boldsymbol{A}_{01} & & \\ & \ddots & \\ & & \boldsymbol{A}_{0N} \end{pmatrix} + \begin{pmatrix} \boldsymbol{B}_{01} & & \\ & \ddots & \\ & & \boldsymbol{B}_{0N} \end{pmatrix} (\boldsymbol{K} \otimes \boldsymbol{I}_r) \begin{pmatrix} \boldsymbol{C}_{01} & & \\ & \ddots & \\ & & \boldsymbol{C}_{0N} \end{pmatrix} \tag{6.9}$$

$$\bar{\boldsymbol{C}} = \begin{pmatrix} \boldsymbol{C}_{01} & & \\ & \ddots & \\ & & \boldsymbol{C}_{0N} \end{pmatrix}. \tag{6.10}$$

Since the extended agents Σ_{0i} are assumed to be completely observable, all pairs $(\boldsymbol{A}_{0i}, \boldsymbol{C}_{0i})$ appearing in Eqs. (6.9) and (6.10) are completely observable. As output feedback cannot alter the observability of a closed loop [246], the pair $(\bar{\boldsymbol{A}}, \bar{\boldsymbol{C}})$ is completely observable as well.

Definition 6.1. (Synchronization)
The system (6.8) with $N \geq 2$ agents is said to be synchronized if the following requirements are satisfied:

1. **Synchronous behavior:** *For specific initial states $\bar{\boldsymbol{x}}_{i0}$ of the agents, all outputs $\boldsymbol{y}_i(t)$ follow a common trajectory $\boldsymbol{y}_{\mathrm{s}}(t)$:*

$$\boldsymbol{y}_1(t) = \boldsymbol{y}_2(t) = ... = \boldsymbol{y}_N(t) = \boldsymbol{y}_{\mathrm{s}}(t), \quad t \geq 0. \tag{6.11}$$

2. **Asymptotic synchronization:** *For all other initial states, the agents asymptotically approach the same trajectory $\boldsymbol{y}_{\mathrm{s}}(t) \in \mathcal{Y}_{\mathrm{s}}$:*

$$\lim_{t\to\infty} \|\boldsymbol{y}_i(t) - \boldsymbol{y}_{\mathrm{s}}(t)\| = 0, \quad i = 1, 2, ..., N. \tag{6.12}$$

Communication Topologies. The communication topology used can be represented by a directed graph $\mathcal{G} = (\mathcal{V}, \mathcal{E}, \mathcal{K})$ with the set $\mathcal{V} = \{1, 2, ..., N\}$ of verticies representing the agents. A directed edge $(i, j) \in \mathcal{E}$ exists if and only if the coupling coefficient k_{ij} used in Eq. (6.2) does not vanish. The Laplacian matrix of the directed graph $\mathcal{G}$ is identical to the matrix $-\boldsymbol{K}$ of the networked controller. A spanning tree of $\mathcal{G}$ is a directed tree formed by graph edges that connect all nodes of the graph. If $\mathcal{G}$ does not have any cycle (closed path), the communication structure is said to be cycle-free.

6.2.3 Internal-Reference Principle

This section shows that synchronization may only occur if all agents have certain dynamic properties in common. These properties are represented by the virtual reference system $\Sigma_{\rm s}$.

System Intersection. The following paragraph defines the notion of system intersection to describe systems having some dynamics in common. It considers the system $\Sigma_{\rm s}$ defined by Eq. (6.5) together with

$$\Sigma_1 : \begin{cases} \dot{\boldsymbol{x}}_1(t) = \boldsymbol{A}_1 \boldsymbol{x}_1(t), \; \boldsymbol{x}_1(0) = \boldsymbol{x}_{10} \\ \boldsymbol{y}_1(t) = \boldsymbol{C}_1 \boldsymbol{x}_1(t) \end{cases} \tag{6.13}$$

$$\Sigma_2 : \begin{cases} \dot{\boldsymbol{x}}_2(t) = \boldsymbol{A}_2 \boldsymbol{x}_2(t), \; \boldsymbol{x}_2(0) = \boldsymbol{x}_{20} \\ \boldsymbol{y}_2(t) = \boldsymbol{C}_2 \boldsymbol{x}_2(t) \end{cases} \tag{6.14}$$

with $\dim \boldsymbol{x}_{\rm s} = n_{\rm s}$, $\dim \boldsymbol{x}_1 = n_1$, $\dim \boldsymbol{x}_2 = n_2$, $\dim \boldsymbol{y}_1 = \dim \boldsymbol{y}_2 = \dim \boldsymbol{y}_{\rm s} = r$ and $n_{\rm s} \leq n_1, n_2$.

Definition 6.2. [243] *The systems Σ_1 and Σ_2 are said to have the intersection $\Sigma_{\rm s}$ (in symbols: $\Sigma_1 \cap \Sigma_2 \supseteq \Sigma_{\rm s}$) if for every initial state $\boldsymbol{x}_{\rm s0} \in \mathbb{R}^{n_{\rm s}}$ there exist initial states $\boldsymbol{x}_{10} \in \mathbb{R}^{n_1}$ and $\boldsymbol{x}_{20} \in \mathbb{R}^{n_2}$ such that the relation*

$$\boldsymbol{y}_1(t) = \boldsymbol{y}_2(t) = \boldsymbol{y}_{\rm s}(t), \quad t \geq 0 \tag{6.15}$$

holds for the solutions of Eqs. (6.5), (6.13) and (6.14).

For the output sets (6.6),

$$\mathcal{Y}_1 = \{\boldsymbol{C}_1 {\rm e}^{\boldsymbol{A}_1 t} \boldsymbol{x}_{10} \,|\, \boldsymbol{x}_{10} \in \mathbb{R}^{n_1}\} \tag{6.16}$$

$$\mathcal{Y}_2 = \{\boldsymbol{C}_2 {\rm e}^{\boldsymbol{A}_2 t} \boldsymbol{x}_{20} \,|\, \boldsymbol{x}_{20} \in \mathbb{R}^{n_2}\} \tag{6.17}$$

system intersection is equivalent to $\mathcal{Y}_{\rm s} \subseteq \mathcal{Y}_1 \cap \mathcal{Y}_2$. If $\mathcal{Y}_1 \cap \mathcal{Y}_2 = \{0\}$ holds, where 0 is the vanishing signal, the intersection is said to be empty: $\Sigma_1 \cap \Sigma_2 = \emptyset$.

Theorem 6.1. (System intersection) [243]
The system $\Sigma_{\rm s}$ *with dimension* $n_{\rm s} \geq 1$ *is an intersection of the systems* Σ_1 *and* Σ_2 *with diagonalizable matrix* $\boldsymbol{A}_i$, $(i = 1, 2)$ *if and only if there exist a regular* $(n_1 \times n_1)$*-matrix* $\boldsymbol{T}_1$ *and a regular* $(n_2 \times n_2)$*-matrix* $\boldsymbol{T}_2$ *such that the relations*

$$\boldsymbol{T}_i^{-1}\boldsymbol{A}_i\boldsymbol{T}_i = \begin{pmatrix} \boldsymbol{A}_{\rm s} & \boldsymbol{O} \\ \boldsymbol{O} & \boldsymbol{A}_{{\rm p}i} \end{pmatrix}, \quad \boldsymbol{C}_i\boldsymbol{T}_i = (\boldsymbol{C}_{\rm s} \;\; \boldsymbol{C}_{{\rm p}i}), \quad i = 1, 2 \tag{6.18}$$

hold, where $\boldsymbol{A}_{\rm p1}$, $\boldsymbol{A}_{\rm p2}$, $\boldsymbol{C}_{\rm p1}$ *and* $\boldsymbol{C}_{\rm p2}$ *are matrices of appropriate dimensions.*

For the synchronization analysis, the *maximum intersection* $\Sigma_{\rm s}^*$ of a given system pair (Σ_1, Σ_2) should be determined, which is the system $\Sigma_{\rm s} = \Sigma_1 \cap \Sigma_2$ with the largest dynamic order $n_{\rm s}$. Obviously, a maximum intersection exists and is unique up to state transformation. The proof of Theorem 6.1 in [243] implies a constructive way for its determination.

If a set of systems Σ_i, $(i = 1, ..., N)$ is considered, a consecutive application of the intersection operation yields the common intersection of all systems of this set: $\Sigma_\cap = \cap_{i=1}^N \Sigma_i$.

Internal-Reference Principle for Synchronization. In the synchronization structure shown in Fig. 6.5 the synchronous trajectory $\boldsymbol{y}_{\rm s}(t)$ occurs as a result of a "consensus" among the agents. It will be shown first that in order to be able to agree about a common trajectory $\boldsymbol{y}_{\rm s} \in \mathcal{Y}_{\rm s}$ the agents must have a common intersection.

Theorem 6.2. (Internal-reference principle for synchronization) [243]
If the overall system $\Sigma_{\rm LL}$ *is synchronized, the extended agents* Σ_{0i}, $(i = 1, 2, ..., N)$ *have a nonempty intersection* $\Sigma_{\rm s}$

$$\Sigma_{\rm s} = \left(\cap_{i=1}^N \Sigma_{0i}\right)^* \neq \emptyset. \tag{6.19}$$

The intersection $\Sigma_{\rm s}$ is said to be the *internal reference model* of the synchronous trajectory because it parametrizes the set $\mathcal{Y}_{\rm s}$ of synchronous trajectories according to Eq. (6.6). Theorem 6.2 requires that all individual agents have to be able to follow the synchronous trajectory $\boldsymbol{y}_{\rm s}(t)$ for $\boldsymbol{e}_i(t) = \boldsymbol{0}$. This principle is similar to the well known internal-model principle of multivariable control, which requires the open-loop system to possess a model of the external signals if the closed-loop system should follow asymptotically a given command signal. However, the difference of both principles lie in the structures considered in both situations. Whereas in multivariable control the external signals are generated by a given model (or are at least assumed to be generated in this way), the reference model generating the synchronous trajectory need not be a given linear system of the form (6.5) but occurs as the intersection of the agents. $\Sigma_{\rm s}$ is, therefore, referred to as a virtual reference system, which does not appear as a system of its own in the network shown in Fig. 6.3.

The theorem has two important consequences for the design of the local units:

- If the agents Σ_i do not have a common intersection ($\cap_{i=1}^{N} \Sigma_i = \emptyset$) they cannot be synchronized by any static networked controller ($\dim x_{{\rm r}i} = 0$).
- If the synchronous trajectory should belong to some specific class $\mathcal{Y}_{\rm s}$ of trajectories, then the local units C_i have to ensure that the corresponding reference model $\Sigma_{\rm s}$, which generates this class of outputs, is included in all extended agents Σ_{0i}, $(i = 1, 2, ...N)$.

6.2.4 Asymptotic Synchronization

As the first step to get a synchronization condition, the following lemma describes the synchronization errors $\boldsymbol{e}_{i1}(t) = \boldsymbol{y}_1(t) - \boldsymbol{y}_i(t)$, $(i = 2, 3, ..., N)$ which form the vector

$$\boldsymbol{e}_{\Delta}(t) = (\boldsymbol{e}_{21}(t)^{\rm T} \;\; \boldsymbol{e}_{31}(t)^{\rm T} \; ... \; \boldsymbol{e}_{N1}(t)^{\rm T})^{\rm T}. \tag{6.20}$$

Lemma 6.1. *Assume that the extended agents Σ_{0i} have a nonempty intersection $\Sigma_{\rm s}$. Then the synchronization error (6.20) appearing in the overall system (6.8) is described by*

$$\Sigma_{\Delta}: \;\; \begin{cases} \frac{\rm d}{{\rm d}t}\breve{\boldsymbol{x}}(t) = \bar{\boldsymbol{A}}_{\rm LL}\breve{\boldsymbol{x}}(t), & \breve{\boldsymbol{x}}(0) = \breve{\boldsymbol{x}}_0 \\ \boldsymbol{e}_{\Delta}(t) = \bar{\boldsymbol{C}}_{\rm LL}\breve{\boldsymbol{x}}(t) \end{cases} \tag{6.21}$$

with

$$\bar{\boldsymbol{A}}_{\rm LL} = \begin{pmatrix} \boldsymbol{A}_{\rm p1} & & & \\ & \boldsymbol{A}_{02} & & \\ & & \ddots & \\ & & & \boldsymbol{A}_{0N} \end{pmatrix} \tag{6.22}$$

$$+ \begin{pmatrix} \boldsymbol{B}_{\rm p1} & \boldsymbol{O} \; ... & \boldsymbol{O} \\ -\boldsymbol{T}_2\hat{\boldsymbol{B}}_{\rm q1} & \boldsymbol{B}_{02} & \\ \vdots & & \ddots \\ -\boldsymbol{T}_N\hat{\boldsymbol{B}}_{\rm q1} & & \boldsymbol{B}_{0N} \end{pmatrix} (\boldsymbol{K} \otimes \boldsymbol{I}_r) \begin{pmatrix} \boldsymbol{C}_{\rm p1} & & & \\ & \boldsymbol{C}_{02} & & \\ & & \ddots & \\ & & & \boldsymbol{C}_{0N} \end{pmatrix}$$

$$\hat{\boldsymbol{B}}_{\rm q1} = \begin{pmatrix} \boldsymbol{B}_{\rm q1} \\ \boldsymbol{O} \end{pmatrix}, \quad \bar{\boldsymbol{C}}_{\rm LL} = \begin{pmatrix} \boldsymbol{C}_{\rm p1} & -\boldsymbol{C}_{02} & & \\ \vdots & & \ddots & \\ \boldsymbol{C}_{\rm p1} & & & -\boldsymbol{C}_{0N} \end{pmatrix}. \tag{6.23}$$

The matrices $\boldsymbol{T}_i$, $\boldsymbol{B}_{\mathrm{p1}}$, $\boldsymbol{B}_{\mathrm{q1}}$ and $\boldsymbol{C}_{\mathrm{p1}}$ result from the decomposition of the agents with respect to their common intersection. Reference [243] gives a relation between the state $\breve{\boldsymbol{x}}(t)$ of the system (6.21) and the state $\bar{\boldsymbol{x}}(t)$ of the overall system (6.8) and shows that the model (6.21) may have any initial state $\breve{\boldsymbol{x}}_0$. The latter fact directly leads to the following synchronization condition:

Theorem 6.3. (Synchronization) [243]
A set of extended agents Σ_{0i} described by Eq. (6.7) is synchronized if and only if

1. *there exists a nonempty intersection*

$$\Sigma_{\mathrm{s}} = \left(\cap_{i=1}^{N} \Sigma_{0i}\right)^* \neq \emptyset \tag{6.24}$$

of all extended agents Σ_{0i} and

2. *the matrix $\bar{\boldsymbol{A}}_{\mathrm{LL}}$ defined in Eq. (6.22) is asymptotically stable.*

If the communication graph includes cycles, the stability of the agent matrices

$$\bar{\boldsymbol{A}}_i = \boldsymbol{A}_{0i} + k_{ii}\boldsymbol{B}_{0i}\boldsymbol{C}_{0i} \tag{6.25}$$

is not sufficient for synchronization, but the matrix $\bar{\boldsymbol{A}}_{\mathrm{LL}}$ given in (6.22) has to be checked. However, the synchronization analysis described in Theorem 6.3 is considerably simplified if the communication graph does not have any cycle. Then the agents can be enumerated such that the Laplacian matrix $-\boldsymbol{K}$ is a lower triangular matrix. Due to Eq. (6.3), the relation $k_{11} = 0$ holds. Consequently, the matrix $\bar{\boldsymbol{A}}_{\mathrm{LL}}$ is upper block-triangular with the diagonal blocks $\boldsymbol{A}_{\mathrm{p1}}$, $\bar{\boldsymbol{A}}_2$, $\bar{\boldsymbol{A}}_3$,..., $\bar{\boldsymbol{A}}_N$ (cf. Eqs. (6.22) and (6.25)). The diagonal blocks have to be asymptotically stable for the second condition of Theorem 6.3 to be satisfied. This condition implies that the diagonal elements k_{22}, k_{33},..., k_{NN} of $\boldsymbol{K}$ must not vanish, which means that the communication graph must have a spanning tree with Σ_1 as root node.

Corollary 6.1. (Synchronization with cycle-free communication) *The overall system (6.8) with cycle-free communication structure is synchronized if and only if*

- *the extended agents Σ_{0i}, $(i = 1, 2, ..., N)$ have a common intersection (6.24),*
- *the communication graph $\mathcal{G}$ has a spanning tree with the root node enumerated as Agent 1,*
- *the matrix $\boldsymbol{A}_{\mathrm{p1}}$ of the root agent is asymptotically stable and*
- *the matrices (6.25), $(i = 2, 3, ..., N)$ are asymptotically stable.*

Corollary 6.1 has the important implication that any set of agents with a non-empty common intersection is synchronizable. As the extended agents Σ_{0i} are controllable and observable, their local units can be chosen by well-known multivariable control methods [246] so as to stabilize them.

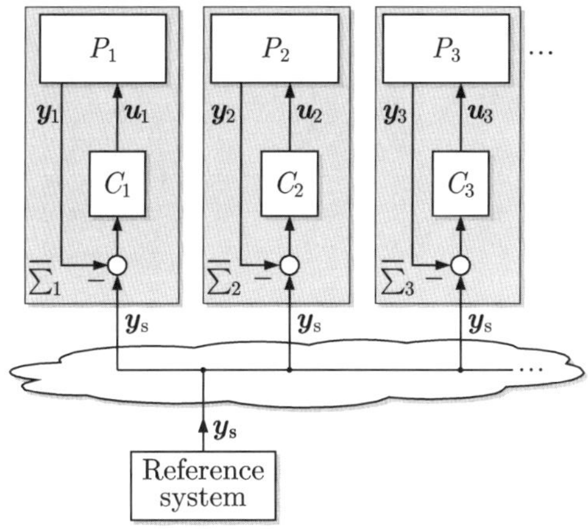

Fig. 6.6 Leader-follower synchronization with decentralized networked controller

The corollary can also be considered from the opposite viewpoint. Given any set of asymptotically stable controlled agents (6.25). If all agents are connected by a cycle-free communication structure including a rooted tree, the overall system synchronizes at a trajectory that has the dynamics of the common intersection of the extended agents Σ_{0i} and is generated by the first agent.

Synchronization by Decentralized Control. Decentralized control has a specific control structure, which satisfies the requirements stated above (Fig. 6.6). All agents get the reference trajectory $\boldsymbol{y}_s(t)$ communicated.

All followers are synchronized with the leader if and only if they include the reference model Σ_s and the matrices

$$\bar{\boldsymbol{A}}_{ii} = \boldsymbol{A}_{0i} - \boldsymbol{B}_{0i}\boldsymbol{C}_{0i}, \quad i = 1, 2, ..., N$$

are asymptotically stable. Note that the local controllers C_i can be designed for the agents Σ_i independently of each other.

Synchronization with Communication among Neighboring Agents. Another specific control structure, which satisfies the requirements on the communication graph described above, is depicted in Fig. 6.7. Only the first agent receives the reference trajectory $\boldsymbol{y}_s(t)$ and all further agents are informed about the output $\boldsymbol{y}_i(t)$ of the preceding agent.

Again, all followers are synchronized with the leader if and only if they include the reference model Σ_s and the matrices

$$\bar{\boldsymbol{A}}_{ii} = \boldsymbol{A}_{0i} - \boldsymbol{B}_{0i}\boldsymbol{C}_{0i}, \quad i = 1, 2, ..., N$$

are asymptotically stable.

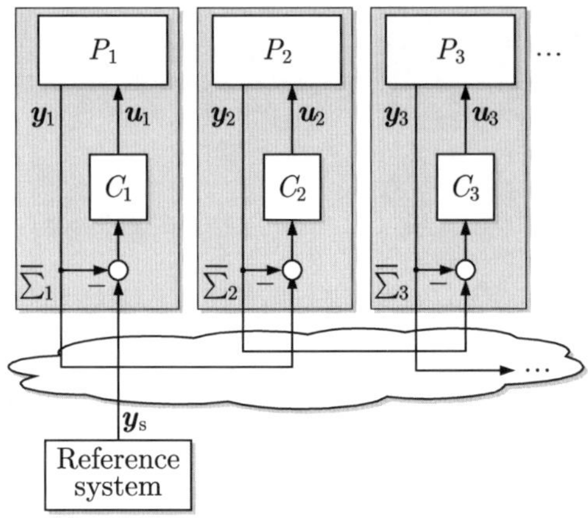

Fig. 6.7 Leader-follower synchronization with information exchange among neighboring agents

Example 6.1 *Distance control in a vehicle platoon*

As an example, consider the vehicle platoon shown in Fig. 6.8. The vehicles should be synchronized with respect the trajectory $y_s(t) = s_0 + \bar{v}t$, which is generated by the virtual reference system

$$\Sigma_s : \begin{cases} \begin{pmatrix} \dot{x}_{s1}(t) \\ \dot{x}_{s2}(t) \end{pmatrix} = \underbrace{\begin{pmatrix} 0 & 1 \\ 0 & 0 \end{pmatrix}}_{\boldsymbol{A}_s} \begin{pmatrix} x_{s1}(t) \\ x_{s2}(t) \end{pmatrix}, \quad \begin{pmatrix} x_{s1}(0) \\ x_{s2}(0) \end{pmatrix} = \begin{pmatrix} s_0 \\ \bar{v} \end{pmatrix} \\ y_s(t) = \underbrace{(1 \quad 0)}_{\boldsymbol{C}_s} \begin{pmatrix} x_{s1}(t) \\ x_{s2}(t) \end{pmatrix}. \end{cases} \tag{6.26}$$

The initial states of the reference system fixes the initial position s_0 and the common velocity $\bar{v}$. The reference model has two vanishing eigenvalues.

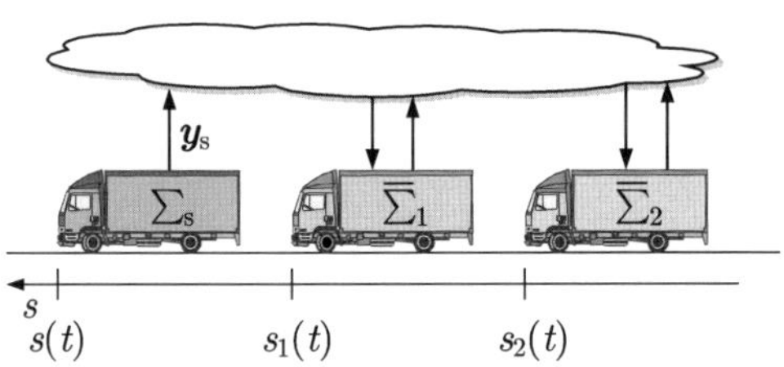

Fig. 6.8 Vehicle platoon

In a platoon, the i-th vehicle should follow this trajectory with the deviation $\bar{s}_i$:

$$y_i(t) = s_i(t) - \bar{s}_i, \quad i = 1, 2, ..., N.$$

The networked controller to be designed should bring the vehicles from their initial positions $s_i(0) = s_{i0}$, $(i = 1, 2, ..., N)$ to the reference trajectory $y_i(t) = y_s(t)$ and, hence, to the position $s_i(t) = y_s(t) + \bar{s}_i$.

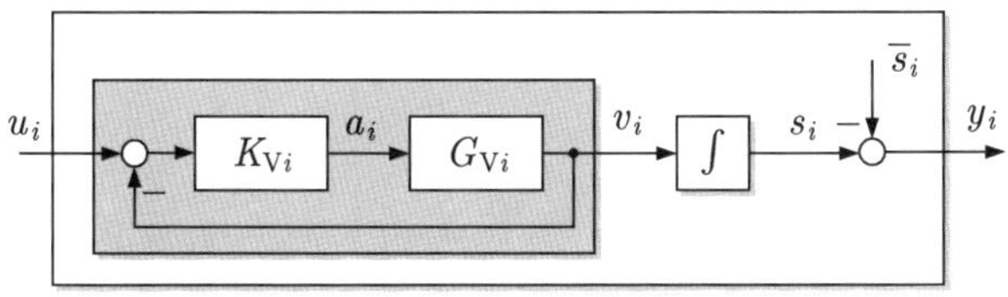

Fig. 6.9 Structure of the vehicle model Σ_i

According to Fig. 6.9, the vehicles considered here have a velocity controller K_{Vi} that adapts the velocity $v_i(t)$ to the reference velocity $u_i(t)$, which is the input used by the networked controller to be designed. The signal $a_i(t)$ determined by the velocity controller K_{Vi} is the acceleration, which is the input to the vehicle model G_{Vi}. The position $s_i(t)$ is obtained by integrating the velocity $v_i(t)$. Hence, the feedback loop of the i-th vehicle with the single input $u_i(t)$ and the single output $v_i(t)$ can be described by some state-space model

$$\begin{aligned} \dot{\boldsymbol{x}}_{ai}(t) &= \boldsymbol{A}_{ai}\boldsymbol{x}_{ai}(t) + \boldsymbol{b}_{ai}u_i(t) \\ v_i(t) &= \boldsymbol{c}_{ai}^{\mathrm{T}}\boldsymbol{x}_{ai}(t) \end{aligned}$$

with appropriate parameters.

As the velocity controller has an integral part, for constant input u_i the relation $|v_i(t) - u_i| \to 0$ holds for $t \to \infty$. For the output $y_i(t)$ of the i-th vehicle the differential equation $\dot{y}_i(t) = \dot{s}_i(t) = \boldsymbol{c}_{ai}^{\mathrm{T}}\boldsymbol{x}_{ai}(t)$ is obtained. Hence, the i-th vehicle has the state-space model

$$\Sigma_i : \left\{ \begin{aligned} \begin{pmatrix} \dot{y}_i(t) \\ \dot{\boldsymbol{x}}_{ai}(t) \end{pmatrix} &= \underbrace{\begin{pmatrix} 0 & \boldsymbol{c}_{ai}^{\mathrm{T}} \\ \boldsymbol{0} & \boldsymbol{A}_{ai} \end{pmatrix}}_{\boldsymbol{A}_i} \underbrace{\begin{pmatrix} y_i(t) \\ \boldsymbol{x}_{ai}(t) \end{pmatrix}}_{\boldsymbol{x}_i(t)} + \underbrace{\begin{pmatrix} 0 \\ \boldsymbol{b}_{ai} \end{pmatrix}}_{\boldsymbol{B}_i} u_i(t) \\ \begin{pmatrix} y_i(0) \\ \boldsymbol{x}_{ai}(0) \end{pmatrix} &= \underbrace{\begin{pmatrix} y_{i0} \\ \boldsymbol{x}_{ai0} \end{pmatrix}}_{\boldsymbol{x}_{i0}} \\ y_i(t) &= \underbrace{(1 \;\; \boldsymbol{0}^{\mathrm{T}})}_{\boldsymbol{C}_i} \begin{pmatrix} y_i(t) \\ \boldsymbol{x}_{ai}(t) \end{pmatrix}. \end{aligned} \right. \tag{6.27}$$

The vehicles distinguish with respect to the dynamics represented by the model G_{Vi} and the velocity controller K_{Vi} and, thus, have different model parameters in the matrix $\boldsymbol{A}_{ai}$ and the vectors $\boldsymbol{b}_{ai}$ and $\boldsymbol{c}_{ai}^{\mathrm{T}}$. However, the structure of the model shown in Fig. 6.9 is the same for all vehicles.

Application of the Internal-Reference Principle. The vehicles do not include the reference system $\Sigma_{\rm s}$ because they have only one vanishing eigenvalue whereas the reference system has two of them. Hence, the local controllers C_i have to include one vanishing eigenvalue. For the control law

$$C_i : \begin{cases} \dot{x}_{{\rm r}i}(t) = b_{{\rm r}i} e_i(t), \quad x_{{\rm r}i}(0) = x_{{\rm r}i0} \\ u_i(t) = k_{{\rm r}i} x_{{\rm r}i}(t) + k_{{\rm e}i} e_i(t) \end{cases} \tag{6.28}$$

with the scalar controller state $x_{{\rm r}i}(t)$, the extended vehicles have the model

$$\Sigma_{0i} : \begin{cases} \begin{pmatrix} \dot{\boldsymbol{x}}_i(t) \\ \dot{x}_{{\rm r}i}(t) \end{pmatrix} = \underbrace{\begin{pmatrix} \boldsymbol{A}_i & \boldsymbol{B}_i k_{{\rm r}i} \\ \boldsymbol{0}^{\rm T} & 0 \end{pmatrix}}_{\boldsymbol{A}_{0i}} \underbrace{\begin{pmatrix} \boldsymbol{x}_i(t) \\ x_{{\rm r}i}(t) \end{pmatrix}}_{\bar{\boldsymbol{x}}_i(t)} + \underbrace{\begin{pmatrix} \boldsymbol{B}_i k_{{\rm e}i} \\ b_{{\rm r}i} \end{pmatrix}}_{\boldsymbol{B}_{0i}} e_i(t) \\ y_i(t) = \underbrace{(\boldsymbol{C}_i \;\; 0)}_{\boldsymbol{C}_{0i}} \begin{pmatrix} \boldsymbol{x}_i(t) \\ x_{{\rm r}i}(t) \end{pmatrix}. \end{cases} \tag{6.29}$$

By means of a state transformation it can be shown that the extended model includes the reference system. Note that the controller structure does not depend upon the vehicle parameters. The structure of the model shown in Fig. 6.9 is responsible for the result that the controller has to include only one of the eigenvalues of the reference system and not both of them. In particular, the internal model remains the same if a more detailed model $(\boldsymbol{A}_{{\rm a}i}, \boldsymbol{b}_{{\rm a}i}, \boldsymbol{c}_{{\rm a}i}^{\rm T})$ is used for the vehicle dynamics.

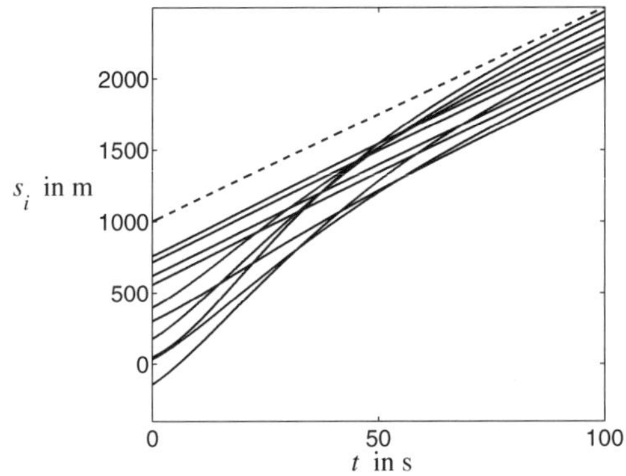

Fig. 6.10 Positions $s_i(t)$ of ten vehicles with decentralized control

Decentralized Networked Controller. For decentralized control with the direct communication of the position of the leader to all vehicles (Fig. 6.6), Fig. 6.10 shows the asymptotic synchronization of the vehicle positions $s_i(t)$. The vehicles simultaneously approach their final position in the platoon. After 100 s the transient behavior is finished. This is roughly the same time interval in which the step response of each vehicle is settled for the parameter values used.

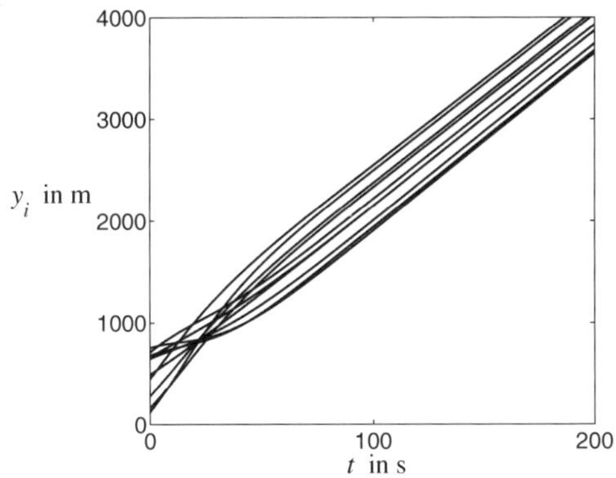

Fig. 6.11 Leaderless synchronization of the vehicles with all-to-all communication

Leaderless Synchronization. In the leaderless synchronization the vehicles have to find a "consensus" about the trajectory on which they synchronize. This means that vehicles have to adapt their velocity to the preceding and to the following vehicles. Each vehicle participates in determining the initial values s_0 and $\bar{v}$ of the trajectory prescribed by the virtual reference system (6.26).

Figure 6.11 shows the outputs $y_i(t)$ of ten vehicles, where all vehicles transfer their output to all others ("all-to-all communication"). The synchronization is ensured, because the matrix $\bar{\boldsymbol{A}}_{\rm LL}$ determined for the ten vehicles according to Eq. (6.22) is asymptotically stable (cf. Theorem 6.3). □

6.2.5 Controller Design for Synchronization

An alternative way of solution to the synchronization problem for agents with individual dynamics results from the following idea proposed in [383, 385]. Every agent is equiped with a dynamic controller C_i that includes the reference system $\Sigma_{\rm s}$, which is called the exosystem. Then all extended agents Σ_{0i}, $(i = 1, 2, .., N)$ have at least the dynamics of $\Sigma_{\rm s}$ in common and trivially satisfy the internal-reference principle. Now the question occurs under what conditions the controlled agents can be synchronized.

An answer to this question can be given in analogy to the well-known internal-model principle of multivariable feedback control. It is known that an asymptotically stable control loop follows an external reference trajectory if the open-loop system includes a model of the reference trajectory. In synchronization, the reference trajectory is given by the weighted sum of the outputs of the neighboring agents. Consequently, the remaining task is to stabilize the extended agents.

Stabilization is possible if the information structure of the controlled agents is extended with respect to Fig. 6.5. For example, if internally a state feedback is allowed for all agents in addition to the output feedback used above, stabilization is possible under the well-known conditions on the exosystem and the agent dynamics.

Alternatively, if Assumption 6.1 on the communications among the agents is released and the state of the exosystem is allowed to be sent to the neighboring agents, synchronization can be ensured in the following way. First, a feedback of the state or some output of the exosystem is applied such that the exosystems of all agents synchronize. Then the exosystem of each agent is interpreted as an individual generator for the reference trajectory of the agent and a local controller is designed such that the agent follows its own exosystem. This way of solution, however, necessitates the exchange of "auxiliary" information including the state of the exosystem rather than the exchange of the agent outputs, which in term make it necessary that all agents are extended by the same exosystem.

Bibliographical Notes

There are several extensions of the synchronization problem with respect to the properties of the agents and the structure of the networked controllers that should be mentioned here. The internal-model principle developed in [385] can be extended to nonlinear systems as shown in [384]. A necessary condition for the synchronizability of agents with individual dynamics is the existence of a synchronous steady state.

Extensions to the synchronization of linear parameter-varying systems (LPV systems) are described in [334, 335].

In references [97, 333] event-based control strategies are developed for the synchronization of autonomous agents. Here the networked controller uses spanning tree a communication link only after an event generator has indicated that the synchronization error exceeds a certain bound. These strategies transfer the idea of event-based control described in detail in Chapter 5 to synchronization problems.

6.3 Synchronizing Subspaces and Geometric Control

6.3.1 Motivation and Earlier Results

Recently, the synchronization problem for networks of systems has drawn increasing attention in the literature, due to the fact that the synchronization phenomenon appears frequently in engineering and natural sciences, both as a design objective as well as an unwanted ancillary effect.

We restrict ourselves to the special case of a homogeneous network of linear systems, e.g., all interconnected systems have the same linear, time-invariant dynamics. The synchronization problem for networks of this type has been studied for instance in [240],[249],[328],[362] and [397]. In these papers, several aspects of networks of linear systems are considered, for example, synchronizability through networks, networks with communication delay, networks that change over time, etc. Our aim is to highlight the link between stabilizability of networks and synchronizability of networks and to employ both frequency domain and state space analysis. Stabilizability of homogenous networks of linear systems is interesting in itself, and some properties of homogenous networks can be conviently described using frequency domain methods, in particular, using generalized frequency method suggested in [165]. A generalization of this approach to heterogeneous networks has been recently given in [129]. Moving into the slightly different, though still related, direction of formation control, we also mention our recent joint work [5], where a full Morse-theoretic analysis of formation control problems of N agents on a line is given.

We consider networks that consist of N node systems, $i = 1, 2, \dots, N$,

$$\begin{aligned} \dot{\boldsymbol{x}}_i(t) &= \boldsymbol{\alpha x}_i(t) + \boldsymbol{\beta u}_i(t) \\ \boldsymbol{y}_i(t) &= \boldsymbol{\gamma x}_i(t) \end{aligned} \tag{6.30}$$

where the matrices $\boldsymbol{\alpha} \in \mathbb{R}^{n\times n}$, $\boldsymbol{\beta} \in \mathbb{R}^{n\times m}$, $\boldsymbol{\gamma} \in \mathbb{R}^{p\times n}$ are assumed to be controllable and observable.

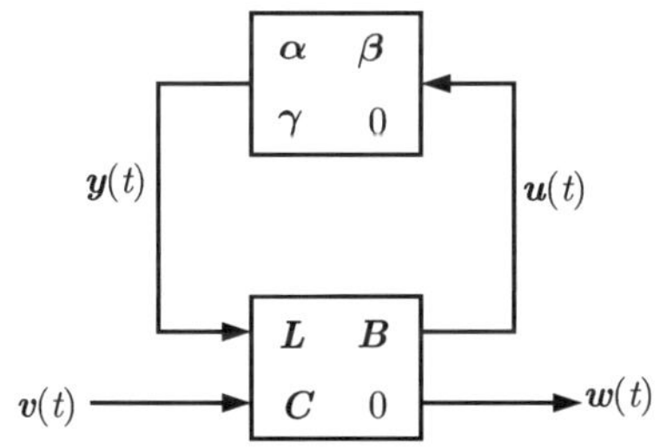

Fig. 6.12 Network structure considered

The control input $\boldsymbol{u}_i$ to each node system depends on the network structure defined through the interconnection matrix $\boldsymbol{L} \in \mathbb{R}^{N\times N}$, and a feedback matrix $\boldsymbol{K} \in \mathbb{R}^{m\times p}$, as well as an external control input. The input to system i from the network is

$$\boldsymbol{\xi}_i(t) = \sum_{j=1}^{N} \boldsymbol{L}_{ij} \boldsymbol{K} \boldsymbol{\gamma} \boldsymbol{x}_j(t). \tag{6.31}$$

Adding *input/output interconnection matrices* $\boldsymbol{B} = (\boldsymbol{B}_1^{\mathrm{T}}, \ldots, \boldsymbol{B}_N^{\mathrm{T}})^{\mathrm{T}} \in \mathbb{R}^{mN \times q}$ with $\boldsymbol{B}_i \in \mathbb{R}^{m \times q}$ and $\boldsymbol{C} = (\boldsymbol{C}_1, \ldots, \boldsymbol{C}_N) \in \mathbb{R}^{l \times pN}$ with $\boldsymbol{C}_i \in \mathbb{R}^{l \times p}$ respectively, (see Fig. 6.12), and external control input $\boldsymbol{v} = (v_1, ..., v_q)^{\mathrm{T}} \in \mathbb{R}^q$ applied to the whole network, the control input to node system i is

$$\boldsymbol{u}_i(t) = \boldsymbol{\xi}_i(t) + \boldsymbol{B}_i \boldsymbol{v}(t).$$

Similarily, the output $\boldsymbol{w} \in \mathbb{R}^l$ of the network is a linear combination of the individual node outputs $\boldsymbol{y}_i$,

$$\boldsymbol{w}(t) = \boldsymbol{C} \boldsymbol{y}(t).$$

with $\boldsymbol{y} = (\boldsymbol{y}_1^{\mathrm{T}}, \ldots, \boldsymbol{y}_N^{\mathrm{T}})^{\mathrm{T}}$. Let $\boldsymbol{x} = (\boldsymbol{x}_1^{\mathrm{T}}, \ldots, \boldsymbol{x}_N^{\mathrm{T}})^{\mathrm{T}} \in \mathbb{R}^{nN}$ denote the global state of the network. Using the interconnection matrices $\boldsymbol{L}, \boldsymbol{B}, \boldsymbol{C}$ and node dynamics $\boldsymbol{\alpha}, \boldsymbol{\beta}, \boldsymbol{\gamma}$ the resulting homogeneous network has the form

$$\begin{aligned} \dot{\boldsymbol{x}}(t) &= \bar{\boldsymbol{A}} \boldsymbol{x}(t) + \bar{\boldsymbol{B}} \boldsymbol{v}(t) \\ \boldsymbol{y}(t) &= \bar{\boldsymbol{C}} \boldsymbol{x}(t). \end{aligned} \tag{6.32}$$

The matrices $\bar{\boldsymbol{A}}, \bar{\boldsymbol{B}}$ and $\bar{\boldsymbol{C}}$ have the tensor product representations

$$\begin{aligned} \bar{\boldsymbol{A}} &= \boldsymbol{I}_N \otimes \boldsymbol{\alpha} + \boldsymbol{L} \otimes \boldsymbol{\beta} \boldsymbol{K} \boldsymbol{\gamma} \in \mathbb{R}^{nN \times nN} \\ \bar{\boldsymbol{B}} &= (\boldsymbol{I}_N \otimes \boldsymbol{\beta}) \boldsymbol{B} \in \mathbb{R}^{nN \times q} \\ \bar{\boldsymbol{C}} &= \boldsymbol{C} (\boldsymbol{I}_N \otimes \boldsymbol{\gamma}) \in \mathbb{R}^{l \times nN}. \end{aligned} \tag{6.33}$$

We discuss several aspects of stability and synchronizability for given node systems $(\boldsymbol{\alpha}, \boldsymbol{\beta}, \boldsymbol{\gamma})$. In particular,

- stabilizability and synchronizability of SISO node systems through a network,
- necessary and sufficient criteria for the existence of a feedback matrix $\boldsymbol{K}$, such that the network (6.33) synchronizes.

Synchronizability involves invariance and stability, in other words, the stability and synchronization problem are not separable. We therefore start with a discussion on stability of networks.

In the first part of this section, the node systems are assumed to be single-input-single-output, i.e., $\boldsymbol{\alpha} \in \mathbb{R}^{n \times n}$, $\boldsymbol{\beta} \in \mathbb{R}^n$ and $\boldsymbol{\gamma} \in \mathbb{R}^{1 \times n}$. The aspect stabilizability of identical node systems through a network is tackled first, which we call *network stabilizable* node systems. To this end, we use frequency domain analysis. Further on, based on the *Bezoutian*, we introduce a *Linear Matrix Inequality formulation* that allows for an algorithmic check of stability. It also reveals some more structure of the problem. In the second part of this section, we formulate necessary and sufficient conditions for synchronizability of networks with MIMO node systems in state space. We extend the analysis to synchronization on subspaces. Last, we consider again networks of SISO systems and link the synchronizability problem to that of stabilizability using the results of the previous subsection. Proofs to the results presented in this section are found in [175].

6.3.2 Stability and Stabilization of Homogenous Networks

In what follows, we discuss networks of SISO systems, that is, $\boldsymbol{\alpha} \in \mathbb{R}^{n\times n}$, $\boldsymbol{\beta} \in \mathbb{R}^{n}$ and $\boldsymbol{\gamma} \in \mathbb{R}^{1\times n}$. Throughout we assume that $(\boldsymbol{\alpha}, \boldsymbol{\beta}, \boldsymbol{\gamma})$ is controllable and observable. Then it is known that the controllability and observability of the overall system $(\bar{\boldsymbol{A}}, \bar{\boldsymbol{B}}, \bar{\boldsymbol{C}})$ is equivalent to $(\boldsymbol{L}, \boldsymbol{B}, \boldsymbol{C})$ being controllable and observable [165], [129]. In the sequel, we assume that $(\boldsymbol{L}, \boldsymbol{B}, \boldsymbol{C})$ are controllable and observable. Networks of SISO node systems are then conveniently described through transfer functions.

The *node transfer function* is the strictly proper McMillan degree n rational function

$$g(s) = \boldsymbol{\gamma}(s\boldsymbol{I}_n - \boldsymbol{\alpha})^{-1}\boldsymbol{\beta}. \tag{6.34}$$

Thus $h(s) := 1/g(s)$ is real rational with $h(\infty) = \infty$.

The *interconnection transfer function* is defined as

$$\boldsymbol{F}_L(s) = \boldsymbol{C}(s\boldsymbol{I}_N - \boldsymbol{L})^{-1}\boldsymbol{B} \in \mathbb{R}^{l\times q}(s). \tag{6.35}$$

We denote by

$$\mathcal{P}_{\boldsymbol{F}_L} := \{s \in \mathbb{C} \mid \det(s\boldsymbol{I}_N - \boldsymbol{L}) = 0\} \tag{6.36}$$

the set of poles of $\boldsymbol{F}_L(s)$.

The *global network transfer function* of (6.33)

$$\boldsymbol{F}(s) := \bar{\boldsymbol{C}}(s\boldsymbol{I}_{nN} - \bar{\boldsymbol{A}})^{-1}\bar{\boldsymbol{B}} = \boldsymbol{C}(h(s)\boldsymbol{I}_N - \boldsymbol{L})^{-1}\boldsymbol{B} \tag{6.37}$$

is strictly proper and is given by the composition of rational functions as $\boldsymbol{F}(s) = (\boldsymbol{F}_L \circ h)(s)$. Therefore, interconnecting N identical linear systems (6.30) to a network of linear systems is equivalent to applying a frequency transformation $h(s)$ into the interconnection transfer function $\boldsymbol{F}_L(s)$. This point of view of studying homogeneous networks of systems via compositions of transfer functions has been proposed in [165].

In the case that we allow for a scalar feedback term $\boldsymbol{K} = k \in \mathbb{R}$, the interconnection transfer function is

$$\boldsymbol{F}_L(s) = \boldsymbol{C}(s\boldsymbol{I}_N - k\boldsymbol{L})^{-1}\boldsymbol{B} \in \mathbb{R}^{l\times q}(s). \tag{6.38}$$

Therefore, the question whether there exists $\boldsymbol{L} \in \mathbb{R}^{N\times N}$ such that the node systems synchronize is independent of the value of k for SISO node systems.

Throughout, let $\mathbb{C}_- \subset \mathbb{C}$ denote the open left-half plane, $\mathbb{C}_+$ the open right-half plane and $\overline{\mathbb{C}_-}$, $\overline{\mathbb{C}_+}$ the corresponding closed half planes. Recall that a proper rational function $f(s)$ is called *stable*, if all its poles are in the open left-half plane $\mathbb{C}_-$.

Stabilization through Networks. We consider necessary and sufficient conditions for the stabilizability of a set of node systems with identical linear dynamics by a network, which we call *network stabilizable*. The following result is taken from [165].

Lemma 6.2. *The global network transfer function* $\boldsymbol{F} = \boldsymbol{F}_L \circ h$ *is stable if and only if*

$$\mathcal{P}_{\boldsymbol{F}_L} \cap h(\overline{\mathbb{C}_+}) = \emptyset. \tag{6.39}$$

Recall that a scalar rational function $g(s)$ is called *positive real* if g is stable and $\mathrm{Real}\,(g(s)) \geq 0$ for all $s \in \overline{\mathbb{C}_+}$. Thus, given that $g(s)$ is a positive real function, $h(s) = \frac{1}{g(s)}$ satisfies $h(\overline{\mathbb{C}_+}) \subset \overline{\mathbb{C}_+} \cup \{\infty\}$. By Lemma 6.2 we conclude

Corollary 6.2. *Let* $g(s)$ *be positive real and* $\boldsymbol{F}_L(s)$ *be stable. Then* $\boldsymbol{F} = \boldsymbol{F}_L \circ h$ *is stable.*

We return to node systems with arbitrary strictly proper transfer function g.

Definition 6.3. *The real transfer function* $g(s) = \frac{1}{h(s)}$ *is* network stabilizable *by an interconnection transfer function* $\boldsymbol{F}_L$*, if* $\boldsymbol{F}_L \circ h$ *is stable and* $\boldsymbol{F}_L$ *is nonconstant.*

We have the following criterion for network stabilizability.

Theorem 6.4. *The node transfer function* g *is network stabilizable by an interconnection transfer function* $\boldsymbol{F}_L$ *if and only if*

$$h(\overline{\mathbb{C}_+} \cup \{\infty\}) \neq \mathbb{C} \cup \{\infty\}.$$

The proof of this theorem relies on the observation that the set

$$\Omega := (\mathbb{C} \cup \{\infty\}) \setminus h(\overline{\mathbb{C}_+} \cup \{\infty\})$$

is self-conjugate, allowing to choose an interconnection matrix $\boldsymbol{L}$ with real entries and eigenvalues in Ω.

Corollary 6.3. *The following statements are equivalent:*
(i) $\Omega \neq \emptyset$,
(ii) g *is network stabilizable,*

(iii) g is network stabilizable by $\boldsymbol{F}_L$ with MacMillan degree $\delta(\boldsymbol{F}_L) = N$, $N \geq 1$ arbitrary,
(iv) g is stabilizable by real or complex static output feedback.

Corollary 6.4. *The following statements are equivalent:*
(i) Ω contains a real point,
(ii) g is network stabilizable by $\boldsymbol{F}_L$ with MacMillan degree $\delta(\boldsymbol{F}_L) = N$, N odd arbitrary,
(iv) g is stabilizable by constant real static output feedback.

Stability of Homogenous Networks Based on the Bezoutian. We are interested in a characterization of the set of node transfer functions g and interconnection matrices $\boldsymbol{L}$ such that the network transfer function $\boldsymbol{F}(s) = \boldsymbol{C}\left(\frac{1}{g(s)}\boldsymbol{I}_N - \boldsymbol{L}\right)^{-1}\boldsymbol{B}$ is stable. To this end, we apply the Bezoutian to decide for a given polynomial if it is Hurwitz (see [172] and [175] for details).

Recall that the Bezoutian of two polynomial $p(s), q(s) \in \mathbb{R}[s]$, deg$p \leq$ deg$q$, is the $(n \times n)$-matrix $\boldsymbol{B}_n(p, q)$ with

$$(1 \ z \ \dots \ z^{n-1})\,\boldsymbol{B}_n(p,q)\begin{pmatrix} 1 \\ w \\ \vdots \\ w^{n-1} \end{pmatrix} = \frac{p(z)q(w) - p(w)q(z)}{z - w}.$$

Some properties of the Bezoutian are the following:

1. For $p(s), q(s) \in \mathbb{R}[s]$, the matrix $\boldsymbol{B}_n(p,q)$ is real symmetric.
2. $\boldsymbol{B}_n : \mathbb{R}[s] \times \mathbb{R}[s] \to \mathbb{R}^{n \times n}$ is a bilinear function.
3. The matrix $\boldsymbol{B}_n(p,q)$ is invertible if and only if the polynomials p and q are coprime.

We refer to [181] for a survey on Bezoutians.

For $n(s), d(s) \in \mathbb{R}[s]$ we define $n(\mathrm{j}s) = n_R(s) + \mathrm{j}n_I(s)$, with $n_R, n_I \in \mathbb{R}[s]$, $d_R, d_I \in \mathbb{R}[s]$ are defined respectively.

Theorem 6.5. *The network transfer function*

$$\boldsymbol{F}(s) = \boldsymbol{C}\left(\frac{1}{g(s)}\boldsymbol{I}_N - \boldsymbol{L}\right)^{-1}\boldsymbol{B},$$

with $g(s) = \frac{n(s)}{d(s)}$, is stable if and only if for all $\lambda_i \in \sigma(\boldsymbol{L})$, the matrix

$$\boldsymbol{M}(\lambda_i) = |\lambda_i|^2\boldsymbol{M}_2 + \mathrm{Real}(\lambda_i)\boldsymbol{M}_1 + \mathrm{Imag}(\lambda_i)\tilde{\boldsymbol{M}}_1 + \boldsymbol{M}_0 \qquad (6.40)$$

is positive definite, where

$$\begin{aligned} \boldsymbol{M}_2 &= \boldsymbol{B}_n(n_I, n_R) \\ \boldsymbol{M}_1 &= \boldsymbol{B}_n(d_R, n_I) + \boldsymbol{B}_n(n_R, d_I) \\ \tilde{\boldsymbol{M}}_1 &= \boldsymbol{B}_n(d_R, n_R) + \boldsymbol{B}_n(d_I, n_I) \\ \boldsymbol{M}_0 &= \boldsymbol{B}_n(d_I, d_R). \end{aligned}$$

Example 6.2 *Honeycomb structured network*

This honeycomb structured network has 13 nodes and 15 edges (Fig. 6.13). The output of each node system $(\boldsymbol{\alpha}, \boldsymbol{\beta}, \boldsymbol{\gamma})$ is input to the adjacent nodes. Suppose the transfer function of the node systems has McMillan degree 2,

$$g(s) = \frac{as + b}{s^2 + cs + d}$$

with $a, b, c, d \in \mathbb{R}$. The 13 eigenvalues of the interconnection matrix $\boldsymbol{L}$ are $\pm\sqrt{6}$ (alg. mult. 1), $\pm\sqrt{3}$ (alg. mult. 2), ± 1 (alg. mult. 3) and 0 (alg. mult. 1). Theorem 6.5 implies that the network is stable if and only if

$$\begin{aligned} &c > 0, d > 0, |\sqrt{6}a| < c, ab + cd > |ad + bc|, \\ &3ab + cd > |\sqrt{3}(ad + bc)|, 6ab + cd > |\sqrt{6}(ad + bc)|. \end{aligned}$$

□

As a consequence of Theorem 6.5 we get the following result.

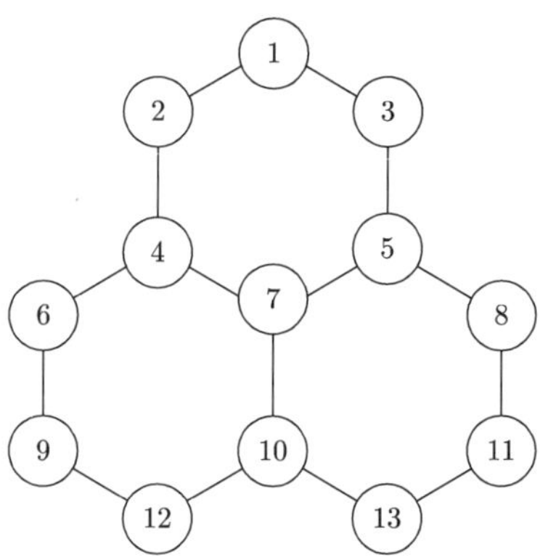

Fig. 6.13 Network with honeycomb structure

Corollary 6.5. *Suppose that all zeros of $g(s)$ are in $\mathbb{C}_+$, then*

$$M_+ := \{\lambda \in \mathbb{C} \mid \boldsymbol{M}(\lambda) > 0\}$$

is convex or empty.

In particular, for any transfer function $g(s) = \frac{1}{d(s)}$, the set M_+ is convex (or empty). Furthermore, it can be shown that M_+ is convex or empty if the nominator of $g(s)$, $n(s)$, is either an even or an odd polynomial [175].

Linear Matrix Inequality Formulation. In view of Theorem 6.5, introducing the variables $z = |\lambda|^2$, $x = \mathrm{Real}(\lambda)$ and $y = \mathrm{Imag}(\lambda)$, the problem of determining those $\lambda \in \mathbb{C}$, such that $\boldsymbol{F}(s)$ is stable, reduces to the following Linear Matrix Inequality problem with nonlinear constraint:

Find $x, y, z \in \mathbb{R}$ such that

$$\begin{aligned} &z\boldsymbol{M}_2 + x\boldsymbol{M}_1 + y\tilde{\boldsymbol{M}}_1 + \boldsymbol{M}_0 > 0 \\ &\text{s. t. } x^2 + y^2 = z. \end{aligned} \tag{6.41}$$

The nonlinear constraint $z = x^2 + y^2$ can equivalently be cast as rank-constrained Linear Matrix Inequality problem, namely $x^2 + y^2 = z$ if and only if

$$\mathrm{rank}\begin{pmatrix} 1 & 0 & x \\ 0 & 1 & y \\ x & y & z \end{pmatrix} = 2.$$

An algorithm for the numerical solution of the latter problem can be found in [279]. We have shown the following result:

Theorem 6.6. *Let $x, y, z \in \mathbb{R}$ be any solution of the rank-constrained LMI*

$$z\boldsymbol{M}_2 + x\boldsymbol{M}_1 + y\tilde{\boldsymbol{M}}_1 + \boldsymbol{M}_0 > 0 \quad s.t. \ \mathrm{rank}\begin{pmatrix} 1 & 0 & x \\ 0 & 1 & y \\ x & y & z \end{pmatrix} = 2.$$

Then the degree 2 interconnection transfer function $f(s) = \frac{1}{s^2 - 2xs + x^2 + y^2}$ stabilizes $g(s)$.

A Line Segmentation Stability Criterion. With normal forms of matrix pencils more structure of the set of interconnection matrices $\boldsymbol{L}$ that stabilize the homogenous network can be revealed. We define $\mathrm{Real}(\lambda) = \rho$ and $\mathrm{Imag}(\lambda) = \sigma\rho$, with $\rho, \sigma \in \mathbb{R}$ and rewrite equation (6.40)

$$\boldsymbol{M}(\rho) = \rho^2(1 + \sigma^2)\boldsymbol{M}_2 + \rho(\boldsymbol{M}_1 + \sigma\tilde{\boldsymbol{M}}_1) + \boldsymbol{M}_0 > 0.$$

The network is stable if both $\boldsymbol{M}(\rho) > 0$ and $\boldsymbol{M}_-(\rho) := \rho^2(1 + \sigma^2)\boldsymbol{M}_2 + \rho(\boldsymbol{M}_1 - \sigma\tilde{\boldsymbol{M}}_1) + \boldsymbol{M}_0 > 0$ hold.

We define the pencil

$$\begin{aligned} \boldsymbol{H}(\rho) &= \boldsymbol{H}_1 + \rho\boldsymbol{H}_2 \\ &= \begin{bmatrix} \boldsymbol{M}_0 + \rho(\boldsymbol{M}_1 + \sigma\tilde{\boldsymbol{M}}_1) & \rho(1 + \sigma^2)\boldsymbol{M}_2 \\ \rho(1 + \sigma^2)\boldsymbol{M}_2 & -(1 + \sigma^2)\boldsymbol{M}_2 \end{bmatrix} \end{aligned} \tag{6.42}$$

with

$$H_1 := \begin{bmatrix} M_0 & 0 \\ 0 & -(1+\sigma^2)M_2 \end{bmatrix}$$

and

$$H_2 := \begin{bmatrix} M_1 + \sigma \tilde{M}_1 & (1+\sigma^2)M_2 \\ (1+\sigma^2)M_2 & 0 \end{bmatrix},$$

$H_-(\rho)$ accordingly.

If M_2 is nonsingular, then H_2 is invertible and we may consider the Jordan normal form of $H_2^{-1}H_1$, to characterize definiteness of $H(\rho)$ depending on ρ. The following result is Theorem 3.4 in [172] applied to the pencil $H(\rho)$, see there for a definition of the sign characteristic. The signature of a symmetric matrix M, $\text{sign}(M)$, is defined as the number of positive eigenvalues of M minus the number of negative eigenvalues of M.

Theorem 6.7. *Let $\rho_1 < \cdots < \rho_m$ denote the real generalized eigenvalues of the pencil $H(\rho)$ (6.42) and n_{ij}, $j = 1, \ldots, \nu_i$, denote the sizes of the Jordan blocks $\mathcal{J}(\mu_i)$, $\mu_i = -\rho_i$, occuring in the Jordan matrix of $H_2^{-1}H_1$.*
Let ε_{ij} denote the corresponding sign characteristics and

$$\delta_i = \sum_{j=1}^{\nu_i} \varepsilon_{ij} \frac{1-(-1)^{n_{ij}}}{2}, \quad i = 1, \ldots, m.$$

Let M_2 be invertible. Then for $\sigma \in \mathbb{R}$ given
a) there exists $\rho \in \mathbb{R}$ such that $M(\rho) > 0$ if and only if

$$n = \text{sign}(M_2) + 2\max\{\delta_1, \delta_1 + \delta_2, \ldots, \delta_1 + \cdots + \delta_m\};$$

b) if $M_2 > 0$ then $M(\rho) > 0$ for $\rho > \rho_m$ or $\rho < \rho_1$;
c) for $\rho_i < \rho < \rho_{i+1}$, $i = 1, \ldots, m-1$ it holds that

$$\text{sign}(M(\rho)) = \text{sign}(M_2) + 2(\delta_1 + \cdots + \delta_i).$$

6.3.3 Synchronization of Homogenous Networks

Regarding the synchronizability problem for networks of identical node systems, we work in state space first and return to the frequency domain for the analysis of SISO node systems.

Synchronizability of MIMO Systems. In this section we tackle the stability and synchronizability problem of networks of identical agents using state space methods. Further, we recapitulate a result on how to synchronize certain networks via appropriate choice of K.

A generalization of the synchronization problem is that of synchronizing subspaces. Let $\mathcal{U} \subseteq \mathbb{R}^n$ be a subspace, $\boldsymbol{U} \in \mathbb{R}^{n\times k}$ denote a matrix whose column vectors span $\mathcal{U}$. Further, let $\mathbf{1}_N := (1,1,\ldots,1)^{\mathrm{T}}$.

Definition 6.4. *The system* (6.33) synchronizes on the subspace $\mathcal{U} \subseteq \mathbb{R}^n$ *if*

a) the subspace spanned by the column vectors of $\mathbf{1}_N \otimes \boldsymbol{U}$ *is* $\boldsymbol{I}_N \otimes \boldsymbol{\alpha} + \boldsymbol{L} \otimes \boldsymbol{\beta K \gamma}$*-invariant;*

b) for all initial values $\boldsymbol{x}(0) \in \mathbb{R}^{nN}$*, the state vectors of all node systems converge to the same state,*

$$\lim_{t\to\infty} \|\boldsymbol{x}_i(t) - \boldsymbol{x}_j(t)\| = 0 \quad \text{for all } i,j = 1,\ldots,N. \tag{6.43}$$

c) for all $i = 1,\ldots,N$,

$$\operatorname{dist}(\mathcal{U}, \boldsymbol{x}_i(t)) \to 0 \quad \text{for } t \to \infty$$

We call $\mathcal{U}$ *a synchronizing subspace. If for* $(\boldsymbol{\alpha}, \boldsymbol{\beta}, \boldsymbol{\gamma}, \boldsymbol{L})$ *given there exists a feedback matrix* $\boldsymbol{K} \in \mathbb{R}^{m\times p}$ *such that* (6.33) *fulfills a)-c) the system* (6.33) *is* synchronizable.

For $\mathcal{U} = \mathbb{R}^n$, Definition 6.4 reduces to that of synchronizability, while for $\mathcal{U} = 0$ Definition 6.4 defines stabilizability for networks.

Necessary and sufficient conditions for synchronization can be found in the literature. A similar version to the following theorem is stated in [397], with $\mathcal{U} = \mathbb{R}^n$ and $\boldsymbol{L} = \boldsymbol{D}\tilde{\boldsymbol{L}}$, where $\boldsymbol{D}$ is a diagonal matrix.

Theorem 6.8. *Let* $\mathcal{U} \subseteq \mathbb{R}^n$ *be a k-dimensional linear subspace. Suppose the interconnection matrix* $\boldsymbol{L} \in \mathbb{R}^{N\times N}$ *has a simple eigenvalue* λ_1 *with eigenvector* $\mathbf{1}_N = (1,1,\ldots,1)^{\mathrm{T}}$. $\mathcal{U}$ *is a synchronizing subspace if and only if the following three conditions hold:*

a) $\mathcal{U}$ *is* $\boldsymbol{\alpha} + \lambda_1 \boldsymbol{\beta K \gamma}$*-invariant.*

b) $\sigma(\boldsymbol{\alpha} + \lambda_i \boldsymbol{\beta K \gamma}) \subseteq \mathbb{C}_-$ *for all* $\lambda_i \in \sigma(\boldsymbol{L}) \setminus \{\lambda_1\}$.

c) $\sigma\left((\boldsymbol{\alpha} + \lambda_1 \boldsymbol{\beta K \gamma})\Big|_{\mathbb{R}^n/\mathcal{U}} \right) \subseteq \mathbb{C}_-$.

The previous theorem unites the stabilizability and synchronizability problem in that the stability and the synchronization problem are covered by setting $\mathcal{U} = 0$ and $\mathcal{U} = \mathbb{R}^n$, respectively.

Corollary 6.6. *Suppose that the assumptions of Theorem 6.8 hold. Then*

(a) there exists a matrix $\boldsymbol{F} \in \mathbb{R}^{m \times n}$ *such that* $\dot{\boldsymbol{x}} = (\boldsymbol{I}_N \otimes \boldsymbol{\alpha} + \boldsymbol{L} \otimes \boldsymbol{\beta}\boldsymbol{F})\boldsymbol{x}$ *synchronizes on* $\mathcal{U}$,

(b) there exists a matrix $\boldsymbol{H} \in \mathbb{R}^{n \times p}$ *such that* $\dot{\boldsymbol{x}} = (\boldsymbol{I}_N \otimes \boldsymbol{\alpha} + \boldsymbol{L} \otimes \boldsymbol{H}\boldsymbol{\gamma})\boldsymbol{x}$ *synchronizes on* $\mathcal{U}$.

Here we make contact with classical concepts from geometric control [391], [360] such as *controlled invariant* and *conditioned invariant* subspaces for $(\boldsymbol{\alpha}, \boldsymbol{\beta}, \boldsymbol{\gamma})$. Recall that a subspace $\mathcal{W}$ is controlled invariant (conditioned invariant) if there exists a matrix $\boldsymbol{F}$ such that $\mathcal{W}$ is $\boldsymbol{\alpha} + \boldsymbol{\beta}\boldsymbol{F}$-invariant ($\mathcal{W}$ is $\boldsymbol{\alpha} + \boldsymbol{F}\boldsymbol{\gamma}$-invariant, respectively).

It is known that if $\mathcal{W}$ is controlled invariant and conditioned invariant, then there exists a matrix $\boldsymbol{K} \in \mathbb{R}^{m \times p}$ such that $\boldsymbol{\alpha} + \boldsymbol{\beta}\boldsymbol{K}\boldsymbol{\gamma}$ is $\mathcal{W}$-invariant. It is an open question whether a similar result holds for the synchronization problem, i.e., if the converse of Corollary 6.6 holds.

A Sufficient Condition for Synchronizability. For fixed node systems and network matrix with certain spectra, the task of calculating $\boldsymbol{K} \in \mathbb{R}^{m \times p}$ such that the node systems synchronize can be done by solving an Algebraic Riccati Equation. This result can be found in [249] for networks where the interconnection matrix $\boldsymbol{L}$ is a real symmetric Laplacian matrix, however, the proof shows that it also holds for more general network structures. Recall that $(\boldsymbol{\alpha}, \boldsymbol{\beta}, \boldsymbol{\gamma})$ is assumed to be stabilizable and observable.

Theorem 6.9. *Suppose that* λ_1 *is a simple eigenvalue of the interconnection matrix* $\boldsymbol{L}$ *with eigenvector* $\boldsymbol{1}_N$. *Let* $0 < \mathrm{Real}(\lambda_2) < \cdots < \mathrm{Real}(\lambda_{\mathrm{N}})$ *for the other* $N-1$ *eigenvalues of* $\boldsymbol{L}$. *Assume that*

$$\mathrm{rank}\boldsymbol{\gamma} = \mathrm{rank}\begin{pmatrix} \boldsymbol{\gamma} \\ \boldsymbol{\beta}^{\mathrm{T}}\boldsymbol{P} \end{pmatrix}$$

holds for the unique solution $\boldsymbol{P} = \boldsymbol{P}^{\mathrm{T}} > 0$ *of the Algebraic Riccati Equation*

$$\boldsymbol{\alpha}^{\mathrm{T}}\boldsymbol{P} + \boldsymbol{P}\boldsymbol{\alpha} - \boldsymbol{P}\boldsymbol{\beta}\boldsymbol{\beta}^{\mathrm{T}}\boldsymbol{P} + \boldsymbol{\gamma}^{\mathrm{T}}\boldsymbol{\gamma} = 0.$$

Then there exists a matrix $\boldsymbol{M} \in \mathbb{R}^{m \times p}$ *such that* $\boldsymbol{\beta}^{\mathrm{T}}\boldsymbol{P} = \boldsymbol{M}\boldsymbol{\gamma}$ *and* $(\boldsymbol{\alpha}, \boldsymbol{\beta}, \boldsymbol{\gamma})$ *is synchronizable by* $\boldsymbol{K} := \tau\boldsymbol{M}$ *with* $\tau = -\frac{1}{\mathrm{Real}(\lambda_2)}$.

Synchronizability of Homogeneous Networks of SISO Systems. The literature on synchronizability usually refers to a state-space formulation of the problem. We find it convenient to work also in the frequency domain and define synchronization as follows.

Definition 6.5. *Let* $g(s) := \boldsymbol{\gamma}(s\boldsymbol{I}_n - \boldsymbol{\alpha})^{-1}\boldsymbol{\beta} \in \mathbb{R}(s)$ *be strictly proper. We say that an interconnection transfer function* $f(s) \in \mathbb{R}(s)$, $f(\infty) = 0$, synchronizes g *if there exists a minimal realization* $(\boldsymbol{L}, \boldsymbol{b}, \boldsymbol{c})$ *of* f,

$$f(s) = \boldsymbol{c}(s\boldsymbol{I}_N - \boldsymbol{L})^{-1}\boldsymbol{b}$$

with $\delta(f) = N$ *and*
(i) 0 *is a simple pole of* f *and* $\boldsymbol{L}\mathbf{1}_N = 0$.
(ii) For the $N-1$ *poles* $\lambda_2, \ldots, \lambda_N \in \mathbb{C}$ *of* f

$$\sigma(\boldsymbol{\alpha} + \lambda_i \boldsymbol{\beta}\boldsymbol{\gamma}) \subset \mathbb{C}_-, \quad i = 2, \ldots, N,$$

holds.

Proposition 6.1. *If the transfer function* $f(s) = \boldsymbol{c}(s\boldsymbol{I}_N - \boldsymbol{L})^{-1}\boldsymbol{b}$ *synchronizes* $g(s)$, *then the network (6.32) achieves synchronization, i.e., the assumptions of Theorem 6.8 are fulfilled.*

Theorem 6.10. *The interconnection transfer function* f *synchronizes* g *if and only if*
(i) 0 *is simple pole of* f.
(ii) $\left(f^{-1}(\infty) \setminus \{0\}\right) \cap h(\overline{\mathbb{C}_+}) = \emptyset$.

Theorem 6.11. N *identical SISO minimal systems*

$$\begin{aligned} \dot{\boldsymbol{x}}_i(t) &= \boldsymbol{\alpha}\boldsymbol{x}_i(t) + \boldsymbol{\beta}\boldsymbol{u}_i(t) \\ \boldsymbol{y}_i(t) &= \boldsymbol{\gamma}\boldsymbol{x}_i(t) \end{aligned} \tag{6.44}$$

are synchronizable with $\boldsymbol{K} = 1$ *through an interconnection if and only if there exists a proper rational real transfer function* $f(s) \in \mathbb{R}(s)$ *of McMillan degree* $N-1$ *with*

$$f^{-1}(\infty) \cap h(\overline{\mathbb{C}_+}) = \emptyset \tag{6.45}$$

and $0 \notin f^{-1}(\infty)$.

A consequence of the Theorem 6.11 is that $N-1$ systems that are network stabilizable by f with $0 \notin f^{-1}(\infty)$ are synchronizable by adding one more identical node system to the network in an appropriate way (such that $\mathbf{1}_N$ is an eigenvector of the new interconnection matrix with eigenvalue 0).

Proposition 6.2. *Let* $\boldsymbol{L} \in \mathbb{R}^{N\times N}$ *with simple eigenvalue* 0 *and eigenvector* $\mathbf{1}_N = (1, \ldots, 1)^{\mathrm{T}}$; *assume that the other* $N-1$ *eigenvalues* $\lambda_2, \ldots, \lambda_N \in \mathbb{C}$ *of* $\boldsymbol{L}$ *satisfy* $\mathrm{Real}(\lambda_\mathrm{i}) > 0$, $\mathrm{i} = 2, \ldots, \mathrm{N}$. *Let* $(\boldsymbol{\alpha}, \boldsymbol{\beta}, \boldsymbol{\gamma})$ *be a minimal realization of a positive real transfer function* $g \in \mathbb{R}(s)$. *Then the spectrum of* $\bar{\boldsymbol{A}} = \boldsymbol{I}_N \otimes \boldsymbol{\alpha} + \boldsymbol{L} \otimes \boldsymbol{\beta}\boldsymbol{\gamma}$ (6.33) *is*

$$\sigma(\bar{A}) = \sigma(\boldsymbol{\alpha}) \ \dot{\cup} \ \underbrace{\bigcup_{i=2}^{N} \sigma(\boldsymbol{\alpha} - \lambda_i \boldsymbol{\beta}\boldsymbol{\gamma})}_{\subseteq \mathbb{C}_-}$$

and $\dot{\boldsymbol{x}} = \bar{A}\,\boldsymbol{x}$ *synchronizes.*

6.4 Optimization Methods for Cluster Synchronization Analysis

6.4.1 Cluster Synchronization

When many dynamic systems mutually interact, complicated and interesting dynamic behaviors can appear. One fundamental emergent behavior is that all systems agree on a common trajectory as it was shown in the preceding section.

In many natural dynamic systems full synchronization of a complete network is a rather special case, and it is more often observed that only parts of the network, which we will call clusters, synchronize [24], while being out of synchrony with the rest of the network. Additionally, it is widely known that real-world networks inherit often very significant structures, such as very strongly connected components, called communities [277]. The existence of communities in graphs is not necessarily reflected in the connectivity of the network and does, in the current understanding, not contribute to the ability of a network to synchronize. An evident question is now whether and how *cluster synchronization* in dynamic networks is related to the existence of community structures in the underlying network. Answering this question promises to provide a profound understanding of complex networked systems, and can lead to fundamental design principles for large-scale networks.

Literature Survey. Cluster synchronization has recently attained significant attention from the dynamic systems perspective [24, 239, 296]. A majority of research has been dedicated to forcing a network of dynamic systems to cluster synchronization using pinning control [250]. A more analysis-oriented approach has been recently pursued in [395] where different mechanisms leading to cluster synchronization of diffusively coupled systems are studied. In [2, 3, 90] a first-order dynamic network model with saturated diffusive couplings is shown to exhibit cluster synchronization behavior.

The literature on community detection is vast and too extensive to be reviewed here. A overview of the field is given in [125]. All community detection methods are based on the same principle. A measure for the "quality" of a community is defined and then computational attractive methods are sought to find the best partitioning according to this measure. The main problem related to this approach is that quality measures for a community are not well defined, which introduces a degree of arbitrariness.

However, considering the analysis of dynamic networks one might search for other quality measures, which incorporate both, the network topology and the dynamic properties of the nodes, while still being efficiently computable. Attempts to specify the quality measures of community detection problems to the functionality of dynamic networks have recently been made in the context of power network analysis [36, 228, 381]. However, a general understanding of how the dynamics of a network and its community structure interact is just beginning to emerge.

Contributions of this Section. This section gives a descriptive overview on how cluster synchronization is related to the hierarchical structures of a network. We consider a class of dynamic networks which eventually exhibit, depending on the dynamic properties of the nodes and the interaction structures, a cluster synchronization behavior. The model has been proposed in [43–45] and is a variant of the network models which are widely studied in the context of full synchronization. We consider networks of nodes with second-order dynamics and saturated diffusive couplings. We show that the condition for complete synchronization of the network can be traced back to a combinatorial optimization problem on a weighted graph. Complex cluster synchronization structures appear if the synchronization condition is not satisfied. For an exact analysis, we relate the clustering structure of the dynamic network to a static saddle-point optimization problem, which provides a computationally attractive tool to predict the clustering structure for a given network configuration.

Having identified the saddle-point problem as a tool to predict the clustering structure of the dynamic network, it is straightforward to see that the same problem can be used for a structural network analysis. In fact, since the saddle-point problem can efficiently compute the clustering structure for one network configuration, it can also be solved for a variety of network parameterizations, uncovering a detailed hierarchical structure of the network. The cluster synchronization directly connects the different fields of synchronization and computational network analysis. We provide in this section an overview on the results originally presented in [43–45], and refer the reader for the detailed proofs to these references.

In the remainder of this section, we present first the considered dynamic network model in Section 6.4.2 and provide some combinatorial conditions for full synchronization on Section 6.4.3. The main analysis tool for predicting the clustering structure, the static saddle-point problem, is presented in Section 6.4.4, and is used for a hierarchical network analysis in Section 6.4.5.

6.4.2 A Dynamic Network Model for Clustering

We start the discussion with a class of dynamic networks defined on undirected graphs. A graph $\mathcal{G} = (\mathcal{V}, \mathcal{E})$, consists of a set of nodes $\mathcal{V} = \{v_1, \ldots, v_{|\mathcal{V}|}\}$ and a set of edges $\mathcal{E} = \{e_1, \ldots, e_{|\mathcal{E}|}\}$ describing the incidence relation between pairs of nodes. Each node of the graph has a dynamic behavior and neighboring nodes are interacting with a diffusive nonlinear coupling.

We consider in particular the following class of dynamic networks

$$\ddot{x}_i(t) = -f_i(\dot{x}_i(t), \omega_i) + \sum_{e_k=(i,j)\in\mathcal{E}} \alpha_k \psi_k(x_j(t) - x_i(t)), \tag{6.46}$$

where $x_i \in \mathbb{R}$ is the dynamic variable associated to a node. The local node dynamics are characterized by the possibly nonlinear function $f_i(\dot{x}_i, \omega_i)$, which is parameterized by a constant ω_i. We assume that $f_i(\omega_i, \omega_i) = 0$ such that ω_i can be understood as a natural "velocity" of node i. Nodes which are neighbors in $\mathcal{G}$ are interacting with each other according to the diffusive nonlinear coupling $\psi_k(x_j(t) - x_i(t))$. The positive scalars α_k indicate the coupling strength between neighboring nodes.

A key observation concerning the dynamic model (6.46) is that for certain problem configurations the network exhibits a cluster synchronization. We first impose the following assumptions on the node dynamics and the diffusive coupling.

Assumption 6.2. *The system* (6.46) *is such that*
1. *for all nodes* $v_i \in \mathcal{V}$

$$(f_i(\dot{x}_i, \omega_i) - f_i(\dot{x}_i', \omega_i))\,(\dot{x}_i - \dot{x}_i') \geq \eta_i\,(\dot{x}_i - \dot{x}_i')$$

for some $\eta_i > 0$ *and all* $\dot{x}_i, \dot{x}_i' \in \mathbb{R}$*;*
2. *for all edges* $e_k \in \mathcal{E}$*,* $\psi_k(\cdot)$ *is odd, monotonically increasing and saturated, i. e.,*

$$\lim_{|x_j - x_i| \to \infty} |\psi_k(x_j - x_i)| \to 1.$$

The two assumptions have explicit interpretations. The first assumption implies that each node, if decoupled from the network, will converge to its "natural" velocity ω_i. Note that if two nodes $v_i, v_j \in \mathcal{V}$ have different parameters $\omega_i \neq \omega_j$, then the two uncoupled systems will approach different "velocities" and will drift apart. Our assumption requires slightly more, i. e., it requires strict monotonicity of $f_i(\cdot)$. Strict monotonicity implies that for any constant input signal $\bar{u}$ the dynamic system

$$\ddot{x}_i(t) = -f_i(\dot{x}_i(t), \omega_i) + \bar{u}$$

will asymptotically approach a unique velocity $\dot{x}_i$.

The second assumption concerns the diffusive coupling functions. It is assumed that the coupling functions are saturated and that neighboring systems are not influencing each other arbitrarily strongly. The interplay between the two factors, i.e., the inhomogeneous node dynamics and the saturated interactions, will cause the clustering of the network.

Clusters. Let us first define a cluster in a purely graph-theoretical way. A graph $\mathcal{G}' = (\mathcal{V}', \mathcal{E}')$ is said to be a subgraph of $\mathcal{G}$ if $\mathcal{V}' \subseteq \mathcal{V}$ and $\mathcal{E}' \subseteq \mathcal{E}$; equivalently, we write $\mathcal{G}' \subseteq \mathcal{G}$. A cluster $\mathcal{P}$ is a connected sub-graph of $\mathcal{G}$. Each partition of the node set $\mathbb{P} = \{\mathcal{P}_1, \ldots, \mathcal{P}_p\}$ with $\mathcal{P}_i \subset \mathcal{V}$, $\bigcup_{i=1}^{p} \mathcal{P}_i = \mathcal{V}$, and $\mathcal{P}_i \cap \mathcal{P}_j = \emptyset$ for all $\mathcal{P}_i, \mathcal{P}_j \in \mathbb{P}$, induces a family of clusters $\mathcal{P}_i \subset \mathcal{G}$, $i \in \{1, \ldots, p\}$, which results after removing all edges from $\mathcal{E}$ which connect nodes in different partitions. The set of edges connecting nodes in different partitions are in the following called the cut-set $\mathcal{Q} \subset \mathcal{E}$ of the respective partition.

Having defined this general concept of a cluster, the main observation related to the dynamic network (6.46) is that the network shows a cluster synchronization behavior [43]. If we include the complete graph, i.e., $\mathbb{P} = \{\mathcal{V}\}$, and the fully decoupled graph, i.e., $\mathbb{P} = \big\{\{v_1\}, \ldots, \{v_{|\mathcal{V}|}\}\big\}$ into the definition of a cluster, then the cluster synchronization of (6.46) can be formalized as follows. For any network configuration, there exists a partition $\mathbb{P} = (\mathcal{P}_1, \ldots, \mathcal{P}_p)$ of the node set $\mathcal{V}$ such that nodes within one partition agree on the same velocity, i.e.,

$$\lim_{t\to\infty} \|\dot{x}_i(t) - \dot{x}_j(t)\| \to 0 \quad \text{if} \quad v_i, v_j \in \mathcal{P}_l,$$

while nodes in different partitions converge to different velocities, i.e.

$$\lim_{t\to\infty} \|\dot{x}_i(t) - \dot{x}_j(t)\| > 0 \quad \text{if} \quad v_i \in \mathcal{P}_l \quad \text{and} \quad v_j \notin \mathcal{P}_l$$

for all initial conditions $\boldsymbol{x}_0 = (x_{10}, \ldots, x_{n0})^{\mathrm{T}}$.

To illustrate the concept of cluster synchronization as it is understood in this section, some example trajectories from a dynamic network with 120 nodes defined according to (6.46) are shown in Fig. 6.14. We will discuss the underlying network model and the parameterization of the system later in Example 6.3. Here, we only want to point out that always groups of nodes in the network reach an agreement on the velocities. How many clusters appear depends on different properties of the dynamic network.

Networks with Linear Node Dynamics. To give a better insight into the model structure (6.46) and how it relates to other models presented for clustering analysis, we want to discuss the special case of linear node dynamics. A particular representation of (6.46) is the model

$$\ddot{x}_i(t) = -A_i(\dot{x}_i(t) - \omega_i) + \sum_{e_k=(i,j)\in\mathcal{E}} \alpha_k \psi_k(x_j(t) - x_i(t)) \tag{6.47}$$

with $A_i \in \mathbb{R}_{>0}$, and $\psi_k(\cdot)$ satisfying Assumption 6.2. The simplified network model (6.47) also provides directly the connection to the clustering model studied in [3, 90]. Considering the asymptotic behavior of (6.47), we can study the dynamics for $\ddot{x}_i = 0$. In this case the dynamics reduces to

$$\dot{x}_i(t) = w_i + A_i^{-1} \sum_{e_k=(i,j)\in\mathcal{E}} \alpha_k \psi_k(x_j(t) - x_i(t)). \tag{6.48}$$

The model (6.48) is in fact exactly the model analyzed in [3, 90] as a substitute for Kuramoto oscillator networks. Surprisingly it turns out that in this particular case the second order system can be more easily analyzed than

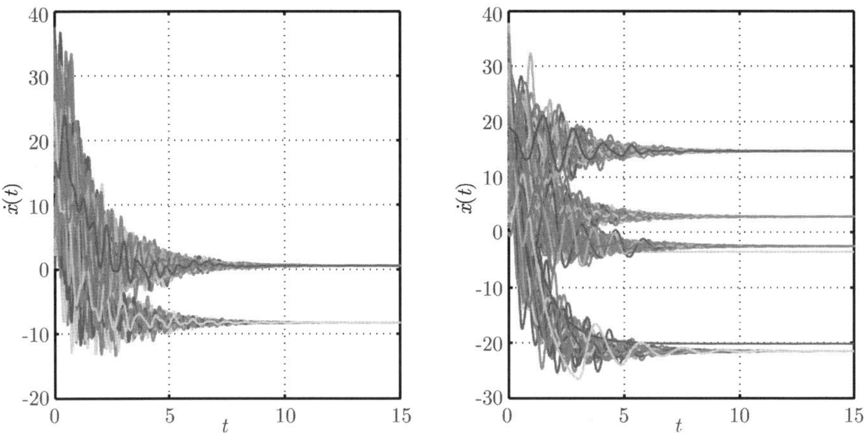

Fig. 6.14 Trajectories of the dynamic network (6.46) with the same topology but different parameterizations. The network forms two (left) or four (right) clusters

the first-order system. In fact, the second-order model can be connected to a static optimization problem, which allows for an explicit characterization of the clustering. Before we discuss this relation here, we want to summarize some results presented in [44, 45] concerning the full synchronization of (6.46).

6.4.3 Combinatorial Conditions for Synchronization

In a first step, we analyze the complete synchronization of the network. We say that the network reaches complete synchronization if all nodes agree on the same velocity. This is, there exists $\nu^* \in \mathbb{R}$, called the agreement velocity, such that

$$\lim_{t\to\infty} \|\dot{x}_i(t) - \nu^*\| \to 0$$

for all $i \in \mathcal{V}$. The agreement velocity ν^* is uniquely defined by the structure of the local node dynamics. To see this, consider the asymptotic behavior of the network, i.e., $\ddot{x}_i = 0$ and sum over all nodes. Due to the symmetric coupling of the nodes, the agreement velocity ν^* must satisfy

$$0 = \sum_{i=1}^{|\mathcal{V}|} f_i(\nu^*, \omega_i). \tag{6.49}$$

Due to the monotonicity of $f_i(\cdot)$ the agreement velocity is uniquely determined by the node dynamics and is, in particular, independent of the underlying network structure.

We are now interested to see under which conditions the network will completely synchronize. Although the value of ν^* is completely determined by the local node dynamics, whether the full network will converge to this solution depends on the interplay of three factors: (i) the dynamics of the local nodes, (ii) the interaction strength, and (iii) the network topology.

To obtain the condition for full synchronization, we interpret the network as a weighted graph. We associate to each node a weight, namely the scalar value $f_i(\nu^*, \omega_i)$. Each edge in the graph is weighted with the capacity α_k. We consider now all possible bi-partitions $\mathbb{P} = (\mathcal{P}_1, \mathcal{P}_2)$ of the network nodes, with $\mathcal{P}_1 \cup \mathcal{P}_2 = \mathcal{V}$. The set of all possible bi-partitions of the node set $\mathcal{V}$ is denoted by $\mathbb{P}^2(\mathcal{G})$. Note that for each $\mathbb{P} \in \mathbb{P}^2(\mathcal{G})$ it follows from (6.49) that

$$0 = \sum_{i \in \mathcal{P}_1} f_i(\nu^*, \omega_i) + \sum_{i \in \mathcal{P}_2} f_i(\nu^*, \omega_i). \tag{6.50}$$

Please note that we use here a short hand notation $i \in \mathcal{P}_1$ to indicate that the node v_i is contained in the set of nodes $\mathcal{P}_1$. Although this notation in not exact, it is unambiguous in the context here. We will similarly write in the following $k \in \mathcal{E}$ to indicate that edge e_k is in the set of edges.

Without loss of generality, we assume that $\mathcal{P}_1$ is chosen such that

$$\sum_{i \in \mathcal{P}_1} f_i(\nu^*, \omega_i) \geq 0.$$

To each bi-partition $\mathbb{P} = (\mathcal{P}_1, \mathcal{P}_2) \in \mathbb{P}^2(\mathcal{G})$ of the node set we associate now the *quality measure*

$$\Psi(\mathbb{P}) = \frac{\sum_{i \in \mathcal{P}_1} f_i(\nu^*, \omega_i)}{\sum_{k \in \mathcal{Q}} \alpha_k}, \tag{6.51}$$

where $\mathcal{Q} \subset \mathcal{E}$ is the set of edges connecting nodes in $\mathcal{P}_1$ and $\mathcal{P}_2$. The "quality" measures the imbalance of the two node sets over the capacity connecting the two partitions.

The quality measure $\Psi(\mathbb{P})$ provides now a condition for the full synchronization of the network (6.46), which is derived in [44].

The network reaches full agreement on the node velocities if and only if

$$\Psi(\mathbb{P}) \leq 1 \quad \text{for all } \mathbb{P} \in \mathbb{P}^2(\mathcal{G}). \tag{6.52}$$

Complementarily, we can also conclude that the network exhibits a (non-trivial) cluster synchronization if and only if there exists a bi-partition of the network nodes for which $\Psi(\mathbb{P}) > 1$.

A particularly clear condition can be derived for networks with linear node dynamics (6.47) and the additional assumptions that all edges have identical capacity $\alpha_1 = \alpha_2 = \cdots = \alpha_{|\mathcal{E}|} = \alpha$ and parameters $A_1 = A_2 = \cdots = A_{|\mathcal{V}|} = 1$. Then the network (6.47) reaches agreement if all bi-partitions of the graph $\mathbb{P} = (\mathcal{P}_1, \mathcal{P}_2)$ with cut-set $\mathcal{Q}$ satisfy

$$\left| r_1 \left(\sum_{i \in \mathcal{P}_2} \omega_i \right) - r_2 \left(\sum_{j \in \mathcal{P}_1} \omega_j \right) \right| \leq \alpha |\mathcal{Q}|, \tag{6.53}$$

where $|\mathcal{Q}|$ denotes the number of edges in the cut-set and $r_i = \frac{|\mathcal{P}_i|}{|\mathcal{V}|}$ is the ratio of nodes in the respective partition.

6.4.4 Static Saddle-Point Problem for Clustering Analysis

The previous discussion showed that the full synchronization of (6.46) is inherently related to combinatorial properties of the weighted network. While it is still tractable to present analytic conditions for the full synchronization in terms of the network properties, it is laborious to explicitly characterize for a given network configuration the exact clustering structure in terms of the combinatorial network properties. Therefore, we pursue another approach. Instead of characterizing every possible clustering structure explicitly, we provide an efficient computational tool which can predict the clustering structure. To analyze the clustering structure, we connect the asymptotic behavior of the dynamic network model to a static optimization problem. The optimization problem can be solved very efficiently and allows to predict the clustering structure for any possible network configuration.

We define first for each node a function $F_i(\nu_i, \omega_i)$ with decision variable ν_i such that

$$\frac{\mathrm{d}}{\mathrm{d}\nu_i} F_i(\nu_i, \omega_i) = f_i(\nu_i, \omega_i),$$

i.e., as the integral function of the local node dynamics. Note that the strong monotonicity assumption on $f_i(\cdot, \omega_i)$ directly implies that $F_i(\nu_i, \omega_i)$ is a strictly convex function.

Additionally, we introduce the incidence matrix of the graph $\mathcal{G}$. Only for the purpose of defining the incidence matrix, we introduce an arbitrary direction on the edges of $\mathcal{G}$. Which direction is chosen for one edge does not influence the results of the analysis. The incidence matrix $\boldsymbol{E}(\mathcal{G}) \in \mathbb{R}^{|\mathcal{V}| \times |\mathcal{E}|}$ of the graph $\mathcal{G}$ is a $\{0, \pm 1\}$-matrix with rows and columns indexed by the nodes and edges of $\mathcal{G}$ such that $[\boldsymbol{E}(\mathcal{G})]_{ik}$ has value '+1' if node i is the initial node of edge k, '-1' if it is the terminal node, and '0' otherwise. This definition implies that for any graph, $\boldsymbol{1}^{\mathrm{T}} \boldsymbol{E}(\mathcal{G}) = \boldsymbol{0}$. The Laplacian matrix $\boldsymbol{L}(\mathcal{G})$ of the undirected graph $\mathcal{G}$ is defined as $\boldsymbol{L}(\mathcal{G}) = \boldsymbol{E}(\mathcal{G}) \boldsymbol{E}(\mathcal{G})^{\mathrm{T}}$.

We consider now the following static *saddle-point problem*

$$\max_{\boldsymbol{\mu}} \min_{\boldsymbol{\nu}} \; L(\boldsymbol{\nu}, \boldsymbol{\mu}) = \sum_{i=1}^{|\mathcal{V}|} F_i(\nu_i, \omega_i) + \boldsymbol{\mu}^{\mathrm{T}} \boldsymbol{E}(\mathcal{G})^{\mathrm{T}} \boldsymbol{\nu}. \qquad (6.54)$$
$$\text{s.t. } |\mu_k| \leq \alpha_k \quad \text{for all } k \in \mathcal{E},$$

where $\boldsymbol{\mu} = (\mu_1, \dots, \mu_{|\mathcal{E}|})^{\mathrm{T}}$ and $\boldsymbol{\nu} = (\nu_1, \dots, \nu_{|\mathcal{V}|})^{\mathrm{T}}$ are the vectors of the static decision variables. For simplicity of notation, we introduce the constraint set of the μ variables as $\Gamma = \{\boldsymbol{\mu} \in \mathbb{R}^{|\mathcal{E}|} : |\mu_k| \leq \alpha_k\}$.

A point $(\bar{\boldsymbol{\nu}}, \bar{\boldsymbol{\mu}})$ is a saddle-point of (6.54) if

$$L(\bar{\boldsymbol{\nu}}, \boldsymbol{\mu}) \leq L(\bar{\boldsymbol{\nu}}, \bar{\boldsymbol{\mu}}) \leq L(\boldsymbol{\nu}, \bar{\boldsymbol{\mu}})$$

for all $\boldsymbol{\nu} \in \mathbb{R}^{|\mathcal{V}|}$ and all $\boldsymbol{\mu} \in \Gamma$. Due to strict convexity of $F_i(\nu_i, \omega_i)$, the point $\bar{\boldsymbol{\nu}}$ at which a saddle-point can be attained is unique. However, given $\bar{\boldsymbol{\nu}}$ there is not necessarily a unique $\bar{\boldsymbol{\mu}}$ which establishes a saddle-point. In fact, given a $\bar{\boldsymbol{\mu}}$ which leads to a saddle-point, one can always add to $\bar{\boldsymbol{\mu}}$ an element from the nullspace of $\boldsymbol{E}(\mathcal{G})$ and, as long as the resulting $\boldsymbol{\mu}$ still satisfies the constraints, obtain another saddle-point. The nullspace of the incidence matrix, however, corresponds directly to the circles of the graph $\mathcal{G}$, see [136]. The structure of the set of saddle-points is, therefore, in fact explicitly connected to the topology of the graph. It seems worthwhile to point out that the existence of cycles in $\mathcal{G}$ leads to additional degrees of freedom in the solution of the saddle-point problem.

Interpretation of the Saddle-Point Problem. To understand the saddle point problem it is helpful to consider the unconstrained case with $\Gamma = \mathbb{R}^{|\mathcal{E}|}$. In this case, the saddle-point problem becomes the Lagrange-dual of the network optimization problem

$$\min_{\boldsymbol{\nu}} \sum_{i=1}^{|\mathcal{V}|} F_i(\nu_i, \omega_i) \quad \text{s.t. } \boldsymbol{E}(\mathcal{G})^{\mathrm{T}} \boldsymbol{\nu} = \mathbf{0}. \qquad (6.55)$$

The equality constraint $\boldsymbol{E}(\mathcal{G})^{\mathrm{T}} \boldsymbol{\nu} = \mathbf{0}$ is in fact a constraint forcing the decision variables of all nodes to be identical, i.e., $\nu_1 = \nu_2 = \cdots = \nu_{|\mathcal{V}|}$. We can understand the network optimization problem (6.55) as the equivalent to the synchronization problem, i.e., a setup where all nodes agree on the same variable. The saddle-point problem (6.54) has then an interpretation as a Lagrange-dual with additional constraints on the dual variables.

We can establish a similar cluster synchronization result as we have established for the dynamic network in Section 6.4.2 for the solution of the saddle-point problem (see [43, 45]). For any network configuration, there exists a partition $\mathbb{P} = \{\mathcal{P}_1, \dots, \mathcal{P}_p\}$, of the node set $\mathcal{V}$ such for all nodes $v_i, v_j \in \mathcal{P}_l$ within one partition the equality

$$\bar{\nu}_i = \bar{\nu}_j$$

holds, while for nodes $v_i \in \mathcal{P}_l, v_j \notin \mathcal{P}_l$ in different partitions

$$\bar{\nu}_i \neq \bar{\nu}_j.$$

The interpretation of the saddle-point problem (6.54) as a variation of the Lagrange-dual to the network optimization problem (6.55) provides insights into the clustering phenomenon. The dual variables in the saddle-point problem (6.54) are constrained and cannot fully compensate for the inhomogeneity of the node's objective functions, which will cause the clustered primal solutions of (6.54). In fact, if two nodes are connected by an edge, say $e_k = (v_i, v_j)$, for which there exists a dual solution such that $|\bar{\mu}_k| < \alpha_k$, i.e., the constraint is inactive, then both nodes will "agree" on the same primal value, i.e., $\bar{\nu}_i = \bar{\nu}_j$. If, however, $|\bar{\mu}_k| = \alpha_k$ and $\bar{\boldsymbol{\mu}}$ cannot be varied in the null-space of $\boldsymbol{E}(\mathcal{G})$ such that the constraint becomes inactive, then the two incident nodes will be separated and have different values, i.e., $\bar{\nu}_i \neq \bar{\nu}_j$. We say then that e_k is a saturated edge. All the appearing clusters will be separated by saturated edges.

Clustering Structures of the Saddle-Point Problem and of a Dynamic Network. The clustering structure resulting from the saddle-point problem is in fact closely related to the clustering structure of the dynamic network (6.46). Let $\bar{\boldsymbol{\nu}}$ be the unique solution to (6.54), then it has been shown in [45] that

$$\lim_{t\to\infty} \dot{x}_i(t) = \bar{\nu}_i, \quad \forall v_i \in \mathcal{V}. \tag{6.56}$$

The saddle-point problem provides us with further information about the dynamic system. Let $\bar{\boldsymbol{\nu}}$ be the unique primal component of the saddle-point, then there exists a vector $\tilde{\boldsymbol{\mu}}$ such that $(\bar{\boldsymbol{\nu}}, \tilde{\boldsymbol{\mu}})$ is a saddle-point of (6.54) and

$$\lim_{t\to\infty} \alpha_k \psi_k(x_i(t) - x_j(t)) \to \tilde{\mu}_k, \quad \forall e_k \in \mathcal{E}, \tag{6.57}$$

i.e., the nonlinear couplings will asymptotically approach a dual solution of the saddle-point problem. Thus, given any network configuration the clustering synchronization structure can be fully predicted by solving the corresponding static problem. Note that under the assumptions used here, the saddle-point problem is convex and can be solved efficiently using standard optimization tools.

Reduction to a Quadratic Optimization Problem. For the network with linear node dynamics (6.47), the functions $F_i(\nu_i, \omega_i)$ are quadratic functions of the form

$$F_i(\nu_i, \omega_i) = \frac{1}{2} A_i (\nu_i - \omega_i)^2.$$

Let in the following $\boldsymbol{A} = \mathrm{diag}(A_1, \ldots, A_{|\mathcal{V}|})$ and $\boldsymbol{\omega} = (\omega_1, \ldots, \omega_{|\mathcal{V}|})^{\mathrm{T}}$. At first note that for a given $\boldsymbol{\mu}$ the minimization over $\boldsymbol{\nu}$ is easy, as the optimal solution to the minimization problem can be directly determined:

$$\boldsymbol{\nu}^*(\boldsymbol{\mu}) = \boldsymbol{\omega} - \boldsymbol{A}^{-1}\boldsymbol{E}(\mathcal{G})\boldsymbol{\mu}.$$

The outer maximization part becomes then simply a quadratic program with linear constraints and reduces, after changing the maximization to a minimization problem, to

$$\begin{aligned} &\min_{\boldsymbol{\mu}} \frac{1}{2}\boldsymbol{\mu}^{\mathrm{T}}\boldsymbol{E}(\mathcal{G})^{\mathrm{T}}\boldsymbol{A}^{-1}\boldsymbol{E}(\mathcal{G})\boldsymbol{\mu} - \boldsymbol{\omega}^{\mathrm{T}}\boldsymbol{E}(\mathcal{G})\boldsymbol{\mu} \\ &\text{s.t. } |\mu_k| \leq \alpha_k, \quad \forall e_k \in \mathcal{E}. \end{aligned} \tag{6.58}$$

Note that $\boldsymbol{L}_A = \boldsymbol{E}(\mathcal{G})^{\mathrm{T}}\boldsymbol{A}^{-1}\boldsymbol{E}(\mathcal{G})$ is a special matrix, which in the literature is called weighted Edge-Laplacian [402]. This matrix can be shown to have the same non-zero eigenvalues as the weighted Laplacian of the network and has in particular only non-negative eigenvalues. The quadratic optimization problem (6.58) is, therefore, convex.

This discussion suggest that the analysis of the dynamic network (6.46) can be significantly simplified by considering a static convex optimization problem. For a given network configuration we can computationally very efficiently predict the clustering structure of the network.

6.4.5 Hierarchical Clustering Analysis and Community Detection

One motivation for studying cluster synchronization is the explicit relation between synchronization phenomena and the network topology and, in particular, communities in the graph. Detecting communities in graphs is one of the major research area in modern network theory and has received tremendous attention in the literature [125, 277]. The problem is to find groups of nodes which are either very similar to each other or strongly connected with each other in order to understand the hierarchical organization structure of the network.

Cluster synchronization is a promising concept to provide further insights into a structural network analysis. A natural idea is to define a community as a group of nodes which agree in their dynamic behavior. We want to address now the question how we can use the dynamic systems (6.46) and the associated static saddle-point problem (6.54) to uncover a hierarchical structure.

We start with the following idea. Given a network configuration, we can vary parameters of the network to enforce different clustering structures. In general, if we assume the graph topology to be fixed, we can either manipulate the node dynamics $f_i(\dot{x}_i, \omega_i)$ or the capacities α_k of the edges to enforce a clustering of the network. This approach provides many degrees of freedom and it would be desirable to have a single parameter such that a variation of this parameter uncovers the hierarchical structure. We restrict our discussion, therefore, to a simplified situation where all edges have identical capacities: $\alpha_1 = \alpha_2 = \cdots = \alpha_{|\mathcal{E}|} = \alpha$. The edge capacity α is now a parameter that can be varied.

Let $\mathbb{P}_\alpha$ denote the clustering structure of the network for a given value α. The network is fully separated for $\alpha = 0$ with $\mathbb{P}_0 = \{\{v_1\}, \ldots, \{v_{|\mathcal{V}|}\}\}$, and is in complete agreement for $\alpha \to \infty$ with $\mathbb{P}_{\alpha\to\infty} = \{\mathcal{V}\}$. If we compare now the clustering structure associated to two different edge capacities with $\alpha' > \alpha$, then it can be directly shown that the clustering structure $\mathbb{P}_{\alpha'}$ can be obtained only by joining clusters of $\mathbb{P}_\alpha$. A variation of the edge capacities will, therefore, reveal a hierarchical clustering structure of the underlying network. We will call the smallest edge capacity, which is required for the network to reach full synchronization the minimal agreement capacity $\alpha_{\min}$, i.e., the network reaches full synchronization for all $\alpha \geq \alpha_{\min}$. In fact, one can compute the clustering structure for a sample set of the parameters and determine the structure. Note that the clustering structure can be exactly computed as the solution to the convex saddle-point problem (6.54) and a hierarchical network analysis can be performed very efficiently from a computational perspective.

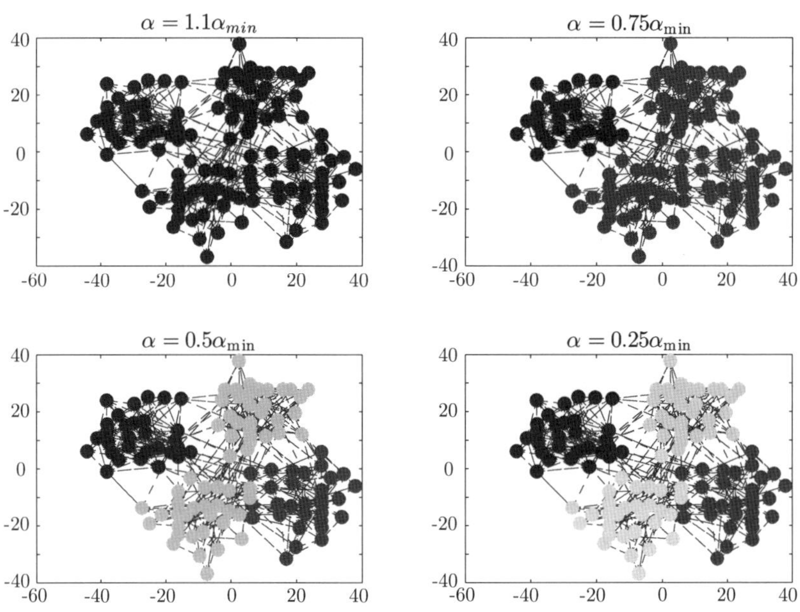

Fig. 6.15 Topology of the hierarchically structured network with Fruchterman-Reingold layout and cluster synchronization structure (node colors). For $\alpha = 1.1\alpha_{\min}$ (top-left) the network reaches complete agreement while different partitions are formed for $\alpha < \alpha_{\min}$

Example 6.3 *Hierarchically structured network*

As an illustrative example we consider the structured network shown in Fig. 6.15. The network has 120 nodes and is by its construction hierarchically structured. It consists of four node groups. Nodes within one group are connected with a probability of $p = 0.3$. Then, two partitions are grouped and nodes in one group are connected with a probability $p = 0.04$, while nodes in different groups are connected with a probability $p = 0.02$. In Fig. 6.15 the resulting network is shown according to a Fruchterman-Reingold layout [127].

We associate now to each node the linear function $f_i = (\dot{x}_i - \omega_i)$ where ω_i is chosen as the horizontal coordinate of the node. The coloring of the nodes in Fig. 6.15 indicates the clustering structure as computed by the corresponding static saddle-point problem. The network forms one ($\alpha = 1.1\alpha_{\min}$), two ($\alpha = 0.75\alpha_{\min}$), three ($\alpha = 0.5\alpha_{\min}$) and more ($\alpha = 0.25\alpha_{\min}$) clusters. The trajectories of the corresponding dynamic network have already been shown in Fig. 6.14. The dynamic simulations are preformed with random initial conditions and $\alpha = 0.75\alpha_{\min}$ and $\alpha = 0.25\alpha_{\min}$, respectively. The dynamic network clearly reflects the predicted clustering structure.

The optimality properties of the resulting clusters and the relation to other graph partitioning problems are discussed explicitly in [45]. We want to point out that the clustering structure depends strongly on the choice of ω_i and is not only related to the graph topology. However, it is a general principle that internally strongly connected components are less likely to partition than weakly connected components. □

Summary. A class of dynamic networks has been considered which shows asymptotically clustering structures. The mechanisms leading to this clustering are the inhomogeneous node dynamics along with the saturated diffusive coupling functions. We discussed that the full synchronization as well as the clustering are related to combinatorial properties of the network. Since the clustering structure is inherently connected to a static optimization problem, the saddle-point problem, one can efficiently use the cluster synchronization concept for a structural network analysis.

6.5 Autonomy and Cooperation in Networked Systems

6.5.1 Information Reduction in the Control of Systems with Identical Subsystems

Communication networks make it possible to exchange information among local control loops whenever an information exchange can improve the performance of the overall system. This section is devoted to interconnected systems consisting of an arbitrary number of identical subsystems, where the subsystem couplings are either caused by the physical relations among the subsystems (physically interconnected systems depicted in Fig. 6.2) or have to be introduced by the networked controller to cope with cooperative control goals (multi-agent systems depicted in Fig. 6.1). In both cases, the subsystems are controlled by associated decentralized controllers. The controllers may exchange some information through the communication network in order to improve the overall system performance. Conditions are elaborated for describing which and when some information exchange should be temporarily invoked among the decentralized controllers.

The subsystems operate in two alternative modes:

- **Autonomous mode:** The subsystems have decentralized control stations that do not interact with each other and are designed to meet individual aims of the subsystems. No communication is invoked among the subsystems that work in the autonomous mode.

- **Cooperative mode:** The control stations exchange information in order to improve the overall system behavior (disturbance attenuation, set-point following).

The main results concern methods for the selection of the information to be exchanged, the description of the situations in which this information has to be temporarily communicated from specific subsystems to specified other subsystems, and methods for the evaluation of the performance gain obtained due to the communication. The results are illustrated by their application to a multizone furnace.

It is obvious that the optimal system performance will be achieved in the cooperative mode with a centralized controller using a continuous communication. In order to reduce the communication load, the information exchange will be only temporarily invoked among the local controllers. The corresponding performance degradation will be analyzed. If this degradation is too large, communication among all or a subset of subsystems will be invoked.

The results for identical subsystems are extended to interconnected subsystems with similar dynamics. The subsystems are decomposed in a nominal model with identical dynamics and individual models that describe the deviation of the subsystems behavior from the common nominal model. Hence, the control design methods for identical subsystems can be applied to the nominal models and afterwards the original subsystem behavior with similar dynamics is analyzed for stability.

Another way to reduce the communication effort is to apply event-based control strategies (cf. Chapter 5) , where the need of an information exchange is reduced by the estimation of the states of the neighbor subsystems. The main idea of the event-based control is to approximate the behavior of the continuous state-feedback controller with a desired precision. The information exchange is only invoked if the deviation of the estimated state from the true state exceeds a communication threshold. Interested readers should consult the references [96, 97, 353] for the event-based implementation of the controllers described in this section.

6.5.2 System Models

Multi-agent Systems. Figure 6.1 on p. 265 shows a multi-agent system consisting of N identical subsystems

$$\begin{aligned} \dot{\boldsymbol{x}}_i(t) &= \bar{\boldsymbol{A}}\boldsymbol{x}_i(t) + \boldsymbol{B}\boldsymbol{u}_{\mathrm{c},i}(t) + \boldsymbol{G}\boldsymbol{d}_i(t), \quad \boldsymbol{x}_i(0) = \boldsymbol{x}_{i0} \\ \boldsymbol{y}_i(t) &= \boldsymbol{C}\boldsymbol{x}_i(t), \quad i = 1, 2, ...N, \end{aligned} \tag{6.59}$$

where

- $\boldsymbol{x}_i(t)$ is the n-dimensional subsystem state,
- $\boldsymbol{y}_i(t)$ is the output vector,
- $\boldsymbol{d}_i(t)$ is the unknown disturbance and
- $\boldsymbol{u}_{\mathrm{c},i}(t)$ is the cooperative part of the control input generated by using the communicated information.

The system matrix $\bar{\boldsymbol{A}} = \boldsymbol{A} - \boldsymbol{B}\boldsymbol{K}_{\mathrm{d}}$ includes the decentralized feedback term, which is effective in the autonomous mode as well as in the cooperative mode. The control input of a single subsystem consists of two parts,

$$\boldsymbol{u}_i(t) = \boldsymbol{u}_{\mathrm{c},i}(t) + \boldsymbol{u}_{\mathrm{a},i}(t),$$

where $\boldsymbol{u}_{\mathrm{a},i}(t)$ denotes the autonomous part of the control input that is generated by the decentralized controllers without communication:

$$\boldsymbol{u}_{\mathrm{a},i}(t) = \boldsymbol{K}_{\mathrm{d}}\boldsymbol{x}_i(t). \tag{6.60}$$

The overall system is represented by

$$\begin{aligned} \dot{\boldsymbol{x}}(t) &= (\boldsymbol{I}_N \otimes \boldsymbol{A})\, \boldsymbol{x}(t) + (\boldsymbol{I}_N \otimes \boldsymbol{B})\, \boldsymbol{u}(t) + (\boldsymbol{I}_N \otimes \boldsymbol{G})\, \boldsymbol{d}(t) \\ &= \boldsymbol{A}_{\mathrm{os}}\boldsymbol{x}(t) + \boldsymbol{B}_{\mathrm{os}}\boldsymbol{u}(t) + \boldsymbol{G}_{\mathrm{os}}\boldsymbol{d}(t) \end{aligned} \tag{6.61}$$

with

$$\boldsymbol{A}_{\mathrm{os}} = (\boldsymbol{I}_N \otimes \boldsymbol{A}), \quad \boldsymbol{B}_{\mathrm{os}} = (\boldsymbol{I}_N \otimes \boldsymbol{B}), \quad \boldsymbol{G}_{\mathrm{os}} = (\boldsymbol{I}_N \otimes \boldsymbol{G}) \tag{6.62}$$

$$\begin{aligned} &\boldsymbol{x}(t) = \left(\boldsymbol{x}_1^{\mathrm{T}}(t), \ldots, \boldsymbol{x}_N^{\mathrm{T}}(t)\right)^{\mathrm{T}}, \quad \boldsymbol{u}_{\mathrm{c}}(t) = \left(\boldsymbol{u}_{\mathrm{c},i}^{\mathrm{T}}(t), \ldots, \boldsymbol{u}_{\mathrm{c,N}}^{\mathrm{T}}(t)\right)^{\mathrm{T}} \text{ and} \\ &\boldsymbol{d}(t) = \left(\boldsymbol{d}_1^{\mathrm{T}}(t), \ldots, \boldsymbol{d}_N^{\mathrm{T}}(t)\right)^{\mathrm{T}}. \end{aligned} \tag{6.63}$$

Physically Interconnected Systems. Figure 6.2 on p. 265 shows a physically interconnected system consisting of N identical subsystems

$$\begin{aligned} \dot{\boldsymbol{x}}_i(t) &= \bar{\boldsymbol{A}}\boldsymbol{x}_i(t) + \boldsymbol{B}\boldsymbol{u}_{\mathrm{c},i}(t) + \boldsymbol{e}s_i(t) + \boldsymbol{G}\boldsymbol{d}_i(t), \quad \boldsymbol{x}_i(0) = \boldsymbol{x}_{i0} \qquad (6.64) \\ \boldsymbol{y}_i(t) &= \boldsymbol{C}\boldsymbol{x}_i(t) \\ z_i(t) &= \boldsymbol{c}_{\mathrm{z}}^{\mathrm{T}}\boldsymbol{x}_i(t) \\ s_i(t) &= \sum_{j=1}^{N} l_{ij} z_j(t), \quad i = 1, 2, ..., N, \qquad (6.65) \end{aligned}$$

where

- $\boldsymbol{x}_i(t)$ is the n-dimensional subsystem state,
- $\boldsymbol{y}_i(t)$ is the output vector,
- $s_i(t)$ is the scalar interconnection input signal,
- $z_i(t)$ is the scalar interconnection output signal,
- $\boldsymbol{d}_i(t)$ is the unknown disturbance,
- $\boldsymbol{u}_{\mathrm{c},i}(t)$ is the cooperative part of the control input generated by using the communicated information.

The physical interconnections within the overall system are described by

$$\boldsymbol{s}(t) = \boldsymbol{L}_{\mathrm{p}}\boldsymbol{z}(t) \tag{6.66}$$

with $\boldsymbol{s}(t) = (s_1(t), \ldots, s_N(t))^{\mathrm{T}}$ and $\boldsymbol{z}(t) = (z_1(t), \ldots, z_N(t))^{\mathrm{T}}$. The interconnection matrix $\boldsymbol{L}_{\mathrm{p}} = (l_{\mathrm{p}ij})$ is assumed to be diagonalizable.

The overall system can be represented as

$$\begin{aligned}\dot{\boldsymbol{x}}(t) &= \left((\boldsymbol{I}_N \otimes \bar{\boldsymbol{A}}) + (\boldsymbol{I}_N \otimes \boldsymbol{e})\left(\boldsymbol{L}_{\mathrm{p}} \otimes \boldsymbol{c}_z^{\mathrm{T}}\right)\right)\boldsymbol{x}(t) + (\boldsymbol{I}_N \otimes \boldsymbol{B})\,\boldsymbol{u}_{\mathrm{c},i}(t) \\ &\quad + (\boldsymbol{I}_N \otimes \boldsymbol{G})\,\boldsymbol{d}(t) \qquad (6.67) \\ &= \left((\boldsymbol{I}_N \otimes \bar{\boldsymbol{A}}) + \left(\boldsymbol{L}_{\mathrm{p}} \otimes \boldsymbol{e}\boldsymbol{c}_{\mathrm{z}}^{\mathrm{T}}\right)\right)\boldsymbol{x}(t) + (\boldsymbol{I}_N \otimes \boldsymbol{B})\,\boldsymbol{u}_{\mathrm{c}}(t) + (\boldsymbol{I}_N \otimes \boldsymbol{G})\,\boldsymbol{d}(t) \\ &= \boldsymbol{A}_{\mathrm{os}}\boldsymbol{x}(t) + \boldsymbol{B}_{\mathrm{os}}\boldsymbol{u}_{\mathrm{c}}(t) + \boldsymbol{G}_{\mathrm{os}}\boldsymbol{d}(t) \qquad (6.68)\end{aligned}$$

with

$$\boldsymbol{A}_{\mathrm{os}} = \left(\boldsymbol{I}_N \otimes \bar{\boldsymbol{A}}\right) + \left(\boldsymbol{L}_{\mathrm{p}} \otimes \boldsymbol{e}\boldsymbol{c}_{\mathrm{z}}^{\mathrm{T}}\right), \quad \boldsymbol{B}_{\mathrm{os}} = (\boldsymbol{I}_N \otimes \boldsymbol{B}), \quad \boldsymbol{G}_{\mathrm{os}} = (\boldsymbol{I}_N \otimes \boldsymbol{G}) \tag{6.69}$$

$$\begin{aligned}\boldsymbol{x}(t) &= \left(\boldsymbol{x}_1^{\mathrm{T}}(t), \ldots, \boldsymbol{x}_N^{\mathrm{T}}(t)\right)^{\mathrm{T}}, \quad \boldsymbol{u}_{\mathrm{c}}(t) = \left(\boldsymbol{u}_{\mathrm{c},i}^{\mathrm{T}}(t), \ldots, \boldsymbol{u}_{\mathrm{c,N}}^{\mathrm{T}}(t)\right)^{\mathrm{T}} \\ \boldsymbol{d}(t) &= \left(\boldsymbol{d}_1^{\mathrm{T}}(t), \ldots, \boldsymbol{d}_N^{\mathrm{T}}(t)\right)^{\mathrm{T}}.\end{aligned}$$

6.5.3 Offline Information Reduction: Decomposition of the Controller Design Problem

A typical linear-quadratic regulator (LQR) design problem for multi-agent systems and physically interconnected systems is considered. In this section, a decomposition will be introduced that reduces the complexity of the design problem in both cases.

LQR Design Problem for Multi-agent Systems. As mentioned before, multi-agent systems consist of isolated subsystems that share some control goals. Consequently, the cost function contains terms that penalize local subsystem states and inputs as well as the differences between the subsystem states. In the cost function

$$J_{\mathrm{M}}(\boldsymbol{u}_{\mathrm{c}}(t), \boldsymbol{x}_0) = \int_0^{\infty} \boldsymbol{x}^{\mathrm{T}}(t)\boldsymbol{Q}_{\mathrm{os}}\boldsymbol{x}(t) + \boldsymbol{u}_{\mathrm{c}}(t)^{\mathrm{T}}\boldsymbol{R}_{\mathrm{os}}\boldsymbol{u}_{\mathrm{c}}(t)\,\mathrm{d}t \tag{6.70}$$

with

$$\boldsymbol{Q}_{\mathrm{os}} = (\boldsymbol{I}_N \otimes \boldsymbol{Q}_{\mathrm{a}}) + (\boldsymbol{L}_{\mathrm{q}} \otimes \boldsymbol{Q}_{\mathrm{c}}) \succeq 0 \quad \text{and} \quad \boldsymbol{R}_{\mathrm{os}} = (\boldsymbol{I}_N \otimes \boldsymbol{R}) \succ 0, \tag{6.71}$$

the interconnection matrix $\boldsymbol{L}_{\mathrm{q}}$ is a symmetric Laplacian matrix describing the couplings among the control goals of the agents. In cost function J_{M}, $\boldsymbol{Q}_{\mathrm{c}} \succ 0$ denotes the cooperative weighting matrix penalizing the differences among the subsystem states while $\boldsymbol{Q}_{\mathrm{a}} \succ 0$ denotes the autonomous weighting matrix for the local subsystem states and $\boldsymbol{R}$ penalizes the cooperative control input, which makes a communication among the subsystems necessary.

Since $\boldsymbol{L}_{\mathrm{q}}$ is symmetric, there exits an orthogonal matrix $\boldsymbol{T}_{\mathrm{q}}$, which diagonalizes $\boldsymbol{L}_{\mathrm{q}}$:

$$\tilde{\boldsymbol{L}}_{\mathrm{q}} := \boldsymbol{T}_{\mathrm{q}}^{-1}\boldsymbol{L}_{\mathrm{q}}\boldsymbol{T}_{\mathrm{q}} = \begin{pmatrix} \lambda_1(\boldsymbol{L}_{\mathrm{q}}) & 0 & \dots & 0 \\ 0 & \lambda_2(\boldsymbol{L}_{\mathrm{q}}) & \ddots & \vdots \\ \vdots & \ddots & \ddots & 0 \\ 0 & \dots & 0 & \lambda_N(\boldsymbol{L}_{\mathrm{q}}) \end{pmatrix}. \tag{6.72}$$

The design problem is given by

$$\min_{\boldsymbol{u}(t)} J_{\mathrm{M}}(\boldsymbol{u}_{\mathrm{c}}(t), \boldsymbol{x}_0) \tag{6.73}$$

$$s.t. \quad \dot{\boldsymbol{x}}(t) = \boldsymbol{A}_{\mathrm{os}}\boldsymbol{x}(t) + \boldsymbol{B}_{\mathrm{os}}\boldsymbol{u}_{\mathrm{c}}(t), \quad \boldsymbol{x}(0) = \boldsymbol{x}_0 \tag{6.74}$$

with $\boldsymbol{A}_{\mathrm{os}}$ and $\boldsymbol{B}_{\mathrm{os}}$ introduced in Eq. (6.62).

It is well known that the control input $\boldsymbol{u}_{\mathrm{c}}^*(t)$, which minimizes the cost function, can be represented as state feedback

$$\boldsymbol{u}_{\mathrm{c}}^*(t) = \boldsymbol{K}_{\mathrm{os}}^*\boldsymbol{x}(t) \tag{6.75}$$

with the optimal controller matrix

$$\boldsymbol{K}_{\mathrm{os}}^* = -\boldsymbol{R}_{\mathrm{os}}^{-1}\boldsymbol{B}_{\mathrm{os}}^{\mathrm{T}}\boldsymbol{P}_{\mathrm{os}}^*, \tag{6.76}$$

where $\boldsymbol{P}_{\mathrm{os}}^*$ is the positive definite solution of the algebraic Riccati equation (ARE)

$$\boldsymbol{A}_{\mathrm{os}}^{\mathrm{T}}\boldsymbol{P}_{\mathrm{os}}^* + \boldsymbol{P}_{\mathrm{os}}^*\boldsymbol{A}_{\mathrm{os}} - \boldsymbol{P}_{\mathrm{os}}^*\boldsymbol{B}_{\mathrm{os}}\boldsymbol{R}_{\mathrm{os}}^{-1}\boldsymbol{B}_{\mathrm{os}}^{\mathrm{T}}\boldsymbol{P}_{\mathrm{os}}^* + \boldsymbol{Q}_{\mathrm{os}} = \boldsymbol{O} \tag{6.77}$$

for the overall system.

LQR Design Problem for Physically Interconnected Systems. In physically interconnected systems the subsystems often have individual control objectives. A reasonable cost function is, therefore,

$$J_{\mathrm{P}} = \int_0^{\infty} \boldsymbol{x}(t)^{\mathrm{T}} \boldsymbol{Q}_{\mathrm{os}} \boldsymbol{x}(t) + \boldsymbol{u}_{\mathrm{c}}(t)^{\mathrm{T}} \boldsymbol{R}_{\mathrm{os}} \boldsymbol{u}_{\mathrm{c}}(t) \, \mathrm{d}t \tag{6.78}$$

with

$$\boldsymbol{Q}_{\mathrm{os}} = (\boldsymbol{I}_N \otimes \boldsymbol{Q}_{\mathrm{a}}) \text{ and } \boldsymbol{R}_{\mathrm{os}} = (\boldsymbol{I}_N \otimes \boldsymbol{R}), \tag{6.79}$$

where $\boldsymbol{R} \succ 0$ and $\boldsymbol{Q}_{\mathrm{a}} \succ 0$ holds. The design problem is given by

$$\min_{\boldsymbol{u}_{\mathrm{c}}(t)} J_{\mathrm{P}}(\boldsymbol{u}_{\mathrm{c}}(t), \boldsymbol{x}_0) \tag{6.80}$$

$$s.t. \quad \dot{\boldsymbol{x}}(t) = \boldsymbol{A}_{\mathrm{os}} \boldsymbol{x}(t) + \boldsymbol{B}_{\mathrm{os}} \boldsymbol{u}_{\mathrm{c}}(t), \quad \boldsymbol{x}(0) = \boldsymbol{x}_0 \tag{6.81}$$

with $\boldsymbol{A}_{\mathrm{os}}$ and $\boldsymbol{B}_{\mathrm{os}}$ introduced in Eq. (6.69). The control input $\boldsymbol{u}_{\mathrm{c}}^*(t)$ that minimizes the cost function can be represented as

$$\boldsymbol{u}_{\mathrm{c}}^*(t) = -\boldsymbol{K}_{\mathrm{os}}^* \boldsymbol{x}(t) \tag{6.82}$$

with the optimal controller matrix

$$\boldsymbol{K}_{\mathrm{os}}^* = \boldsymbol{R}_{\mathrm{os}}^{-1} \boldsymbol{B}_{\mathrm{os}}^{\mathrm{T}} \boldsymbol{P}_{\mathrm{os}}^*, \tag{6.83}$$

where $\boldsymbol{P}_{\mathrm{os}}^*$ is the positive definite solution of the following algebraic Riccati equation (ARE) for the overall system

$$\boldsymbol{A}_{\mathrm{os}}^{\mathrm{T}} \boldsymbol{P}_{\mathrm{os}}^* + \boldsymbol{P}_{\mathrm{os}}^* \boldsymbol{A}_{\mathrm{os}} - \boldsymbol{P}_{\mathrm{os}}^* \boldsymbol{B}_{\mathrm{os}} \boldsymbol{R}_{\mathrm{os}}^{-1} \boldsymbol{B}_{\mathrm{os}}^{\mathrm{T}} \boldsymbol{P}_{\mathrm{os}}^* + \boldsymbol{Q}_{\mathrm{os}} = \boldsymbol{O}. \tag{6.84}$$

Although $\boldsymbol{Q}_{\mathrm{os}}$ is a block-diagonal matrix the solution $\boldsymbol{P}_{\mathrm{os}}^*$ of the optimization problem is (generally) a fully occupied matrix. Hence, the implementation of the optimal controller $\boldsymbol{K}_{\mathrm{os}}^*$ requires an information exchange among all the subsystems.

For both multi-agent systems and physically interconnected systems the complexity of this design problem increases with the number N of subsystems. However, the following investigation will show that, first, the design problem can be reduced to N optimal control problems for auxiliary subsystems and, second, the controller can be implemented with situation-dependent rather than continuous communication among the subsystems.

Decomposition of the Design Problem for Multi-agent Systems. Consider the state transformation

$$\tilde{\boldsymbol{x}}(t) = (\boldsymbol{T}_{\mathrm{q}} \otimes \boldsymbol{I}_{\mathrm{n}})^{-1} \boldsymbol{x}(t) = \boldsymbol{T}_{\mathrm{q,os}}^{-1} \boldsymbol{x}(t) \tag{6.85}$$

with the transformation matrix $\boldsymbol{T}_{\rm q}$ that diagonalizes the Laplacian matrix $\boldsymbol{L}_{\rm q}$:

$$\boldsymbol{T}_{\rm q}^{-1}\boldsymbol{L}_{\rm q}\boldsymbol{T}_{\rm q} = {\rm diag}(\lambda_i(\boldsymbol{L}_{\rm q})).$$

The state transformation (6.85) is applied to the model (6.61) to get the system matrix

$$\begin{aligned}\tilde{\boldsymbol{A}}_{\rm os} = \boldsymbol{T}_{\rm q,os}^{-1}\boldsymbol{A}_{\rm os}\boldsymbol{T}_{\rm q,os} &= (\boldsymbol{T}_{\rm q}\otimes\boldsymbol{I}_n)^{-1}(\boldsymbol{I}_N\otimes\boldsymbol{A})(\boldsymbol{T}_{\rm q}\otimes\boldsymbol{I}_n)\\ &= (\boldsymbol{T}_{\rm q}^{-1}\otimes\boldsymbol{I}_n)(\boldsymbol{T}_{\rm q}\otimes\boldsymbol{A})\\ &= (\boldsymbol{I}_N\otimes\boldsymbol{A}) \qquad (6.86)\end{aligned}$$

and to the weighting matrix $\boldsymbol{Q}_{\rm os}$ to get

$$\begin{aligned}\tilde{\boldsymbol{Q}}_{\rm os} &= \boldsymbol{T}_{\rm q,os}^{-1}\boldsymbol{Q}_{\rm os}\boldsymbol{T}_{\rm q,os}\\ &= \left(\boldsymbol{T}_{\rm q}^{-1}\otimes\boldsymbol{I}_n\right)\boldsymbol{Q}_{\rm os}\left(\boldsymbol{T}_{\rm q}\otimes\boldsymbol{I}_n\right) \qquad (6.87)\\ &= \left(\boldsymbol{T}_{\rm q}^{-1}\otimes\boldsymbol{I}_n\right)\left(\boldsymbol{I}_N\otimes\boldsymbol{Q}_{\rm a}\right)\left(\boldsymbol{T}_{\rm q}\otimes\boldsymbol{I}_n\right) + \left(\boldsymbol{T}_{\rm q}^{-1}\otimes\boldsymbol{I}_n\right)\left(\boldsymbol{L}_{\rm q}\otimes\boldsymbol{Q}_{\rm c}\right)\left(\boldsymbol{T}_{\rm q}\otimes\boldsymbol{I}_n\right)\\ &= \left(\boldsymbol{I}_N\otimes\boldsymbol{Q}_{\rm a}\right) + \left(\boldsymbol{T}_{\rm q}^{-1}\boldsymbol{L}_{\rm q}\boldsymbol{T}_{\rm q}\otimes\boldsymbol{Q}_{\rm c}\right) = \left(\boldsymbol{I}_N\otimes\boldsymbol{Q}_{\rm a}\right) + \left(\tilde{\boldsymbol{L}}_{\rm q}\otimes\boldsymbol{Q}_{\rm c}\right)\\ &= \left(\boldsymbol{I}_N\otimes\boldsymbol{Q}_{\rm a}\right) + \left({\rm diag}\left(\lambda_i(\boldsymbol{L}_{\rm q})\right)\otimes\boldsymbol{Q}_{\rm c}\right). \qquad (6.88)\end{aligned}$$

If the state transformation (6.85) is applied to the ARE (6.77), the ARE for the overall system can be decomposed into N ARE's for the modified subsystems. This result is summarized in the following theorem.

Theorem 6.12. *The solution of the LQR design problem* (6.73) – (6.74) *for the multi-agent system* (6.61) *can be represented by* (6.75) *and* (6.76) *with*

$$\boldsymbol{P}_{\rm os}^{*} = (\boldsymbol{T}_{\rm q}\otimes\boldsymbol{I}_n)\ {\rm diag}(\tilde{\boldsymbol{P}}_i^{*})\ (\boldsymbol{T}_{\rm q}\otimes\boldsymbol{I}_n)^{-1}$$

in terms of the solutions $\tilde{\boldsymbol{P}}_i^{*}$ of the following N separate ARE's

$$\boldsymbol{A}^{\rm T}\tilde{\boldsymbol{P}}_i^{*} + \tilde{\boldsymbol{P}}_i^{*}\boldsymbol{A} - \tilde{\boldsymbol{P}}_i^{*}\boldsymbol{B}\boldsymbol{R}^{-1}\boldsymbol{B}^{\rm T}\tilde{\boldsymbol{P}}_i^{*} + \tilde{\boldsymbol{Q}}_i = \boldsymbol{O} \qquad (6.89)$$

with

$$\tilde{\boldsymbol{Q}}_i = \boldsymbol{Q}_{\rm a} + \lambda_i(\boldsymbol{L}_{\rm q})\boldsymbol{Q}_{\rm c}, \quad i = 1, \ldots, N. \qquad (6.90)$$

Decomposition of the Design Problem for Physically Interconnected Systems. Similarly to the previous case, consider now the state transformation

$$\tilde{\boldsymbol{x}}(t) = (\boldsymbol{T}_{\rm p}\otimes\boldsymbol{I}_{\rm n})^{-1}\boldsymbol{x}(t) = \boldsymbol{T}_{\rm p,os}^{-1}\boldsymbol{x}(t) \qquad (6.91)$$

with the transformation matrix $\boldsymbol{T}_\mathrm{p}$ that diagonalizes the interconnection matrix $\boldsymbol{L}_\mathrm{p}$:

$$\boldsymbol{T}_\mathrm{p}^{-1}\boldsymbol{L}_\mathrm{p}\boldsymbol{T}_\mathrm{p} = \mathrm{diag}(\lambda_i(\boldsymbol{L}_\mathrm{p})).$$

If this state transformation (6.91) is applied to the overall system, the system matrix $\boldsymbol{A}_\mathrm{os}$ gets the following block-diagonal form:

$$\begin{aligned}\tilde{\boldsymbol{A}}_\mathrm{os} =&(\boldsymbol{T}_\mathrm{p} \otimes \boldsymbol{I}_n)^{-1}\boldsymbol{A}_\mathrm{os}(\boldsymbol{T}_\mathrm{p} \otimes \boldsymbol{I}_n) && (6.92)\\ =&(\boldsymbol{T}_\mathrm{p} \otimes \boldsymbol{I}_n)^{-1}((\boldsymbol{I}_N \otimes \bar{\boldsymbol{A}}) + (\boldsymbol{L}_\mathrm{p} \otimes \boldsymbol{e}\boldsymbol{c}_\mathrm{z}^\mathrm{T}))(\boldsymbol{T}_\mathrm{p} \otimes \boldsymbol{I}_n)\\ =&(\boldsymbol{T}_\mathrm{p}^{-1} \otimes \bar{\boldsymbol{A}})(\boldsymbol{T}_\mathrm{p} \otimes \boldsymbol{I}_n) + (\boldsymbol{T}_\mathrm{p}^{-1}\boldsymbol{L}_\mathrm{p} \otimes \boldsymbol{e}\boldsymbol{c}_\mathrm{z}^\mathrm{T})(\boldsymbol{T}_\mathrm{p} \otimes \boldsymbol{I}_n)\\ =&(\boldsymbol{I}_N \otimes \bar{\boldsymbol{A}}) + (\tilde{\boldsymbol{L}}_\mathrm{p} \otimes \boldsymbol{e}\boldsymbol{c}_\mathrm{z}^\mathrm{T})\\ =&(\boldsymbol{I}_N \otimes \bar{\boldsymbol{A}}) + \mathrm{diag}\left(\lambda_i(\boldsymbol{L}_\mathrm{p})\right) \otimes \boldsymbol{e}\boldsymbol{c}_\mathrm{z}^\mathrm{T} && (6.93)\end{aligned}$$

Accordingly, the stability of the interconnected system is equivalent to the stability of the N modified subsystems given in Eq. (6.93). The deviation of the modified subsystems from each other can be seen as a bounded uncertainty. Hence, the decentralized controller design problem for the interconnected system can be considered as controller design problem for a single uncertain system and strategies from robust control can be applied [94, 299].

If the state transformation (6.91) is applied to the ARE (6.84), the ARE for the overall system can be decomposed into N ARE's for modified subsystems. This result is summarized in the following theorem.

Theorem 6.13. *The solution $\boldsymbol{P}_\mathrm{os}^*$ of the LQR design problem* (6.80) – (6.81) *for the physically interconnected system* (6.68) *can be represented by Eqs.* (6.82) *and* (6.83) *with*

$$\boldsymbol{P}_\mathrm{os}^* = (\boldsymbol{T}_\mathrm{p} \otimes \boldsymbol{I}_n)\,\mathrm{diag}(\tilde{\boldsymbol{P}}_i^*)(\boldsymbol{T}_\mathrm{p} \otimes \boldsymbol{I}_n)^{-1} \qquad (6.94)$$

in terms of the solutions $\tilde{\boldsymbol{P}}_i^*$ of the following N separate ARE's

$$\tilde{\boldsymbol{A}}_i^\mathrm{T}\tilde{\boldsymbol{P}}_i^* + \tilde{\boldsymbol{P}}_i^*\tilde{\boldsymbol{A}}_i - \tilde{\boldsymbol{P}}_i^*\boldsymbol{B}\boldsymbol{R}^{-1}\boldsymbol{B}^\mathrm{T}\tilde{\boldsymbol{P}}_i^* + \boldsymbol{Q}_\mathrm{a} = \boldsymbol{O} \qquad (6.95)$$

with

$$\tilde{\boldsymbol{A}}_i = \bar{\boldsymbol{A}} + \lambda_i(\boldsymbol{L}_\mathrm{p})\boldsymbol{e}\boldsymbol{c}_\mathrm{z}^\mathrm{T}, \quad i = 1, 2, ..., N.$$

Theorems 6.12 and 6.13 deal with different interconnection structures. However, in both cases, a state transformation can be used to decompose the controller design problem into N separate design problems for modified subsystems, which are given in Eq. (6.89) for multi-agent systems and in Eq. (6.95) for physically interconnected systems. This step provides an important simplification.

In Eq. (6.89), the term $\tilde{\boldsymbol{Q}}_i$ contains information on the interconnection of the control goals, while in Eq. (6.95), the term $\tilde{\boldsymbol{A}}_i$ contains information on the physical interconnections of the subsystems. If the matrix has some multiple eigenvalues, the corresponding ARE's (6.95) are identical with each other. The same decomposition approach can also be applied to discrete-time systems to solve the corresponding discrete-time LQR design problem [98].

Extension I: Physical Input and Output Vectors. In Section 6.5.2, it has been assumed that the physical input and output signals can be described as scalars. Now, a more general system class will be considered, where the physical interconnection input and output signals are vectors of the same dimension: $\boldsymbol{s}_i, \boldsymbol{z}_i \in \mathbb{R}^m$. Consequently, $\boldsymbol{L}_\mathrm{p} \in \mathbb{R}^{Nm\times Nm}$, $\boldsymbol{E} \in \mathbb{R}^{n\times m}$ and $\boldsymbol{C}_\mathrm{z} \in \mathbb{R}^{m\times n}$ hold. Then the state equation can be given by

$$\dot{\boldsymbol{x}}(t) = ((\boldsymbol{I}_N \otimes \bar{\boldsymbol{A}}) + (\boldsymbol{I}_N \otimes \boldsymbol{E})\boldsymbol{L}_\mathrm{p}(\boldsymbol{I}_N \otimes \boldsymbol{C}_\mathrm{z}))\boldsymbol{x}(t) \tag{6.96}$$

and the interconnection matrix $\boldsymbol{L}_\mathrm{p}$ can be represented as

$$\boldsymbol{L}_\mathrm{p} = \sum_{t=1}^{N_l} \boldsymbol{L}_t \otimes \boldsymbol{L}^t \tag{6.97}$$

where $\boldsymbol{L}_t \in \mathbb{R}^{N\times N}$, $\boldsymbol{L}^t \in \mathbb{R}^{m\times m}$ and j is a finite number. Note that the interconnection structure considered in (6.65) is a special case of (6.97), where $m = 1$, $\boldsymbol{L}_1 = \boldsymbol{L}_\mathrm{p}$, $\boldsymbol{L}^1 = 1$ and $\boldsymbol{L}^t = 0$ for $t \neq 1$. If a transformation matrix $\boldsymbol{T}$ exists, which simultaneously diagonalizes the interconnection matrices $\boldsymbol{L}_t$, $t = 1, 2, ..., N_l$, after the state transformation (6.91), the system matrix gets again a block-diagonal form. The matrices in the main diagonal are given by

$$\tilde{\boldsymbol{A}}_i = \bar{\boldsymbol{A}} + \sum_{t=1}^{N_l} \lambda_i(\boldsymbol{L}_t)\boldsymbol{E}\boldsymbol{L}^t\boldsymbol{C}_\mathrm{z}, \quad i = 1, 2, ..., N. \tag{6.98}$$

Consequently, the ARE (6.84) can be decomposed in the following N ARE's for the modified subsystems

$$\tilde{\boldsymbol{A}}_i^\mathrm{T}\tilde{\boldsymbol{P}}_i^* + \tilde{\boldsymbol{P}}_i^*\tilde{\boldsymbol{A}}_i - \tilde{\boldsymbol{P}}_i^*\boldsymbol{B}\boldsymbol{R}^{-1}\boldsymbol{B}^\mathrm{T}\tilde{\boldsymbol{P}}_i^* + \boldsymbol{Q}_\mathrm{a} = \boldsymbol{O}, \tag{6.99}$$

with

$$\tilde{\boldsymbol{A}}_i = \bar{\boldsymbol{A}} + \sum_{t=1}^{j} \lambda_t(\boldsymbol{L}_\mathrm{t})\boldsymbol{E}\boldsymbol{L}^t\boldsymbol{C}_\mathrm{z}. \tag{6.100}$$

Extension II: Cooperative Control Goals. If the physically interconnected systems share some control goals, the weighting matrix in the objective function (6.78) has the following structure

$$\boldsymbol{Q}_\mathrm{os} = (\boldsymbol{I}_N \otimes \boldsymbol{Q}_\mathrm{a}) + (\boldsymbol{L}_\mathrm{q} \otimes \boldsymbol{Q}_\mathrm{c}),$$

where $\boldsymbol{L}_\mathrm{q}$ is a symmetric matrix describing the couplings in the control goals, while $\boldsymbol{Q}_\mathrm{c}$ denotes the cooperative weighting matrix and penalizes the difference state of the subsystems. If the matrices $\boldsymbol{L}_\mathrm{q}$ and $\boldsymbol{L}_\mathrm{p}$ are simultaneously diagonalizable, the same simplifications for the controller design is achieved, so that the optimal controller design for the overall system can be performed for the modified subsystems separately. The ARE's for the modified subsystems are given by

$$\tilde{\boldsymbol{A}}_i^\mathrm{T}\tilde{\boldsymbol{P}}_i^* + \tilde{\boldsymbol{P}}_i^*\tilde{\boldsymbol{A}}_i - \tilde{\boldsymbol{P}}_i^*\boldsymbol{B}\boldsymbol{R}^{-1}\boldsymbol{B}^\mathrm{T}\tilde{\boldsymbol{P}}_i^* + \tilde{\boldsymbol{Q}}_i = \boldsymbol{O}$$

with

$$\tilde{\boldsymbol{A}}_i = \bar{\boldsymbol{A}} + \lambda_i(\boldsymbol{L}_\mathrm{p})\boldsymbol{e}\boldsymbol{c}_\mathrm{z}^\mathrm{T} \quad \text{and} \quad \tilde{\boldsymbol{Q}}_i = \boldsymbol{Q}_\mathrm{a} + \lambda_i(\boldsymbol{L}_\mathrm{q})\boldsymbol{Q}_\mathrm{c}.$$

6.5.4 Online Information Reduction: Situation-Dependent Communication

In order to reduce the communication load, while maintaining the controller performance close to the optimal performance, it is reasonable to establish the communication only between subsystems that are directly interconnected. The distributed controller structure is identical with the interconnection structure of the overall system:

$$\boldsymbol{K}^\mathrm{v} = \boldsymbol{L}_\mathrm{q} \otimes \boldsymbol{K}_\mathrm{v}. \tag{6.101}$$

The distributed controller has to ensure the stability of the interconnected system and to improve the performance compared to the decentralized controller by using the additional information obtained by the communication.

Design of the Distributed Controller. The value of the cost function (6.78) reflects the overall system performance. Hence, in order to guarantee that the distributed controller provides a better performance than the decentralized controller, the inequality

$$J^\mathrm{v} \le J^\mathrm{d} \tag{6.102}$$

must hold for an arbitrary initial state $\boldsymbol{x}_0$, where J^v denotes the value of the cost function with the distributed controller and J^d with the decentralized controller. The controller parameter $\boldsymbol{K}_\mathrm{v}$ is determined by solving the ARE for the i-th modified subsystems

$$\boldsymbol{K}_\mathrm{v} = \lambda_i(\boldsymbol{L}_\mathrm{q})^{-1}\boldsymbol{R}^{-1}\boldsymbol{B}^\mathrm{T}\tilde{\boldsymbol{P}}_i^*. \tag{6.103}$$

Theorem 6.14. [94] *Consider the N ARE's* (6.95). *If there exists a matrix* $\tilde{\boldsymbol{P}}_i^*$ *that fulfills the condition*

$$\tilde{\boldsymbol{P}}_i^* \frac{\lambda_j(\boldsymbol{L}_{\mathrm{q}})}{2\lambda_i(\boldsymbol{L}_{\mathrm{q}})} \prec \tilde{\boldsymbol{P}}_j^* \text{ for } j = 1, 2, \ldots, N, \tag{6.104}$$

then the distributed controller (6.101) *with the controller matrix introduced in* (6.103) *provides a better performance than the decentralized controller. Consequently, the inequality in* (6.102) *holds for an arbitrary initial state* $\boldsymbol{x}_0$.

In virtue of Theorem 6.14, a distributed controller can be designed by solving an ARE of the single subsystem order. However, in order to determine the ARE to be solved, the condition (6.104) has to be checked. If the condition is fulfilled by an arbitrary Riccati matrix $\tilde{\boldsymbol{P}}_i^*$, the controller matrix $\boldsymbol{K}_{\mathrm{v}}$ has to be determined using the relation given in (6.103). If the condition is fulfilled by more than one Riccati matrices, then, one of these matrices has to be considered by the determination of the controller matrix $\boldsymbol{K}_{\mathrm{v}}$.

Control of the Interconnected System with Situation-Dependent Communication. In this part of the section, locally interconnected subsystems are considered, i.e. it is assumed that only the nearest subsystems are physically interconnected. Such an interconnection structure arises in many technological examples like vehicle platoons or multizone crystal furnace which will be considered as an example below. A similar control strategy is applied in [95] to a symmetrically interconnected system.

The control input with the distributed controller for a single subsystem can be given by

$$\boldsymbol{u}_{\mathrm{c},i}(t) = \boldsymbol{K}_{\mathrm{v}}(\boldsymbol{x}_{i-1}(t) - \boldsymbol{x}_i(t)) + \boldsymbol{K}_{\mathrm{v}}(\boldsymbol{x}_{i+1}(t) - \boldsymbol{x}_i(t)). \tag{6.105}$$

In order to further reduce the communication effort, a situation-dependent communication is considered. The control input for an arbitrary subsystem with the proposed switching controller is given by

$$\boldsymbol{u}_{\mathrm{c},i}(t) = \begin{cases} \boldsymbol{0} & \text{if } \exists j : ||\boldsymbol{K}_{\mathrm{v}} \Delta \boldsymbol{x}_j(t)|| \leq \epsilon \\ \boldsymbol{K}_{\mathrm{v}} \left((\boldsymbol{x}_{i-1}(t) - \boldsymbol{x}_i(t)) + (\boldsymbol{x}_{i+1}(t) - \boldsymbol{x}_i(t)) \right) & \text{else} \end{cases}$$

with $\Delta \boldsymbol{x}_j(t) = \boldsymbol{x}_j(t) - \boldsymbol{x}_{j-1}(t)$, $(j = 2, \ldots, N)$.

Note that the subsystems are in the autonomous mode if the deviation of the states of all pairs of neighbouring subsystems is less then a given value. If the norm of the vector $||\boldsymbol{K}_{\mathrm{v}} \Delta \boldsymbol{x}_j(t)||$ exceeds for an arbitrary subsystem pair the communication threshold, the subsystems switch in cooperative mode and an information exchange among the nearest subsystems is invoked. A coordinator is required in order to check the communication condition and activate the information exchange if necessary.

Example 6.4 *Control of a multizone furnace*

System model. The multizone furnace presented in Fig. 6.16 is an application example with locally interconnected subsystems [94]. It is used to grow GaAs crystals with the highest possible purity. It consists of a large number of identical heating zones, which have separate sensing and actuating units so that they can be independently controlled. The neighboring heating zones are physically interconnected and cannot be seen as separate subsystems because of the heat transfer between them. The heat transfer between distant heating zones via infrared radiation is neglected.

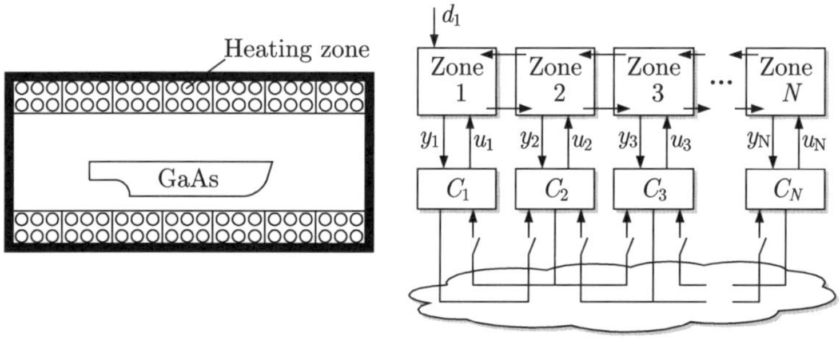

Fig. 6.16 Multizone furnace

The dynamics of a heating zone is described by equations (6.64) – (6.65) with the matrices and vectors

$$\boldsymbol{A} = \begin{pmatrix} -0.3 & 0 \\ 0 & -0.03 \end{pmatrix}, \quad \boldsymbol{c}^{\mathrm{T}} = \boldsymbol{c}_z^{\mathrm{T}} = \begin{pmatrix} 0.2 & 0.8 \end{pmatrix}, \quad \boldsymbol{e} = \boldsymbol{b} = \boldsymbol{g} = \begin{pmatrix} 1.3 \\ 0.1 \end{pmatrix}$$

and with the interconnection matrix

$$\boldsymbol{L} = \begin{pmatrix} -5 & 5 & 0 & \dots & 0 \\ 5 & -10 & 5 & \ddots & \vdots \\ 0 & \ddots & \ddots & \ddots & 0 \\ \vdots & \ddots & 5 & -10 & 5 \\ 0 & \dots & 0 & 5 & -5 \end{pmatrix}.$$

It is assumed that the overall system is composed of 15 identical subsystems and the cost function (6.78) has the weighting matrix $\boldsymbol{Q}_{\mathrm{a}} = 100\boldsymbol{c}^{\mathrm{T}}\boldsymbol{c}$ and the scalar $r = 1$ reflects the performance of the overall system.

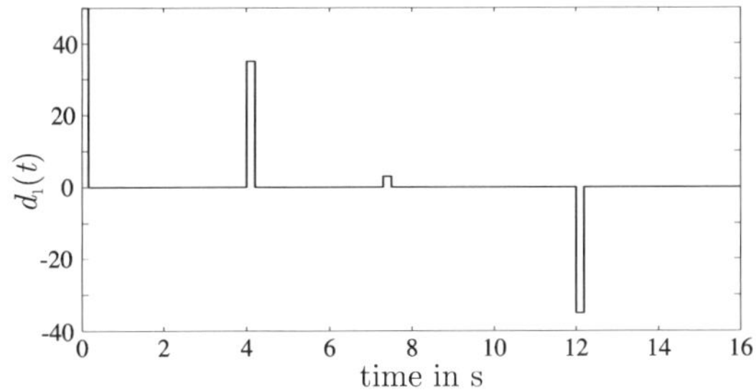

Fig. 6.17 Impulse disturbances affect the first subsystem

Design of the Networked Controller with Situation-Dependent Communication. According to the solution of the optimization problem in the transformed domain, the existence condition (6.104) is fulfilled for the matrices $\tilde{\boldsymbol{P}}_i$ for $i > 10$. The distributed controller is determined as

$$\boldsymbol{K}^{\mathrm{v}} = (\boldsymbol{L} \otimes \boldsymbol{K}_{\mathrm{v}}) = (\boldsymbol{L} \otimes (\lambda_{15}(\boldsymbol{L}_{\mathrm{q}}))^{-1} \boldsymbol{B}^{\mathrm{T}} \tilde{\boldsymbol{P}}_{15}) .$$

The switching condition is set to be

$$||\boldsymbol{K}_{\mathrm{v}} (\boldsymbol{x}_j(t) - \boldsymbol{x}_{j-1}(t)) || = 3, \quad j = 2, \ldots, 15.$$

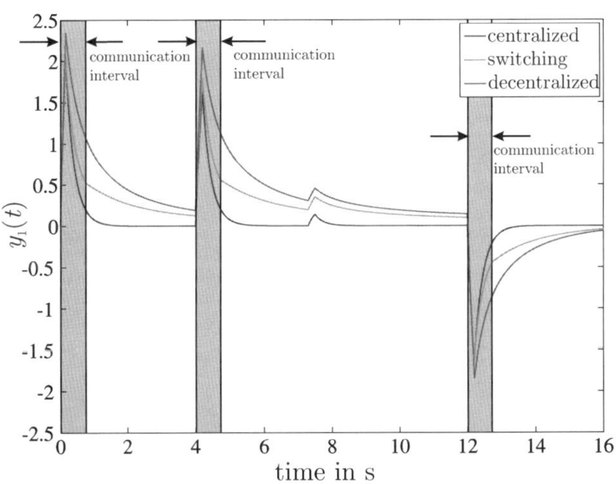

Fig. 6.18 Behavior of the furnace

It is assumed that the disturbance shown in Fig. 6.17 affects the first subsystem. The deviation of the disturbed subsystem from the operation point with decentralized, networked and centralized controllers is presented in Fig. 6.18. The simulation results show that the disturbance behavior of the overall system is clearly improved by using the exchanged information, although the information exchange is performed only in the short time periods indicated by the grey rectangles. It can be further noticed that the communication time depends on the effect of the acting disturbance, which in turn depends on the magnitude and the duration of the disturbance signal. The disturbance signal in the time period [8, 8.1] does not cause a communication, since the effects do not exceed the switching threshold. □

6.5.5 Stability Analysis of Subsystems with Similar Dynamics

This section concerns systems that are composed of subsystems with similar (rather than identical) dynamics. A method for analyzing the stability of such systems is developed, where the subsystems are decomposed into nominal models with identical dynamics and individual models describing the deviation of the subsystems from the common nominal models.

The analysis is divided into three steps. First, the stability of the controlled and physically interconnected nominal subsystems is checked by a state transformation based on the structure of the interconnection (cf. Eq. (6.91)). The second step is to find an upper bound for the deviation of the subsystem models from the common approximate model. The third step is to prove the stability of the overall system by considering the stabilized nominal system and the deviation bound. While considering the autonomous mode and cooperative mode, the stability analysis has to be performed for both modes separately.

Way of Solution. The idea of the stability analysis is to decompose all subsystem models into a nominal model $\bar{\Sigma}_{\mathrm{n}i}$, $(i \in \mathcal{N} := \{1, \ldots, N\})$, which is identical for all subsystems, and into an error model $\Sigma_{\mathrm{e}i}$ that describes the deviation of the subsystem dynamics from this common nominal model $\bar{\Sigma}_{\mathrm{n}i}$. Thus, every nominal subsystem $\bar{\Sigma}_{\mathrm{n}i}$, $(i \in \mathcal{N})$ has a corresponding error subsystem $\Sigma_{\mathrm{e}i}$, which together build the controlled decomposed subsystem $\bar{\Sigma}_{\mathrm{d}i}$. The subsystems $\bar{\Sigma}_{\mathrm{d}i}$, $(i \in \mathcal{N})$ are physically interconnected (denoted by $\boldsymbol{L}_{\mathrm{p}}$) and interconnected over the local controllers C_i which use the communication network (denoted by $\boldsymbol{L}_{\mathrm{q}}$) as shown in Fig. 6.19a. When coupling these controlled nominal subsystems $\bar{\Sigma}_{\mathrm{n}i}$ through $\boldsymbol{L}_{\mathrm{p}}$ and $\boldsymbol{L}_{\mathrm{q}}$, a nominal overall subsystem $\bar{\Sigma}_{\mathrm{n}}$ results (Fig. 6.19b).

The benefit of this decomposition is twofold. First, the controllers C_i, $(i \in \mathcal{N})$ can be designed for the identical subsystems $\bar{\Sigma}_{\mathrm{n}i}$ by using the methods that have been developed in the preceding section and in the literature for systems composed of identical subsystems (cf. [94, 117]). Second, the behavior of the overall system can be analyzed by separately considering the system $\bar{\Sigma}_{\mathrm{n}}$, which is composed of identical controlled subsystems $\bar{\Sigma}_{\mathrm{n}i}$, $(i \in \mathcal{N})$, and the error subsystems $\Sigma_{\mathrm{e}i}$ describing the deviation bounds. If the deviation bounds satisfy the stability test elaborated in the following, this analysis shows that the subsystems are similar enough to be controlled by the identical controllers.

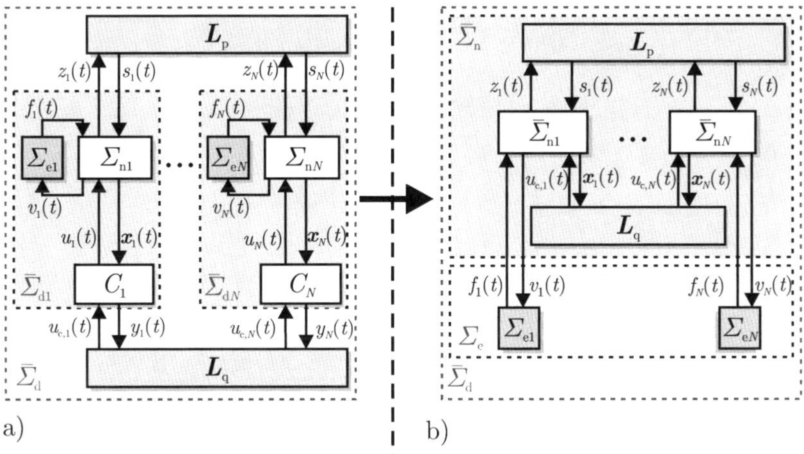

Fig. 6.19 Decomposition approach (a); Stability analysis for $\bar{\Sigma}_d$ consisting of the controlled nominal subsystems $\bar{\Sigma}_{ni}$, $(i \in \mathcal{N})$ coupled over the interconnection matrices $\boldsymbol{L}_p$ and $\boldsymbol{L}_q$ (b)

Note that the stability analysis can be performed for the following three control tasks:

1. **Control of multi-agent systems:** no physical interconnection and an arbitrary communication exchange of the controllers ($\boldsymbol{L}_p = \boldsymbol{0}$)

2. **Decentralized control of physically interconnected systems:** an arbitrary physical interconnection and no communication exchange of the controllers ($\boldsymbol{L}_q = \boldsymbol{0}$)

3. **Networked control of physically interconnected systems:** an arbitrary physical interconnection and an arbitrary communication exchange of the controllers

The first and third task concern the cooperative mode for multi-agent systems and physically interconnected systems, the second task the autonomous mode for physical interconnected systems. For the autonomous control of multi-agent systems only the stability of all separate subsystem has to be tested. In the following the stability analysis is performed for physically interconnected systems, where the controllers communicate over the network (task 3). For tasks 1 and 2 the analysis is similar.

Model. According to Fig. 6.19a, the overall system $\bar{\Sigma}_{\rm d}$ consists of the subsystems $\bar{\Sigma}_{{\rm d}i}$, $(i \in \mathcal{N})$, each of which is controlled by an associated controller C_i. The subsystems are physically coupled ($\boldsymbol{L}_{\rm p}$), whereas the controllers communicate over the network ($\boldsymbol{L}_{\rm q}$). In the following model the subsystems are described by a controlled nominal model $\bar{\Sigma}_{{\rm n}i}$, $(i \in \mathcal{N})$ together with a deviation model $\Sigma_{{\rm e}i}$. The similarity assumption on the subsystems is represented by the fact that all nominal subsystems $\bar{\Sigma}_{{\rm n}i}$ have identical properties.

The subsystems $\bar{\Sigma}_{{\rm d}i}$, $(i \in \mathcal{N})$ with the decentralized controller (6.60) are composed of identical nominal subsystems $\bar{\Sigma}_{{\rm n}i}$ and corresponding error subsystems $\Sigma_{{\rm e}i}$:

$$\bar{\Sigma}_{{\rm d}i}: \begin{cases} \bar{\Sigma}_{{\rm n}i}: \begin{cases} \dot{\boldsymbol{x}}_i(t) = \bar{\boldsymbol{A}}\boldsymbol{x}_i(t) + \boldsymbol{b}u_{{\rm c},i}(t) + \boldsymbol{e}s_i(t) + \boldsymbol{p}f_i(t), \quad \boldsymbol{x}_i(0) = \boldsymbol{x}_{0i} \\ y_i(t) = \boldsymbol{c}^{\rm T}\boldsymbol{x}_i(t) \\ z_i(t) = \boldsymbol{c}_{\rm z}^{\rm T}\boldsymbol{x}_i(t) \\ v_i(t) = \bar{\boldsymbol{c}}_{\rm v}^{\rm T}\boldsymbol{x}_i(t) + d_{\rm v}u_{{\rm c},i}(t) + e_{\rm v}s_i(t) \end{cases} \\ \Sigma_{{\rm e}i}: \begin{cases} \dot{\boldsymbol{x}}_{{\rm e}i}(t) = \boldsymbol{A}_{{\rm e}i}\boldsymbol{x}_{{\rm e}i}(t) + \boldsymbol{b}_{{\rm e}i}v_i(t) \\ f_i(t) = \boldsymbol{c}_{{\rm e}i}^{\rm T}\boldsymbol{x}_{{\rm e}i}(t) + d_{{\rm e}i}v_i(t) \end{cases} \end{cases} \tag{6.106}$$

with the matrices

$$\bar{\boldsymbol{A}} = \boldsymbol{A} - \boldsymbol{k}_{\rm v}^{\rm T}\boldsymbol{b}$$
$$\bar{\boldsymbol{c}}_{\rm v}^{\rm T} = \boldsymbol{c}_{\rm v}^{\rm T} - \boldsymbol{k}_{\rm v}^{\rm T}d_{\rm v}.$$

$\boldsymbol{x}_i(t)$, $(i \in \mathcal{N})$ is the nominal subsystem state and its notation is identical to the subsystem state in Eqs. (6.59) and (6.64) to show that the nominal subsystems $\bar{\Sigma}_{{\rm n}i}$ are identical. The error subsystem state $\boldsymbol{x}_{{\rm e}i}(t)$ is a vector of the dimension $n_{{\rm e}i}$. The control input $u_i(t)$, the control output $y_i(t)$, the error input $f_i(t)$, the error output $v_i(t)$, the interconnection input $s_i(t)$ and the interconnection output $z_i(t)$ are scalars.

The physical interconnection between the nominal subsystem $\bar{\Sigma}_{{\rm n}i}$, $(i \in \mathcal{N})$ is described by $\boldsymbol{L}_{\rm p}$ as in Eq. (6.66). The subsystems have the distributed controller (6.101), where the matrices $\boldsymbol{L}_{\rm p}$ in (6.66) and $\boldsymbol{L}_{\rm q}$ in (6.101) are assumed to be simultaneously diagonalizable (e.g. if $\boldsymbol{L}_{\rm p}$ and $\boldsymbol{L}_{\rm q}$ are identical). From Eqns. (6.106), (6.66), (6.101) the model of the nominal overall system

$$\bar{\Sigma}_{\rm n}: \begin{cases} \dot{\boldsymbol{x}}(t) = \boldsymbol{A}_{\rm L}\boldsymbol{x}(t) + \boldsymbol{P}_{\rm L}\boldsymbol{f}(t), \quad \boldsymbol{x}(0) = \boldsymbol{x}_0 \\ \boldsymbol{y}(t) = \boldsymbol{C}_{\rm L}\boldsymbol{x}(t) \\ \boldsymbol{v}(t) = \boldsymbol{C}_{\rm vL}\boldsymbol{x}(t) \end{cases} \tag{6.107}$$

follows with the matrices

$$\boldsymbol{A}_{\rm L} = (\boldsymbol{I}_N \otimes \bar{\boldsymbol{A}}) + (\boldsymbol{L}_{\rm p} \otimes \boldsymbol{e}\boldsymbol{c}_{\rm z}^{\rm T}) + (\boldsymbol{L}_{\rm q} \otimes \boldsymbol{k}_{\rm v}^{\rm T}\boldsymbol{b})$$
$$\boldsymbol{P}_{\rm L} = (\boldsymbol{I}_N \otimes \boldsymbol{p}) + (\boldsymbol{L}_{\rm p} \otimes \boldsymbol{e})$$

$$\boldsymbol{C}_{\rm L} = (\boldsymbol{I}_N \otimes \boldsymbol{c}^{\rm T})$$
$$\boldsymbol{C}_{\rm vL} = (\boldsymbol{I}_N \otimes \bar{\boldsymbol{c}}_{\rm v}^{\rm T}) + (\boldsymbol{L}_{\rm p} \otimes e_{\rm v}\boldsymbol{c}_{\rm z}^{\rm T}) + (\boldsymbol{L}_{\rm q} \otimes \boldsymbol{k}_{\rm v}^{\rm T} d_{\rm v}) .$$

The signals of $\bar{\Sigma}_{\rm n}$ are $\boldsymbol{x}(t) = (\boldsymbol{x}_1^{\rm T}(t), \dots, \boldsymbol{x}_N^{\rm T}(t))^{\rm T}$, $\boldsymbol{f}(t) = (f_1(t), \dots, f_N(t))^{\rm T}$ and $\boldsymbol{v}(t) = (v_1(t), \dots, v_N(t))^{\rm T}$.

Note that no restrictions are imposed on the structure of the interconnection matrix $\boldsymbol{L}_{\rm p}$ besides the diagonalizability. The following investigations hold for arbitrary strength and direction of the interconnections as well as for an arbitrary number N of subsystems.

Stability Analysis. At the first step, the stability of $\bar{\Sigma}_{\rm n}$ described by Eq. (6.107) is analyzed by extending ideas of [94] on symmetrically interconnected subsystems with identical dynamics and of [117] on the cooperative control of vehicle formations.

Lemma 6.3. *The controller* (6.101) *stabilizes the nominal overall system* $\bar{\Sigma}_{\rm n}$ *in* (6.107), *if and only if the matrices*

$$\tilde{\boldsymbol{A}}_{{\rm L}ii} = \bar{\boldsymbol{A}} + \lambda_i(\boldsymbol{L}_{\rm p})\boldsymbol{e}\boldsymbol{c}_{\rm z}^{\rm T} + \lambda_i(\boldsymbol{L}_{\rm q})\boldsymbol{b}\boldsymbol{c}^{\rm T}, \quad i = 1, 2, ..., N \tag{6.108}$$

are Hurwitz, where $\lambda_i(\boldsymbol{L}_{\rm p})$, $(i \in \mathcal{N})$ *and* $\lambda_i(\boldsymbol{L}_{\rm q})$ *are the eigenvalues of the interconnection matrix* $\boldsymbol{L}_{\rm p}$ *in* (6.66) *and* $\boldsymbol{L}_{\rm q}$ *in* (6.101), *respectively.*

The proof of Lemma 6.3 is given in [331], where $\bar{\Sigma}_{\rm n}$ is transformed by the relation (6.91). Lemma 6.3 reveals that the stability of the nominal overall system $\bar{\Sigma}_{\rm n}$ of the order nN can be analyzed by considering the stability of the N transformed single subsystems $\tilde{\Sigma}_{{\rm n}i}$, $(i \in \mathcal{N})$ of order n separately. The system matrices $\tilde{\boldsymbol{A}}_{{\rm L}ii}$, $(i \in \mathcal{N})$ only differ from each other by the eigenvalues $\lambda_i(\boldsymbol{L}_{\rm p})$ and $\lambda_i(\boldsymbol{L}_{\rm q})$. The case that a complex value for $\lambda_i(\boldsymbol{L}_{\rm p})$ leads to complex-valued subsystems $\tilde{\Sigma}_{{\rm n}i}$ is discussed in [117] for vehicle formations. If $\lambda_i(\boldsymbol{L}_{\rm p})$ is equal to zero, the stability of the corresponding subsystem $\tilde{\Sigma}_{{\rm n}i}$, $(i \in \mathcal{N})$ is independent of the interconnection with the other subsystems $\tilde{\Sigma}_{{\rm n}j}$, $(j \in \mathcal{N}, j \neq i)$.

In the second step, the I/O behavior of the stable system $\Sigma_{\rm e}$ is investigated, which is described by the I/O model

$$\Sigma_{\rm e}: \quad \boldsymbol{f} = \boldsymbol{G}_{\rm e} * \boldsymbol{v}, \tag{6.109}$$

where "$*$" symbolizes the convolution operation

$$\boldsymbol{G} * \boldsymbol{u} = \int_0^t \boldsymbol{G}(t-\tau)\boldsymbol{u}(\tau)\,{\rm d}\tau.$$

$\boldsymbol{G}_{\rm e}(t) = {\rm diag}(g_{{\rm e}i}(t))$ is the impulse response matrix of $\Sigma_{\rm e}$ with the elements $g_{{\rm e}i}(t) = \boldsymbol{c}_{{\rm e}i}^{\rm T} {\rm e}^{\boldsymbol{A}_{{\rm e}i}t}\boldsymbol{b}_{{\rm e}i} + d_{{\rm e}i}$ $i \in \mathcal{N}$ that represents the impulse response matrices of the error subsystems $\Sigma_{{\rm e}i}$.

$\boldsymbol{G}_{\rm e}(t)$ is a block diagonal matrix, hence in $\Sigma_{\rm e}$ each signal $v_i(t)$, $(i \in \mathcal{N})$ has only some effect on the signal $f_i(t)$ of the same subsystem $\Sigma_{\mathrm{e}i}$.

For the stability analysis of $\bar{\Sigma}_{\rm d}$, the I/O behavior of the error subsystems $\Sigma_{\mathrm{e}i}$, $(i \in \mathcal{N})$ are described by upper bounds $\bar{g}_{\mathrm{e}i}(t)$

$$|g_{\mathrm{e}i}(t)| \leq \bar{g}_{\mathrm{e}i}(t) \quad \forall t, \tag{6.110}$$

which show that a bounded error output leads to a bounded error input $|f_i(t)| \leq \bar{g}_{\mathrm{e}i}(t) * |v_i(t)|$.

With

$$\boldsymbol{M} = \int_0^\infty |\boldsymbol{G}(t)| \, \mathrm{d}t, \tag{6.111}$$

the matrix $\bar{\boldsymbol{M}}_{\rm e}$ (that follows from Eq. (6.110) and (6.111)) is finite, since the error overall system $\Sigma_{\rm e}$ is I/O stable.

The I/O behavior of the decomposed overall system $\bar{\Sigma}_{\rm d}$ is described by the model

$$\bar{\Sigma}_{\rm d} : \begin{cases} \bar{\Sigma}_{\rm n} : \begin{cases} \boldsymbol{y} = \boldsymbol{G}_{\rm yf} * \boldsymbol{f} \\ \boldsymbol{v} = \boldsymbol{G}_{\rm vf} * \boldsymbol{f} \end{cases} \\ \Sigma_{\rm e} \;:\; \boldsymbol{f} = \boldsymbol{G}_{\rm e} * \boldsymbol{v} \;. \end{cases} \tag{6.112}$$

The impulse response matrices of $\bar{\Sigma}_{\rm n}$ are defined by:

$$\begin{aligned} \boldsymbol{G}_{\rm yf}(t) &= \boldsymbol{C}_{\rm L} \mathrm{e}^{\boldsymbol{A}_{\rm L} t} \boldsymbol{P}_{\rm L} \\ \boldsymbol{G}_{\rm vf}(t) &= \boldsymbol{C}_{\rm vL} \mathrm{e}^{\boldsymbol{A}_{\rm L} t} \boldsymbol{P}_{\rm L} \,. \end{aligned}$$

The third step of the stability analysis is to apply the following I/O stability criterion to $\bar{\Sigma}_{\rm d}$.

Lemma 6.4. *Consider the decomposed overall system* $\bar{\Sigma}_{\rm d}$ *given in Eq.* (6.112)*, where the impulse response matrix* $\boldsymbol{G}_{\rm e}(t)$ *of the overall error system* $\Sigma_{\rm e}$ *satisfies the inequality*

$$|\boldsymbol{G}_{\rm e}(t)| \leq \bar{\boldsymbol{G}}_{\rm e}(t) \quad \forall t \tag{6.113}$$

for some matrix $\bar{\boldsymbol{G}}_{\rm e}(t)$*. Suppose that the nominal overall system* $\bar{\Sigma}_{\rm n}$ *and the error overall system* $\Sigma_{\rm e}$ *with* $\boldsymbol{G}_{\rm e}(t) = \bar{\boldsymbol{G}}_{\rm e}(t)$ *are I/O stable, i.e. the matrices* $\boldsymbol{M}_{\rm yf}$, $\boldsymbol{M}_{\rm vf}$ *and* $\bar{\boldsymbol{M}}_{\rm e}$*, which are obtained by Eq.* (6.111) *with the same indices have finite elements. Then the decomposed overall system* $\bar{\Sigma}_{\rm d}$ (6.112) *is I/O stable, if*

$$\lambda_{\rm p}(\bar{\boldsymbol{M}}_{\rm e} \boldsymbol{M}_{\rm vf}) < 1, \tag{6.114}$$

where $\lambda_{\rm p}$ *denotes the maximum eigenvalue (Perron root) of the non-negative matrix* $\bar{\boldsymbol{M}}_{\rm e} \boldsymbol{M}_{\rm vf}$.

Lemma 6.4 is based on the results in [244]. The analysis is extended here to an arbitrary interconnection matrix $\boldsymbol{L}_\mathrm{p}$. To use the stability criterion (6.114), the scalars $m_{\mathrm{vf}ij}$, which are the elements of the matrix $\boldsymbol{M}_\mathrm{vf}$, have to be determined. For this, the I/O behavior of the transformed subsystem $\tilde{\Sigma}_{\mathrm{n}i}$ is investigated.

Lemma 6.5. *Transform the system* $\bar{\Sigma}_\mathrm{n}$ *in* (6.112) *by the relation* (6.91). *Then, the elements of the impulse response matrix* $\boldsymbol{G}_\mathrm{nvf}$ *are given by*

$$g_{\mathrm{vf}ij}(t) = \sum_{k=1}^{N} \tilde{\boldsymbol{c}}_{\mathrm{vL}ik}^\mathrm{T} \mathrm{e}^{\tilde{\boldsymbol{A}}_{\mathrm{L}kk} t} \tilde{\boldsymbol{p}}_{\mathrm{L}kj} \,. \tag{6.115}$$

Lemma 6.5 shows that the non-negative matrix $\boldsymbol{M}_\mathrm{vf}$ in (6.114) can be determined just by analyzing the n-th order system $\tilde{\Sigma}_{\mathrm{n}i}$, $(i \in \mathcal{N})$. The stability analysis described in Lemma 6.4 can be further simplified, if multiple eigenvalues of the interconnection matrices $\boldsymbol{L}_\mathrm{p}$ (6.66) $\boldsymbol{L}_\mathrm{q}$ (6.71) lead to multiple subsystems $\tilde{\Sigma}_{\mathrm{n}i}$ with identical dynamics. Hence the I/O behavior $g_{\mathrm{vf}ij}(t)$ from the error input $f_i(t)$ to the error output $v_i(t)$ in (6.115) can be merged.

Main Result. With the results presented above the stability analysis can be performed as in Algorithm 6.1

Algorithm 6.1. Stability of interconnected subsystems with similar dynamics

---

Given: Decentralized controlled subsystems $\bar{\Sigma}_{\mathrm{d}i}$ (6.106) with similar dynamics that are composed of identical nominal models $\bar{\Sigma}_{\mathrm{n}i}$ and corresponding error models $\Sigma_{\mathrm{e}i}$. The interconnection between the subsystems $\bar{\Sigma}_{\mathrm{d}i}$ is described by the matrices $\boldsymbol{L}_\mathrm{p}$ from (6.66) and $\boldsymbol{L}_\mathrm{q}$ from (6.71).

1: Check the stability of $\bar{\Sigma}_\mathrm{n}$ (6.107) with the controller in (6.101) with Lemma 6.3.

2: Define an upper bound for the I/O behavior of the error overall system Σ_e (6.109), with the result that the matrix $\bar{\boldsymbol{M}}_\mathrm{e}$ given by (6.111) and (6.113) is finite.

3: Check the I/O stability of $\bar{\Sigma}_\mathrm{d}$ in (6.112) with Lemma 6.4 while taking into account the simplification made in Lemma 6.5.

Hence, the stability analysis for the overall system can be summarized in the following theorem.

Theorem 6.15. *Consider systems that consist of subsystems with similar dynamics. Decompose the subsystem models* $\bar{\Sigma}_{\mathrm{d}i}$*,* $(i \in \mathcal{N})$ *into identical nominal models* $\bar{\Sigma}_{\mathrm{n}i}$ *and individual error models* $\Sigma_{\mathrm{e}i}$ *as shown in Eq.* (6.106)*. The subsystems are controlled by identical controllers* (6.101)*. Assume that the interconnection matrix* $\boldsymbol{L}_{\mathrm{p}}$ *from* (6.66) *and* $\boldsymbol{L}_{\mathrm{q}}$ *from* (6.71) *can be diagonalized by the transformation* (6.91)*. Then the decomposed overall system* $\bar{\Sigma}_{\mathrm{d}}$ *is I/O stable, if the following conditions hold:*

1. *The matrices* $\tilde{\boldsymbol{A}}_{\mathrm{L}ii}$*,* $(i \in \mathcal{N})$ *defined in Eq.* (6.108) *are Hurwitz.*
2. *The elements of* $\bar{m}_{\mathrm{e}i}$*,* $(i \in \mathcal{N})$*, given by Eqns.* (6.110) *and* (6.111) *are finite.*
3. *The condition* (6.114) *holds, where* $\boldsymbol{M}_{\mathrm{vf}} = (m_{\mathrm{vf}ij})$*,* $(i, j \in \mathcal{N})$ *is given by Eq.* (6.111) *and its entries are defined by Eq.* (6.115)*.*

Theorem 6.15 extends the stability analysis of networked systems composed of identical subsystems to systems composed of similar subsystems. For this, the simplifications on the stability analysis of identical subsystems based on the state transformation (6.91) as shown can also be applied to the nominal overall system $\bar{\Sigma}_{\mathrm{n}}$ described by Eq. (6.107). Furthermore, the determination of the matrix $\boldsymbol{M}_{\mathrm{vf}}$, which is needed for the stability criterion (6.114), can be simplified by the same state transformation.

The elements $\bar{m}_{\mathrm{e}i}$ can be interpreted as "degree of similarity". In general, greater elements $\bar{m}_{\mathrm{e}i}$ of the test matrix lead to a greater value of $\lambda_{\mathrm{p}}(\bar{\boldsymbol{M}}_{\mathrm{e}}\boldsymbol{M}_{\mathrm{vf}})$ due to the monotonicity property of the Perron root [29]. Hence, if the condition in (6.114) holds, then the subsystem $\bar{\Sigma}_{\mathrm{d}i}$, $(i \in \mathcal{N})$ can be called similar.

7

Wireless Networking for Control

A. Haupt, F. Allgöwer, R. Blind, A. Chamaken, R. Gotzhein, M. Krämer, and L. Litz

This chapter is concerned with two important aspects of the implementation of distributed controllers. The first part shows how in control and communication co-design the control law and the scheduling method, which is used to decide which communication node gets access to the network, are chosen simultaneously. In the second part, the suitability of event-based control for different network protocols is evaluated.

Chapter Contents

7.1 Control Implications of Wireless Networks

Wireless Networked Control Systems (WNCS) (Fig. 7.1) constitute a class of networked control systems where sensors, controllers and actuators exchange information over a wireless digital communication network. WNCS are based on networked wireless devices which have many advantages compared to their wired counterparts such as easy deployment, flexible architecture, low installation and maintenance cost, lack of cabling, and high mobility. There is not yet a widespread use of wireless technologies in industrial control applications despite their great potentials. One of the main reasons is that currently available wireless technologies do not have a framework that is applicable across several application domains. Expert knowledge about the whole protocol stack from the communication layer to the application layer is often required for each specific application development.

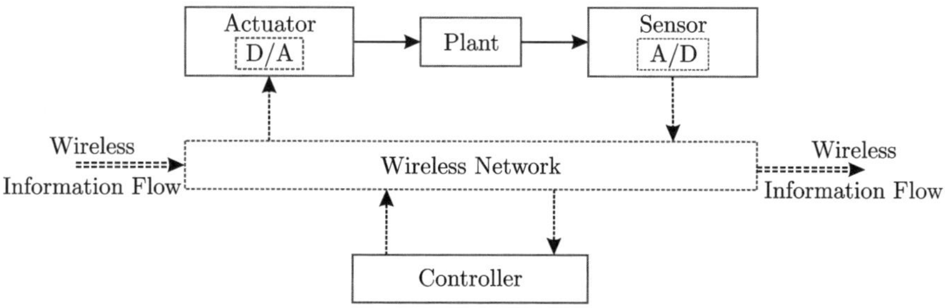

Fig. 7.1 Networked control system with wireless information flow between components and control loops

The multidisciplinary research field of WNCSs combines control theory, communication technology, and computer science into a new dynamic environment for real-time control systems that features both resource limitations and workload variability, resulting in uncertainties about the availability of the computing (e.g. CPU time, energy) and communication (e.g. bandwidth) resources (Fig. 7.2). Traditional control theory focuses on dynamic systems connected through ideal channels assuming equidistant sampling, negligible or constant communication delays and no data loss, whereas communication theory studies the transmission of information over imperfect channels with variable transmission delays and possible data loss. Moreover, systems engineers always assume that control and communication tasks are periodic with a set of fixed periods, having hard deadlines and known worst-case execution time (WCET).

These assumptions are not always true, especially in dynamic environments with changing resource availability. Therefore, in order to design and implement control algorithms and communication protocols for real-time WNCSs, principles and methods from computer science and communication technology should be included into the design and implementation of control systems, thus, leading to the co-design of control, communication and computing.

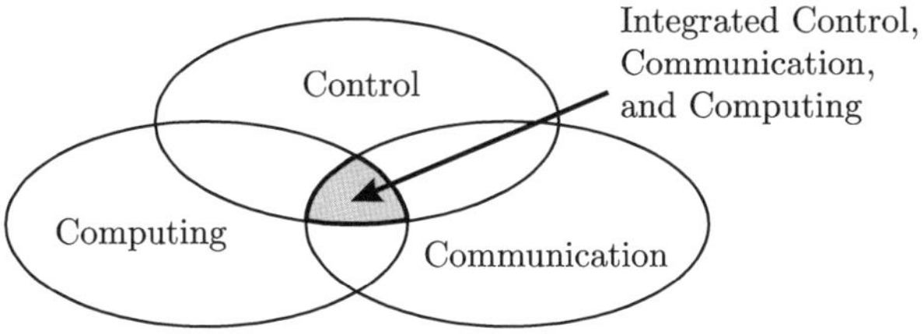

Fig. 7.2 WNCS at the intersection between control, communication and computing

Wireless networks have several implications for the design and the performance of control:

- **Reliability:** Sensor measurements and control action loss can prevent the correct execution of control algorithms. Reliability can be maximized by using retransmissions in the event of packet loss but can result in an increase of the energy consumption of the network since nodes have to be awake for longer periods of time. Therefore, a tradeoff between reliability and network energy consumption has to be considered.

- **Latency:** Sensor measurements and control actions must reach their destinations within a given deadline since the resulting time delay influences control system performance and stability. The retransmission of packets to ameliorate the reliability may increase the delay. Therefore, a tradeoff between reliability and latency must be considered.

- **Energy:** Energy efficient operation of wireless nodes is required to prevent frequent battery replacement and increase the network lifetime. Maximizing the reliability and minimizing the latency may result in a better performance of the control system. However, the drawback is an increase of the energy consumption of the network. A tradeoff between a tolerable control performance and an acceptable network lifetime has to be considered.

- **Flexibility:** Communication protocol parameters and control algorithm parameters should be adaptable to changes of the control application requirements, to time varying wireless channels, and to network topology variations. This requires analytical models describing the relation between protocol parameters, controller parameters, and selected performance indices such as the Quality of Control (QoC), the Quality of Services (QoS), and the energy consumption.

- **Data rate:** Advanced control strategies such as model predictive control (MPC) require much more computation time and memory than standard control approaches. The type and amount of data to be transmitted also depends on the selected control strategy. For scheduled control, sensor measurements and actuation commands must be received periodically within a specified deadline. Therefore, the network must be able to guarantee the delay of a signal within the specified time deadline. For event-based control, sensor measurements and actuation commands do not have a time deadline. However, the network must guarantee access when events occur.
- **Scalability:** Due to the limited processing capabilities of the wireless nodes, control algorithms and communication protocols must be computationally light.

As a result, the design of WNCS must take into consideration a large number of competing tradeoffs. In particular, a tradeoff exists between control performance, communication performance, and network lifetime. Therefore, control and communication should be jointly designed.

7.2 Control and Communication Co-design

7.2.1 Cross-Layering

Control systems and computer networks are scientific disciplines that use different kinds of mathematical models and design approaches. Control theory is based on assumptions regarding sampling instances (e.g. synchronization), performance (e.g. delay), and reliability (e.g. loss) that are difficult to satisfy by off-the-shelf or even specialized communication technologies. Vice versa, many computer networks have properties and constraints that are not suitable for control systems. To resolve this mismatch, the joint design of the control and communication system, referred to as cross-layer design, has been proposed [337].

This section presents the approach of cross-layer design of wireless networked control systems described in detail in [53–55] and [140]. We start by introducing an abstract architecture, identifying aspects to be addressed during the design, and in particular identifying joint parameters of control and communication that can be used both in control theory and as network abstractions. We then present two cross-layer approaches. The static approach is used at design time, exploiting various degrees of freedom to devise matching control algorithms and communication protocols. The dynamic approach exploits possibilities to adapt control and communication to varying load situations at runtime.

From an abstract, architectural point of view, we decompose wireless networked control systems into two layers (Fig. 7.3): the control layer comprising application functionality of controllers, sensors, and actuators, and the communication layer providing all kinds of networking functionality. Both layers use computing resources such as CPU, memory, transceiver, and energy, imposing constraints on the complexity of control algorithms and communication protocols, on throughput and on reaction times.

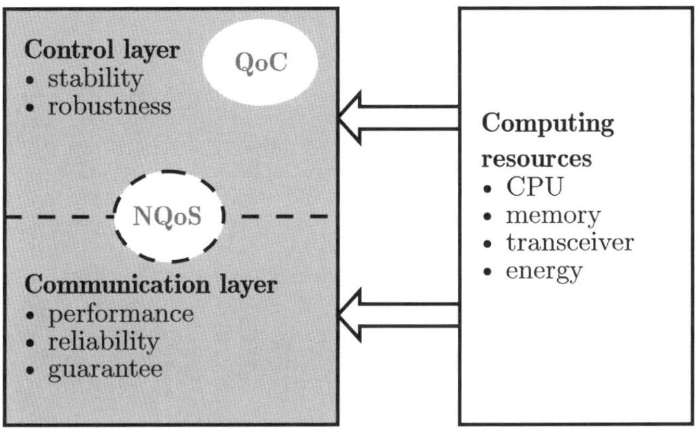

Fig. 7.3 Abstract cross-layering architecture of WNCS

To define an abstract interface between control layer and communication layer, we apply the notion of Network Quality-of-Service (NQoS) specification [330], capturing the abstract communication requirements of the control system (Fig. 7.3). For each data flow between sensors, controllers, and actuators, the NQoS specification defines performance, reliability, and guarantee parameter values in a mathematically precise way [382]. Here, performance comprises efficiency aspects of a data flow, concerning, for instance, latency, bandwidth, and synchronization. Reliability refers to operational safety of data flows, for instance, loss rates and burstiness. Finally, guarantee captures the degree of commitment regarding performance and reliability, ranging from best effort (i.e. no guarantee) to deterministic (i.e. full guarantee). To allow for dynamic adaptation to varying load situations and resource availability, we specify, for each data flow, a minimum NQoS $nqos_{base}$ sufficient to grant a minimum QoC and a preferred NQoS $nqos_{pref} \geq nqos_{base}$ that is useful in case of dynamic disturbances. We note that the NQoS domain is multi-dimensional. Therefore, comparing two NQoS values or determining their difference (see below), which is rather intuitive on an informal level, is a non-trivial step that requires careful formal treatment, which can be found, e.g., in [382]. Our intention is to give strong guarantees for the minimum QoS requirements, and to support the preferred QoS requirements as far as possible.

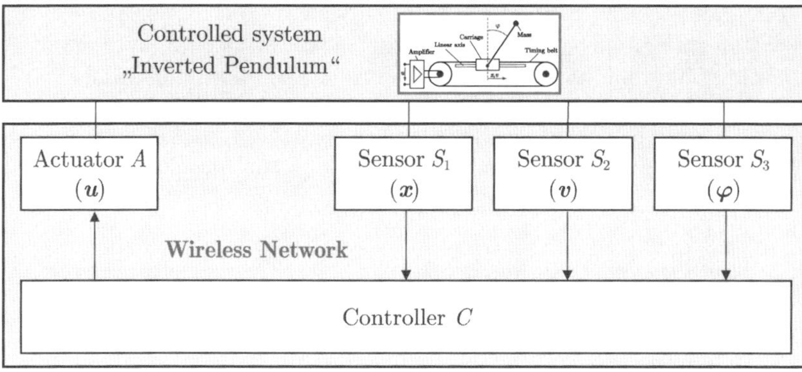

Fig. 7.4 Topology of the inverted pendulum system

Example 7.1 *Networked control of an inverted pendulum*

As an example, we consider the inverted pendulum. The topology of the control system is shown in Fig. 7.4. The symbols x, v, φ and u refer to cart position, cart velocity, pendulum angle, and motor voltage (i.e. cart acceleration), respectively. From the topology, it follows that data flows between the sensors S_i and the controller C as well as between C and actuator A have to be supported. For each data flow, NQoS parameters are to be specified for the required QoC. For instance, we specify a sampling period of 80 ms (or 40 ms), a maximum transmission delay of 16 ms, and a data rate of 6 bytes per sampling period as minimum (preferred) performance for data flows between sensors and controller. In addition, we require at most one lost packet per 5 sampling periods and a maximum clock offset of 1 ms as minimum and preferred reliability, and enhanced best effort guarantees at all times. □

The NQoS specification serves several purposes. First, it provides a suitable abstraction for control, by identifying parameters that are used in existing control theories. Secondly, it forms the starting point for the configuration and development of matching communication technologies. Thirdly, it is the formal basis for static and dynamic cross-layer design, where the NQoS specification is incrementally adapted until it matches both the needs of the control system and the capabilites of a real communication system.

As an example, consider the *static cross-layer design* of the inverted pendulum [53–55]. To provide enhanced best-effort guarantees in multi-hop wireless networks, deterministic communication protocols supporting the specified performance and reliability during periods of sufficient channel quality are needed. This rules out commercial communication technologies such as WLAN, ZigBee, or Bluetooth, as they are prone to packet loss due to collisions, exhibit arbitrary transmission delays due to statistical contention schemes, and have no (deterministic) multi-hop capabilities. Instead, a specialized communication system matching the NQoS requirements of the inverted pendulum and other WNCS such as rate flow control is to be designed. If this is not feasible, the NQoS requirements of the control system can be modified within the available degrees of freedom, e.g., by reducing QoC or by using more powerful control algorithms.

By *dynamic cross-layer design*, we refer to measures for the adaptation of network services to control demands at runtime, and vice versa, and to the adaptation of computing resources such as energy [140]. For instance, if a controlled system is disturbed during operation, switching between control algorithms may be an option for fast stabilization. Alternatively, an increase of the sampling frequency can be considered, leading to additional communication. In the NQoS specification, the degree of freedom is captured by the difference between preferred and minimum NQoS, i.e. $nqos_{pref} - nqos_{base}$. To give strong guarantees for $nqos_{base}$, we use TDMA techniques with exclusive reservations. To support the QoS exceeding $nqos_{base}$, we apply a new method of controlled contention called *mode-based scheduling* [139].

Time slot	s_1	s_2	s_3	s_4	s_5	s_6	s_7	s_8	s_9	s_{10}	s_{11}	s_{12}	s_{13}	s_{14}	s_{15}	s_{16}	s_{17}	s_{18}	s_{19}	s_{20}
$mode_1$ (prio = 1)			$\mathcal{S}_{1,1}$	$\mathcal{S}_{1,2}$	$\mathcal{S}_{1,3}$	$\mathcal{S}_{A,1}$	$\mathcal{C}_1$	$\mathcal{S}_{2,1}$	$\mathcal{S}_{2,2}$	$\mathcal{S}_{2,3}$	$\mathcal{S}_{A,2}$	$\mathcal{C}_2$	$\mathcal{S}_{1,1}$	$\mathcal{S}_{1,2}$	$\mathcal{S}_{1,3}$	$\mathcal{C}_A$	$\mathcal{C}_1$	$\mathcal{S}_{B,1}$	$\mathcal{S}_{B,2}$	
$mode_2$ (prio = 2)													$\mathcal{S}_{2,1}$	$\mathcal{S}_{2,2}$	$\mathcal{S}_{2,3}$		$\mathcal{C}_2$			

Time slot	s_{21}	s_{22}	s_{23}	s_{24}	s_{25}	s_{26}	s_{27}	s_{28}	s_{29}	s_{30}	s_{31}	s_{32}	s_{33}	s_{34}	s_{35}	s_{36}	s_{37}	s_{38}	s_{39}	s_{40}
$mode_1$ (prio = 1)		$\mathcal{C}_B$	$\mathcal{S}_{1,1}$	$\mathcal{S}_{1,2}$	$\mathcal{S}_{1,3}$		$\mathcal{C}_1$	$\mathcal{S}_{2,1}$	$\mathcal{S}_{2,2}$	$\mathcal{S}_{2,3}$		$\mathcal{C}_2$	$\mathcal{S}_{2,1}$	$\mathcal{S}_{2,2}$	$\mathcal{S}_{2,3}$		$\mathcal{C}_2$			
$mode_2$ (prio = 2)													$\mathcal{S}_{1,1}$	$\mathcal{S}_{1,2}$	$\mathcal{S}_{1,3}$		$\mathcal{C}_1$			
$mode_3$ (prio = 3)													$\mathcal{S}_{A,1}$	$\mathcal{S}_{A,2}$			$\mathcal{C}_A$			
$mode_4$ (prio = 4)													$\mathcal{S}_{B,1}$	$\mathcal{S}_{B,2}$			$\mathcal{C}_B$			

Fig. 7.5 Mode-based schedule for inverted pendulums and flow rate controls

Example 7.1 (cont.) *Networked control of an inverted pendulum*

To support the NQoS requirements of four WNCS - two inverted pendulums 1 and 2, and two flow rate controls A and B - operating on top of a joint communication system, we structure time into macro slots having a constant length of 160 ms. Each macro slot is subdivided into 40 time slots s_i with a length of 4 ms each (Fig. 7.5). For deterministic, time-triggered communication, we assign time slots exclusively such that the minimum NQoS $nqos_{base}$ is provided to all data flows. In Fig. 7.5, slots with only a single node assignment (one entry per column) are reserved exclusively. For instance, slots s_3, s_4, s_5 are assigned to sensors $S_{1,1}$, $S_{1,2}$, and $S_{1,3}$ of inverted pendulum 1. The same holds for slots s_{23}, s_{24}, s_{25}, which realizes the minimum sampling period of 80 ms. Assuming that all sensors synchronously determine values, the maximum transmission delay resulting from medium access delay, sending delay, and propagation delay is 12 ms, which satisfies the performance requirements of 16 ms.

To support $nqos_{pref} - nqos_{base}$, we use a new method of controlled, deterministic contention [139] called *mode-based scheduling* combined with a highly efficient method for *fast mode-signaling* (cf. Section 7.2.3). With mode-based scheduling, a time slot can be assigned to more than one node, at most one per mode. Modes are introduced to distinguish between different transmission purposes, and are associated with a priority. In Fig. 7.5, four modes and their priorities are shown. Here, slot s_{13} is assigned to sensors $S_{1,1}$ and $S_{2,1}$ of the inverted pendulums 1 and 2, with the intention that if both sensors have a packet to be transmitted, pendulum 1 gains preference. Note that the preferred NQoS is not guaranteed deterministically as the minimum NQoS, but granted on top for faster stabilization in case of disturbances. Similarly, slot s_{33} is assigned to sensors $S_{1,1}$ and $S_{2,1}$ of the inverted pendums 1 and 2, and to sensors $S_{A,1}$ and $S_{B,1}$ of the flow rate controls A and B. Different from slot s_{13}, preference is given to pendulum 2. If in a particular macro slot, both pendulums do not use slot s_{33}, the flow rate controls A or B may send packets on top of their minimum NQoS, with preference for A. □

We point out that mode-based scheduling solves the problem of uncontrolled contention, where all nodes may contend for unreserved time slots. In wireless communication systems, uncontrolled contention may lead to collisions and therefore is not suitable for WNCS.

7.2.2 Control Design

In a wireless networked control system, the design of a stable controller with good performance is more challenging compared to the wired case. Several reliability issues arise with the introduction of real non-deterministic wireless network architecture with communication imperfections in the control feedback loop such as random packet loss, varying jitter, random delays and signal interferences. In order to design a practical usable controller, these imperfections have to be considered during the controller design phase as well as real-life resource limitations such as limited node energy, computing and transceiver power. These parameters negatively affect the QoC and, therefore, the control architecture has to consider a diametrical tradeoff between energy efficient communication and good QoC. An introduction on co-design or cross-design of control, communication and computing (CCC) is referred to in Section 7.2.1. In this section, we will focus on the choice of high performance and computationally efficient control structures, the design and parameterization of them and some further important aspects on wireless control under resource limitation, e.g. the interoperability between distributed control loops. Further details about the used TDMA communication protocol (WNCS_CoM) can be found in Section 7.2.3. This section is based upon [51, 52, 54, 55, 140].

Problem Formulation. The problem we aim to solve is that of performing control over a dedicated wireless communication network (Section 7.2.3). The main goal is to find solutions for control loops closed over wireless communication networks that guarantee not only control performance and stability, but also minimize the total energy consumption of the wireless network, thus, increasing the network lifetime. Abstractions of the mathematical models of the communication protocols proposed in [51] are used in conjunction with optimal control design techniques to investigate the essential issues of the WNCS cross-design approach. Optimal control design techniques are applied in order to derive a closed-loop control performance metric for WNCSs as a function of delay and packet loss requirements. To this end, the achievable control performance is obtained from a combination of hardware-in-the-loop simulations and experiments.

The proposed cross-design optimization approach is based on the constrained multi-objective optimization problem formulated in [51] for which the objective functions are the control performance J_{con} and the average wireless node lifetime $\overline{T}$ both constrained by delay and packet loss probability requirements, which are derived from the desired control cost. Moving the objective function of the control application J_{con} to the constraints leads to the following single-objective optimization problem

$$\begin{aligned} \max_{\boldsymbol{x}} \quad & \overline{T}(\boldsymbol{x}) \\ s.t. \quad & J_{\text{con}}(\boldsymbol{x}) \leq J_{\text{max}} \\ & R(\boldsymbol{x}) \leq R_{\text{max}} \\ & \mathrm{E}[D(\boldsymbol{x})] \leq D_{\text{max}} \\ & \boldsymbol{x}_{\text{min}} \leq \boldsymbol{x} \leq \boldsymbol{x}_{\text{max}}, \end{aligned}$$

where J_{max} is the desired maximum control cost of the WNCS. J_{max} is chosen so as to guarantee control system stability and an acceptable wireless network lifetime.

The proposed cross-design approach is illustrated through simulations and experiments on the inverted pendulum benchmark with following main contributions:

- Cross-design investigation of the effects of network induced random delays and packet loss on the control performance and lifetime of WNCSs
- Proposition of a cross-design approach based on the design of suitable control algorithms and dedicated communication protocols (refer to Section 7.2.3) in order to meet a desired control performance while minimizing the energy consumption of the wireless network at the same time

Control Design Methodology. The proposed control design methodology is based on the joint design of control and communication with optimization of control performance and network lifetime using available control and communication parameters. Control structures and corresponding controller parameters can be adapted offline or online to changing network conditions in order to obtain best control performance. At the same time, communication protocols and corresponding protocol parameters can be tuned both offline or online to meet a desired control performance and maximize the lifetime of the WNCS. With the goal of optimizing the control performance while minimizing network lifetime, the communication cross-layering architecture in Fig. 7.9 is considered.

The proposed control design approach is carried out in two phases including a design phase and an optimization phase. During the design phase, control system and communication system are designed for a given control application taking into consideration the limited communication, computing, and energy resources of the wireless nodes. This allows resulting control algorithms and communication protocols to be implemented on low energy wireless platforms with limited computational capabilities. At the control layer, the design phase includes the mathematical modeling and identification of the control system followed by the selection, implementation and evaluation of suitable control structures. These control structures are selected based on their performance, computational and energy requirements. Closed-loop performance evaluation of selected control algorithms is done via simulation of real plants with suitable wireless network models and results in acceptable QoC metrics for the control application (e.g. system responses, robustness) from which QoS requirements to the underlying communication system (e.g. transmission rate, average delay, packet dropout rate) are derived.

The degrees of freedom at the control layer are the control structure, the sampling period, and the controller gains. At the communication layers, the design phase includes the implementation of communication protocols that guarantee the QoS requirements from the control application in order to achieve acceptable control performance. The communication system is either capable of delivering the QoS required by the control application (i.e. $nqoS_{pref}$) or, in case this turns out to be infeasible, offers guaranteed QoS (i.e. $nqoS_{base}$) so that the control application can for example change the sampling period, adapt controller parameters or switch controllers to meet the desired QoC. The degrees of freedom at the communication layer are the MAC protocols with corresponding MAC protocol parameters and the nodes transmit powers.

The proposed structure for the control design phase is depicted in Fig. 7.6 and an excerpt of the corresponding workflow is presented in Algorithm 7.1.

The design workflow is aimed at finding optimal controller parameters $\boldsymbol{K}_{opt}$ for a given control structure and selected combinations of network QoS (e.g. delay τ^{rt} and packet loss rate ρ^{rt}) such that the control performance index J_{con} is minimized. For a given control structure, combinations of controller parameters and network QoS are selected within the stability bounds (e.g. MATI, MAD, MAPL) derived in [51]. Both heuristics based optimization algorithms (e.g. Generic Algorithms [130]) and optimization algorithms using optimal control techniques (Section 7.2.2) are applied to find the optimal controller parameters $\boldsymbol{K}_{opt}$. Results are stored in a lookup table for later use during the control optimization phase.

Algorithm 7.1. Control design phase

Init: Select controller (e.g. LQR)
Init: Select sampling period h
Init: Select network QoS τ^{rt} and ρ^{rt} with corresponding distributions $\{\tau_k^{sc}\}_0^{\infty}$, $\{\tau_k^{ca}\}_0^{\infty}$, $\{\alpha_k\}_0^{\infty}$, and $\{\beta_k\}_0^{\infty}$
Init: Select initial controller parameter (e.g. $\boldsymbol{K}$ for the wired case)
Init: $k_f, \boldsymbol{Q}_{k_f}, \boldsymbol{Q}_x, \boldsymbol{Q}_u, J_{\text{con}}$
Init: $k \leftarrow 0$

loop
  while $k \leq k_f$ **do**
    get τ_k^{sc}, τ_k^{ca} from the given distributions
    get β_k^{sc} and α_k^{sc} from the given distributions
    measure $\boldsymbol{x}_k$
    compute $\boldsymbol{u}_k = -\boldsymbol{K}\boldsymbol{x}_k$
    compute $J_{con_k} = \boldsymbol{x}_{k_f}^{\mathrm{T}} \boldsymbol{Q}_{k_f} \boldsymbol{x}_{k_f} + \sum_{k=0}^{k_f-1} \left\{ \boldsymbol{x}_k^{\mathrm{T}} \boldsymbol{Q}_x \boldsymbol{x}_k + (1-\alpha_k)\, \boldsymbol{u}_k^{\mathrm{T}} \boldsymbol{Q}_u \boldsymbol{u}_k \right\}$
    $J_{\text{con}} \Leftarrow J_{\text{con}} + J_{con_k}$
    $k \Leftarrow k+1$
    if $k = k_f$ **then**
      $k \Leftarrow 0$
    end if
  end while
  adapt controller parameters $\boldsymbol{K}$ to minimize J_{con}
end loop{Use of Genetic Algorithms and minimum search algorithms}.
Result: $\boldsymbol{K}_{opt}(h, \tau^{rt}, \rho^{rt})$ for the selected network QoS τ^{rt} and ρ^{rt} are stored in a lookup table for use during the optimization phase.

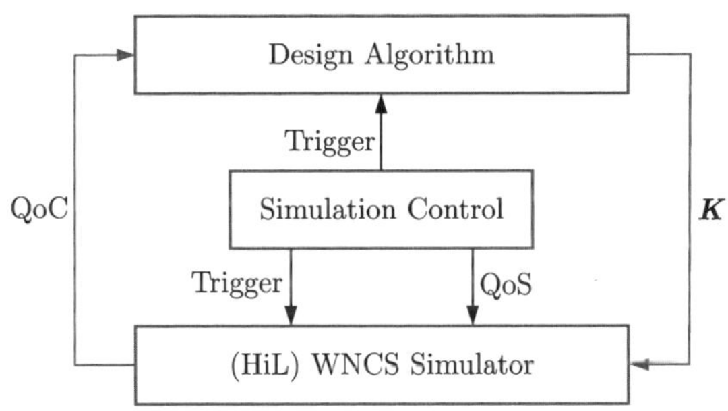

Fig. 7.6 Proposed structure for the control design phase

The proposed structure for the control optimization phase is depicted in Fig. 7.7 and an excerpt of the underlying workflow is presented below:

During the optimization phase, the total energy consumption of the wireless network is minimized while meeting a desired control performance using defined degrees of freedom at both the control layer and the communication layers. The control performance index J_{con} and network QoS parameters delay τ^{rt} and packet loss rate ρ^{rt} and performance metric J_{net} are measured at runtime. Results are then used to estimate the node lifetime. Suitable

Algorithm 7.2. Control optimization phase

Init: Select MAC scheme (e.g. TDMA)
Init: select MAC parameters (e.g. macro slot length S, macro slot time s_i)
Init: Select controller (e.g. LQR)
Init: Select controller parameter (from lookup table)

loop
  while $k \leq k_f$ **do**
    measure τ_k^{sc}, τ_k^{ca}
    get β_k^{sc} and α_k^{sc}
    measure $\boldsymbol{x}_k$
    compute $\boldsymbol{u}_k = -\boldsymbol{K}\boldsymbol{x}_k$
    compute $J_{con_k} = \boldsymbol{x}_{k_f}^{\mathrm{T}} \boldsymbol{Q}_{k_f} \boldsymbol{x}_{k_f} + \sum_{k=0}^{k_f-1} \left\{ \boldsymbol{x}_k^{\mathrm{T}} \boldsymbol{Q}_x \boldsymbol{x}_k + (1-\alpha_k)\, \boldsymbol{u}_k^{\mathrm{T}} \boldsymbol{Q}_u \boldsymbol{u}_k \right\}$
    $J_{\mathrm{con}} \Leftarrow J_{\mathrm{con}} + J_{con_k}$
    $k \Leftarrow k+1$
    if $k = k_f$ **then**
      estimate network lifetime
      $\overline{T} = \frac{\mathrm{E}_{Tot}}{\left[\bar{P}_W(\alpha,l)T_W + P_{M_1}(h-T_W) + (\bar{P}_W(\alpha,l) - P_{M_1})\bar{\epsilon} + \mathrm{E}_{M_{12}}\right]} h$
      $k \Leftarrow 0$
    end if
  end while
  select network and controller parameters (from lookup table) in order to maximize $\overline{T}$ while meeting $J_{\mathrm{con}} \leq J_{\max}$
end loop{Refer to [51] for the derivation of $\overline{T}$}.

Result: $\boldsymbol{K}_{opt}(h, \tau^{rt}, \rho^{rt})$ for measured network QoS, $J_{\mathrm{con}} \leq J_{\max}$, and $\overline{T}$ is maximized

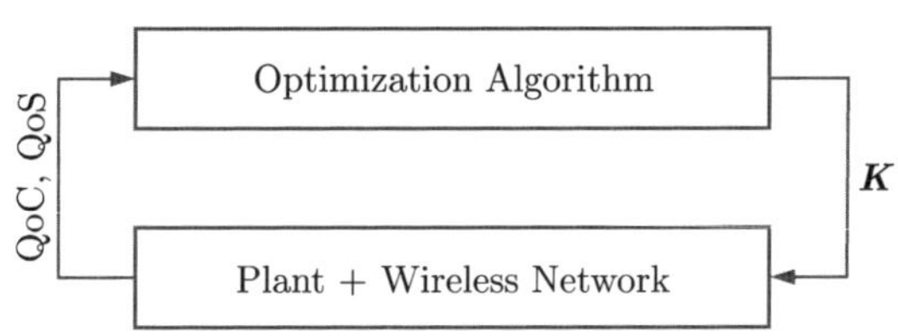

Fig. 7.7 Proposed structure for the control optimization phase

combinations of controller and communication protocol parameters are then selected from a lookup table and applied at runtime. Both the design phase and the optimization phase are extensively addressed in [51].

Control Model. The control design model considered is depicted in Fig. 7.8 where a plant is controlled over a wireless network using the dedicated wireless protocols developed in Section 7.2.3. Plant sensors transmit their measurements to the controller over the wireless network using scheduled (e.g. TDMA) or contention-based (e.g. CSMA/CA) MAC protocols. The wireless network may introduce time-varying delays and packet loss depending on the selected MAC protocol.

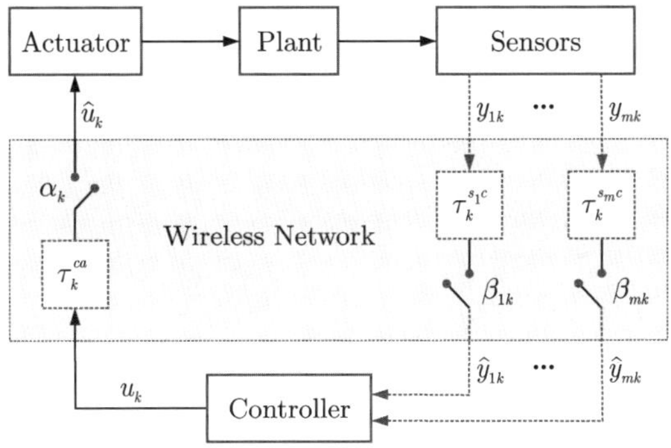

Fig. 7.8 Overview of the control model

The plant dynamics are modeled using the following differential equation

$$\begin{aligned}\dot{\boldsymbol{x}}(t) &= \boldsymbol{A}\boldsymbol{x}(t) + \boldsymbol{B}u(t) + \boldsymbol{v}(t) \\ \boldsymbol{y}(t) &= \boldsymbol{C}\boldsymbol{x}(t) + \boldsymbol{w}(t),\end{aligned} \tag{7.1}$$

where $\boldsymbol{x}(t) \in \mathbb{R}^n$ is the state vector, $\boldsymbol{y}(t) \in \mathbb{R}^m$ the output vector, $u(t) \in \mathbb{R}$ the control input, $\boldsymbol{v}(t) \in \mathbb{R}^n$ the process noise with zero mean and uncorrelated increments and $\boldsymbol{w}(t) \in \mathbb{R}^m$ the measurement noise with zero mean and uncorrelated increments. $\boldsymbol{A}$, $\boldsymbol{B}$, $\boldsymbol{C}$ are matrices of appropriate dimensions. Sampling the plant with time-varying sampling period $h_k = t_{k+1} - t_k$ and expressing the sampling period as $h_k = h_0 + \tau_k^{rt}$ leads to the following discrete-time formulation of the continuous-time plant dynamics (7.1) assuming zero-order hold

$$\begin{aligned}\boldsymbol{x}_{k+1} &= \boldsymbol{\Phi}\boldsymbol{x}_k + \boldsymbol{\Gamma}u_k + \boldsymbol{v}_k \\ \boldsymbol{y}_k &= \boldsymbol{C}\boldsymbol{x}_k + \boldsymbol{w}_k,\end{aligned}$$

where h_0 is constant, $0 < \tau_k^{sc} < h_k$ and $0 < h_{\min} \leq h_k \leq h_{\max}$. $\boldsymbol{\Phi} = e^{\boldsymbol{A}h_k}$, $\boldsymbol{\Gamma} = \int_0^{h_k} e^{\boldsymbol{A}s}\, ds\boldsymbol{B}$. We assume that the random sequences $\{\tau_k^{rt}\}_0^\infty$ and $\{h_k\}_0^\infty$ are independent with known distributions.

Modeling the packet loss sequences $\{\beta_k\}_0^\infty$ and $\{\alpha_k\}_0^\infty$ in Fig. 7.8 as random processes leads to the following expression for the sensor measurements at the controller input and the control value at the actuator input

$$\begin{aligned}\hat{\boldsymbol{y}}_k &= (1 - \beta_k)\,(\boldsymbol{C}\boldsymbol{x}_k + \boldsymbol{w}_k) \\ \hat{u}_k &= (1 - \alpha_k)u_k,\end{aligned}$$

where $\beta_k = \max\{\beta_{ik}\}_{i\in\{1,\ldots,m\}} \in \{0, 1\}$, $\beta_{ik} \in \{0, 1\}$ and $\alpha_k \in \{0, 1\}$, are Bernoulli random variables with packet loss probabilities $\text{Prob}\,[\beta_k = 1] = \rho^{sc}$ and $\text{Prob}\,[\alpha_k = 1] = \rho^{ca}$ modeling the packet loss between the sensor and the controller and between the controller and the actuator respectively.

β_k is known to the controller which expects periodic sensor measurements. However, only α_{k-1} is known to the controller if acknowledgements are used. Therefore, the following set of information $\boldsymbol{I}_k$ is available to the controller node at each sampling time instant k:

$$\boldsymbol{I}_k = \left\{ \hat{\boldsymbol{y}}_0, \cdots, \hat{\boldsymbol{y}}_k; \beta_0, \cdots, \beta_k; \alpha_0, \cdots, \alpha_{k-1}; \tau_0^{sc}, \cdots, \tau_k^{sc}; \tau_0^{ca}, \cdots, \tau_{k-1}^{ca} \right\}.$$

In the event of control packet loss between the controller and the actuator, the actuator does nothing ($\hat{u}_k = 0$).

The following linear difference equations with intermittent sensor measurements and control commands are obtained:

$$\begin{aligned} \boldsymbol{x}_{k+1} &= \boldsymbol{\Phi} \boldsymbol{x}_k + \boldsymbol{\Gamma} \hat{\boldsymbol{u}}_k + \boldsymbol{v}_k \\ \hat{\boldsymbol{y}}_k &= (1 - \beta_k)\, \boldsymbol{y}_k, \end{aligned}$$

where ($\boldsymbol{x}_0$, $\boldsymbol{v}_k$, $\boldsymbol{w}_k$) are Gaussian, uncorrelated, white, with mean ($\bar{\boldsymbol{x}}_0$, 0, 0) and covariance ($\boldsymbol{P}_0$, $\boldsymbol{Q}$, $\boldsymbol{R}$) respectively. In the event of a packet loss between the sensor and the controller, the input to the controller is $\hat{\boldsymbol{y}}_k = \boldsymbol{0}$. However, state estimates must be updated correctly.

State Estimator. Since measurements can be dropped out between sensors and controller, an estimator is required to reconstruct missing plant states. The Kalman filter state estimator used to estimate plant states in the event of packet loss between sensors and controller is of the following form:

$$\begin{aligned} \hat{\boldsymbol{x}}_{k+1|k} &= \boldsymbol{\Phi} \hat{\boldsymbol{x}}_{k|k} + \boldsymbol{\Gamma} \hat{u}_k && (7.2) \\ \boldsymbol{e}_{k+1|k} &= \boldsymbol{\Phi} \boldsymbol{e}_{k|k} + \boldsymbol{v}_k \\ \boldsymbol{P}_{k+1|k} &= \boldsymbol{\Phi} \boldsymbol{P}_{k|k} \boldsymbol{\Phi}^{\mathrm{T}} + \boldsymbol{Q} && (7.3) \end{aligned}$$

$$\begin{aligned} \hat{\boldsymbol{x}}_{k+1|k+1} &= \hat{\boldsymbol{x}}_{k+1|k} + (1 - \beta_{k+1})\, \boldsymbol{K}_{k+1} \left[\boldsymbol{y}_{k+1} - \boldsymbol{C} \hat{\boldsymbol{x}}_{k+1|k} \right] && (7.4) \\ \boldsymbol{e}_{k+1|k+1} &= [\boldsymbol{I} - (1 - \beta_{k+1})\, \boldsymbol{K}_{k+1} \boldsymbol{C}] \boldsymbol{e}_{k+1|k} - (1 - \beta_{k+1})\, \boldsymbol{K}_{k+1} \boldsymbol{v}_{k+1} \\ \boldsymbol{P}_{k+1|k+1} &= \boldsymbol{P}_{k+1|k} - (1 - \beta_{k+1})\, \boldsymbol{K}_{k+1} \boldsymbol{C} \boldsymbol{P}_{k+1|k} \\ \boldsymbol{K}_{k+1} &= \boldsymbol{P}_{k+1|k} \boldsymbol{C}^{\mathrm{T}} \left(\boldsymbol{C} \boldsymbol{P}_{k+1|k} \boldsymbol{C}^{\mathrm{T}} + \boldsymbol{R} \right)^{-1}, && (7.5) \end{aligned}$$

where $\boldsymbol{y}_k$ is the actual sensor measurement, $\boldsymbol{C}\hat{\boldsymbol{x}}_{k|k-1}$ the measurement prediction, $\boldsymbol{K}_k$ the time-varying Kalman filter gain matrix, $\boldsymbol{P}_{k|k-1}$ the error covariance matrix which is function of the packet loss between sensors and controller. The derivation of the above equations is similar to the derivation for the time varying Kalman filter equations using the time varying matrices $\boldsymbol{A}_k = \boldsymbol{\Phi}$ and $\boldsymbol{C}_k = (1 - \beta_k)\boldsymbol{C}$. Equations (7.2) to (7.3) are the time update equations and (7.4) to (7.5) the measurement update equations.

If sensor measurements are dropped on one or several consecutive sampling periods, the estimator must be run in open loop up to the time step $k - 1$ after successful arrival of all sensor packets at the controller at time step k

until the estimate $\hat{x}_{k-1|k-1}$ is obtained before applying Eq. (7.2). Assuming for example that, $\hat{x}_{\epsilon|\epsilon}$, $\hat{u}_\epsilon$, and τ_ϵ^{sc} are known at sampling time instant ϵ, the estimator is run in open loop using the iteration

$$\begin{aligned}\hat{\boldsymbol{x}}_{\epsilon+1|\epsilon+1} &= \hat{\boldsymbol{x}}_{\epsilon+1|\epsilon} = \boldsymbol{\Phi}\hat{\boldsymbol{x}}_{\epsilon|\epsilon} + \boldsymbol{\Gamma} u_\epsilon \\ \hat{u}_{\epsilon+1} &= 0 \\ \boldsymbol{P}_{\epsilon+1|\epsilon} &= \boldsymbol{\Phi}\boldsymbol{P}_{\epsilon|\epsilon-1}\boldsymbol{\Phi}^{\mathrm{T}} + \boldsymbol{Q}\end{aligned}$$

until $\hat{\boldsymbol{x}}_{k-1|k-1}$ is obtained.

There is a limit to how many sensor measurements can be lost before the state estimator loses stability. According to [340], there is a critical sensor to controller packet loss probability value ρ_{crti}^{sc} and controller to actuator packet loss probability ρ_{crit}^{ca} below which the estimated mean state covariance $\mathrm{E}_{\rho^{ca}}[\boldsymbol{P}_{k+1|k}]$ is bounded for all initial conditions (provided that the usually assumed stabilizability and detectability hypotheses are satisfied). Above this critical probability value, the estimated mean state covariance diverges for some initial conditions. The existence of this critical probability values was shown in [340] with bounds $\rho_{\min} \le \rho_{crit}^{sc} \le \rho_{\max}$ and $\rho_{\min} \le \rho_{crit}^{ca} \le \rho_{\max}$.

A close form solution for the lower bound is $\rho_{\min} = \frac{1}{\max_i (\phi_i)^2}$ and $\rho_{\max} = \frac{1}{\prod_i (\phi_i)^2}$ for the upper bound where $\boldsymbol{\phi} = |\lambda_i^u(\Phi)|^2$ is the spectral radius of the matrix $\boldsymbol{\Phi}$ and $\lambda_i^u(\boldsymbol{\Phi})$ the unstable eigenvalues of the matrix $\boldsymbol{\Phi}$. As shown by the authors, this upper bound is the solution of the following linear matrix inequality:

$$\rho_{cri}^{sc} = \arg\min_{\rho^{\bar{s}c}} \boldsymbol{\Psi}_{\rho^{sc}}(\boldsymbol{Y}, \boldsymbol{Z}) > 0, \quad 0 \le \boldsymbol{Y} \le \boldsymbol{I}$$

$$\boldsymbol{\Psi}_{\rho^{sc}}(\boldsymbol{Y}, \boldsymbol{Z}) = \begin{pmatrix} \boldsymbol{Y} & \boldsymbol{Y} & \tilde{\rho}\boldsymbol{Z}\boldsymbol{Q}_u^{\frac{1}{2}} & \tilde{\rho}(\boldsymbol{Y}\boldsymbol{\Phi}^{\mathrm{T}} + \boldsymbol{Z}\boldsymbol{\Gamma}^{\mathrm{T}}) & \sqrt{\rho^{sc}}\boldsymbol{Y}\boldsymbol{\Phi}^{\mathrm{T}} \\ \boldsymbol{Y} & \boldsymbol{Q}^{-1} & 0 & 0 & 0 \\ \tilde{\rho}\boldsymbol{Z}\boldsymbol{Q}_u^{\frac{1}{2}}\boldsymbol{Z}^{\mathrm{T}} & 0 & \boldsymbol{I} & 0 & 0 \\ \tilde{\rho}(\boldsymbol{\Phi}\boldsymbol{Y} + \boldsymbol{\Gamma}\boldsymbol{Z}^{\mathrm{T}}) & 0 & 0 & \boldsymbol{Y} & 0 \\ \sqrt{\rho^{sc}}\boldsymbol{\Phi}\boldsymbol{Y} & 0 & 0 & 0 & \boldsymbol{Y} \end{pmatrix}$$

where $\rho^{\bar{s}c}$ is the mean packet loss probability between the sensors and the controller and

$$\tilde{\rho} = \sqrt{1 - \rho^{sc}}.$$

This linear matrix inequality also applies for ρ_{crit}^{ca}.

Optimal Controller Design. The control objective is to minimize the following cost function

$$J_p(\boldsymbol{U}^{p-1}, \bar{\boldsymbol{x}}_0, \boldsymbol{P}_0) = \tag{7.6}$$
$$\mathrm{E}\left[\boldsymbol{x}_p^{\mathrm{T}}\boldsymbol{Q}_p\boldsymbol{x}_p + \sum_{k=0}^{p-1}\left\{\boldsymbol{x}_k^{\mathrm{T}}\boldsymbol{Q}_{xk}\boldsymbol{x}_k + (1-\alpha_k)\,\boldsymbol{u}_k^{\mathrm{T}}\boldsymbol{Q}_{uk}\boldsymbol{u}_k\right\}|\boldsymbol{U}^{p-1}, \bar{\boldsymbol{x}}_0, \boldsymbol{P}_0\right],$$

where $\boldsymbol{U}^{p-1} = (\boldsymbol{u}_{p-1}, \boldsymbol{u}_{p-2}, \cdots, \boldsymbol{u}_1)$, p is the optimization horizon, $\boldsymbol{Q}_x$ is positive semi-definite, $\boldsymbol{Q}_u$ positive definite, and $\boldsymbol{Q}_p$ positive semi-definite. $\boldsymbol{u}_k$ is the control value computed by the controller upon reception of sensor measurements and $(1-\alpha_k)\,\boldsymbol{u}_k$ is the control value applied by the actuator at sampling instant k. The control input at the plant is only weighted in the control cost if a new control command is received at the controller. Since no control action is taken by the actuator in the event of packet loss between the sensors and the controller, the actuator applies zero input, thus, saving energy. In order to minimize the cost function given by Eq. (7.6) a control input sequence $\boldsymbol{U}^{*p-1}$ is required, which is a function of the available set of information I_k at the controller at sampling time instant k, i.e. $\boldsymbol{u}_k = \boldsymbol{g}_k(\boldsymbol{I}_k)$. This leads to the following equation:

$$J_p^*(\bar{\boldsymbol{x}}_0, \boldsymbol{P}_0) = \min_{\boldsymbol{u}_k = \boldsymbol{g}_k(\boldsymbol{I}_k)} J_p(\boldsymbol{U}^{p-1}, \bar{\boldsymbol{x}}_0, \boldsymbol{P}_0).$$

The derivation of the optimal control law and the corresponding control cost are based on the dynamic programming approach using the cost-to-go iterative procedure. Defining the final cost function to be $V_p = \mathrm{E}[\boldsymbol{x}_p^{\mathrm{T}} \boldsymbol{Q}_p \boldsymbol{x}_p | \boldsymbol{I}_p]$ leads to the following cost-to-go function:

$$V_k = \min_{u_k} \mathrm{E}[\boldsymbol{x}_k^{\mathrm{T}} \boldsymbol{Q}_{xk} \boldsymbol{x}_k + (1-\alpha_k)\boldsymbol{u}_k^{\mathrm{T}} \boldsymbol{Q}_{uk} \boldsymbol{u}_k + V_{k+1}(\boldsymbol{x}_{k+1}) | \boldsymbol{I}_p] \tag{7.7}$$

for $k = p-1, \cdots, 1$. $V_k(\boldsymbol{x}_k)$ in Eq. (7.7) can be written as

$$\begin{aligned}
V_k(\boldsymbol{x}_k) &= \mathrm{E}[\boldsymbol{x}_k^{\mathrm{T}} \boldsymbol{S}_k \boldsymbol{x}_k | \boldsymbol{I}_k] + \boldsymbol{\nu}_k, k = p, \cdots, 0 \\
\boldsymbol{S}_k &= \boldsymbol{\Phi}^{\mathrm{T}} \boldsymbol{S}_{k+1} \boldsymbol{\Phi} + \boldsymbol{Q}_{xk} \\
&\quad - (1-\rho^{ca}) \boldsymbol{\Phi}^{\mathrm{T}} \boldsymbol{S}_{k+1} \boldsymbol{\Gamma} (\boldsymbol{\Gamma}^{\mathrm{T}} \boldsymbol{S}_{k+1} \boldsymbol{\Gamma} + \boldsymbol{Q}_{uk})^{-1} \boldsymbol{\Gamma}^{\mathrm{T}} \boldsymbol{S}_{k+1} \boldsymbol{\Phi} \\
\boldsymbol{\nu}_k &= \mathrm{trace}((\boldsymbol{\Phi}^{\mathrm{T}} \boldsymbol{S}_{k+1} \boldsymbol{\Phi} + \boldsymbol{Q}_{xk} - \boldsymbol{S}_k) \boldsymbol{P}_{k|k}) \\
&\quad + \mathrm{trace}(\boldsymbol{S}_{k+1} \boldsymbol{Q}) + \mathrm{E}[\boldsymbol{\nu}_{k+1} | \boldsymbol{I}_k]
\end{aligned}$$

with initial values $\boldsymbol{S}_p = \boldsymbol{Q}_p$ and $\boldsymbol{\nu}_p = 0$. The optimal control input $\boldsymbol{u}_k$ is given by

$$\begin{aligned}
\boldsymbol{u}_k &= -\boldsymbol{L}_{\rho^{sc}}\left(\tau_k^{sc}, \boldsymbol{S}_\infty\right) \hat{\boldsymbol{x}}_{k|k} \\
&= -\left[\boldsymbol{\Gamma}^{\mathrm{T}} \boldsymbol{S}_{k+1} \boldsymbol{\Gamma} + \boldsymbol{Q}_{uk}\right]^{-1} \boldsymbol{\Gamma}^{\mathrm{T}} \boldsymbol{S}_{k+1} \boldsymbol{\Phi} \hat{\boldsymbol{x}}_{k|k},
\end{aligned}$$

where $\boldsymbol{L}_{\rho^{sc}}\left(\tau_k^{sc}\right) = \boldsymbol{L}_{\rho^{sc}}\left(\tau_k^{sc}, S_\infty\right)$ is a function of the packet loss probability ρ^{sc} as well as the delay τ^{sc} between the sensors and the controller. Since $J_p^*(\bar{\boldsymbol{x}}_0, \boldsymbol{P}_0) = V_0(\boldsymbol{x}_0)$, the cost function of the optimal LQG is given by

$$\begin{aligned}
J_p^*(\bar{\boldsymbol{x}}_0, \boldsymbol{P}_0) &= \bar{\boldsymbol{x}}_0^{\mathrm{T}} \boldsymbol{S}_0 \bar{\boldsymbol{x}}_0 + \mathrm{trace}(\boldsymbol{S}_0 \boldsymbol{P}_0) + \\
&\sum_{k=0}^{p-1} \left\{ \mathrm{trace}(\boldsymbol{\Phi}^{\mathrm{T}} \boldsymbol{S}_{k+1} \boldsymbol{\Phi} + \boldsymbol{Q}_{xk} - \boldsymbol{S}_k) \mathrm{E}_{\rho^{sc}}[\boldsymbol{P}_{k|k}] \right\}
\end{aligned} \tag{7.8}$$

where $\mathrm{E}_{\rho^{sc}}[\boldsymbol{P}_{k|k}]$ explicitly shows that the expectation of the error covariance matrix is computed taking into consideration the packet loss probability between the sensors and the controller. Therefore, the error covariance matrices are stochastic and cannot be computed analytically. However, deterministic bounds on the error covariance matrices can be computed as shown in [340] resulting in the following equations

$$\begin{aligned}
\underline{\boldsymbol{P}_{k|k}} &\leq \mathrm{E}_{\rho^{sc}}[\boldsymbol{P}_{k|k}] \leq \overline{\boldsymbol{P}_{k|k}}, \forall k \geq 0 \\
\overline{\boldsymbol{P}_{k+1|k}} &= \boldsymbol{\Phi}\overline{\boldsymbol{P}_{k|k-1}}\boldsymbol{\Phi}^{\mathrm{T}} + \boldsymbol{Q} - \\
&\quad (1-\rho^{sc})\boldsymbol{\Phi}\overline{\boldsymbol{P}_{k|k-1}}\boldsymbol{C}^{\mathrm{T}}(\boldsymbol{C}\overline{\boldsymbol{P}_{k|k-1}}\boldsymbol{C}^{\mathrm{T}} + \boldsymbol{R})^{-1}\boldsymbol{C}\overline{\boldsymbol{P}_{k|k-1}}\boldsymbol{C}^{\mathrm{T}}\boldsymbol{\Phi}^{\mathrm{T}} \\
\overline{\boldsymbol{P}_{k|k}} &= \overline{\boldsymbol{P}_{k|k-1}} - (1-\rho^{sc})\Phi\overline{\boldsymbol{P}_{k|k-1}}\boldsymbol{C}^{\mathrm{T}}(C\overline{\boldsymbol{P}_{k|k-1}}\boldsymbol{C}^{\mathrm{T}} + \boldsymbol{R})^{-1}\boldsymbol{C}\overline{\boldsymbol{P}_{k|k-1}}\boldsymbol{C}^{\mathrm{T}}\boldsymbol{\Phi}^{\mathrm{T}} \\
\underline{\boldsymbol{P}_{k+1|k}} &= \rho^{sc}\Phi\underline{\boldsymbol{P}_{k|k-1}}\boldsymbol{\Phi}^{\mathrm{T}} + \boldsymbol{Q} \\
\underline{\boldsymbol{P}_{k|k}} &= \rho^{sc}\underline{\boldsymbol{P}_{k|k-1}},
\end{aligned}$$

where the initial conditions are $\overline{\boldsymbol{P}_{0|0}} = \underline{\boldsymbol{P}_{0|0}} = \boldsymbol{P}_0$. The minimum achievable control cost J_p^* in Eq. (7.8) cannot be computed analytically, but bounds can be found as follows:

$$\begin{aligned}
J_p^{\min} &\leq J_p^* \leq J_p^{\max} \\
J_p^{\max} &= \bar{\boldsymbol{x}}_0^{\mathrm{T}}\boldsymbol{S}_0\bar{\boldsymbol{x}}_0 + \mathrm{trace}(\boldsymbol{S}_0\boldsymbol{P}_0) + \sum_{k=0}^{p-1}\mathrm{trace}(\boldsymbol{S}_{k+1}\boldsymbol{Q}) + \\
&\quad \sum_{k=0}^{p-1}\left\{\mathrm{trace}(\boldsymbol{\Phi}^{\mathrm{T}}\boldsymbol{S}_{k+1}\boldsymbol{\Phi} + \boldsymbol{Q}_{xk} - \boldsymbol{S}_k)\overline{\boldsymbol{P}_{k|k}}\right\} \\
J_p^{\min} &= \bar{\boldsymbol{x}}_0^{\mathrm{T}}\boldsymbol{S}_0\bar{\boldsymbol{x}}_0 + \mathrm{trace}(\boldsymbol{S}_0\boldsymbol{P}_0) + \sum_{k=0}^{p-1}\mathrm{trace}(\boldsymbol{S}_{k+1}\boldsymbol{Q}) + \\
&\quad \sum_{k=0}^{p-1}\left\{\mathrm{trace}(\boldsymbol{\Phi}^{\mathrm{T}}\boldsymbol{S}_{k+1}\boldsymbol{\Phi} + \boldsymbol{Q}_{xk} - \boldsymbol{S}_k)\underline{\boldsymbol{P}_{k|k}}\right\}.
\end{aligned}$$

Following the same procedure as for the finite horizon case presented above, the optimal controller gain for the infinite horizon case is given by

$$\lim_{k\to\infty} \boldsymbol{L}_k = \boldsymbol{L}_\infty = \left[\boldsymbol{\Gamma}^{\mathrm{T}}\boldsymbol{S}_\infty\boldsymbol{\Gamma} + \boldsymbol{Q}_{uk}\right]^{-1}\boldsymbol{\Gamma}^{\mathrm{T}}\boldsymbol{S}_\infty\boldsymbol{\Phi}$$

and the expected minimum cost is bounded by the following two deterministic sequences

$$\frac{1}{p}J_p^{\min}\frac{1}{p} \leq J_p^* \leq \frac{1}{p}J_p^{\max}$$

$$\begin{aligned}
J_{\infty}^{\max} &= \lim_{p\to\infty} J_p^{\max} \\
&= \operatorname{trace}(\boldsymbol{S}_\infty \boldsymbol{Q}) + \operatorname{trace}((\boldsymbol{\Phi}^{\mathrm{T}} \boldsymbol{S}_\infty \boldsymbol{\Phi} + \boldsymbol{Q}_x - \boldsymbol{S}_\infty)(\overline{\boldsymbol{P}_\infty} - \\
&\quad (1-\rho^{sc})\overline{\boldsymbol{P}_\infty}\boldsymbol{C}^{\mathrm{T}}(\boldsymbol{C}\overline{\boldsymbol{P}_\infty}\boldsymbol{C}^{\mathrm{T}} + \boldsymbol{R})^{-1}\boldsymbol{C}\overline{\boldsymbol{P}_\infty}\boldsymbol{C}^{\mathrm{T}})) \\
J_{\infty}^{\min} &= \lim_{p\to\infty} J_p^{\max} \\
&= \operatorname{trace}(\boldsymbol{S}_\infty \boldsymbol{Q}) + \rho^{sc}\operatorname{trace}(\boldsymbol{\Phi}^{\mathrm{T}} \boldsymbol{S}_\infty \boldsymbol{\Phi} + \boldsymbol{Q}_x - \boldsymbol{S}_\infty),
\end{aligned}$$

where the matrices $\boldsymbol{S}_\infty$, $\overline{\boldsymbol{P}_\infty}$, and $\underline{\boldsymbol{P}_\infty}$ are positive definite solutions to the following set of equations

$$\begin{aligned}
\boldsymbol{S}_\infty &= \boldsymbol{\Phi}^{\mathrm{T}} \boldsymbol{S}_\infty \boldsymbol{\Phi} + \boldsymbol{Q}_x - (1-\rho^{ca})\boldsymbol{\Phi}^{\mathrm{T}} \boldsymbol{S}_\infty \boldsymbol{\Gamma}(\boldsymbol{\Gamma}^{\mathrm{T}} \boldsymbol{S}_\infty \boldsymbol{\Gamma} + \boldsymbol{Q}_u)^{-1}\boldsymbol{\Gamma}^{\mathrm{T}} \boldsymbol{S}_\infty \boldsymbol{\Phi} \\
\overline{\boldsymbol{P}_\infty} &= \boldsymbol{\Phi}\overline{\boldsymbol{P}_\infty}\boldsymbol{\Phi}^{\mathrm{T}} + \boldsymbol{Q} - (1-\rho^{sc})\boldsymbol{\Phi}^{\mathrm{T}}\overline{\boldsymbol{P}_\infty}\boldsymbol{C}(\boldsymbol{C}^{\mathrm{T}} \boldsymbol{S}_\infty \boldsymbol{C} + \boldsymbol{Q}_u)^{-1}\boldsymbol{C}^{\mathrm{T}}\overline{\boldsymbol{P}_\infty}\boldsymbol{\Phi}^{\mathrm{T}} \\
\underline{\boldsymbol{P}_\infty} &= \rho^{sc}\boldsymbol{\Phi}\underline{\boldsymbol{P}_\infty}\boldsymbol{\Phi}^{\mathrm{T}} + \boldsymbol{Q}.
\end{aligned}$$

7.2.3 Communication Design

The NQoS specification defines an abstract interface between control layer and communication layer, which forms the starting point for the joint design of control and communication system (Section 7.2.1). Refining the abstract architecture of WNCS shown in Fig. 7.3, we decompose the communication layer into four sublayers (Fig. 7.9). With each sublayer, core communication functionalities of WNCS are associated. In addition, WNCS protocols jointly designed or selected, configured, and integrated to provide a cross-layered communication system for inverted pendulum and flow rate control are shown in the figure. In this section, we will argue why certain communication functionalities are relevant to WNCS, and survey our protocol solutions. In addition, we will report on our model-driven development, which supports the high-level design of communication systems.

To provide enhanced best-effort guarantees in multi-hop wireless networks, deterministic protocols supporting time-triggered communication are needed. This means, among other things, that medium access based on statistical contention schemes or without predictable timing is not sufficient. Instead, reservation-based schemes with exclusive access and predictable, controlled contention are mandatory. Also, synchronous sampling and timestamping of sensor values is required, which can only be achieved through a joint time base of all network nodes. These constraints substantially reduce the choice of technologies and protocols, in particular in the context of wireless networks.

Physical Sublayer. An important cross-layer design decision concerns the selection of a suitable transceiver technology for the physical sublayer, which on the one hand has to be powerful enough to support the performance requirements of the control system; on the other hand, it has to be sufficiently

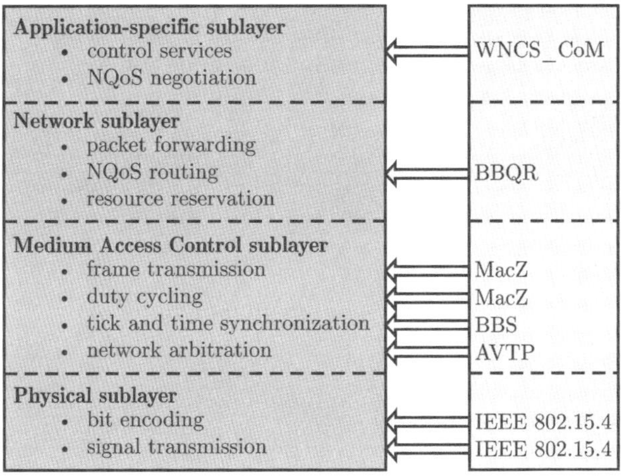

Fig. 7.9 Communication cross-layering architecture and related WNCS protocols

rich and flexible to support the communication functionalities of the upper sublayers. A transceiver satisfying these constraints is the ChipCon CC2420 [57], which follows the IEEE 802.15.4 PHY standard for wireless communication. The CC2420 has been designed for low power and low voltage applications, and provides an effective data rate of 250 Kbps. In addition, there is hardware support for packet handling, data buffering, clear channel assessment (CCA), link quality indication, energy detection and packet timing information. In the project, we have decided to use this off-the-shelf technology as the basis for specialized WNCS protocols on the upper sublayers of the communication cross-layering architecture (Fig. 7.9).

Medium Access Control Sublayer. The MAC sublayer is in charge of controlling medium access, which is crucial for the support of NQoS requirements of WNCS in multi-hop wireless networks. A core functionality is the deterministic network-wide synchronization of nodes. In the project, we have decided to use *Black Burst Synchronization* (BBS) [138], which provides both tick and time synchronization. Tick (heartbeat) synchronization determines global reference points in time, which are needed on MAC level to realize Time Division Multiple Access (TDMA) schemes. Time (clock) synchronization also associates clock values with reference points in time, which is needed on control level for synchronous sampling and timestamping.

BBS is integrated into *MacZ* [18], a MAC protocol providing flexible network-wide medium slotting, several kinds of frame transmission, network-wide deterministic contention schemes, and flexible duty cycling. Based on network-wide tick synchronization, time is decomposed into *macro slots*, which are structured into a resynchronization interval followed by a number of *micro slots* (Fig. 7.10). Micro slots can then be grouped into several

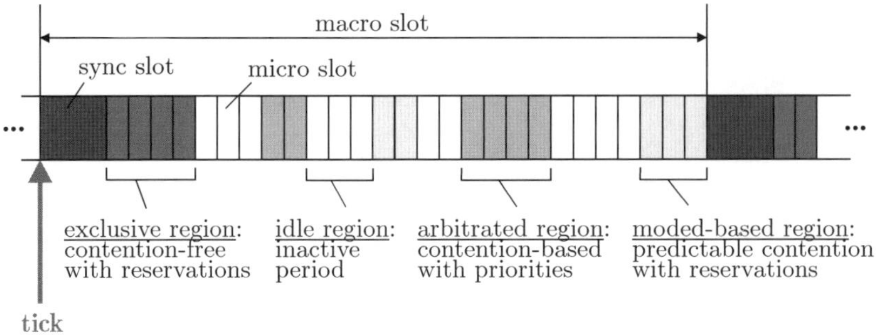

Fig. 7.10 Network-wide medium slotting for WNCS

kinds of virtual slot regions, supporting several kinds of frame transmission as well as flexible duty cycling:

- In *exclusive regions*, medium access is by exclusive reservations and thus is contention-free. Therefore, frame transmissions start at determined points in time, without prior carrier sensing. In multi-hop networks, frame collisions are safely avoided if reservations are network-wide, which means that at most one node can use an exclusive slot. A more effective use of exclusive slots is possible by applying a Spatial Division Multiple Access (SDMA) scheme, where exclusive reservations are not network-wide, but respect the interference range of sender and receiver only [185]. To realize the mode-based schedule shown in Fig. 7.5, slots s_3 to s_{12}, for instance, are grouped into an exclusive region.

- In *arbitrated regions*, medium access is by general contention with strict or statistical priorities. As in existing wireless technologies, an arbitrary number of nodes may contend for the medium, based on a random backoff scheme. After an initial medium idle period, nodes draw values from their contention windows, which are then counted backward every time the medium is observed as idle for the duration of a backoff slot. As soon as the count reaches zero, frame transmission is started. Different from existing wireless technologies, MacZ can be configured to support strict priorities, too, which is a step towards a more predictable contention, however, at the expense of larger contention windows (for details, see [18]). To realize the mode-based schedule shown in Fig. 7.5, no arbitrated regions are configured.

- In *mode-based regions*, medium access is by non-exclusive reservations with unique priorities, resulting in predictable, controlled contention. These regions are especially devised to support mode-based scheduling. Here, frames are to be transmitted according to their unique priority. To implement this efficiently, *fast mode-signaling* [139] is used, to reach consensus on the current transmission mode with the highest priority extremely fast and reliably among all network nodes. In wireless networks where all nodes are in sensing range, fast mode-signaling can be implemented by using a very small contention window of a size equal to the number of used modes minus 1 in a given slot. To realize the mode-based schedule shown in Fig. 7.5, slots s_{13} to s_{15}, for instance, are grouped into a mode-based region, using a contention window of size 1.

- *Idle regions* are inactive periods without any scheduled transmissions. Here, transceivers can change to sleep mode, to save energy. As idle regions may be inserted anywhere in the macro slot, this supports very flexible duty cycling. To realize the mode-based schedule shown in Fig. 7.5, slots s_{20}, s_{21}, s_{26}, s_{31}, s_{36}, and s_{38} to s_{40} are grouped into idle regions.

This medium structuring of MacZ supports a high degree of flexibility especially needed in WNCS, where slot reservations must be in line with sampling periods.

A crucial functionality of wireless networks formed of battery-powered nodes is *duty cycling*. In communication systems, a duty cycle is the fraction of time a transceiver is in an active state. Given a duty cycle of, for example, 2%, the transceiver can be switched to power-down mode for the remaining 98% of the runtime, thereby increasing the networks lifetime by a factor that depends on the activity and power consumption of further hardware components such as CPU, memory, or A/D converters. Energy is wasted, for instance, due to *idle listening*, which occurs when a node listens on the medium without detecting any transmissions. Other reasons are *overhearing* (reception of frames with different destinations), *collisions* (overlapping transmissions of nodes within interference range of a receiver), and *overhead* (exchange of non-payload data, e.g. synchronization frames).

With MacZ, we have designed and realized a very flexible and energy-efficient solution for duty cycling [58, 61, 386]. As explained before, MacZ supports the definition of weakly periodic activity and inactivity periods (virtual slot regions), which are repeated in each macro slot, thereby strengthening flexibility. Energy efficiency is achieved by reducing idle listening, overhearing, collisions, and overhead significantly, especially in exclusive regions. In an experiment, we have shown that the duty cycles of well-known duty cycling protocols are 10-20 times longer than the duty cycles of MacZ, and that their waste factor is 60-80 times higher [62].

With the *Arbitrating Value Transfer Protocol (AVTP)* [60], MacZ integrates a crucial functionality that enables deterministic node arbitration in multi-hop WNCS. AVTP can be classified as a deterministic binary countdown algorithm for wireless multi-hop networks. It achieves arbitration among contending nodes within a given hop radius, which may be chosen as the network diameter for network-wide arbitration. This goes beyond the capabilities of other protocols, which are limited to an arbitration radius of 2 hops. In addition, AVTP can be used to reach network-wide agreement on data values. Under reasonable assumptions, the protocol achieves a small and constant arbitration delay, has low and bounded complexity regarding computation, storage, time, and structure, and is robust against topology changes at runtime.

In the project, we use AVTP to realize deterministic NQoS routing with BBQR [21] (see below). In addition, we have realized the Try-Once-Discard (TOD) protocol, which is of high interest in networked control systems because its properties can be characterized by Lyapunov functions, with AVTP [61].

Network Sublayer. The network sublayer is in charge of packet forwarding, NQoS routing, and resource reservation, which are functionalities especially required in multi-hop networks with dynamic NQoS negotiations. For WNCS, it is usually not sufficient to find some path between communicating nodes, but paths satisfying a specified NQoS. While this is a hard problem in general, it can be solved on the basis of TDMA schemes and dynamic slot reservations. With *Black-Burst-based Quality-of-Service Routing (BBQR)* [21], we have devised a deterministic NQoS routing protocol for WNCS that solves two core problems. First, there is the problem that concurrent route discoveries may fail due to mutual blocking of resources. BBQR solves this problem by serializing route discoveries, using AVTP for deterministic arbitration. Secondly, there is the problem that routing packets may be lost due to frame collisions on the wireless medium, which may lead to failure of route discoveries. BBQR avoids this problem through synchronized route discovery phases and the use of collision-resistant frame encodings with black bursts.

Application-Specific Sublayer. The application-specific sublayer is in charge of providing control services, and of supporting NQoS negotiations. In the project, we have devised the *WNCS Communication Middleware (WNCS_CoM)* [55], which realizes a service interface supporting control applications on a high level of abstraction, while granting a high degree of freedom to the cross-layer design of the lower layers of the communication system. The idea of WNCS_CoM is that sensors and actuators register specific services (e.g., periodical value delivery, event notification), to which controllers can subscribe.

Example 7.1 (cont.) *Networked control of an inverted pendulum*

Figure 7.11 shows a typical usage scenario. The control application `SensorAngle` registers a service called `angle` by a confirmed interaction with `WNCS_CoM`. Afterwards, the controller `InvPendulum` subscribes to this service, specifying the base (preferred) periodicity and transmission delay to be 80ms (40ms) and 16ms (16ms), respectively. This causes `WNCS_CoM` to obtain a timestamped pendulum angle value from `SensorAngle` periodically (internal action `getAngle`), and to forward this value and the timestamp ts to `InvPendulum`. In the figure, this is shown as a periodical message exchange triggered by a timer. We note that `WNCS_CoM` is in fact distributed, with an instance on each node of the control system, but is shown as a single agent in the diagram. □

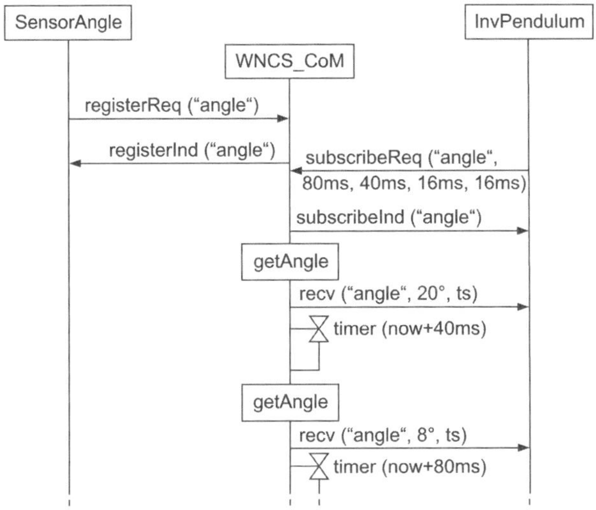

Fig. 7.11 Usage scenario of the WNCS communication middleware (WNCS_CoM)

Model-Driven Development of WNCS. Model-driven engineering is a software development methodology that places the abstract, formal system model in the center of the development activity. The objective is that abstract models guide and direct all development activities, ranging from system design over code generation and deployment to system maintenance, resulting both in quality improvements and productivity increases. In order to cross-layer design communication systems of WNCS, we apply SDL-MDD [213], our model-driven development approach with SDL (Specification and Description Language) as modeling language. SDL supports the high-level design specification of distributed systems, and, in particular, communication protocols. The SDL specification is the starting point for model-driven implementation, where code is generated automatically, instrumented by platform-specific driver routines, deployed and executed under the control of an SDL engine. Furthermore, it can be used as the starting point for model-driven performance simulations [19].

WNCS can be categorized as real-time systems, i.e. reactive systems in which the correctness of the system behavior depends on the correct ordering of events and their occurrence in time. Furthermore, they usually operate under severe limitations regarding computation, communication, and energy. Therefore, aspects such as real-time scheduling, energy control and resource management are crucial for their development, and have to be covered by the specification language [20]. In the course of the project, we have proposed SDL extensions and specification styles to meet these demands: tasks, real-time signaling, priority scheduling, energy-mode signaling, and energy scheduling.

The execution of real-time systems is usually decomposed into processing units called *tasks*, which are scheduled according to their urgency. To specify tasks, we have proposed to add the concept of task to SDL [38]. An SDL task is identified by a unique task id, which is associated with a transition trigger and assigned when the task is created. Tasks can be created by sending SDL signals or by starting SDL timers marked as new task. For scheduling purposes, a priority can be assigned to each task. When a transition is fired, it is associated with the task id of its trigger. In case further triggers are generated during transition execution, by sending SDL signals or by setting timers, these triggers implicitly inherit the current task id of the transition unless marked as new task. Tasks terminate as soon as no further triggers exist. With these stipulations, it is possible to specify complex tasks in SDL, formed by a set of partially ordered SDL transitions.

In real-time systems, tasks may be triggered at predefined points in time, for instance, to sample and forward sensor values. In SDL, timers could be used to specify time-triggered task scheduling. For several reasons, this design solution has major drawbacks [38]. To solve this problem, we have proposed another design solution called *real-time signaling* in [212]. A real-time signal is an SDL signal for which an arrival time is specified when the signal is sent. The signal is transferred to its SDL destination process as usual, and is appended to its input queue. However, consumption of the signal is postponed until the specified arrival time. Thus, a task scheduler can send real-time signals triggering SDL tasks well ahead of the actual point in time at which the task is to be started, by setting the arrival time to the scheduled execution time. We have defined a syntactical extension of SDL and its formal semantics, have implemented real-time signals in our SDL-MDD tool chain, and have applied it to the inverted pendulum.

The capability to achieve short or even predictable reaction times is an essential capability of real-time systems. In [59], we have taken a pragmatic approach by adding language constructs for the specification of SDL process priorities, combined with a mechanism to suspend and resume SDL processes. All SDL extensions were implemented in the SDL-MDD tool chain, and evaluated in a series of experiments. Compared to customary SDL transition scheduling policies, priority scheduling with process suspension indeed achieves significantly shortened reaction times. However, as it turned out, SDL processes do no reflect the task structure of real-time systems. Therefore, we have decided to drop this extension, and to use the concept of SDL tasks [38] together with task priorities instead.

In systems with battery-powered nodes such as WNCS, energy consumption is a major concern, constraining the uptime of nodes and therefore of control systems. To incorporate energy aspects in SDL, we have identified two complementary conceptual approaches [140]. The first approach, called *energy-mode signaling*, requires that the system developer explicitly specifies transitions of the energy model as part of the SDL design. In our solution, this is done by sending special SDL signals called energy-mode signals to the SDL environment. These signals are then mapped to the native operating system, and finally to the hardware platform. We have applied energy-mode signaling to switch the transceiver and CPU between active and idle mode. The second approach, called *energy scheduling*, incorporates additional functionality into the SDL engine, which controls the execution of SDL systems. This functionality consists, for each hardware component (e.g. CPU, transceiver, memory, A/D converter), of an energy scheduler, which is responsible for triggering transitions of the energy model. This scheduler uses state information available to the SDL engine in order to determine when and how the mode of operation is to be changed.To derive this information, signal queues and active SDL timers are inspected. As this does not require modifications of the SDL design specification, the high abstraction level of the model is maintained.

7.3 Control Requirements on Network Protocols

7.3.1 Load, Loss and Control Performance

When designing a networked control system (NCS) , we are mainly interested in the control performance and would like to model the communication system as a transparent tool that allows data exchange whenever necessary but can be neglected when analyzing the control system. Unfortunately, no real communication system is ideal and thus a limited data rate, packet loss, and delay must be accepted. As a consequence thereof, the choice of the optimal communication rate is not straight forward due to the complex interaction between control and communication. Figure 7.12 depicts how load, loss, and delay affect the control performance.

In general, the control performance increases with the network load, i.e., the more data exchanged, the better the control performance. On the other hand, the control performance decreases with loss and delay, i.e., the more loss and delay, the worse the control performance. Moreover, in the field of communication systems it is well known that loss and delay increase with the network load but also depend on several other factors like the communication protocol and the traffic pattern. These relationships are relatively simple and well known when considered separately. However, when taking all three into account, the effect of the network load on the control performance is not that simple anymore. Increasing the network load increases the control performance but also increases loss and delay, which decrease the control performance. Thus, the choice of the optimal network load becomes a challenging problem, which is up to now not well understood.

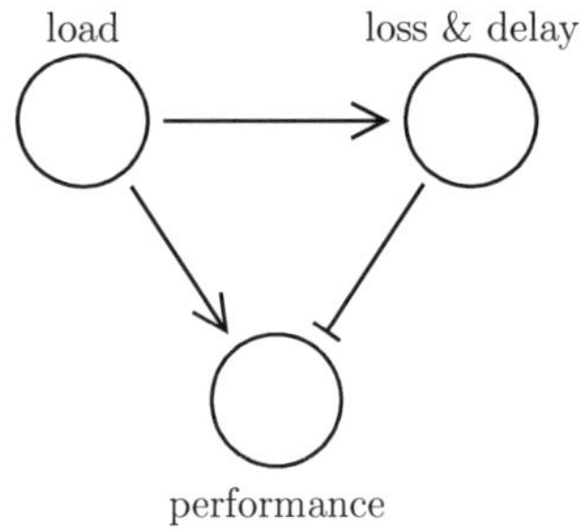

Fig. 7.12 The effect of load, loss, and delay on the control performance

One of the challenges of a detailed analysis of a networked control system is the fact that different layers of the ISO/OSI stack must be taken into account. The controller is located within the application layer but loss and delay is determined within the medium access control (MAC) layer. The control strategy determines the traffic pattern and thereby affects the choice of the MAC protocol. On the other hand, loss and delay of the communication system affect the controller design. Thus, these two layers can not be designed separately.

7.3.2 Control Problem (Application Layer)

In this section, we consider the control of N integrator systems

$$\dot{x}(t) = u(t) + w(t), \tag{7.9}$$

where $x \in \mathbb{R}$ is the state of an individual system and u its input. Since all systems are identical, we omit an index to distinguish between the different systems. The disturbance w is assumed to be white Gaussian noise with unit incremental variance. Each system is controlled by an agent whose goal is to minimize the cost

$$J = \lim_{M \to \infty} \sup \frac{1}{M} \int_0^M \mathrm{E}[x(t)^2] \mathrm{d}t. \tag{7.10}$$

Moreover, all agents uses the same communication system to close the loop.

The model of the communication system is relatively simple. Each packet takes some time to be transmitted, the *packet duration* τ. Moreover, packets collide and are lost whenever two or more agents send simultaneously. To avoid collisions, there might be an additional delay caused by the MAC protocol. Thus, d denotes the expected end-to-end delay of a packet, i.e., the time between packet generation and reception. Moreover, p denotes the packet loss probability. As already stated, the end-to-end delay d and the loss probability p are a result of the control and communication strategy. Hence, loss and delay depend on the choice of the communication protocol, the packet duration τ, the control strategy, and the network load.

When the communication is ideal, i.e., when there is no loss and delay, the control input u is a sequence of impulses

$$u(t) = \sum_{k} -\delta(t - t_k)x(t_k), \tag{7.11}$$

where t_k, $(k = 0, 1, ...)$ are the times of event generation. Each impulse is such that it resets the state to the origin when applied immediately. Between the events, the input is zero and the system just integrates the noise.

Since packets are delayed or even lost, the control input is slightly different. Obviously, if a packet is lost, the corresponding impulse will not be applied. Hence, we use $\mathcal{A}$ to denote the index set of arrived packets. Moreover, if the packet is delayed by d_k, the impulse is applied at time $t_k + d_k$ and will not reset the state exactly to the origin. Thus, the control input becomes

$$u(t) = \sum_{k \in \mathcal{A}} -\delta(t - t_k - d_k)x(t_k). \tag{7.12}$$

Combining (7.9) and (7.12), we get

$$x^+(t_k + d_k) = x(t_k + d_k) - x(t_k) \qquad \text{for } t = t_k + d_k, k \in \mathcal{A} \tag{7.13}$$

$$\dot{x}(t) = w(t),\ u(t) = 0 \qquad \text{else,} \tag{7.14}$$

where x^+ is the state of the system directly after applying the impulse.

Now, the interesting question is when to apply an impulse. The most well known scheme is time-triggered control, where the time between control updates is constant, i. e.,

$$t_{k+1} = t_k + T_{\mathrm{TT}}, \tag{7.15}$$

where T_{TT} is the sampling time. Since the control updates are applied without taking the current state of the plant into account this might result in control updates at times when they are not essential and a lack of control updates at times when they would be important.

This problem can be solved by explicitly taking the state of the system into account when determining the time for updating the control action. One such approach is event-based control, (cf. Chapter 5). Here, events are generated whenever the state x exceeds a bound, i. e., when $x \leq \underline{\Delta}$ or $x \geq \overline{\Delta}$. To guarantee that this scheme works properly even when events are delayed or lost, the bounds must be changed as follows. Whenever a bound is exceeded the bounds are shifted by the boundary increment Δ such that the state is exactly between the new bounds. Whenever a packet arrives and the corresponding event is applied, the bounds must be changed accordingly.

Example 7.2 *Event-based control of an integrator system*

Figure 7.13 shows an example of event-based control with packet loss and delay. In this example, the state of the system starts at the origin and the bounds are $\pm\Delta$. At time t_1, the state reaches the upper bound $\overline{\Delta} = \Delta$ and an event occurs. Thus, a packet is sent over the communication system and the bounds are shifted by $+\Delta$ to $\underline{\Delta} = 0$ and $\overline{\Delta} = 2\Delta$. At time $\tilde{t}_1 = t_1 + d_1$, this packet arrives and the corresponding impulse is applied, which changes the state by $-\Delta$. Due to the delay between the packet generation and the application of the corresponding impulse the state is not reset exactly to the origin. Moreover, the bounds are also shifted by $-\Delta$ back to $\pm\Delta$. At time t_2, the state reaches the upper bound. As a consequence thereof, the bounds are shifted by $+\Delta$ and a packet is sent over the network. Now, we assume that this packet is lost. At time t_3, the state reaches the upper bound $\overline{\Delta} = 2\Delta$. Again, the bounds are shifted by $+\Delta$ and a packet is generated. Since the previous packet was lost, this packet contains the command to change the state and bounds by -2Δ. When this packet arrives at time $\tilde{t}_3 = t_3 + d_3$, the corresponding impulse is applied and the bounds are shifted by -2Δ back to $\underline{\Delta} = -\Delta$ and $\overline{\Delta} = +\Delta$. □

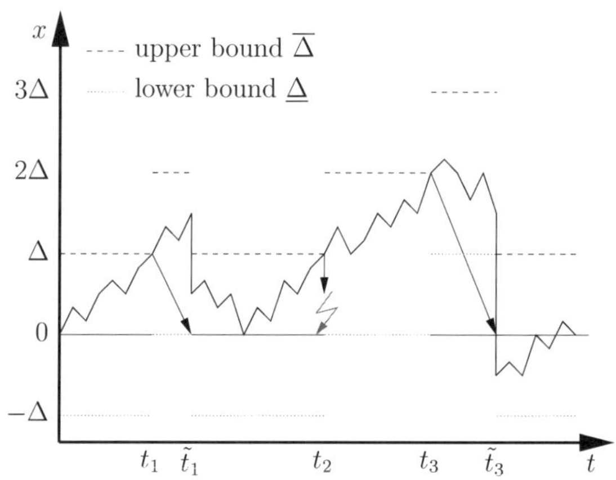

Fig. 7.13 An example of event-based control

7.3.3 Comparison of Event-Based and Time-Triggered Control

For event-based control, the expected inter-event time T_{EB} can be calculated from the boundary increments as $T_{\text{EB}} = \Delta^2$. However, for the analysis of communication systems, it is more appropriate to use the inter-arrival rate $\lambda := 1/T$, where T is the sampling time or expected inter-event time, respectively, instead of the sampling time or expected inter-event time. This is mainly due to the fact that the accumulated rate of all agents is just the sum of the individual rates, i.e., $\lambda_\Sigma = \sum_i \lambda_i$. Moreover, the offered load ρ of an agent is the product of the packet duration τ and its inter-arrival rate λ, i.e., $\rho := \tau\lambda = \tau/\Delta^2$. Since there are N identical agents, the network load becomes $\rho_\Sigma = \sum_i \rho_i = N\rho$.

For both control schemes, it turns out that the cost directly scales with the speed of the communication system, i.e., with the packet duration τ. To compare the different control and communication schemes fairly, we hence use the normalized cost J/τ as performance measure. Using the previous definitions, the normalized cost of time-triggered control becomes

$$\frac{J_{\mathrm{TT}}}{\tau} = \frac{1}{2\rho} + \frac{p}{\rho(1-p)} + \frac{d}{\tau}, \tag{7.16}$$

whereas the normalized cost of event-based control becomes

$$\frac{J_{\mathrm{EB}}}{\tau} = \frac{1}{6\rho} + \frac{p}{\rho(1-p)} + \frac{d}{\tau}. \tag{7.17}$$

Comparing the cost function of event-based and time-triggered control, we see that the first term is the cost with ideal communication, i.e., without loss and delay. Assuming that the load ρ is identical, event-based control outperforms time-triggered control. The second term is an additional cost due to loss, which is similar for both schemes. This additional cost is zero for $p = 0$ and approaches infinity for $p \to 1$. This is not surprising since for $p = 0$ the communication is ideal and for $p = 1$ all packets are lost and thus no control action takes place. Finally, the third term is an additional cost due to delay, which is identical for both control schemes.

When comparing time-triggered and event-based control, it is important to take into account that the traffic pattern is different. This difference affects the choice of the communication protocol and thereby also the loss and delay of the communication system. As already stated, the time between two events is constant in time-triggered control. Thus, the event times are known in advance and the communication system can be optimized accordingly. Obviously, this is not the case for event-based control since the events are generated by a random process. However, the probability distribution function (PDF) of the inter-event times can be derived from the Wiener first passage time with two bounds. When the bounds are $\pm\Delta$, the PDF of the inter-event times of event-based control is

$$f(t|\Delta) = \Delta\sqrt{\frac{2}{\pi t^3}} \sum_{k=-\infty}^{\infty} (4k+1)\mathrm{e}^{-\frac{(4k+1)^2\Delta^2}{2t}}, \tag{7.18}$$

which can be written as a function of a normalized PDF $f(t|1)$:

$$f(t|\Delta) = \frac{1}{\Delta^2} f(\frac{t}{\Delta^2}|1) = \lambda f(\lambda t|1). \tag{7.19}$$

Figure 7.14 shows the PDF of the normalized inter-arrival time $f(t|1)$, which reflects our intuition. For very small values of t, i.e., directly after an event, it is very unlikely that there will be a new event. Then, there is a period where most events will happen. Finally, it is also unlikely that the time between two events is large.

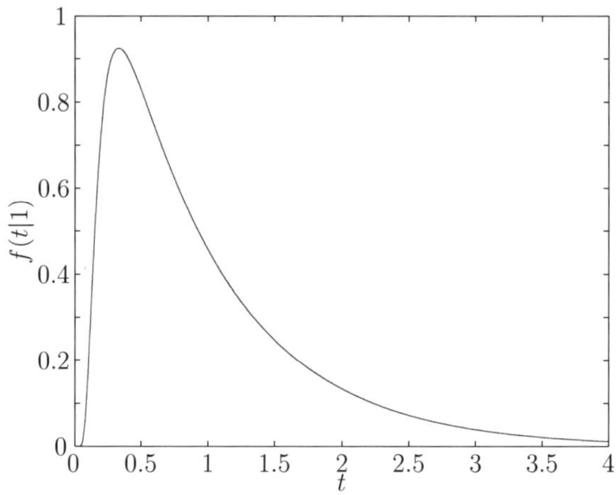

Fig. 7.14 The PDF $f(t|1)$ of the normalized inter-arrival time

From (7.18) and Fig. 7.14, we see that the event times of event-based control are neither deterministic nor a Poisson process. Thus, most results found in the communication literature can not be applied directly because a Poisson arrival process is assumed in most cases. However, it is possible to show that the arrival process converges to a Poisson process for $N \to \infty$ as stated in the following theorem.

Theorem 7.1. *Suppose all N agents together send with a certain rate $\lambda_\Sigma < \infty$, i.e., each agent sends with the rate λ_Σ/N. As $N \to \infty$, the superposition of the arrival processes of all agents approaches a Poisson process with rate λ_Σ.*

This theorem allows us to use the classical results from communication literature when analyzing networked event-based control.

7.3.4 Properties of Communication Systems (MAC Layer)

In this section, we give the expected delay and loss probability for different communication protocols. Due to the significant difference in the traffic generated by time-triggered and event-based control, we distinguish between two classes of protocols. In time-triggered control the event times are known in advance and can be tuned accordingly. This allows the usage of *deterministic protocols*, where the resources must be assigned in advance. Hence, we analyze time-triggered control with the two most important deterministic protocols:

TDMA and FDMA. In contrast, when using event-based control, the event times are not known in advance and thus the resources can not be reserved. Whenever an agent has to send a packet, it has to compete with other agents for the medium. Hence, we analyze event-based control with the most basic *contention-based protocols.*

Time Division Multiple Access (TDMA). Here, every agent gets its own timeslot. Thus, there will be no collisions and the delay is just the packet duration, i.e.,

$$d = \tau, \tag{7.20}$$

$$p = 0. \tag{7.21}$$

Nevertheless, it must be guaranteed that the entire network load does not exceed one:

$$\rho_\Sigma \leq 1. \tag{7.22}$$

Frequency Division Multiple Access (FDMA). Here, each agents gets its own frequency, thereby avoiding collisions. However, since every agent gets only $1/N$-th of the available frequency, the packets take N times longer to be transmitted. Thus, the end-to-end delay and loss probability is

$$d = N\tau, \tag{7.23}$$

$$p = 0. \tag{7.24}$$

Again, it must be guaranteed that the entire network load does not exceed one:

$$\rho_\Sigma \leq 1. \tag{7.25}$$

Pure ALOHA. Here, every agent sends a packet whenever an event is generated without checking the medium. Thus, the end-to-end delay is just the packet duration, i.e.,

$$d = \tau. \tag{7.26}$$

Since all agents send without checking the medium, many collisions occur and thus the loss probability is relatively high.

Theorem 7.2. *Suppose that the offered load of each agent is* ρ *and all packets involved in a self-interference are lost. Then the loss probability of pure ALOHA is:*

$$p = 1 - \left(1 - \int_0^{\rho} f(x|1)\mathrm{d}x\right)^2 \times \left(1 - \int_0^{2\rho} [1 - \int_0^{x} f(t|1)\mathrm{d}t]\mathrm{d}x\right)^{N-1}, \quad (7.27)$$

where $f(t|1)$ *is the PDF of the inter-arrival times given by* (7.19).

Unfortunately, the loss probability given in Theorem 7.2 is a relatively complex function. Thus, it is difficult to derive further analytical results like the optimal load or minimal cost based on (7.27). However, Theorem 7.1 shows that the arrival process of all agents together approaches a Poisson process for $N \to \infty$. Hence, we will derive results based on the assumption that the arrival process is a Poisson process. Under this assumption, the loss probability simplifies to

$$p = 1 - \mathrm{e}^{-2\rho_\Sigma} = 1 - \mathrm{e}^{-2\rho N}. \quad (7.28)$$

Figure 7.15 shows the exact loss probability obtained from Theorem 7.2 for $N = 2$, 4, 8, 16, and 32 agents with the loss probability for a Poisson arrival process. For a small number of agents, there is a significant difference in the loss probability. In this case, it is important to use (7.27) to calculate the loss probability. On the other hand, we also see that for a moderate number of agents, the difference is already relatively small.

Slotted ALOHA. In slotted ALOHA, the collision probability is reduced by restricting the times an agent is allowed to start sending. Therefore, the time is divided into slots and every agent is only allowed to start sending at the begin of a slot. We assume that the slot length is identical to the packet duration. Thus, the expected end-to-end delay becomes the packet duration plus the expected time until the begin of the next slot:

$$d = \tau + 0.5\tau = 1.5\tau. \quad (7.29)$$

Note that loss happens not only due to collisions with other agents, but also due to self-interference, i.e., when a new packet is generated while the agent has another packet to send. Depending on the resolution mechanism, either both packets involved in a self-interference, the newer, or the older packet might be lost. Since in slotted ALOHA the agents wait for the begin of the next slot before sending, self-interference can be resolved such that

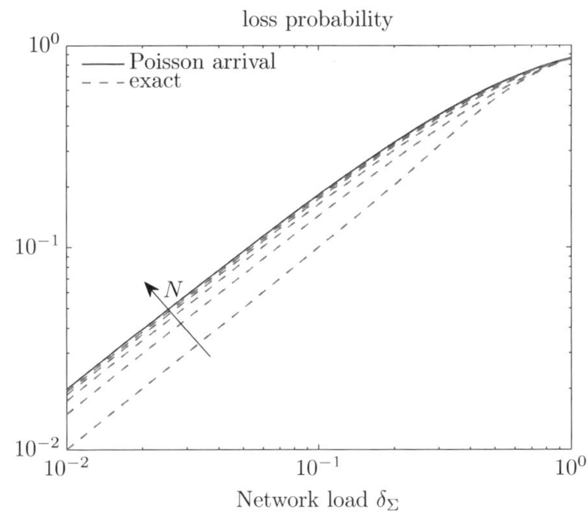

Fig. 7.15 Comparison of the exact loss probability of Theorem 7.2 for $N = 2$, 4, 8, 16, and 32 agents with the loss probability for a Poisson arrival process

the newest packet is kept. For this case, the loss probability is given in the following theorem.

Theorem 7.3. *Suppose, the offered load of each agent is ρ and the newest packet involved in a self-interference is kept. Then the lss probability of slotted ALOHA is:*

$$p = 1 - \frac{1}{\rho} \int_0^{\rho} \left(1 - \int_0^{\rho - x} f(t|1) \mathrm{d}t\right) \mathrm{d}x \\ \times \left(1 - \int_0^{\rho} \left[1 - \int_0^{x} f(t|1) \mathrm{d}t\right] \mathrm{d}x\right)^{N-1}, \quad (7.30)$$

where $f(t|1)$ is the PDF of the inter-arrival times, given by (7.19).

Again, (7.30) is a relatively complex function and it is difficult to derive analytical results based on this equation. Hence, we will analyze event-based control with slotted ALOHA under the assumption of Poisson arrival, where the loss probability is

$$p = 1 - \mathrm{e}^{-\rho_\Sigma} = 1 - \mathrm{e}^{-\rho N}. \quad (7.31)$$

Carrier Sense Multiple Access (CSMA). In the following, we present two models for analyzing event-based control with a CSMA communication system. As the name suggests, the medium is sensed before an agent starts to send. When the medium is found idle, the agent starts to send. The CSMA protocols differ for the case that the medium is found busy. The two most common approaches are *1-persistent CSMA* and *non-persistent CSMA*. For 1-persistent CSMA, the agent waits until the medium becomes idle and then starts to send. However, if two agents find the medium busy, both will wait until the medium becomes idle, then simultaneously start to send and collide. This problem is avoided in non-persistent CSMA, where the agent waits a random time and then starts the arbitration process again. Unfortunately, in both approaches, collisions are still possible and there exist many different CSMA protocols, which mainly differ in the details of collision detection and avoidance.

However, modern Ethernet is no longer built by connecting all users to a shared wire and thereby forming a global collision domain. Instead, each user is connected to a switch, which resolves collisions by buffering the packets in a first-in-first-out (FIFO) queue. For larger networks, the switches are connected with each other, and a distributed algorithm is used to find a spanning tree. Hence, we use a queueing system to model an Ethernet-like communication system.

For real-time applications, it does not make much sense to wait a random time and then try to retransmit the original packet, as done in non-persistent CSMA. Instead, it makes more sense to drop this packet and let the event-based control scheme generate the next packet. Note that this scheme is identical to Erlang's loss process.

Queueing System. In general, a queue is characterized by its arrival process, its service process, its number of service units, and its queue size. By Theorem 7.1, the arrival process of all agents together approaches a Poisson process. Hence, we use a Markovian (abbreviated by M) arrival process. Moreover, the packet duration is constant, i.e., the service process is deterministic (D). Finally, all agents share the same medium, i.e., there is 1 service unit. Hence, we use an M/D/1 queue to model an Ethernet-like communication system.

For an infinite M/D/1 queue, the delay and loss is

$$d = \frac{2 - \rho_\Sigma}{2(1 - \rho_\Sigma)}\tau, \tag{7.32}$$

$$p = 0. \tag{7.33}$$

Note that there are no losses due to collisions or an overloaded queue; the packets are just delayed.

For a finite M/D/1 queue of size M (waiting room plus processing unit) the delay and loss is

$$d = \left(M - \frac{\sum_{k=0}^{M-1} b_k - M}{\rho_\Sigma b_{M-1}} \right) \tau, \tag{7.34}$$

$$p = 1 - \frac{b_{M-1}}{1 + \rho_\Sigma b_{M-1}}, \tag{7.35}$$

where the coefficients b_n are

$$b_n = \sum_{k=0}^{n} \frac{(-1)^k}{k!} (n-k)^k \mathrm{e}^{(n-k)\rho_\Sigma} \rho_\Sigma^k. \tag{7.36}$$

Erlang's Loss Model. As already stated, the shared medium is sensed before an agent starts to send. If the medium is idle, the packet is immediately sent. If the medium is found busy, the packet is dropped. Thus, the delay is just the packet duration

$$d = \tau \tag{7.37}$$

and the loss probability is

$$p = \frac{\rho_\Sigma}{1 + \rho_\Sigma}. \tag{7.38}$$

7.3.5 Control Performance for Different Communication Schemes

Figure 7.16 shows the normalized cost for $N = 8$ agents. For calculating the loss and delay of the different contention based MAC protocols, the arrival process is assumed to be a Poisson process. From Fig. 7.16, we see that the additional cost due to loss and delay is not significant if the network load is relatively low. Consequently, the cost is determined by the sampling scheme and does not depend on the communication protocol. If the network load is large, the additional cost due to loss and delay is significant and the effects of the different communication protocols become dominant.

An important aspect of the analysis of networked control systems is the minimal cost achievable with a certain control and communication strategy, which is summarized in Table 7.1. Although event-based control outperforms time-triggered control under ideal conditions, i.e., when there is no loss and delay, we get the highest minimal cost for event-based control with an ALOHA communication scheme. Not surprisingly, for TDMA and FDMA the optimal load is $\rho_\Sigma = 1$, where the cost is smaller than the minimal cost of event-based control with an ALOHA communication system. However, the best performance is achieved with event-based control with an infinite queue.

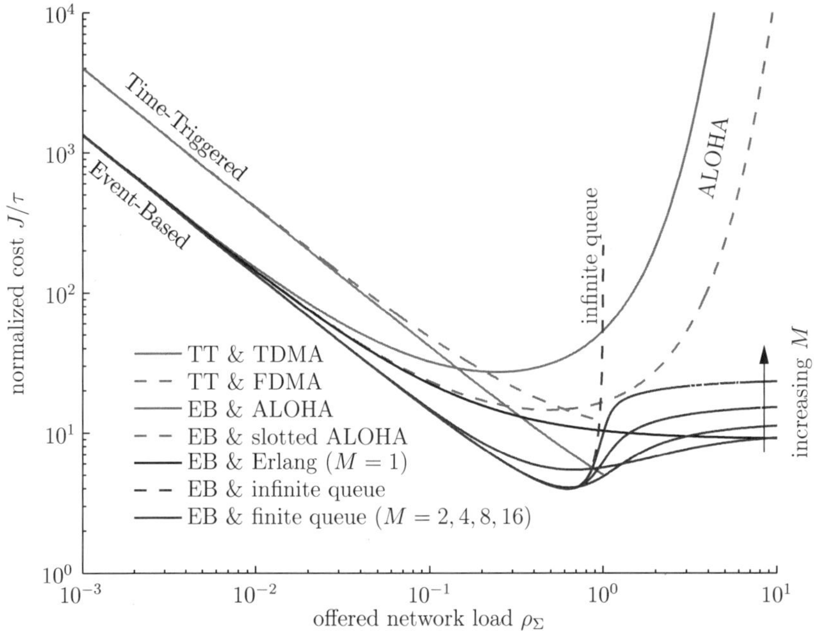

Fig. 7.16 The normalized cost for $N = 8$ agents

Table 7.1 The minimal normalized cost of the different schemes

control scheme	communication scheme	minimal normalized cost J^*/τ
time-triggered	TDMA	$0.5N + 1$
	FDMA	$1.5N$
event-based	unslotted ALOHA	$2e^{2\rho_\Sigma^{u*}} N + 1 \approx 3.26N + 1$
	slotted ALOHA	$e^{\rho_\Sigma^{s*}} N + 1.5 \approx 1.63N + 1.5$
	infinite queue	$\frac{N-3}{6(1-\sqrt{3/N})} + \frac{\sqrt{N}+\sqrt{3}-6/\sqrt{N}}{2(-3/\sqrt{N}+\sqrt{3})}$
		$\approx \frac{1}{6}N + \frac{\sqrt{N}}{2\sqrt{3}}$ for large N
	Erlang's loss model	$N + 1$

Interestingly, it turns out that event-based control with Erlang's loss model is optimal for $\rho_\Sigma = \infty$. In this case, all agents try to send all the time. Thereby, the choice of the sending agent is random, resulting in a random sampling of the agents. Obviously, this is not the optimal strategy and we see that the minimal cost with this approach is larger than the minimal cost of time-triggered control with TDMA.

Event-based control with an infinite queue gives the minimal cost of all analyzed control and communication schemes but might be risky. If the offered network load ρ_Σ exceeds one, the cost becomes infinite. Fortunately, the cost remains finite when the queue is finite. If the network load is not too high, we see no significant difference in Fig. 7.16 between the cost with an infinite queue and the cost with a properly chosen finite queue. However, the choice of the queue size is not a simple task. If the queue is relatively small, e.g., $M = 2$, the minimal cost is larger than the minimal cost with an infinite queue but the cost remains relatively small when the network is overloaded. On the other hand, if the queue is relatively large, e.g., $M = 16$, then the minimal cost is similar to the minimal cost achieved with an infinite queue but the cost is relatively large, but still finite, when the network is overloaded.

Finally, except for event-based control with Erlang's loss model, the network load should always be limited. For time-triggered control, $\rho_\Sigma = 1$ is a hard limit that cannot be exceeded. On the other hand, the limit for event-based control is soft and depends on the protocol. Increasing the network load beyond the optimal load is possible but should be avoided because the cost due to loss and delay become significant. Moreover, for event-based control with an infinite queue, the cost becomes even infinite for $\rho_\Sigma > 1$. Obviously, this must be avoided but it is nevertheless possible to overload the communication system for a short time.

Another important aspect, not considered so far, is the flexibility of the configuration, i.e., the ability to add another agent to an already running system. For time-triggered control this is relatively difficult because it requires to change the preassigned sending times and thus also the sampling times of all agents. For event-based control it is much simpler because a contention based communication system that allows arbitrary access times must be used anyhow. However, for event-based control with an infinite queue, this can be dangerous because adding another agent might overload the communication system, resulting in an infinite cost.

Summary. In this section, we considered the optimal communication rate for networked control systems. Thereby, we showed that different layers of the ISO/OSI stack must be taken into account. The control scheme, located within the application layer, determines the traffic pattern and thereby affects the choice of the MAC protocol. Both, the MAC protocol and the traffic pattern determine the loss and delay of the communication system. Thus, a comparison of time-triggered and event-based control without considering the details of the communication strategy might be misleading. Under ideal communication, event-based control gives a better performance than time-triggered control. However, when a simple communication system like ALOHA is used to transmit the packets, the performance of event-based control is significantly degraded and becomes worse than the one achievable with time-triggered control with a TDMA or FDMA communication protocol. However, when using an adequate communication system for event-based control, like a CSMA based communication system, a better control performance is achieved than possible with time-triggered control.

Bibliographical Notes

Event-based and time-triggered control were first compared in [13] under the assumption that the communication is ideal. In [302], this comparison is extended to include the effect of packet loss; the effect of delay is studied in [33]. Event-based control with pure ALOHA and slotted ALOHA is compared to time-triggered control with TDMA in [33] and [34], respectively. In [35], it is shown that the Palm–Khintchine Theorem holds for event-based traffic, i.e., the arrival process of all agents together converges to a Poisson process. Based on this result, event-based control with different CSMA protocols are compared to time-triggered control with TDMA and FDMA in [35].

References

[1] Aeyels, D.: Global observability of Morse-Smale vector fields. *Differential Equations* 45, 1–15 (1982)

[2] Aeyels, D., De Smet, F.: A mathematical model for the dynamics of clustering. *Physica D: Nonlinear Phenomena* 237, 2517–2530 (2008)

[3] Aeyels, D., De Smet, F.: Cluster formation in a time-varying multi-agent system. *Automatica* 47, 2481–2487 (2011)

[4] Ames, A. D., Zheng, H., Gregg, R. D., Sastry, S.: Is there life after Zeno? Taking executions past the breaking (Zeno) point. In: *Proc. 2006 American Control Conf.*, Minneapolis, USA, pp. 2652–2657 (2006)

[5] Anderson, B. D. O., Helmke, U.: Counting critical formations on a line. *SIAM J. Control Optim.* (submitted, 2013)

[6] Anta, A., Tabuada, P.: To sample or not to sample: Self-triggered control of nonlinear systems. *IEEE Trans. Automat. Control* 55, 2030–2042 (2010)

[7] Antsaklis, P., Baillieul, J.: Guest editorial special issue on networked control systems. *IEEE Trans. Automat. Control* 49, 1421–1423 (2004)

[8] Antsaklis, P., Baillieul, J.: Special issue on technology of networked control systems. *Proc. IEEE* 95, 5–8 (2007)

[9] Antunes, D. J., Heemels, W. P.M. H., Hespanha, J. P., Silvestre, C. J.: Scheduling measurements and controls over networks – Part I: Rollout strategies for protocol design. In: *Proc. 2012 American Control Conf.*, pp. 2036–2041 (2012)

[10] Antunes, D. J., Heemels, W. P. M. H., Hespanha, J. P., Silvestre, C. J.: Scheduling measurements and controls over networks – Part II: Rollout strategies for simultaneous protocol and controller design. In: *Proc. 2012 American Control Conf.*, pp. 2042–2047 (2012)

[11] Åström, K. J.: Event based control. In: *Analysis and Design of Nonlinear Control Systems*, pp. 127–147. Springer, Berlin (2008)

[12] Åström, K. J.: *Introduction to Stochastic Control Theory.* Dover Publications (2006)

[13] Åström, K. J., Bernhardsson, B. M.: Comparison of Riemann and Lebesgue sampling for first order stochastic systems. In: *Proc. 41st IEEE Conf. on Decision and Control*, Las Vegas, USA, pp. 2011–2016 (2002)

[14] Baillieul, J.: Feedback designs in information-based control. In: *Workshop on Stochastic Theory and Control* (2001)

[15] Bamieh, B., Paganini, F., Dahleh, M. A.: Distributed control of spatially invariant systems. *IEEE Trans. Automat. Control* 47, 1091–1107 (2002)

[16] Bar-Shalom, Y., Li, X. R.: *Estimation and Tracking-Principles, Techniques, and Software*. Artech House, Inc., Norwood (1993)

[17] Bardi, M., Capuzzo-Dolcetta, I.: *Optimal Control and Viscosity Solutions of Hamilton-Jacobi-Bellman Equations*. Systems & Control: Foundations & Appl. Birkhäuser, Boston (1997)

[18] Becker, P., Gotzhein, R., Kuhn, T.: Macz - a quality-of-service mac layer for ad-hoc networks. In: König, A., Köppen, M., Abraham, A., Igel, C., Kasabov, N. (eds.) *Proc. 7th Int. Conf. on Hybrid Intelligent Systems*, Kaiserslautern, Germany, pp. 277–282 (2007)

[19] Becker, P., Gotzhein, R., Kuhn, T.: Model-driven performance simulation of self-organizing systems with partssim. *PIK - Praxis der Informationsverarbeitung und Kommunikation* 31, 45–50 (2008)

[20] Becker, P., Christmann, D., Gotzhein, R.: Model-driven development of time-critical protocols with SDL-MDD. In: Reed, R., Bilgic, A., Gotzhein, R. (eds.) SDL 2009. LNCS, vol. 5719, pp. 34–52. Springer, Heidelberg (2009)

[21] Becker, P., Birtel, M., Christmann, D., Gotzhein, R.: Black-Burst-based quality-of-service routing (BBQR) for wireless ad-hoc networks. In: *Proc. 11th Int. Conf. on New Technologies of Distributed Systems*, Paris, France, pp. 1–8 (2011)

[22] Beker, O., Hollot, C. V., Chait, Y., Han, H.: Fundamental properties of reset control systems. *Automatica* 40, 905–915 (2004)

[23] Bellman, R.: *Dynamic Programming*. Princeton University Press (1957)

[24] Belykh, V. N., Belykh, I. V., Mosekilde, E.: Cluster synchronization modes in an ensemble of coupled chaotic oscillators. *Phys. Rev. E* 63, 036216 (2001)

[25] Bemporad, A.: Predictive control of teleoperated constrained systems with unbounded communication delays. In: *Proc. 37th IEEE Conf. on Decision and Control*, Tampa, USA, pp. 2133–2138 (1998)

[26] Bemporad, A., Heemels, W. P. M. H., Johansson, M.: *Networked Control Systems*. Springer, Berlin (2010)

[27] Ben-Tal, A., Goryashko, A., Guslitzer, E., Nemirovski, A.: Adjustable robust solutions of uncertain linear programs. *Math. Programming* 99, 351–376 (2004)

[28] Berger, L. T., Iniewski, K.: *Smart Grid: Applications, Communications, and Security*. John Wiley & Sons, New Jersey (2012)

[29] Berman, A., Plemmons, R. J.: *Nonnegative Matrices in the Mathematical Sciences*. Classics in Applied Mathematics. SIAM (1994)

[30] Bertsekas, D. P.: *Dynamic Programming and Optimal Control*, 3rd edn., vol. 1. Athena Scientific, Belmont (2005)

[31] Bertsekas, D. P., Tsitsiklis, J. N.: *Parallel and Distributed Computation: Numerical Methods*. Prentice-Hall, Inc., Englewood Cliffs (1989)

[32] Bitmead, R. R., Gevers, M. R., Petersen, I. R., Kaye, R. J.: Monotonicity and stabilizability properties of solutions of the Riccati difference equation: Propositions, lemmas, theorems, fallacious conjectures and counterexamples. *Systems & Control Lett.* 5, 309–315 (1985)

[33] Blind, R., Allgöwer, F.: Analysis of networked event-based control with a shared communication medium: Part I - pure ALOHA. In: *Proc. 18th IFAC World Congress*, Milan, Italy, pp. 10092–10097 (2011)

[34] Blind, R., Allgöwer, F.: Analysis of networked event-based control with a shared communication medium: Part II - slotted ALOHA. In: *Proc. 18th IFAC World Congress*, Milan, Italy, pp. 8830–8835 (2011)
[35] Blind, R., Allgöwer, F.: On the optimal sending rate for networked control systems with a shared communication medium. In: *Proc. Joint 50th IEEE Conf. on Decision and Control and European Control Conf.*, Orlando, USA, pp. 4704–4709 (2011)
[36] Bompard, E., Wu, D., Xue, F.: The concept of betweenness in the analysis of power grid vulnerability. In: *Proc. 2010 Complexity in Engineering*, Rome, Italy, pp. 52–54 (2010)
[37] Braun, R. D., Kroo, I. M.: Development and application of the collaborative optimization architecture in a multidisciplinary design environment. In: Hussaini, M. Y., Alexandrov, N. M. (eds.) *Multidisciplinary Design Optimization: State of the Art*, pp. 98–116. SIAM, Philadelphia (1997)
[38] Braun, T., Christmann, D., Gotzhein, R., Igel, A.: Model-driven engineering of networked ambient systems with SDL-MDD. In: *Proc. 3th Int. Conf. on Ambient Systems, Networks and Technologies, Procedia Computer Science*, Niagara Falls, Ontario, Canada, pp. 490–498. Elsevier (2012)
[39] Brockett, R. W.: On the control of a flock by a leader. *Proc. Steklov Institute of Mathematics* 268, 49–57 (2010)
[40] Brockett, R. W.: The Lie groups of simple feedback systems. In: *Proc. 15th IEEE Conf. on Decision and Control*, Clearwater, USA (1976)
[41] Brockett, R. W., Liberzon, D.: Quantized feedback stabilization of linear systems. *IEEE Trans. Automat. Control* 45, 1279–1289 (2000)
[42] Bungartz, H.- J., Griebel, M.: Sparse grids. *Acta Numerica* 13, 1–123 (2004)
[43] Bürger, M., Zelazo, D., Allgöwer, F.: Network clustering: A dynamical systems and saddle-point perspective. In: *Proc. Joint 50th IEEE Conf. on Decision and Control and European Control Conf.*, Orlando, USA, pp. 7825–7830 (2011)
[44] Bürger, M., Zelazo, D., Allgöwer, F.: Combinatorial insights and robustness analysis for clustering in dynamical networks. In: *Proc. 2012 American Control Conf.*, Montreal, Canada, pp. 454–459 (2012)
[45] Bürger, M., Zelazo, D., Allgöwer, F.: Hierarchical clustering of dynamical networks using a saddle-point analysis. *IEEE Trans. Automat. Control* 58, 113–124 (2013)
[46] Callier, F. M., Nahum, C. D.: Necessary and sufficient conditions for the complete controllability and observability of systems in series using the coprime decomposition of a rational matrix. *IEEE Trans. Circuits and Systems* 22, 90–95 (1975)
[47] Camacho, E. F., Bordons, C.: *Model Predictive Control.* Springer, London (2004)
[48] Candès, E. J., Tao, T.: Near-optimal signal recovery from random projections: Universal encoding strategies? *IEEE Trans. Information Theory* 52, 5406–5425 (2006)
[49] Cervin, A., Henriksson, D., Lincoln, B., Eker, J., Arzen, K. -E.: How does control timing affect performance? Analysis and simulation of timing using jitterbug and TrueTime. *IEEE Control Systems Magazine* 23, 16–30 (2003)
[50] Cervin, A., Henriksson, D., Ohlin, M.: *TrueTime Reference Manual.* Lund University, Sweden (2010)

[51] Chamaken, A.: *Model-Based Cross-Design for Wireless Networked Control Systems with Limited Resources.* PhD thesis, TU Kaiserslautern, Germany (2012)

[52] Chamaken, A., Litz, L.: Joint design of control and communication for wireless networked control systems: A case study. In: *Proc. 2010 American Control Conf.*, Baltimore, Maryland, USA, pp. 1834–1840 (2010)

[53] Chamaken, A., Krämer, M., Litz, L., Gotzhein, R.: Model-based c^3-cross-design for wireless networked control systems. In: *Proc. of the Symp. on Recent Trends in Networked Systems and Cooperative Control*, Stuttgart, Germany (2009)

[54] Chamaken, A., Litz, L., Krämer, M., Gotzhein, R.: A new approach to the joint design of control and communication in wireless networked control systems. In: AUTOMATION 2009, Baden-Baden, Germany, pp. 251–255 (2009)

[55] Chamaken, A., Litz, L., Krämer, M., Gotzhein, R.: Cross-layer design of wireless networked control systems with energy limitations. In: *Proc. 10th European Control Conf.*, Budapest, Hungary, pp. 2325–2330 (2009)

[56] Chert, G., Teboulle, M.: A proximal-based decomposition method for convex minimization problems. *Math. Programming* 64, 81–101 (1994)

[57] ChipCon. Cc2420 2.4 GHz IEEE 802.15.4 transceiver (2012), `http://www.ti.com/lit/ds/symlink/cc2420.pdf`

[58] Christmann, D., Gotzhein, R., Krämer, M., Winkler, M.: Flexible and energy-efficient duty cycling in wireless networks with MacZ. In: *Proc. 10th Int. Conf. on New Technologies of Distributed Systems*, Tozeur, Tunisia, pp. 121–128 (2010)

[59] Christmann, D., Becker, P., Gotzhein, R.: Priority scheduling in SDL. In: Ober, I., Ober, I. (eds.) SDL 2011. LNCS, vol. 7083, pp. 202–217. Springer, Heidelberg (2011)

[60] Christmann, D., Gotzhein, R., Rohr, S.: The arbitrating value transfer protocol (avtp) - deterministic binary countdown in wireless multi-hop networks. In: *Proc. 21st Int. Conf. on Computer Communications and Networks*, Munich, Germany, pp. 1–9 (2012)

[61] Christmann, D., Gotzhein, R., Siegmund, S., Wirth, F.: Realization of try-once-discard in wireless multi-hop networks (2012) (submitted)

[62] Christmann, D., Gotzhein, R., Krämer, M., Winkler, M.: Flexible and energy-efficient duty cycling in wireless networks with MacZ (extended version). *Concurrency and Computation: Practice and Experience* 25, 218–233 (2013)

[63] Cloosterman, M. B. G., van de Wouw, N., Heemels, W. P. M. H., Nijmeijer, H.: Stability of networked control systems with uncertain time-varying delays. *IEEE Trans. Automat. Control* 54, 1575–1580 (2009)

[64] Cogill, R.: Event-based control using quadratic approximate value functions. In: *Proc. Joint 48th IEEE Conf. on Decision and Control and 28th Chinese Control Conf.*, Shanghai, China, pp. 5883–5888 (2009)

[65] Cohen, A., Dahmen, W., DeVore, R.: Compressed sensing and best k-term approximation. *J. American Mathematical Society* 22, 211–231 (2009)

[66] Colonius, F.: Minimal data rates and invariance entropy. In: *Proc. 19th Int. Symp. on Mathematical Theory of Networks and Systems*, Budapest, Hungary, pp. 5–9 (2010)

[67] Colonius, F.: Minimal bit rates and entropy for stabilization. *SIAM J. Control Optim.* 50, 2988–3010 (2012)

[68] Colonius, F.: Subspace entropy and controlled invariant subspaces. In: Hüper, K., Trumpf, J. (eds.) *Mathematical Systems Theory – Festschrift in Honor of Uwe Helmke on the Occasion of his 60th birthday* (to appear, 2013)
[69] Colonius, F., Helmke, U.: Entropy of controlled-invariant subspaces. *J. of Applied Mathematics and Mechanics (ZAMM)* (2011) (submitted)
[70] Colonius, F., Kawan, C.: Invariance entropy for control systems. *SIAM J. Control Optim.* 48, 1701–1721 (2009)
[71] Colonius, F., Kawan, C.: Invariance entropy for outputs. *Math. Control Signals Syst.* 22, 203–227 (2011)
[72] Colonius, F., Kliemann, W.: *The Dynamics of Control.* Birkhäuser, Boston (2000)
[73] Colonius, F., Kawan, C., Nair, G. N.: A note on topological feedback entropy and invariance entropy. *Systems & Control Lett.* (to appear, 2013)
[74] Crouch, P.: Dynamical realizations of finite Voltera series. *SIAM J. Control Optim.* 19, 177–202 (1981)
[75] Dacic, D. B., Nesic, D.: Quadratic stabilization of linear networked control systems via simultaneous protocol and controller design. *Automatica* 43, 1145–1155 (2007)
[76] Dashkovskiy, S. N., Rüffer, B. S., Wirth, F.: An ISS small gain theorem for general networks. *Math. Control Signals Syst.* 19, 93–122 (2007)
[77] Dashkovskiy, S. N., Rüffer, B. S., Wirth, F.: An ISS small gain theorem for general networks. *Math. Control Signals Syst.* 19, 93–122 (2007)
[78] Dashkovskiy, S. N., Rüffer, B. S., Wirth, F.: Small gain theorems for large scale systems and construction of ISS Lyapunov functions. *SIAM J. Control Optim.* 48, 4089–4118 (2010)
[79] Dashkovskiy, S. N., Ito, H., Wirth, F.: On a small gain theorem for ISS networks in dissipative Lyapunov form. *Eur. J. Control* 17, 357–365 (2011)
[80] Dashkovskiy, S. N., Kosmykov, M., Wirth, F.: A small gain condition for interconnections of ISS systems with mixed ISS characterizations. *IEEE Trans. Automat. Control* 56, 1247–1258 (2011)
[81] De Benedetto, M. D., D'Innocenzo, A., Pola, G., Rinaldi, C., Santucci, F.: A theoretical framework for control over wireless networks. In: *Proc. 17th Int. Symp. on Mathematical Theory of Networks and Systems*, Kyoto, Japan, pp. 2431–2437 (2006)
[82] De Leone, R., Meyer, R. R., Kontogiorgis, S.: Alternating direction splittings for block-angular parallel optimization. *J. Optim. Theory and Appl.* 90, 2–9 (1996)
[83] De Persis, C.: n-bit stabilization of n-dimensional nonlinear systems in feedforward form. *IEEE Trans. Automat. Control* 50, 299–311 (2005)
[84] De Persis, C.: Minimal data rate stabilization of nonlinear systems over networks with large delays. In: *Proc. 5th Int. Symp. on Modeling and Optimization in Mobile, Ad Hoc and Wireless Networks and Workshops*, Limassol, Cyprus, pp. 1–9 (2007)
[85] De Persis, C.: Robustness of quantized continuous-time nonlinear systems to encoder/decoder mismatch. In: *Proc. Joint 48th IEEE Conf. on Decision and Control and 28th Chinese Control Conf.*, Shanghai, China, pp. 13–18 (2009)
[86] De Persis, C., Isidori, A.: Stabilizability by state feedback implies stabilizability by encoded state feedback. *Systems & Control Lett.* 53, 249–258 (2004)

[87] De Persis, C., Sailer, R., Wirth, F.: On a small-gain approach to distributed event-triggered control. In: *Proc. 18th IFAC World Congress*, Milan, Italy, pp. 2401–2406 (2011)

[88] De Persis, C., Sailer, R., Wirth, F.: Event-triggering of large-scale systems without Zeno behaviour. In: *Proc. 20th Int. Symp. on Mathematical Theory of Networks and Systems MTNS 2012*, Melbourne, Australia (2012)

[89] De Persis, C., Sailer, R., Wirth, F.: Parsimonious event-triggered distributed control: A zeno free approach. *Automatica* (2012) (accepted)

[90] De Smet, F., Aeyels, D.: Clustering in a network of non-identical and mutually interacting agents. *Proc. Royal Society A* 45, 745–768 (2009)

[91] de Souza, C., Gevers, M., Goodwin, G.: Riccati equations in optimal filtering of nonstabilizable systems having singular state transition matrices. *IEEE Trans. Automat. Control* 31, 831–838 (1986)

[92] Delchamps, D.: Stabilizing a linear system with quantized state feedback. *IEEE Trans. Automat. Control* 35, 916–924 (1990)

[93] Delvenne, J. -C.: An optimal quantized feedback strategy for scalar linear systems. *IEEE Trans. Automat. Control* 51, 298–303 (2006)

[94] Demir, O., Lunze, J.: A decomposition approach to decentralized and distributed control of spatially interconnected systems. In: *Proc. 18th IFAC World Congress*, Milan, Italy (2011)

[95] Demir, O., Lunze, J.: Networked control of symmetrically interconnected systems. In: *Proc. 19th Mediterranean Conf. on Control and Automation*, Corfu, Greece (2011)

[96] Demir, O., Lunze, J.: Event-based synchronisation of multi-agent systems. In: *Proc. 4th IFAC Conf. on Analysis and Design of Hybrid Systems*, Eindhoven, Netherlands (2012)

[97] Demir, O., Lunze, J.: Cooperative control of multi-agent systems with event-based communication. In: *Proc. 2012 American Control Conf.*, Montreal, Canada (2012)

[98] Demir, O., Lunze, J., Otte, P.: Analogies in the controller design for multi-agent and physically interconnected systems with identical subsystems. In: *Proc. 20th Mediterranean Conf. on Control and Automation*, Barcelona, Spain (2012)

[99] Deroo, F., Ulbrich, M., Anderson, B. D. O., Hirche, S.: Accelerated iterative distributed controller synthesis with a Barzilai-Borwein step size. In: *Proc. 51st IEEE Conf. on Decision and Control*, Maui, Hawaii, USA (2012)

[100] Dion, J. -M., Commault, C., van der Woude, J.: Generic properties and control of linear structured systems: a survey. *Automatica* 39, 1125–1144 (2003)

[101] Dirr, G., Helmke, U., Jordan, J.: Control and observation of the matrix Riccati differential equation. In: *Emergent Problems in Nonlinear Systems and Control*, vol. 393, pp. 169–184. Springer, Berlin (2009)

[102] Dirr, G., Helmke, U., Jordan, J.: Global observability of real analytic systems. In: *Proc. 10th European Control Conf.*, Budapest, Hungary, pp. 4582–4586 (2009)

[103] Dirr, G., Helmke, U., Rüppel, F.: The system Lie algebras on networked control systems (submitted, 2013)

[104] do Valle Costa, O. L., Fragoso, M. D., Marques, R. P.: *Discrete-Time Markov Jump Linear Systems*. Springer, Berlin (2005)

[105] Dold, J., Stursberg, O.: Distributed predictive control of communicating and platooning vehicles. In: *Proc. 49th IEEE Conf. on Decision and Control*, Atlanta, USA, pp. 561–566 (2009)

[106] Dold, J., Stursberg, O.: Robust predictive control of platooning vehicles using communication. *Automatisierungstechnik* 58, 207–216 (2010)

[107] Donkers, M. C. F., Heemels, W. P. M. H.: Output-based event-triggered control with guaranteed $\mathcal{L}_2$-gain and improved event-triggering. In: *Proc. 49th IEEE Conf. on Decision and Control*, Atlanta, USA, pp. 3246–3251 (2010)

[108] Donkers, M. C. F., Hetel, L., Heemels, W. P.M. H.: Stability analysis of networked control systems using a switched linear systems approach. In: *Hybrid Systems: Computation and Control*, pp. 150–164. Springer, Berlin (2010)

[109] Donoho, D. L.: Compressed sensing. *IEEE Trans. Information Theory* 52, 1289–1306 (2006)

[110] Downarowicz, T.: *Entropy in Dynamical Systems.* Cambridge University Press (2010)

[111] Dunbar, W. B., Murray, R. M.: Distributed receding horizon control for multi-vehicle formation stabilization. *Automatica* 42, 549–558 (2006)

[112] Ekanayake, J., Liyanage, K., Wu, J., Yokoyama, A., Jenkins, N.: *Smart Grid: Technology and Applications.* John Wiley & Sons, New Jersey (2012)

[113] Encarnacao, J.: Ambient intelligence. *IEEE Trans. Information Theory* 50, 5–6 (2008)

[114] Fagnani, F., Zampieri, S. A.: A symbolic approach to performance analysis of quantized feedback systems: The scalar case. *SIAM J. Control Optim.* 44, 816–866 (2005)

[115] Fardad, M., Lin, F., Jovanovic, M. R.: On the dual decomposition of linear quadratic optimal control problems for vehicular formations. In: *Proc. 49th IEEE Conf. on Decision and Control*, Atlanta, USA, pp. 6287–6292 (2010)

[116] Farokhi, F., Langbort, C., Johansson, K. H.: Control design with limited model information. In: *Proc. 2011 American Control Conf.*, San Francisco, USA (2011)

[117] Fax, J. A., Murray, R. M.: Information flow and cooperative control of vehicle formations. *IEEE Trans. Automat. Control* 49, 1465–1476 (2004)

[118] Ferris, M. C., Mangasarian, O. L.: Parallel variable distribution. *SIAM J. Optim.* 4, 815–832 (1994)

[119] Findeisen, R., Varutti, P.: Stabilizing nonlinear predictive control over non-deterministic communication networks. In: Magni, L., Raimondo, D.M., Allgöwer, F. (eds.) Nonlinear Model Predictive Control. LNCIS, vol. 384, pp. 167–179. Springer, Heidelberg (2009)

[120] Findeisen, R., Grüne, L., Pannek, J., Varutti, P.: Robustness of prediction based delay compensation for nonlinear systems. In: *Proc. 18th IFAC World Congress*, pp. 203–208 (2011)

[121] Fischer, J., Hekler, A., Dolgov, M., Hanebeck, U. D.: Optimal sequence-based LQG control over TCP-like networks subject to random transmission delays and packet losses. In: *Proc. 2013 American Control Conf.* (2012) (accepted)

[122] Fischer, J., Hekler, A., Hanebeck, U. D.: State estimation in networked control systems. In: *Proc. 15th Int. Conf. on Information Fusion*, Singapore (2012)

[123] Fischer, J., Reinhardt, M., Hanebeck, U. D.: Optimal sequence-based control and estimation of networked linear systems. *Arxiv preprint, arXiv:1211.5086* (2012)

[124] Fornasier, M., Schnass, K., Vybíral, J.: Learning functions of few arbitrary linear parameters in high dimensions (2012) (to appear)
[125] Fortunato, S.: Community detection in graphs. *Physics Reports* 486, 75–174 (2010)
[126] Franco, E., Magni, L., Parisini, T., Polycarpou, M. M., Raimondo, D. M.: Cooperative constrained control of distributed agents with nonlinear dynamics and delayed information exchange: A stabilizing receding-horizon approach. *IEEE Trans. Automat. Control* 53, 324–338 (2008)
[127] Fruchterman, T. M., Reingold, E. M.: Graph drawing by force-directed placement. *Software - Practice and Experience* 21, 1129–1164 (1991)
[128] Fuhrmann, P. A.: On controllability and observability of systems connected in parallel. *IEEE Trans. Circuits and Systems* 22, 57 (1975)
[129] Fuhrmann, P. A., Helmke, U.: Reachability, observability and strict equivalence of networked linear systems. *Math. Control Signals Syst.* 2, 299–306 (2013)
[130] Gabel, O.: *Design of QoS-adaptive Control for AmI-Systems with heuristic methods.* PhD thesis, Departement of Automatic Control, University of Kaiserslautern (2008)
[131] Geiselhart, R., Wirth, F.: Numerical construction of LISS Lyapunov functions under a small gain condition. *Math. Control Signals Syst.* 24, 3–32 (2012)
[132] Geiselhart, R., Wirth, F.: Numerical construction of LISS Lyapunov functions under a small gain condition. In: *Proc. Joint 50th IEEE Conf. on Decision and Control and European Control Conf.*, Orlando, USA, pp. 6967–6972 (2011)
[133] Gevers, M., Li, G.: *Parametrizations in Control, Estimation and Filtering Problems: Accuracy Aspects.* Communications and Control Engineering Series. Springer, Berlin (1993)
[134] Ghosh, B., Rosenthal, J.: A generalized Popov-Belevitch-Hautus test of observability. *IEEE Trans. Automat. Control Engineering Practice* 40, 176–180 (1995)
[135] Giselsson, P., Rantzer, A.: Distributed model predictive control with suboptimality and stability guarantees. In: *Proc. 49th IEEE Conf. on Decision and Control*, Atlanta, USA, pp. 7272–7277 (2010)
[136] Godsil, C., Royle, G.: *Algebraic Graph Theory.* Springer, Berlin (2001)
[137] Goodman, D., Gersho, A.: Theory of an adaptive quantizer. *IEEE Trans. Automat. Control* 22, 1037–1045 (1974)
[138] Gotzhein, R., Kuhn, T.: Black Burst Synchronization (BBS) – A protocol for deterministic tick and time synchronization in wireless networks. *Computer Networks* 55, 3015–3031 (2011)
[139] Gotzhein, R., Kuhn, T.: Zeit- und Prioritäts-gesteuerter Sende/ empfangsknoten. Patent 10, 039 488, Deutsches Patent- und Markenamt, erteilt am 29.3 (2010)
[140] Gotzhein, R., Krämer, M., Litz, L., Chamaken, A.: Energy-aware system design with SDL. In: Reed, R., Bilgic, A., Gotzhein, R. (eds.) SDL 2009. LNCS, vol. 5719, pp. 19–33. Springer, Heidelberg (2009)
[141] Goulart, P. J., Kerrigan, E. C., Maciejowski, J. M.: Optimization over state feedback policies for robust control with constraints. *Automatica* 42, 523–533 (2006)

[142] Grimm, G., Messina, M. J., Tuna, S. E., Teel, A. R.: Model predictive control: for want of a local control Lyapunov function, all is not lost. *IEEE Trans. Automat. Control* 50, 546–558 (2005)

[143] Groß, D., Stursberg, O.: Robust distributed predictive control of communicating and constrained systems. In: *Proc. 18th IFAC World Congress*, Milan, Italy, pp. 8926–8932 (2011)

[144] Groß, D., Stursberg, O.: Distributed predictive control of communicating and constrained systems (2011) (submitted)

[145] Grüne, L.: An adaptive grid scheme for the discrete Hamilton-Jacobi-Bellman equation. *Numerische Mathematik* 75, 319–337 (1997)

[146] Grüne, L.: Analysis and design of unconstrained nonlinear MPC schemes for finite and infinite dimensional systems. *SIAM J. Control Optim.* 48, 1206–1228 (2009)

[147] Grüne, L., Junge, O.: A set oriented approach to optimal feedback stabilization. *Systems & Control Lett.* 54, 169–180 (2005)

[148] Grüne, L., Junge, O.: Optimal stabilization of hybrid systems using a set oriented approach. In: *Proc. 17th Int. Symp. on Mathematical Theory of Networks and Systems*, Kyoto, Japan, pp. 2089–2093 (2006)

[149] Grüne, L., Junge, O.: Approximately optimal nonlinear stabilization with preservation of the Lyapunov function property. In: *Proc. 46th IEEE Conf. on Decision and Control*, New Orleans, USA, pp. 702–707 (2007)

[150] Grüne, L., Junge, O.: Global optimal control of perturbed systems. *J. Optim. Theory Appl.* 136, 411–429 (2008)

[151] Grüne, L., Junge, O.: *Gewöhnliche Differentialgleichungen. Eine Einführung aus der Perspektive der dynamischen Systeme.* Vieweg+Teubner Verlag (2009)

[152] Grüne, L., Müller, F.: Set oriented optimal control using past information. In: *Proc. 18th Int. Symp. on Mathematical Theory of Networks and Systems*, Blacksburg, Virginia, USA (2008)

[153] Grüne, L., Müller, F.: An algorithm for event-based optimal feedback control. In: *Proc. Joint 48th IEEE Conf. on Decision and Control and 28th Chinese Control Conf.*, Shanghai, China, pp. 5311–5316 (2009)

[154] Grüne, L., Pannek, J.: *Nonlinear Model Predictive Control: Theory and Algorithms.* Springer, Berlin (2011)

[155] Grüne, L., Worthmann, K.: A distributed NMPC scheme without stabilizing terminal constraints. In: Johansson, R., Rantzer, A. (eds.) Distributed Decision Making and Control. LNCIS, vol. 417, pp. 259–285. Springer, Berlin (2012)

[156] Grüne, L., Pannek, J., Worthmann, K.: A prediction based control scheme for networked systems with delays and packet dropouts. In: *Proc. Joint 48th IEEE Conf. on Decision and Control and 28th Chinese Control Conf.*, Shanghai, China, pp. 537–542 (2009)

[157] Grüne, L., Pannek, J., Worthmann, K.: A networked unconstrained nonlinear MPC scheme. In: *Proc. 10th European Control Conf.*, Budapest, Hungary, pp. 371–376 (2009)

[158] Grüne, L., Jerg, S., Junge, O., Lehmann, D., Lunze, J., Müller, F., Post, M.: Two complementary approaches to event-based control. *Automatisierungstechnik* 58, 173–182 (2010)

[159] Grüne, L., Litz, L., Lunze, J.: Schwerpunktheft "Digital vernetzte Systeme". *Automatisierungstechnik* 58 (2010)
[160] Grüne, L., Pannek, J., Seehafer, M., Worthmann, K.: Analysis of unconstrained nonlinear MPC schemes with time varying control horizon. *SIAM J. Control Optim.* 48, 4938–4962 (2010)
[161] Grüne, L., Pannek, J., Worthmann, K.: Ensuring stability in networked systems with nonlinear MPC for continuous time systems. In: *Proc. 51st IEEE Conf. on Decision and Control*, Maui, Hawaii, USA, pp. 14–19 (2012)
[162] Guo, W., Chen, S., Lü, J., Yu, X.: Consensus of multi-agent systems with an active leader and asymmetric adjacency matrix. In: *Proc. Joint 48th IEEE Conf. on Decision and Control and 28th Chinese Control Conf.*, Shanghai, China, pp. 3697–3702 (2009)
[163] Gurt, A., Nair, G. N.: Internal stability of dynamic quantised control for stochastic linear plants. *Automatica* 45, 1387–1396 (2009)
[164] Gusrialdi, A., Hirche, S.: Performance-oriented communication topology design for large-scale interconnected systems. In: *Proc. 49th IEEE Conf. on Decision and Control*, Atlanta, USA, pp. 5707–5713 (2010)
[165] Hara, S.: A unified approach to decentralized cooperative control for large-scale networked dynamical systems. In: Willems, J.C., Hara, S., Ohta, Y., Fujioka, H. (eds.) Persp. in Math. Sys. Theory, Ctrl., & Sign. Pro. LNCIS, vol. 398, pp. 61–72. Springer, Heidelberg (2010)
[166] Hara, S., Hayakawa, T., Sugata, H.: LTI systems with generalized frequency variables: A unified framework for homogeneous multi-agent dynamical systems. *SICE J. of Control, Measurement, and System Integration* 2, 299–306 (2009)
[167] He, C.: On the distance to uncontrollability and the distance to instability and their relation to some condition numbers in control. *Numerische Mathematik* 76, 463–477 (1997)
[168] Heemels, W. P.M. H., Teel, A. R., van de Wouw, N., Nesic, D.: Networked control systems with communication constraints: Tradeoffs between transmission intervals, delays and performance. *IEEE Trans. Automat. Control* 55, 1781–1796 (2010)
[169] Hekler, A., Fischer, J., Hanebeck, U. D.: Control over unreliable networks based on control input densities. In: *Proc. 15th Int. Conf. on Information Fusion*, Singapore, pp. 1277–1283 (2012)
[170] Hekler, A., Fischer, J., Hanebeck, U. D.: Sequence-based control for networked control systems based on virtual control inputs. In: *Proc. 51th IEEE Conf. on Decision and Control*, Maui, Hawaii, USA (2012)
[171] Hekler, A., Fischer, J., Hanebeck, U. D.: Sequence-based control for networked control systems based on virtual control inputs. *Arxiv preprint, arXiv:1206.0549* (2012)
[172] Helmke, U., Anderson, B. D. O.: Hermitian pencils and output feedback stabilization of scalar systems. *Int. J. Control* 56, 857–876 (1992)
[173] Helmke, U., Moore, J. B.: L^2 sensitivity minimization of linear system representations via gradient flows. *J. Math. Syst. Estim. Control* 5, 79–98 (1995)
[174] Helmke, U., Schönlein, M.: Approximate controllability of parameter-dependent linear systems. *Systems & Control Lett.* (2012) (submitted)
[175] Helmke, U., von Heusinger, A.: Synchronizing subspaces of networks of homogeneous linear control systems. *Automatisierungstechnik* (accepted, 2013)

[176] Helmke, U., Kurniawan, I., Lang, P., Schönlein, M.: Sensitivity optimal design of networks of identical linear systems. In: *Proc. 20th Int. Symp. on Mathematical Theory of Networks and Systems*, Melbourne, Australia (2012)

[177] Helmke, U., Kurniawan, I., Lang, P., Schönlein, M.: Interconnection sensitivity of networks of linear systems (in preparation, 2013)

[178] Hespanha, J. P., Naghshtabrizi, P.: A survey of recent results in networked control systems. *Proc. IEEE* 95, 138–162 (2007)

[179] Hespanha, J. P., Liberzon, D., Teel, A. R.: Lyapunov conditions for input-to-state stability of impulsive systems. *Automatica* 44, 2735–2744 (2008)

[180] Hewer, G., Kenney, C.: The sensitivity of the stable Lyapunov equation. *SIAM J. Control Optim.* 26, 321–344 (1988)

[181] Hinrichsen, D., Pritchard, A. J.: *Mathematical systems theory. I.* Texts in Applied Mathematics, vol. 48. Springer, Berlin (2005)

[182] Hirche, S., Lunze, J.: Digital vernetzte Regelungssysteme. *Automatisierungstechnik* 56, 1–3 (2008)

[183] Hoock, A. -M.: Topological entropy and invariance entropy for infinite-dimensional linear systems. *J. Dyn. Control Syst.* (2011) (submitted)

[184] Hu, M., Li, C.: Design smart city based on 3S, internet of things, grid computing and cloud computing technology. In: Wang, Y., Zhang, X. (eds.) IOT 2012. CCIS, vol. 312, pp. 466–472. Springer, Heidelberg (2012)

[185] Igel, A., Gotzhein, R.: An analysis of the interference problem in wireless networks. In: *Proc. 8th Int. Conf. on Wireless and Mobile Communications*, Venice, Italy, pp. 187–194 (2012)

[186] Ito, H., Dashkovskiy, S., Wirth, F.: On a small gain theorem for networks of iISS systems. In: *Proc. Joint 48th IEEE Conf. on Decision and Control and 28th Chinese Control Conf.*, Shanghai, China, pp. 4210–4215 (2009)

[187] Jadbabaie, A., Lin, J., Morse, S.: Coordination of groups of mobile autonomous subsystems using nearest neighbor rules. *IEEE Trans. Automat. Control* 48, 988–1001 (2003)

[188] Jamshidi, M.: *Large-Scale Systems: Modeling and Control.* North-Holland, New York (1983)

[189] Javidi, T., Franceschetti, M., Kumar, P. R., Mitter, S., Teneketzis, D.: Guest editorial control and communications. *IEEE J. Selected Areas in Communications* 26, 577–579 (2008)

[190] Jerg, S., Junge, O.: Lazy global feedbacks for quantized nonlinear event systems. In: *Proc. 7th IFAC Symp. on Robust Control Design*, Aalborg, Denmark, pp. 294–301 (2012)

[191] Jerg, S., Junge, O., Post, M.: Global optimal feedbacks for stochastic quantized nonlinear event systems (2012) (preprint)

[192] Jia, D., Krogh, B.: Min-max feedback model predictive control for distributed control with communication. In: *Proc. 2002 American Control Conf.*, Anchorage, Alaska, USA, pp. 4507–4512 (2002)

[193] Jiang, Z. -P., Wang, Y.: Input-to-state stability for discrete-time nonlinear systems. *Automatica* 37, 857–869 (2001)

[194] Jiang, Z. -P., Sontag, E. D., Wang, Y.: Input-to-state stability for discrete-time nonlinear systems. In: *Proc. 14th IFAC World Congress*, Beijing, China, pp. 277–282 (1999)

[195] Junge, O., Osinga, H. M.: A set oriented approach to global optimal control. *ESAIM Control Optim. Calc. Var.* 10, 259–270 (2004)

[196] Kaszkurewicz, E., Bhaya, A.: *Matrix Diagonal Stability in Systems and Computation*. Birkäuser, Boston (2000)

[197] Katok, A., Hasselblatt, B.: *Introduction to the Modern Theory of Dynamical Systems*. Cambridge University Press (1995)

[198] Kawan, C.: *Invariance Entropy for Control Systems*. PhD thesis, Institut für Mathematik, Universität Augsburg (2010)

[199] Kawan, C.: Invariance entropy of control sets. *SIAM J. Control Optim.* 49, 732–751 (2011)

[200] Kawan, C.: Lower bounds for the strict invariance entropy. *Nonlinearity* 24, 1909–1935 (2011)

[201] Kawan, C.: Upper and lower estimates for invariance entropy. *Discrete Contin. Dyn. Syst.* 30, 169–186 (2011)

[202] Kawan, C.: Minimal data rates for invariance of sets – an introduction to invariance entropy for finite-dimensional deterministic systems (2012) (submitted)

[203] Kenney, C., Hewer, G.: The sensitivity of the algebraic and differential Riccati equations. *SIAM J. Control Optim.* 28, 50–69 (1990)

[204] Kerrigan, E. C., Mayne, D. Q.: Optimal control of constrained, piecewise affine systems with bounded disturbances. In: *Proc. 41st IEEE Conf. on Decision and Control*, Las Vegas, USA, pp. 1552–1557 (2002)

[205] Khalil, H. K.: *Nonlinear Systems*. Prentice Hall, New Jersey (2002)

[206] Kim, K. -D., Kumar, P. R.: Cyber-physical systems: A perspective at the centennial. *Proc. IEEE* 100, 1287–1308 (2012)

[207] Kiwiel, K. C.: Convergence of approximate and incremental subgradient methods for convex optimization. *SIAM J. Optim.* 14, 807–840 (2004)

[208] Koltai, P., Junge, O.: Optimal value functions for weakly coupled systems: a posteriori estimates. *Arxiv preprint, arXiv:1110.3616* (2012)

[209] Koltai, P., Junge, O.: Split dynamic programming and ignorant policies in distributed control. *J. of Applied Mathematics and Mechanics (ZAMM)*, Arxiv preprint, arXiv:1110.3616 (to appear, 2013)

[210] Konstantinov, M., Gu, D., Mehrmann, V., Petkov, P.: Perturbation theory for matrix equations. *Studies in Computational Mathematics*. Elsevier, North Holland (2003)

[211] Koshita, S., Abe, M., Kawamata, M.: Analysis of second-order modes of linear discrete-time systems under bounded-real transformations. *IEICE Trans. Fundamentals of Electronics, Communications and Computer Science* 90, 2510–2515 (2007)

[212] Krämer, M., Braun, T., Christmann, D., Gotzhein, R.: Real-time signaling in SDL. In: Ober, I., Ober, I. (eds.) SDL 2011. LNCS, vol. 7083, pp. 186–201. Springer, Heidelberg (2011)

[213] Kuhn, T., Gotzhein, R., Webel, C.: Model-driven development with sdl – process, tools, and experiences. In: *Proc. 9th Int. Conf. on Model Driven Engineering Languages and Systems*, Genova, Italy, pp. 83–97 (2006)

[214] Laila, D. S., Astolfi, A.: Input-to-state stability for discrete-time time-varying systems with applications to robust stabilization of systems in power form. *Automatica* 41, 1891–1903 (2005)

[215] Langbort, C., Chandra, R. S., D'Andrea, R.: Distributed control design for systems interconnected over an arbitrary graph. *IEEE Trans. Automat. Control* 49, 1502–1519 (2004)

[216] Lazar, M., Heemels, W. P.M. H.: A new dual-mode hybrid mpc algorithm with a robust stability guarantee. In: *Proc. 2nd IFAC Conf. on Analysis and Design of Hybrid Systems*, Alghero, Sardinia, Italy (2006)

[217] Lazar, M., Muñoz de la Peña, D., Heemels, W. P. M. H., Alamo, T.: On input-to-state stability of min-max nonlinear model predictive control. *Systems & Control Lett.* 57, 39–48 (2008)

[218] Lee, E. A.: CPS foundations. In: *Design and Automation Conf.*, pp. 742–747 (2010)

[219] Lehmann, D.: *Event-based state-feedback control.* Logos Verlag, Berlin (2011)

[220] Lehmann, D.: *Event-based state-feedback control.* PhD thesis, Ruhr-Universität Bochum (2011)

[221] Lehmann, D., Lunze, J.: Event-based control using quantized state information. In: *Proc. 2nd IFAC Workshop on Distributed Estimation and Control in Networked Systems*, Annecy, France, pp. 1–6 (2010)

[222] Lehmann, D., Lunze, J.: Extension and experimental evaluation of an event-based state-feedback approach. *Control Engineering Practice* 19, 101–112 (2011)

[223] Lehmann, D., Lunze, J.: Event-based control with communication delays. In: *Proc. 18th IFAC World Congress*, Milan, Italy, pp. 3262–3267 (2011)

[224] Lehmann, D., Lunze, J.: Event-based output-feedback control. In: *Proc. 19th Mediterranean Conf. on Control and Automation*, Corfu, Greece, pp. 982–987 (2011)

[225] Lehmann, D., Lunze, J.: Event-based control with communication dealys and packet losses. *Int. J. Control* 85, 563–577 (2012)

[226] Lehmann, D., Lunze, J., Johansson, K. H.: Comparison between sampled-data control, deadband control and model-based event-triggered control. In: *Proc. 4th IFAC Conf. on Analysis and Design of Hybrid Systems*, Eindhoven, Netherlands (2012)

[227] Leonard, N. E., Krishnaprasad, P. S.: Control of switched electrical networks using averaging on Lie groups. In: *Proc. 33rd IEEE Conf. on Decision and Control*, Lake Buena Vista, USA, pp. 1919–1924 (1994)

[228] Lesieutre, B., Roy, S., Pinar, A.: Power system extreme event screening using graph partitioning. In: *Proc. 38th North American Power Symp.*, Carbondale, Illinois, USA, pp. 503–510 (2006)

[229] Li, X. R., Jilkov, V. P.: Survey of maneuvering target tracking. part v. multiple-model methods. In: *Conf. on Signal and Data Processing of Small Targets*, San Diego, California, USA, vol. 4473, pp. 559–581 (2003)

[230] Lian, F. -L., Moyne, J. R., Tilbury, D.M.: Performance evaluation of control networks: Ethernet, controlnet, and devicenet. *IEEE Control Systems Magazine* 21, 66–83 (2001)

[231] Liberzon, D., Hespanha, J. P.: Stabilization of nonlinear systems with limited information feedback. *IEEE Trans. Automat. Control* 50, 910–915 (2005)

[232] Lin, C. -T.: Structural controllability. *IEEE Trans. Automat. Control* 19, 2001–2008 (1974)

[233] Lincoln, B., Rantzer, A.: Relaxing dynamic programming. *IEEE Trans. Automat. Control* 51, 1249–1260 (2006)

[234] Lipsa, G. M., Martins, N. C.: Remote state estimation with communication costs for first-order LTI systems. *IEEE Trans. Automat. Control* 56, 2013–2025 (2011)

[235] Littman, M. L.: Markov games as a framework for multi-agent reinforcement learning. In: *Proc. 11th Int. Conf. on Machine Learning*, New Brunswick, New Jersey, USA, pp. 157–163 (1994)

[236] Litz, L., Gabriel, T., Groß, M., Gabel, O.: Networked control systems - Stand und Ausblick. *Automatisierungtechnik* 56, 4–19 (2008)

[237] Liu, J., de la Peña, D. M., Christofides, P. D.: Distributed model predictive control system design using lyapunov techniques. In: Magni, L., Raimondo, D.M., Allgöwer, F. (eds.) Nonlinear Model Predictive Control. LNCIS, vol. 384, pp. 181–194. Springer, Heidelberg (2009)

[238] Löfberg, J.: *Minimax Approaches to Robust Model Predictive Control.* PhD thesis, Linköping University (2003)

[239] Lu, W., Liu, B., Chen, T.: Cluster synchronization in networks of coupled nonidentical dynamical systems. *Chaos* 20, 013120 (2010)

[240] Lunze, J.: Synchronizable nodes in networked systems. *J. Phys. A, Math. Theor.* 44 (2011)

[241] Lunze, J.: Event-based control: a tutorial introduction. *J. the Society of Instrument and Control Engineers* 49, 783–788 (2010)

[242] Lunze, J.: An internal-model principle for the synchronisation of autonomous agents with individual dynamics. In: *Proc. Joint 50th IEEE Conf. on Decision and Control and European Control Conf.*, Orlando, USA, pp. 2106–2111 (2011)

[243] Lunze, J.: Synchronization of heterogeneous agents. *IEEE Trans. Automat. Control* 47, 2885–2890 (2012)

[244] Lunze, J.: Stability analysis of large-scale systems composed of strongly coupled similar subsystems. *Automatica* 25, 561–570 (1989)

[245] Lunze, J.: *Feedback Control of Large-Scale Systems.* Prentice Hall (1992)

[246] Lunze, J.: *Regelungstechnik 2.* Springer (2008)

[247] Lunze, J., Lamnabhi-Lagarrigue, F.: *Handbook of Hybrid Systems Control.* Cambridge University Press (2009)

[248] Lunze, J., Lehmann, D.: A state-feedback approach to event-based control. *Automatica* 46, 211–215 (2010)

[249] Ma, C. -Q., Zhang, J. -F.: Necessary and sufficient conditions for consensusabilityof linear multi-agent systems. *IEEE Trans. Automat. Control* 55, 1263–1268 (2010)

[250] Ma, Z., Liu, Z., Zhang, G.: A new method to realize cluster synchronization in connected chaotic networks. *Chaos* 16, 023103 (2006)

[251] Maciejowski, J. M.: *Predictive Control with Constraints.* Prentice-Hall, Englewood Cliffs (2002)

[252] Mahboobi Esfanjani, R., Reble, M., Münz, U., Nikravesh, S. K. Y., Allgöwer, F.: Model predictive control of constrained nonlinear time-delay systems. In: *Proc. Joint 48th IEEE Conf. on Decision and Control and 28th Chinese Control Conf.*, Shanghai, China, pp. 1324–1329 (2009)

[253] Martensson, K., Rantzer, A.: A scalable method for continuous-time distributed control synthesis. In: *Proc. 2012 American Control Conf.*, Montreal, Canada, pp. 6308–6313 (2012)

[254] Matveev, A., Savkin, A.: *Estimation and Control over Communication Networks.* Birkhäuser, Boston (2009)

[255] Mazo, M., Tabuada, P.: Decentralized event-triggered control over wireless sensor/actuator networks. *IEEE Trans. Automat. Control* 58, 2456–2461 (2011)
[256] Meinel, M., Ulbrich, M.: A class of distributed optimization methods with event-triggered communication. Preprint (2012)
[257] De Miguel, V., Murray, W.: A local convergence analysis of bilevel decomposition algorithms. *Optim. and Engineering* 7, 99–133 (2006)
[258] De Miguel, V., Nogales, F. J.: On decomposition methods for a class of partially separable nonlinear programs. *Math. Operations Research* 33, 119–139 (2008)
[259] Molin, A., Hirche, S.: An iterative algorithm for optimal event-triggered estimation. In: *Proc. 4th IFAC Conf. on Analysis and Design of Hybrid Systems*, Eindhoven, Netherlands, pp. 64–69 (2012)
[260] Molin, A., Hirche, S.: On the optimality of certainty equivalence for event-triggered control systems. *IEEE Trans. Automat. Control* 58, 470–474 (2013)
[261] Molin, A., Hirche, S.: On the optimal design of decentralized event-triggered controllers for large-scale systems with contention-based communication. In: *Proc. Joint 50th IEEE Conf. on Decision and Control and European Control Conf.*, Orlando, USA (2011)
[262] Momoh, J.: *Smart Grid: Fundamentals of Design and Analysis*. John Wiley & Sons, New Jersey (2012)
[263] Moreau, L.: Stability of multi-agent systems with time-dependent communication links. *IEEE Trans. Automat. Control* 50, 169–182 (2005)
[264] Moschik, S., Stadler, M., Dourdoumas, N.: Zur Überprüfung der perspektivischen Beobachtbarkeit zeitkontinuierlicher linearer zeitinvarianter Systeme. *Automatisierungstechnik* 60, 735–742 (2012)
[265] Müller, M. A., Reble, M., Allgöwer, F.: A general distributed MPC framework for cooperative control. In: *Proc. 18th IFAC World Congress*, Milan, Italy, pp. 7987–7992 (2011)
[266] Müller, M. A., Reble, M., Allgöwer, F.: Cooperative control of dynamically decoupled systems via distributed model predictive control. *Int. J. Robust and Nonlinear Control* 22, 1376–1397 (2012)
[267] Müller, M. A., Schürmann, B., Allgöwer, F.: Robust cooperative control of dynamically decoupled systems via distributed MPC. In: *Proc. IFAC Conf. on Nonlinear Model Predictive Control*, Noordwijkerhout, Netherlands, pp. 412–417 (2012)
[268] Mullis, C. T., Roberts, R. A.: Roundoff noise in digital filters: frequency transformations and invariants. *IEEE Trans. Acoust. Speech Signal Processing* 24, 538–550 (1976)
[269] Nair, G. N., Evans, R. J.: Stabilizability of stochastic linear systems with finite feedback data rates. *SIAM J. Control Optim.* 43, 413–436 (2004)
[270] Nair, G. N., Evans, R. J., Mareels, I. M. Y., Moran, W.: Topological feedback entropy and nonlinear stabilization. *IEEE Trans. Automat. Control* 49, 1585–1597 (2004)
[271] Nair, G. N., Fagnani, F., Zampieri, S., Evans, R. J.: Feedback control under data rate constraints: an overview. *Proc. IEEE* 95, 108–137 (2007)
[272] Necoara, I., Suykens, J.: Application of a smoothing technique to decomposition in convex optimization. *IEEE Trans. Automat. Control* 53, 2674–2679 (2008)

[273] Necoara, I., Suykens, J. A. K.: Interior-point lagrangian decomposition method for separable convex optimization. *J. Optim. Theory and Appl.* 143, 567–588 (2009)
[274] Nedic, A., Bertsekas, D. P.: The effect of deterministic noise in subgradient methods. *Math. Programming* 125, 75–99 (2010)
[275] Nesterov, Y.: Smooth minimization of non-smooth functions. *Math. Programmming* 103, 127–152 (2005)
[276] Nešić, D., Liberzon, D.: A unified framework for design and analysis of networked and quantized control systems. *IEEE Trans. Automat. Control* 54, 732–747 (2009)
[277] Newman, M. E., Girvan, M.: Finding and evaluating community structures in networks. *Phys. Rev. E* 69, 026113 (2004)
[278] Olfati-Saber, R., Fax, J. A., Murray, R. M.: Consensus and cooperation in networked multi-agent systems. *Proc. IEEE* 95, 215–233 (2007)
[279] Orsi, R., Helmke, U., Moore, J. B.: A Newton-like method for solving rank constrained linear matrix inequalities. *Automatica* 42, 1875–1882 (2006)
[280] Osborne, M.J., Rubinstein, A.: A Course in Game Theory. The MIT Press (1994)
[281] Osipov, G., Kurths, J., Zhou, C.: Synchronization in Oscillatory Networks. Springer, Berlin (2007)
[282] Palomar, D.P., Chiang, M.: A tutorial on decomposition methods for network utility maximization. IEEE J. Selected Areas in Communication 24, 1439–1451 (2006)
[283] Papadimitriou, C.H., Tsitsiklis, J.: Intractable problems in control theory. SIAM J. Control Optim. 24, 639–654 (1986)
[284] Papavassiliou, S., Kato, N., Liu, Y., Xu, C.-Z., Wang, X.: Guest editors' introduction: Special issue on cyber-physical systems. IEEE Trans. Parallel and Distributed Systems 23, 1569–1571 (2012)
[285] Philipp, P.: Zustands- und Parameterschätzung für nichtlineare Networked Control Systems. In: Roppenecker, G., Lohmann, B. (eds.) Methoden und Anwendung der Regelungstechnik, pp. 1–13. Shaker-Verlag, Aachen (2009)
[286] Philipp, P.: Structure exploiting derivative computation for moving horizon estimation. In: Proc. 2011 American Control Conf., San Francisco, USA, pp. 4263–4268 (2011)
[287] Philipp, P.: Exact state sensitivity calculation and its role in moving horizon estimation for nonlinear networked control systems. In: Roppenecker, G., Lohmann, B. (eds.) Methoden und Anwendung der Regelungstechnik, pp. 27–39. Shaker-Verlag, Aachen (2011)
[288] Philipp, P.: Observability of nonlinear NCS with unsynchronized sensor clocks. In: Proc. 51st IEEE Conf. on Decision and Control, Maui, Hawaii, USA, pp. 1349–1355 (2012)
[289] Philipp, P., Altmannshofer, S.: Experimental validation of a new moving horizon estimator approach for networked control systems with unsynchronized clocks. In: Proc. 2012 American Control Conf., Montreal, Canada, pp. 4939–4944 (2012)
[290] Philipp, P., Lohmann, B.: Gauss-newton moving horizon observer for state and parameter estimation of nonlinear networked control systems. In: Proc. 10th European Control Conf., Budapest, Hungary, pp. 1740–1745 (2009)

[291] Philipp, P., Lohmann, B.: Moving horizon estimation for nonlinear networked control systems with unsynchronized timescales. In: Proc. 18th IFAC World Congress, Milan, Italy, pp. 12457–12464 (2011)

[292] Philipp, P., Schmid-Zurek, T.: Distributed moving horizon estimation via dual decomposition. In: Proc. 51st IEEE Conf. on Decision and Control, Maui, Hawaii, USA, pp. 4792–4798 (2012)

[293] Philipp, P., Schneider, M.: A decentralized moving horizon observer for distributed implementation of centralized controllers. In: Proc. 2013 American Control Conf., Washington D.C., USA (2013)

[294] Pikovsky, A., Rosenblum, M., Kurths, J.: Synchronization: A Universal Concept in Nonlinear Sciences. Cambridge Univ. Press, Cambridge (2001)

[295] Pin, G., Parisini, T.: Stabilization of networked control systems by nonlinear model predictive control: a set invariance approach. In: Magni, L., Raimondo, D.M., Allgöwer, F. (eds.) Int. Workshop on Assessment and Future Directions of NMPC. Springer, Berlin (2008)

[296] Pogromsky, A., Santoboni, G., Nijmeijer, H.: Partial synchronization: from symmetry towards stability. Physica D: Nonlinear Phenomena 172, 65–87 (2002)

[297] Polushin, I.G., Liu, P.X., Lung, C.-H.: On the model-based approach to nonlinear networked control systems. Automatica 44, 2409–2414 (2008)

[298] Poovendran, R., Sampigethaya, K., Gupta, S.K.S., Lee, I., Prasad, K.V., Corman, D., Paunicka, J.: Special issue on cyber - physical systems [scanning the issue]. Proc. IEEE 100, 6–12 (2012)

[299] Popov, A., Werner, H.: A robust control approach to formation control. In: Proc. 10th European Control Conf., Budapest, Hungary, pp. 4428–4433 (2009)

[300] Purchala, K., Meeus, L., Van Dommelen, D., Belmans, R.: Usefulness of dc power flow for active power flow analysis. In: 2005 IEEE Power Engineering Society General Meeting, San Francisco, USA, pp. 454–459 (2005)

[301] Quevedo, D. E., Nešić, D.: Robust stability of packetized predictive control of nonlinear systems with disturbances and Markovian packet losses. Automatica 48, 1803–1811 (2012)

[302] Rabi, M., Johansson, K. H.: Scheduling packets for event-triggered control. In: *Proc. 10th European Control Conf.*, Budapest, Hungary, pp. 3779–3784 (2009)

[303] Rabi, M., Johansson, K. H., Johansson, M.: Optimal stopping for event-triggered sensing and actuation. In: *Proc. 47th IEEE Conf. on Decision and Control*, Cancun, Mexico, pp. 3607–3612 (2008)

[304] Raff, T., Ebenbauer, C., Allgower, F.: Passivity-based nonlinear dynamic output feedback design: a semidefinite programming approach. In: *Proc. 43rd IEEE Conf. on Decision and Control*, Paradise Islands, The Bahamas, vol. 5, pp. 5409–5414 (2004)

[305] Raff, T., Ebenbauer, C., Allgöwer, F.: Nonlinear model predictive control: a passivity-based approach. In: Findeisen, R., Allgöwer, F., Biegler, L.T. (eds.) Assesments and Future Directions of Nonlinear Model Predictive Control. LNCIS, vol. 358, pp. 151–162. Springer, Heidelberg (2007)

[306] Rakovic, S. V., Grieder, P., Kvasnica, M., Mayne, D. Q., Morari, M.: Computation of invariant sets for piecewise affine discrete time systems subject to bounded disturbances. In: *Proc. 43rd IEEE Conf. on Decision and Control*, Paradise Islands, The Bahamas, pp. 1418–1423 (2004)

[307] Rawlings, J. B., Mayne, D. Q.: *Model Predictive Control: Theory and Design*. Nob Hill Publishing, Madison (2009)

[308] Reble, M.: *Model Predictive Control for Nonlinear Continuous-Time Systems with and without Time-Delays*. PhD thesis, University of Stuttgart (2012) (submitted)

[309] Reble, M., Allgöwer, F.: Unconstrained model predictive control and suboptimality estimates for nonlinear continuous-time systems. *Automatica* 48, 1812–1817 (2012)

[310] Reble, M., Allgöwer, F.: Stabilizing design parameters for model predictive control of constrained nonlinear time-delay systems. In: *Proc. 9th IFAC Workshop on Time Delay Systems*, Prague, Czech Republic, pp. 361–366 (2010)

[311] Reble, M., Allgöwer, F.: General design parameters of model predictive control for nonlinear time-delay systems. In: *Proc. 49th IEEE Conf. on Decision and Control*, Atlanta, USA, pp. 176–181 (2010)

[312] Reble, M., Allgöwer, F.: Design of terminal cost functionals and terminal regions for model predictive control of nonlinear time-delay systems. In: Sipahi, R., Vyhlidal, T., Niculescu, S.-I., Pepe, P. (eds.) Time Delay Sys.: Methods, Appli. and New Trends. LNCIS, vol. 423, pp. 355–366. Springer, Heidelberg (2012)

[313] Reble, M., Brunner, F. D., Allgöwer, F.: Model predictive control for nonlinear time-delay systems without terminal constraint. In: *Proc. 18th IFAC World Congress*, Milan, Italy, pp. 9254–9259 (2011)

[314] Reble, M., Mahboobi Esfanjani, R., Nikravesh, S. K. Y., Allgöwer, F.: Model predictive control of constrained nonlinear time-delay systems. *IMA J. Mathematical Control and Information* 28, 183–201 (2011)

[315] Reble, M., Müller, M. A., Allgöwer, F.: Unconstrained model predictive control and suboptimality estimates for nonlinear time-delay systems. In: *Proc. Joint 50th IEEE Conf. on Decision and Control and European Control Conf.*, Orlando, USA, pp. 7599–7604 (2011)

[316] Reble, M., Quevedo, D. E., Allgöwer, F.: Stochastic stability and performance estimates of packetized unconstrained model predictive control for networked control systems. In: *Proc. 9th IEEE Int. Conf. on Control and Automation*, Santiago, Chile, pp. 171–176 (2011)

[317] Reble, M., Quevedo, D. E., Allgöwer, F.: Control over erasure channels: Stochastic stability and performance of packetized unconstrained model predictive control. *Int. J. Robust and Nonlinear Control* (2012)

[318] Remagnino, P., Foresti, G. L., Ellis, T.: *Ambient Intelligence - A Novel Paradigm*. Springer, Berlin (2004)

[319] Ren, W., Beard, R. W.: Consensus seeking in multiagent systems under dynamically changing interaction topologies. *IEEE Trans. Automat. Control* 50, 655–661 (2005)

[320] Richards, A., How, J.: A decentralized algorithm for robust constrained model predictive control. In: *Proc. 2004 American Control Conf.*, Boston, USA, pp. 4261–4266 (2004)

[321] Richards, A., How, J.: Robust distributed model predictive control. *Int. J. Control* 80, 1517–1531 (2007)

[322] Rotkowitz, M., Lall, S.: A characterization of convex problems in decentralized control. *IEEE Trans. Automat. Control* 51, 274–286 (2006)

[323] Rüffer, B., Wirth, F.: Stability verification for monotone systems using homotopy algorithms. *Numer. Algorithms* 58, 529–543 (2011)

[324] Sagastizabal, C. A., Solodov, M. V.: Parallel variable distribution for constrained optimization. *Computational Optim. and Appl.* 22, 111–131 (2002)

[325] Sailer, R., Wirth, F.: Stabilization by encoded feedback with Markovian communication channels. In: *Proc. 8th IFAC Nonlinear Control Systems Design Symp.*, Bologna, Italy, pp. 84–89 (2010)

[326] Sailer, R., Wirth, F.: Stabilization of nonlinear systems with delayed data-rate-limited feedback. In: *Proc. 10th European Control Conf.*, Budapest, Hungary, pp. 1734–1739 (2009)

[327] Samar, S., Boyd, S., Gorinevsky, D.: Distributed estimation via dual decomposition. In: *Proc. 9th European Control Conf.*, Kos, Greece, pp. 1511–1516 (2007)

[328] Scardovi, L., Sepulchre, R.: Synchronization in networks of identical linear systems. *Automatica* 45, 2557–2562 (2009)

[329] Scattolini, R.: Architectures for distributed and hierarchical model predictive control – a review. *J. Process Control* 19, 723–731 (2009)

[330] Schmitt, J. B.: *Heterogeneous Network Quality of Service Systems.* Kluwer Academic Publishers, Boston (2001)

[331] Schuh, R., Lunze, J.: Stability analysis of networked systems with similar dynamics. In: *Proc. 12th European Control Conf.*, Zurich, Switzerland (accepted, 2013)

[332] Schuler, S., Zhou, W., Münz, U., Allgöwer, F.: Controller structure design for decentralized control of coupled higher order subsystems. In: *Proc. 2nd IFAC Workshop on Distributed Estimation and Control in Networked Systems*, Annecy, France, pp. 269–274 (2010)

[333] Seyboth, G., Dimarogonas, D. V., Johansson, K. H.: Control of multi-agent systems via event-based communication. In: *Proc. 18th IFAC World Congress*, Milan, Italy, pp. 10086–10091 (2011)

[334] Seyboth, G., Schmidt, G. S., Allgöwer, F.: Cooperative control of linear parameter-varying systems. In: *Proc. 2012 American Control Conf.*, Montreal, Canada, pp. 2407–2412 (2012)

[335] Seyboth, G., Schmidt, G. S., Allgöwer, F.: Output synchronization of linear parameter-varying systems via dynamic couplings. In: *Proc. 51st IEEE Conf. on Decision and Control*, Maui, Hawaii, USA (2012)

[336] Shah, P., Parrilo, P.A.: H2-optimal decentralized control over posets: A state space solution for state-feedback. In: *Proc. 49th IEEE Conf. on Decision and Control*, Atlanta, USA (2010)

[337] Shakkottai, S., Rappaport, T. S., Karlsson, P. C.: Cross-layer design for wireless networks. *IEEE Communications Magazine* 41, 74–80 (2003)

[338] Shor, N. Z.: *Minimization Methods for Non-Differentiable Functions.* Springer, New York (1985)

[339] Siljak, D. D.: *Large-Scale Dynamic Systems: Stability and Structure.* North-Holland, Amsterdam (1978)

[340] Sinopoli, B., Schenato, L., Franceschetti, M., Poolla, K., Jordan, M. I., Sastry, S. S.: Kalman filtering with intermittent observations. *IEEE Trans. Automat. Control* 49, 1453–1464 (2004)

[341] Solodov, M. V.: New inexact parallel variable distribution algorithms. *Computational Optim. and Appl.* 7, 165–182 (1997)

[342] Solodov, M. V.: On the convergence of constrained parallel variable distribution algorithms. *SIAM J. Optim.* 8, 187–196 (1998)

[343] Sontag, E. D.: The ISS philosophy as a unifying framework for stability-like behavior. In: Isidori, A., Lamnabhi-Lagarrigue, F., Respondek, W. (eds.) Nonlinear Control in the Year 2000. LNCIS, vol. 259, pp. 443–467. Springer, Heidelberg (2001)

[344] Sontag, E. D.: Finite-dimensional open-loop control generators for nonlinear systems. *Int. J. Control* 47, 537–556 (1988)

[345] Sontag, E. D.: Universal nonsingular controls. *Systems & Control Lett.* 19, 221–224 (1992)

[346] Sontag, E. D.: *Mathematical Control Theory.* Springer, New York (1998)

[347] Sontag, E. D., Wang, Y.: On characterizations of the input-to-state stability property. *Systems & Control Lett.* 24, 351–359 (1995)

[348] Stöcker, C., Lunze, J.: Event-based control of input-output linearizable systems. In: *Proc. 18th IFAC World Congress*, Milan, Italy, pp. 10062–10067 (2011)

[349] Stöcker, C., Lunze, J.: Event-based control of nonlinear systems: an input-output linearization approach. In: *Proc. Joint 50th IEEE Conf. on Decision and Control and European Control Conf.*, Orlando, USA, pp. 2541–2546 (2011)

[350] Stöcker, C., Lunze, J.: Event-based control with incomplete state measurement and guaranteed performance. In: *Proc. 3rd IFAC Workshop on Distributed Estimation and Control in Networked Systems*, Santa Barbara, USA, pp. 49–54 (2012)

[351] Stöcker, C., Lunze, J.: Input-to-state stability of event-based state-feedback control. In: *Proc. 13th European Control Conf.*, Zurich, Switzerland, pp. 49–54 (accepted, 2013)

[352] Stöcker, C., Lunze, J., Ngo, C.: Two methods for the event-based control of interconnected systems and their experimental evaluation. *Automatisierungstechnik* 60, 724–734 (2012)

[353] Stöcker, C., Lunze, J., Vey, D.: Stability analysis of interconnected event-based control loops. In: *Proc. 4th IFAC Conf. on Analysis and Design of Hybrid Systems*, Eindhoven, Netherlands, pp. 58–63 (2012)

[354] Tabuada, P.: Event-triggered real-time scheduling of stabilizing control tasks. *IEEE Trans. Automat. Control* 52, 1680–1685 (2007)

[355] Tammer, K.: The application of parametric optimization and imbedding for the foundation and realization of a generalized primal decomposition approach. In: Guddat, J., Jongen, H., Kummer, B., Nozicka, F. (eds.) *Parametric Optimization and Related Topics, Mathematical Research*, pp. 376–386. Akademie-Verlag, Berlin (1989)

[356] Tang, P. L., de Silva, C. W.: Stability validation of a constrained model predictive networked control system with future input buffering. *Int. J. Control* 80, 1954–1970 (2007)

[357] Tatikonda, S., Mitter, S.: Control under communication constraints. *IEEE Trans. Automat. Control*,1056–1068 (2004)
[358] Tatikonda, S., Sahai, A., Mitter, S. K.: Stochastic linear control over a communication channel. *IEEE Trans. Automat. Control*, 1549–1561 (2004)
[359] Tatikonda, S. C.: *Control Under Communication Constraints*. PhD thesis, Department of EECS, Masachussets Institute of Technology (2000)
[360] Trentelman, H., Stoorvogel, A., Hautus, M.: *Control Theory for Linear Systems*. Springer, London (2001)
[361] Tseng, P.: Dual coordinate ascent methods for non-strictly convex minimization. *Math. Programming* 59, 231–248 (1993)
[362] Tuna, S. E.: Conditions for synchronizability in arrays of coupled linear systems. *IEEE Trans. Automat. Control* 54, 2416–2420 (2009)
[363] Tuna, S. E.: Synchronizing linear systems via partial-state coupling. *Automatica* 44, 2179–2184 (2008)
[364] Tuna, S. E., Messina, M. J., Teel, A. R.: Shorter horizons for model predictive control. In: *Proc. 2006 American Control Conf.*, Minneapolis, USA (2006)
[365] Vamsi, A. S. M., Elia, N.: Design of distributed controllers realizable over arbitrary directed networks. In: *Proc. 49th IEEE Conf. on Decision and Control*, Atlanta, USA, pp. 4795–4800 (2010)
[366] Van Loan, C.: How near is a stable matrix to an unstable matrix? *Contemp. Math.* 47, 465–477 (1985)
[367] Varutti, P.: *Model Predictive Control for Nonlinear Networked Control Systems: A Model-based Compensation Approach for Nondeterministic Communication Networks*. PhD thesis, Otto-von-Guericke University Magdeburg (2012)
[368] Varutti, P., Findeisen, R.: Predictive control of nonlinear chemical processes under asynchronous measurements and controls. In: *Proc. 7th IFAC Int. Symp. on Advanced Control of Chemical Processes*, Instanbul, Turkey, pp. 156–161 (2009)
[369] Varutti, P., Findeisen, R.: Event-based NMPC for networked control systems over UDP-like communication channels. In: *Proc. 2011 American Control Conf.*, San Francisco, USA, pp. 3166–3171 (2011)
[370] Varutti, P., Findeisen, R.: Compensating network delays and information loss by predictive control methods. In: *Proc. 10th European Control Conf.*, Budapest, Hungary, pp. 1722–1727 (2009)
[371] Varutti, P., Kern, B., Faulwasser, T., Findeisen, R.: Event-based model predictive control for networked control systems. In: *Proc. Joint 48th IEEE Conf. on Decision and Control and 28th Chinese Control Conf.*, Shanghai, China, pp. 567–572 (2009)
[372] Varutti, P., Kern, B., Findeisen, R.: Dissipativity-based distributed nonlinear predictive control for cascaded nonlinear systems. In: *Proc. 8th IFAC Int. Symp. Advanced Control of Chemical Processes*, Singapore, vol. 8, pp. 439–444 (2012)
[373] Venkat, A. N., Rawlings, J. B., Wright, S.J.: Distributed model predictive control of large-scale systems. In: *Assessment and Future Directions of Nonlinear Model Predictive Control*, pp. 591–605. Springer, Berlin (2007)
[374] Venkat, A. N., Hiskens, I. A., Rawlings, J. B., Wright, S. J.: Distributed MPC strategies with application to power system automatic generation control. *IEEE Trans. Contr. Sys. Tech.*, 1192–1206 (2008)

[375] von Lossow, M.: A min-max version of Dijkstra's algorithm with application to perturbed optimal control problems. In: *Proc, 6th Int. Congress on Industrial and Applied Mathematics and GAMM Annual Meeting*, Zurich, Switzerland, pp. 4130027–4130028 (2007)

[376] Walters, P.: *An Introduction to Ergodic Theory*. Springer (1982)

[377] Wan, P., Lemmon, M. D.: Distributed network utility maximization using event-triggered augmented lagrangian methods. In: *Proc. 2009 American Control Conf.*, St. Louis, Missouri, USA, pp. 3298–3303 (2009)

[378] Wang, J., Elia, N.: A control perspective for centralized and distributed convex optimization. In: *Proc. Joint 50th IEEE Conf. on Decision and Control and European Control Conf.*, Orlando, USA, pp. 3800–3805 (2011)

[379] Wang, X., Lemmon, M.: Self-triggered feedback control systems with finite-gain L_2 stability. *IEEE Trans. Automat. Control* 54, 452–467 (2009)

[380] Wang, X., Lemmon, M.: Event-triggering in distributed networked control systems. *IEEE Trans. Automat. Control* 56, 586–601 (2011)

[381] Wang, Z., Scaglione, A., Thomas, R. J.: Electrical centrality measures for electric power grid vulnerability. In: *Proc. of the 49th IEEE Conf. on Decision and Control*, Atlanta, USA, pp. 5792–5797 (2010)

[382] Webel, C., Gotzhein, R.: Formalization of network quality-of-service requirements. In: Derrick, J., Vain, J. (eds.) FORTE 2007. LNCS, vol. 4574, pp. 309–324. Springer, Heidelberg (2007)

[383] Wieland, P., Allgöwer, F.: An internal model principle for synchronization. In: *Proc. 7th IEEE Int. Conf. Control and Automation*, Christchurch, New Zealand, pp. 285–290 (2009)

[384] Wieland, P., Allgöwer, F.: On synchronous steady states and internal models of diffusively coupled systems. In: *Proc. 4th IFAC Symp. on System, Structure and Control*, Ancona, Italy, pp. 1–10 (2010)

[385] Wieland, P., Sepulchre, R., Allgöwer, F.: An internal model principle is necessary and sufficient for linear output synchronization. *Automatica* 47, 1068–1074 (2010)

[386] Winkler, M., Christmann, D., Krämer, M.: Customized duty cycling with MacZ. In: *Proc. 23rd Int. Conf. on Architecture of Computing Systems - Workshop Proc.*, Hannover, Germany, pp. 1–9 (2010)

[387] Witsenhausen, H. S.: A counterexample in stochastic optimum control. *SIAM J. Control* 6, 131 (1968)

[388] Witsenhausen, H. S.: A standard form for sequential stochastic control. *Theory of Computing Systems* 7, 5–11 (1973)

[389] Wong, W. S., Brockett, R.: ystems with finite communication bandwidth constraints. part I. state estimation problems. *IEEE Trans. Automat. Control* AC-42, 1294–1299 (1997)

[390] Wong, W. S., Brockett, R.: Systems with finite communication bandwidth constraints. part II. stabilization with limited information feedback. *IEEE Trans. Automat. Control* AC-44, 1049–1053 (1999)

[391] Wonham, W. M.: *Linear Multivariable Control. A Geometric Approach*, 3rd edn., New York. Applications of Mathematics, p. 10 (1985)

[392] Worthmann, K.: *Stability Analysis of Unconstrained RHC*. PhD thesis, University of Bayreuth (2011)

[393] Worthmann, K.: Estimates on the prediction horizon length in MPC. In: *Proc. 19th Int. Symp. on Mathematical Theory of Networks and Systems*, Melbourne, Australia (2012)
[394] Xia, F., Sun, Y.: *Control and Scheduling Codesign - Flexible Resource Management in Real-Time Control Systems.* Springer, Berlin (2008)
[395] Xia, W., Cao, M.: Clustering in diffusively coupled networks. *Automatica* 47, 2395–2405 (2011)
[396] Xie, L.: Topological entropy and data rate for practical stability: A scalar case. *Asian J. Control* 11, 376–385 (2009)
[397] Yang, T., Stoorvogel, A .A., Saberi, A.: Consensus for multi-agent systems-synchronization and regulation for complex networks. In: *Proc. 2011 American Control Conf.*, San Francisco, USA, pp. 5312–5317 (2011)
[398] You, K., Xie, L.: Necessary and sufficient conditions for consensusability of discrete-time multi-agent systems. In: *Proc. 2nd IFAC Workshop on Distributed Estimation and Control in Networked Systems*, Annecy, France, pp. 109–114 (2010)
[399] Yu, H., Antsaklis, P. J.: *Passivity of cascade systems based on the analysis of passivity indices.* Technical report, ISIS Group, Department of Electrical Engineering, University of Notre Dame (2009)
[400] Yu, H., Antsaklis, P. J.: A passivity measure of systems in cascade based on passivity indices. In: *Proc. 49th IEEE Conf. on Decision and Control*, Atlanta, USA, pp. 2186–2191 (2010)
[401] Zamani, M., Anderson, B. D. O., Helmke, U., Chen, W.: On the zeros of blocked time-invariant systems. *Systems & Control Lett.* (to appear, 2013)
[402] Zelazo, D., Mesbahi, M.: Edge agreement: Graph-theoretic performance bounds and passivity analysis. *IEEE Trans. Automat. Control* 56, 544–555 (2010)
[403] Zenger, C.: Sparse grids. In: *Parallel Algorithms for Partial Differential Equations*, pp. 241–251. Vieweg-Verlag, Braunschweig (1991)
[404] Zhang, L., Hristu-Varsakelis, D.: Communication and control co-design for networked control systems. *Automatica* 42, 953–958 (2006)
[405] Zhang, W., Branicky, M. S., Phillips, S. M.: Stability of networked control systems. *IEEE Control Systems Magazine* 21, 84–99 (2001)

Index

Printed by Printforce, the Netherlands